# Lecture Notes in Mathematics

Edited by A. Dold and B. Eckmann

883


# Cylindric Set Algebras

Cylindric Set Algebras and Related Structures
By L. Henkin, J. D. Monk, and A. Tarski

On Cylindric-Relativized Set Algebras
By H. Andréka and I. Németi


Springer-Verlag
Berlin Heidelberg New York 1981

**Authors**

Leon Henkin
Department of Mathematics, University of California
Berkeley, CA 94720, USA

J. Donald Monk
Department of Mathematics, University of Colorado
Boulder, CO 80309, USA

Alfred Tarski
462 Michigan Ave.
Berkeley, CA 94707, USA

Hajnalka Andréka
István Németi
Mathematical Institute, Hungarian Academy of Sciences
Reáltanoda u. 13–15, 1053 Budapest, Hungary

AMS Subject Classifications (1980): 03C55, 03G15

ISBN 978-3-540-10881-8 Springer-Verlag Berlin Heidelberg New York
ISBN 978-0-387-10881-0 Springer-Verlag New York Heidelberg Berlin



2141/3140-543210

# Introduction

This volume is devoted to a comprehensive treatment of certain set-theoretical structures which consist of fields of sets enhanced by additional fundamental operations and distinguished elements. The treatment is largely self-contained.

Each of these structures has an associated dimension $\alpha$, a finite or infinite ordinal; their basic form is well illustrated in the case $\alpha = 3$. Let $R$ be an arbitrary set, and let $\mathfrak{F}$ be a field of subsets of the set ${}^3R$ of all triples of elements of $R$. Thus $\mathfrak{F}$ is a non-empty collection of subsets of ${}^3R$ closed under union, intersection, and complementation relative to ${}^3R$. We shall assume that $\mathfrak{F}$ is closed under the three operations $C_0, C_1, C_2$ of cylindrification, where $C_0$, for example, is the operation given by:

$$C_0X = \{\langle x,y,z\rangle : \langle u,y,z\rangle \in X \text{ for some } u, \text{ with } x \in R\};$$

$C_0X$ is the cylinder formed by moving $X$ parallel to the first axis. $C_1X$ and $C_2X$ are similarly related to the second and third axes. We also assume that the diagonal planes $D_{01}, D_{02}, D_{12}$ are in $\mathfrak{F}$; here, for example,

$$D_{01} = \{\langle x,x,y\rangle : x,y \in R\}.$$

Similarly $D_{02}$ (resp., $D_{12}$) consists of all triples of ${}^3R$ whose first and third (resp., second and third) coordinates coincide. A collection $\mathfrak{F}$ satisfying all of these conditions is called a cylindric field of sets (of dimension 3). Cylindric fields of sets and certain closely related structures are the objects of study in this volume. Considered not merely as collections of sets, but as algebraic objects endowed with fundamental operations and distinguished elements, cylindric fields of sets are called cylindric set algebras. $Cs_\alpha$ is the class of all cylindric set algebras of dimension $\alpha$, and $ICs_\alpha$ is the class of algebras isomorphic to them.

In much of the work, general algebraic notions are studied in their application to cylindric set algebras. We consider subalgebras, homomorphisms, products, and ultraproducts of them, paying special attention, for example, to the closure of $ICs_\alpha$ and related classes under these operations. In addition, there are natural operations upon these structures which are specific to their form as certain Boolean algebras with operators, such as relativization to subsets of ${}^3R$ and isomorphism to algebras of subsets of ${}^3S$ with $S \neq R$, and there are relationships between set algebras of different dimensions.

Although, as mentioned, the volume is largely self-contained, we shall often refer to the book Cylindric Algebras, Part I, by Henkin, Monk, and Tarski. Many notions touched on briefly in the present volume are treated in detail in that one, and motivation for considering certain questions can be found there. Indeed, the present work had its genesis in the decision by Henkin, Monk, and Tarski to publish a series of papers which would form the bulk of Part II of their earlier work. Their contribution to the present volume is, in fact, the first of this proposed series. As their writing proceeded, they learned of the closely related results obtained by Andréka and Németi, and invited the latter to publish jointly with themselves.

Thus, the present volume consists of two parts. The first, by Henkin, Monk, and Tarski, contains the basic defintions and results on various kinds of cylindric set algebras. The second, by Andréka and Németi, is organized parallel to the first. In it, certain aspects of the theory are investigated more thoroughly; in particular, many results which are merely formulated in Part I, are provided with proofs in Part II. In both parts, many open problems concerning the structures considered are presented.

The authors

Berkeley
Boulder
Budapest

Table of contents

# Cylindric set algebras and related structures [1)]

by L. Henkin, J.D. Monk, and A. Tarski

The abstract theory of cylindric algebras is extensively developed in the book [HMT] by the authors. Several kinds of special set algebras were mentioned, primarily for motivational purposes, in that book. It is the purpose of this article to begin the examination of these set algebras in more detail. The simplest and most important kind of set algebras are the cylindric set algebras introduced in 1.1.5. (Throughout this article we shall refer to items from [HMT] by number without explicitly mentioning that book). Recall that the unit element of any $\alpha$-dimensional cylindric set algebra $(Cs_\alpha)$ $\mathfrak{A}$ is the Cartesian power ${}^\alpha U$ of a set $U$ (the base), and the other elements of $A$ are subsets of ${}^\alpha U$. The diagonal element $D_{\kappa\lambda}$ of $\mathfrak{A}$ is the set $\{x \in {}^\alpha U: x_\kappa = x_\lambda\}$ for each $\kappa, \lambda < \alpha$ ; the fundamental Boolean operations of $\mathfrak{A}$ are union, intersection, and complementation; and for each $\kappa < \alpha$ the fundamental operation $C_\kappa$ consists of cylindrification by translation parallel to the $\kappa$th axis of the space. Several other kinds of set algebras were briefly considered in [HMT], and their definition is similar to that of a cylindric set algebra: weak cylindric set algebras (cf. 2.2.11), generalized cylindric set algebras (cf. 1.1.13), and what we shall now call generalized weak cylindric set algebras (cf. 2.2.11). The algebras of each of these kinds have for their unit elements subsets of a special kind of some Cartesian space ${}^\alpha U$, while

---

1) This article is the first in a series intended to form a large portion of the second volume of the work Cylindric Algebras, of which Part I has appeared ([HMT] in the bibliography). The research and writing were supported in part by NSF grants MPS 75-03583, MCS 77-22913.

the fundamental operations of any such algebra are obtained from those of a $Cs_\alpha$ , with unit element ${}^\alpha U$ , by relativization (in the sense of 2.2) to the unit element of the algebra discussed. To unify our treatment of these several classes of cylindric set algebras we use here as the most general class of set algebras that of the cylindric-relativized set algebras, in which the unit elements may be arbitrary subsets of a Cartesian space. These algebras are simply subalgebras of those algebras that are obtained by arbitrary relativizations from full cylindric set algebras. We shall not discuss here, however, the class of all cylindric-relativized set algebras in any detail, restricting ourselves to the aspects of relativization directly relevant to our discussion of those set algebras which are CA's .

Much of the importance of cylindric set algebras stems from the following construction. Given any relational structure $\mathfrak{C}$ and any first-order discourse language $\Lambda$ for $\mathfrak{C}$ , the collection $\{\widetilde{\varphi}^{\mathfrak{C}} : \varphi$ a formula of $\Lambda\}$ forms an $\alpha$ - dimensional cylindric field of sets, where $\alpha$ is the length of the sequence of variables of $\Lambda$ (for the notation used here, see the Preliminaries of [HMT]). Thus the above collection is the universe of a cylindric set algebra $\mathfrak{B}$ . This algebra is locally finite dimensional in the sense of 1.11.1 , since $c_\kappa \widetilde{\varphi}^{\mathfrak{C}} = \widetilde{\varphi}^{\mathfrak{C}}$ except possibly for the finitely many $\kappa < \alpha$ such that the $\kappa$th variable of $\Lambda$ occurs free in $\varphi$ . Furthermore, $\mathfrak{B}$ has an additional property of regularity: if $x \in B$ , $f \in x$ , $g \in {}^\alpha C$ , and $\Delta x 1 f = \Delta x 1 g$ , then $g \in x$ . (Here $\Delta x$ , the dimension set of $x$ , is $\{\kappa \in \alpha : c_\kappa x \neq x\}$ ; see 1.6.1 .) Regular set algebras will be discussed extensively later.

The article has nine sections. In section 1 we give formal definitions of the classes of set algebras which are studied in this article and we state the simplest relationships between them; the proofs are found in later sections of the paper. In section 2 some deeper relationships are established, using the notion of relativization. Section 3 is concerned with change of base, treating the question of conditions under which a set algebra with base U is isomorphic to one with a different base W ; the main results are algebraic versions of the Löwenheim, Skolem, Tarski theorems (some results on change of base are also found in section 7). In section 4 the algebraic notion of subalgebra is investigated for our various set algebras, paying particular attention to the problem about the minimum number of generators for a set algebra. Homomorphisms of set algebras are discussed in section 5, and products, along with the related indecomposability notions, are studied in their application to set algebras in section 6. Section 7, devoted to ultraproducts of set algebras, gives perhaps the deepest results in the paper. In particular, it is in this section that the less trivial of the relationships between the classes of set algebras described in section 1 are established. Reducts and neat embeddings of set algebras are discussed in section 8. Finally, in section 9 we list the most important problems concerning set algebras which are open at this time, and we also take this opportunity to describe the status of the problems stated in [HMT] .

For reference in later articles, we refer to theorems, definitions, etc., by three figures, e.g. I.2.2 for the second item in section 2 of paper number I, which is just the present paper (see the initial footnote).

The very most basic results on set algebras were first described in the paper Henkin, Tarski [HT]. Other major results were obtained in Henkin, Monk [HM] . In preparing the present comprehensive discussion of set algebras many natural questions arose. Some of these questions were solved by the authors, and their solutions are found here. A large number of the questions were solved by H. Andréka and I. Németi. Where their solutions were short we have usually included the results here, with their permission, and we have indicated that the results are theirs. Many of their longer solutions will be found in the paper [AN3] following this one, which is organized parallel to our paper; a few of their related results are found in [AN2], [AN4], or [N]. In the course of our article we shall have occasion to mention explicitly most of their related results. We are indebted to Andréka and Németi for their considerable help in preparing this paper for publication.

The following set-theoretical notation not in [HMT] will be useful. If $f \in {}^{\alpha}U$, $\kappa < \alpha$, and $u \in U$, then $f^{\kappa}_{u}$ is the member of ${}^{\alpha}U$ such that $(f^{\kappa}_{u})\lambda = f\lambda$ if $\lambda \neq \kappa$, while $(f^{\kappa}_{u})\kappa = u$. For typographical reasons we sometimes write $f(\kappa/u)$ in place of $f^{\kappa}_{u}$.

## 1. Various set algebras

Definition I.1.1. (i) Let $U$ be a set, $\alpha$ an ordinal, and $V \subseteq {}^{\alpha}U$. For all $\kappa,\lambda < \alpha$ we set

$$D^{[V]}_{\kappa\lambda} = \{y \in V : y_{\kappa} = y_{\lambda}\} ,$$

and we let $C^{[V]}_{\kappa}$ be the mapping from $SbV$ into $SbV$ such that, for every $X \subseteq V$,

$$C^{[V]}_{\kappa}X = \{y \in V : y^{\kappa}_{u} \in X \text{ for some } u \in U\} .$$

(When $V$ is implicitly understood we shall write simply $D_{\kappa\lambda}$ or $C_{\kappa}$.)

(ii) $A$ is an $\alpha$-dimensional cylindric-relativized field of sets iff there is a set $U$ and a set $V \subseteq {}^{\alpha}U$ such that $A$ is a non-empty family of subsets of $V$ closed under all the operations $\cup$, $\cap$, ${}_{V}\sim$ and $C^{[V]}_{\kappa}$ (for each $\kappa < \alpha$), and containing as elements the subsets $D^{[V]}_{\kappa\lambda}$ (for all $\kappa,\lambda < \alpha$). The base of $A$ is the set $\bigcup_{x \in V} Rgx$.

(iii) $\mathfrak{A}$ is a cylindric-relativized set algebra of dimension $\alpha$ with base $U$ iff there is a set $V \subseteq {}^{\alpha}U$ such that

$\mathfrak{A} = \langle A, \cup, \cap, {}_V\sim, 0, V, C_\kappa^{[V]}, D_{\kappa\lambda}^{[V]}\rangle_{\kappa,\lambda<\alpha}$ , where A is an $\alpha$-dimensional cylindric-relativized field of sets with unit element V and base U . $Crs_\alpha$ is the class of all cylindric-relativized set algebras of dimension $\alpha$ . In case $A = SbV$ , the set A and the algebra $\mathfrak{A}$ are called respectively a <u>full cylindric-relativized field of sets</u> and a <u>full cylindric-relativized set algebra</u>.

(iv) Let U be a set and $\alpha$ an ordinal. ${}^\alpha U$ is then called the <u>Cartesian space with base</u> U <u>and dimension</u> $\alpha$ . Moreover, for every $p \in {}^\alpha U$ we set

$${}^\alpha U^{(p)} = \{x \in {}^\alpha U : \{\xi < \alpha : x_\xi \neq p_\xi\} \text{ is finite}\} ,$$

and we call ${}^\alpha U^{(p)}$ the <u>weak Cartesian space with base</u> U <u>and dimension</u> $\alpha$ <u>determined by</u> p .

For the following parts (v) - (ix) of this definition we assume that A is a cylindric-relativized field of sets and $\mathfrak{A}$ is a cylindric-relativized set algebra, both with dimension $\alpha$ , base U , and unit element V .

(v) A , respectively $\mathfrak{A}$ , is called an $\alpha$-<u>dimensional cylindric field of sets</u>, respectively <u>set algebra</u>, if $V = {}^\alpha U$ . The class of all cylindric set algebras of dimension $\alpha$ is denoted by $Cs_\alpha$ .

(vi) A , respectively $\mathfrak{A}$ , is called an $\alpha$-<u>dimensional weak cylindric field of sets</u>, respectively <u>set algebra</u>, if there is a $p \in {}^\alpha U$ such that $V = {}^\alpha U^{(p)}$ . The class of all $\alpha$-dimensional weak cylindric set algebras is denoted by $Ws_\alpha$ .

(vii) A , respectively $\mathfrak{A}$ , is called an $\alpha$-<u>dimensional generalized cylindric field of sets</u>, respectively <u>set algebra</u>, if V has the form $\bigcup_{i\in I} {}^\alpha Y_i$ , where $Y_i \neq 0$ for each $i \in I$ , and $Y_i \cap Y_j = 0$ for any two distinct $i, j \in I$ . The sets $Y_i$ are called the <u>subbases</u> of $\mathfrak{A}$ . The symbol $Gs_\alpha$ denotes the class of all generalized cylindric set algebras of dimension $\alpha$ .

(viii) A , respectively $\mathfrak{A}$ , is called an $\alpha$-dimensional generalized weak cylindric field of sets, respectively set algebra, if V has the form $\bigcup_{i\in I}{}^{\alpha}Y_i^{(pi)}$ , where $pi \in {}^{\alpha}Y_i$ for each $i \in I$ and ${}^{\alpha}Y_i^{(pi)} \cap {}^{\alpha}Y_j^{(pj)} = 0$ for any two distinct $i,j \in I$ . The sets $Y_i$ are called the subbases of $\mathfrak{A}$ . We use $Gws_\alpha$ for the class of all generalized weak cylindric set algebras of dimension $\alpha$.

(ix) An element $x \in A$ is regular provided that

$g \in x$ whenever $f \in x$ , $g \in V$ , and $(\Delta x \cup 1)1 f = (\Delta x \cup 1)1 g$ . A and $\mathfrak{A}$ are called regular if each $x \in A$ is regular. (We assume here $\alpha > 0$ ; if $\alpha = 0$ , all elements, as well as A and $\mathfrak{A}$ , are called regular.) K being any class included in $Crs_\alpha$ , we denote by $K^{reg}$ the class of all regular algebras which belong to K .

(x) Let $\mathfrak{A}$ be a cylindric algebra and $b \in A$ . We denote by $\mathfrak{Rl}_b\mathfrak{A}$ the algebra obtained by relativizing $\mathfrak{A}$ to b (see 2.2.1 for a formal definition). K being a class of cylindric algebras, we let $R\ell K = \{\mathfrak{Rl}_b\mathfrak{A} : \mathfrak{A} \in K , b \in A\}$ .

Remark I.1.2. The definition of cylindric set algebras in I.1.1(v) coincides with definition I.1.5. Other parts of I.1.1 are consistent with the informal definitions in 1.1.13 and 2.2.11 . Originally, regularity was defined only for cylindric set algebras, in the form given in the introduction. The general definition in I.1.1 (ix) is due to H. Andréka and I. Németi and turns out to have the desired meaning for cylindric set algebras as well as $Ws_\alpha$'s and $Gs_\alpha$'s ; see I.1.13 - I.1.16.

The inclusions holding among the various classes of set algebras introduced in I.1.1 are indicated in Figure I.1.3 for $\alpha > 0$ ; if we consider the classes of isomorphic images of the various set algebras, then the diagram collapses as indicated in Figure I.1.4 . The inclusions in each case are proper inclusions.

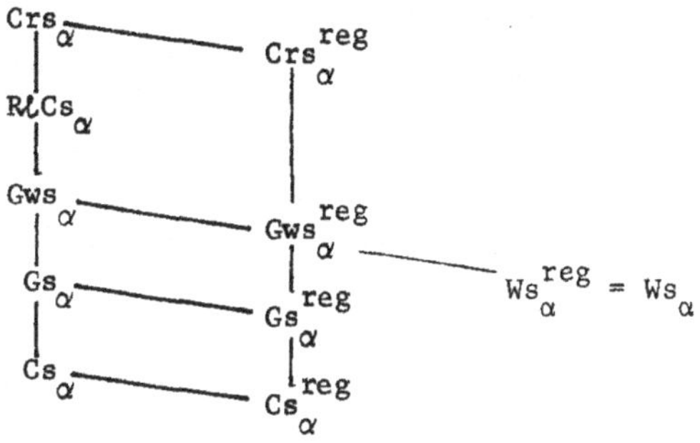


Figure I.1.3

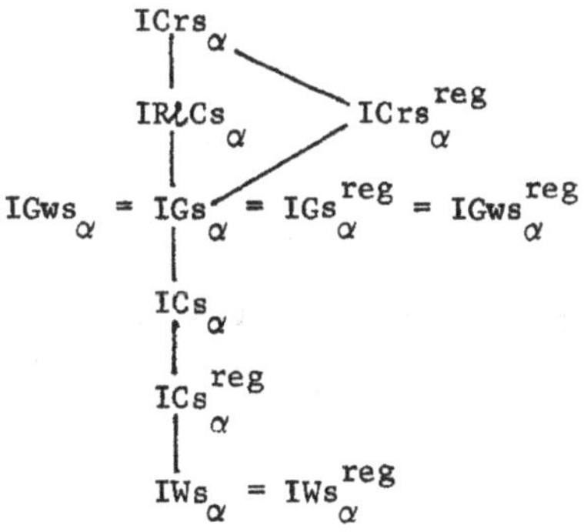


Figure I.1.4

In case $\alpha < \omega$, the classes $Ws_\alpha$ and $Cs_\alpha$ coincide, and so do $Gws_\alpha$ and $Gs_\alpha$ ; furthermore, under this assumption each member of any of these classes is regular. In the general case, $Gs_\alpha$'s are isomorphic to subdirect products of $Cs_\alpha$'s and conversely; similarly for $Gws_\alpha$'s and $Ws_\alpha$'s . Every $Ws_\alpha$ is regular. Every $Gs_\alpha$ is isomorphic to a regular $Gs_\alpha$ , and to a subdirect product of regular $Cs_\alpha$'s. Proofs of these facts and the relationships in the diagrams will be found at the appropriate places in this paper.

We begin our discussion by describing some degenerate cases of the notions in I.1.1 , and giving those inclusions between the classes which follow easily from the definitions. Throughout the paper we omit proofs which seem trivial.

<u>Corollary</u> I.1.5. Let $\mathfrak{A}$ be a $Crs_\alpha$ with base $U$ and unit element $V$ .

(i) $V = 0$ iff $|A| = 1$ ; if $V = \{0\}$ then $|A| = 2$ .

(ii) If $\alpha = 0$ , then $V \subseteq \{0\}$ .

(iii) If $\alpha > 0$ and $U = 0$ , then $V = 0$ .

Because of this theorem we will frequently make such assumptions as $\alpha > 0$ , $U \neq 0$ , or $V \neq 0$ .

<u>Corollary</u> I.1.6. If $\alpha < \omega$ , $U$ is any set, and $p \in {}^\alpha U$ , then ${}^\alpha U^{(p)} = {}^\alpha U$ . Hence for $\alpha < \omega$ we have $Gs_\alpha = Gws_\alpha$ , if $0 < \alpha < \omega$ we have $Cs_\alpha = Ws_\alpha \cup \{\mathfrak{A}_\alpha\}$ , where $\mathfrak{A}_\alpha$ is the unique $Cs_\alpha$ with universe 1 , and finally $Cs_0 = Ws_0$ .

<u>Corollary</u> I.1.7. If $\mathfrak{A}$ is a $Gws_\alpha$ with every subbase having only one element, then $\mathfrak{A}$ is a discrete $Gs_\alpha$ .

<u>Corollary</u> I.1.8. $Crs_1 = Gws_1 = Gs_1 = Cs_1 = Ws_1 \cup \{\mathfrak{B}\}$ , where $\mathfrak{B}$ is the $Crs_1$ with universe 1 ; $Crs_0 = Gws_0 = Gs_0 = \{\mathfrak{A}_1, \mathfrak{A}_2\}$ and $Cs_0 = Ws_0 = \{\mathfrak{A}_2\}$ , where $\mathfrak{A}_1$ and $\mathfrak{A}_2$ are the unique $Crs_0$'s with universes 1 and 2 respectively. Furthermore, every $Crs_0$ is full, and the base of any $Crs_0$ is 0 .

<u>Corollary</u> I.1.9. (i) For any $\alpha$ , $Cs_\alpha \cup Gs_\alpha \cup Ws_\alpha \cup Gws_\alpha \subseteq CA_\alpha$ .

(ii) For $\alpha \leq 1$ , $Crs_\alpha \subseteq CA_\alpha$ .

(iii) For $\alpha \geq 2$ , $Crs_\alpha \nsubseteq CA_\alpha$ .

Proof. Both (i) and (ii) are trivial. To establish (iii), we construct $\mathfrak{A} \in Crs_\alpha \sim CA_\alpha$ for $\alpha \geq 2$ by choosing any set $U$ with $|U| > 1$, taking $V = {}^\alpha U \sim D_{01}^{[{}^\alpha U]}$, and letting $\mathfrak{A}$ be the full $Crs_\alpha$ with unit element $V$ (I.1.1(iii)). We have $D_{01}^{[V]} = 0$, so that $V = D_{00}^{[V]} \neq C_1^{[V]}(D_{01}^{[V]} \cap D_{10}^{[V]}) = 0$. Thus $\mathfrak{A}$ fails to satisfy axiom $(C_6)$, whence $\mathfrak{A} \notin CA_\alpha$ and the proof is complete.

Corollary I.1.9(iii) explains why we shall not give many results concerning the class $Crs_\alpha$ in these papers; as indicated in the introduction, this class is introduced just to unify some definitions and results, and plays an auxiliary role in our discussion.

Corollary I.1.10. If $\alpha \geq 2$, then $Cs_\alpha \subset Gs_\alpha$ and $R\ell Cs_\alpha \subseteq Crs_\alpha = SR\ell Cs_\alpha$.

We shall be able to strengthen these results below by showing that $Gws_\alpha \subset R\ell Cs_\alpha$ for $\alpha \geq 2$ (see I.2.12 - I.2.13). It is known that although $R\ell Cs_0 = Crs_0$ and $R\ell Cs_1 = Cs_1$, we have $R\ell Cs_\alpha \subset Crs_\alpha$ for $\alpha > 1$, but we shall not give an example here; see [HR] and Prop.2.3 of [AN3] (p.155).

Corollary I.1.11. If $\alpha \geq \omega$, then $Ws_\alpha \subset Gws_\alpha \subset Crs_\alpha$ and $Cs_\alpha \subset Gs_\alpha \subset Gws_\alpha$.

Theorem I.1.12. Let $\mathfrak{A}$ be a $Gws_\alpha$ with unit element $\bigcup_{i\in I} {}^\alpha U_i^{(pi)}$, where ${}^\alpha U_i^{(pi)} \cap {}^\alpha U_j^{(pj)} = 0$ for all distinct $i, j \in I$. Assume that ${}^\alpha U_i^{(pi)} \in A$ for all $i \in I$. Also assume $\alpha \neq 1$.

Then for any $X \in A$ the following conditions are equivalent:

(i) $X = {}^{\alpha}U_i^{(pi)}$ for some $i \in I$ ;

(ii) $X$ is a minimal element of $\mathfrak{A}$ (under $\subseteq$ ) such that $X \neq 0$ and $\underset{\sim}{\Delta} X = 0$ .

Proof. (i) ⇒ (ii) . Clearly for any $i \in I$ we have $0 \neq {}^{\alpha}U_i^{(pi)}$ and $\underset{\sim}{\Delta}{}^{\alpha}U_i^{(pi)} = 0$ . Now suppose that $0 \neq Y \subseteq {}^{\alpha}U_i^{(pi)}$ and $\underset{\sim}{\Delta} Y = 0$ . Fix $y \in Y$ , and let $x \in {}^{\alpha}U_i^{(pi)}$ be arbitrary. There is then a finite $\Gamma \subseteq \alpha$ such that $(\alpha\sim\Gamma)1y = (\alpha\sim\Gamma)1x$ . Thus $x \in c_{(\Gamma)}Y = Y$, and hence $Y = {}^{\alpha}U_i^{(pi)}$.

(ii) ⇒ (i) . Assume (ii), and choose $i \in I$ such that $X \cap {}^{\alpha}U_i^{(pi)} \neq 0$ . Since $\underset{\sim}{\Delta}{}^{\alpha}U_i^{(pi)} = 0$ , we have $\underset{\sim}{\Delta}(X \cap {}^{\alpha}U_i^{(pi)}) = 0$ by 1.6.6 . Hence by (ii) , $X = X \cap {}^{\alpha}U_i^{(pi)}$ , while by the implication (i) ⇒ (ii) already established, ${}^{\alpha}U_i^{(pi)} = X \cap {}^{\alpha}U_i^{(pi)}$ . Thus (i) holds.

Now we discuss regular set algebras. In the case of the classes $Cs_\alpha$ , $Ws_\alpha$ , $Gs_\alpha$ the definition assumes a simpler form mentioned in the introduction.

Corollary I.1.13. Let $\mathfrak{A}$ be a $Cs_\alpha$ with base $U$ , and let $X \in A$ . Then the following conditions are equivalent:

(i) $X$ is regular;

(ii) for all $f \in X$ and all $g \in {}^{\alpha}U$ , if $\underset{\sim}{\Delta} X 1 f = \underset{\sim}{\Delta} X 1 g$ then $g \in X$ .

Proof. (ii) ⇒ (i) . Trivial. (i) ⇒ (ii) . Assume (i) and the hypotheses of (ii) . If $0 \in \underset{\sim}{\Delta} X$ , the desired conclusion $g \in X$ is obvious. Suppose therefore $0 \notin \underset{\sim}{\Delta} X$. Then $f^0_{g0} \in X$ since $0 \notin \underset{\sim}{\Delta} X$ and $(\underset{\sim}{\Delta} X \cup 1) 1 f^0_{g0} = (\underset{\sim}{\Delta} X \cup 1) 1 g$ . Hence $g \in X$ by (i) .

The next two Corollaries are proved in the same way as I.1.13.

<u>Corollary</u> I.1.14. Let $\mathfrak{A}$ be a $Ws_\alpha$ with unit element ${}^{\alpha}U^{(p)}$, and let $X \in A$. Then the following conditions are equivalent:

(i) $X$ is regular;

(ii) for all $f \in X$ and all $g \in {}^{\alpha}U^{(p)}$, if $\Delta X \upharpoonright f = \Delta X \upharpoonright g$ then $g \in X$.

<u>Corollary</u> I.1.15. Let $\mathfrak{A}$ be a $Gs_\alpha$ with unit element $\bigcup_{i \in I} {}^{\alpha}Y_i$, where $Y_i \cap Y_j = 0$ for $i \neq j$, and let $X \in A$. Then the following conditions are equivalent:

(i) $X$ is regular;

(ii) for all $i \in I$, all $f \in X \cap {}^{\alpha}Y_i$, and all $g \in {}^{\alpha}Y_i$, if $\Delta X \upharpoonright f = \Delta X \upharpoonright g$ then $g \in X$.

No analogous simplification of the notion of regularity for arbitrary $Gws_\alpha$'s is known. Weak cylindric set algebras are always regular:

<u>Corollary</u> I.1.16. $Ws_\alpha^{reg} = Ws_\alpha$.

<u>Proof</u>. Suppose that $\mathfrak{A}$ is a $Ws_\alpha$ with unit element ${}^{\alpha}U^{(p)}$. Also assume $X \in A$, $f \in X$, $g \in {}^{\alpha}U^{(p)}$, and $\Delta X \upharpoonright f = \Delta X \upharpoonright g$. Since $f, g \in {}^{\alpha}U^{(p)}$, there is a finite $\Gamma \subseteq \alpha$ with $(\alpha \sim \Gamma) \upharpoonright f = (\alpha \sim \Gamma) \upharpoonright g$. Hence $[\alpha \sim (\Gamma \sim \Delta X)] \upharpoonright f = [\alpha \sim (\Gamma \sim \Delta X)] \upharpoonright g$, so that $g \in c_{(\Gamma \sim \Delta X)} X = X$. Thus by I.1.14, $X$ is regular.

<u>Corollary</u> I.1.17. If $\alpha < \omega$, then $Cs_\alpha = Cs_\alpha^{reg}$, $Ws_\alpha = Ws_\alpha^{reg}$, $Gs_\alpha = Gs_\alpha^{reg}$, and $Gws_\alpha = Gws_\alpha^{reg}$.

<u>Corollary</u> I.1.18. If $\alpha \geq \omega$, then $Ws_\alpha^{reg} \subset Gws_\alpha^{reg} \subset Crs_\alpha^{reg}$ and

$Cs_\alpha^{reg} \subset Gs_\alpha^{reg} \subset Gws_\alpha^{reg}$ .

Proof. To produce a member of $Gws_\alpha^{reg} \sim Ws_\alpha^{reg}$ , let U and V be two disjoint sets with at least two elements, and let $p \in {}^\alpha U$ , $q \in {}^\alpha V$ be arbitrary. Let $\mathfrak{A}$ be the full $Gws_\alpha$ with unit element ${}^\alpha U^{(p)} \cup {}^\alpha V^{(q)}$ . Then it is easily checked, along the lines of the proof of I.1.16, that $\mathfrak{A} \in Gws_\alpha^{reg} \sim Ws_\alpha$ . It is also clear that $\mathfrak{A} \notin Gs_\alpha$ . If we let $\mathfrak{B}$ be the two-element $Crs_\alpha$ with unit element $\{p\}$ , we see that $\mathfrak{B} \in Crs_\alpha^{reg} \sim Gws_\alpha$ (provided that p has more than one element in its range). Finally, if U and V are as above and $\mathfrak{C}$ is the minimal subalgebra of the full $Gs_\alpha$ with unit element ${}^\alpha U \cup {}^\alpha V$ , then $\mathfrak{C} \in Gs_\alpha^{reg} \sim Cs_\alpha$ .

Corollary I.1.19. For $\alpha \geq \omega$ we have $Cs_\alpha^{reg} \subset Cs_\alpha$ , $Gs_\alpha^{reg} \subset Gs_\alpha$ , and $Gws_\alpha^{reg} \subset Gws_\alpha$ .

Proof. By I.1.11 it suffices to exhibit a $Cs_\alpha$ which is not regular. Let $\mathfrak{A}$ be the full $Cs_\alpha$ with base 2 . Let $p = \langle 0 : \kappa < \alpha \rangle$ . Then ${}^\alpha 2^{(p)} \in A$ and $\underset{\sim}{\Delta}{}^\alpha 2^{(p)} = 0$ . Hence ${}^\alpha 2^{(p)}$ is not regular, since in a regular $Cs_\alpha$ the only elements X such that $\underset{\sim}{\Delta} X = 0$ are the zero element and the unit element.

## 2. Relativization

Our basic kinds of set algebras have been defined in terms of relativization of full cylindric set algebras. We want to present here some results, I.2.5, I.2.9, and I.2.11, which involve relativization and exhibit important connections between $Cs_\alpha$'s and $Gws_\alpha$'s. First we establish an elementary characterization of unit sets of $Gws_\alpha$'s, which was mentioned in 2.2.11.

<u>Theorem</u> I.2.1. Let $\mathfrak{A}$ be a $Cs_\alpha$ with base $U$, and let

$$F = \{X \in A : s^\kappa_\lambda X \cap s^\lambda_\kappa X = X \text{ for all } \kappa,\lambda < \alpha\} .$$

Under these premises, dependent on an additional assumption imposed on $\alpha$, each of the following conditions (i)-(iii) is necessary and sufficient for any given set $V$ to belong to $F$ :

(i) $V$ belongs to $A$, assuming $\alpha \leq 1$ (so that in this case $F = A$ );

(ii) $V$ belongs to $A$ and is a Cartesian space, assuming $\alpha = 2$ ;

(iii) $V$ belongs to $A$ and is the unit element of a $Gws_\alpha$, assuming $\alpha \geq 3$ .

Under the same premises we obtain an additional conclusion:

(iv) assuming $2 \leq \alpha < \omega$, the conditions $C_{(\alpha\sim 1)}V = V$ and $V \in F$ are jointly equivalent to the condition that $V$ is the unit element of a $Cs_\alpha$ in case $\alpha = 2$, or a $Gs_\alpha$ in case $\alpha \geq 3$, with base $U$ .

<u>Proof</u>. The necessity and sufficiency of (i) is obvious, and so is the sufficiency of (ii) . To establish the sufficiency of (iii) , assume $\alpha \geq 3$ and consider a set $V \in A$ which is the unit element of a $Gws_\alpha$, say $V = \bigcup_{i\in I} {}^\alpha W_i^{(pi)}$ where ${}^\alpha W_i^{(pi)} \cap {}^\alpha W_j^{(pj)} = 0$ for any $i,j \in I$ with $i \neq j$ . Given any $\kappa,\lambda < \alpha$, we clearly have $V \subseteq s^\kappa_\lambda V \cap s^\lambda_\kappa V$ . Now suppose $f \in s^\kappa_\lambda V \cap s^\lambda_\kappa V$ ; thus, say, $f^\kappa_{f\lambda} \in {}^\alpha W_i^{(pi)}$ and $f^\lambda_{f\kappa} \in {}^\alpha W_j^{(pj)}$ . Pick a $\mu \in \alpha \sim \{\kappa,\lambda\}$ and notice that $f_\mu \in W_i \cap W_j$ . There is a finite $\Gamma \subseteq \alpha$ such that $(\alpha\sim\Gamma)1 f = (\alpha\sim\Gamma)1 pi = (\alpha\sim\Gamma)1 pj$ . If therefore we define $g$ by stipulating that $g\nu = f\mu$ for all $\nu \in \Gamma$, and $g\nu = f\nu$ for every $\nu \in \alpha \sim \Gamma$, we get $g \in {}^\alpha W_i^{(pi)} \cap {}^\alpha W_j^{(pj)}$ . Consequently $i = j$ and $f \in {}^\alpha W_i^{(pi)} \subseteq V$ . Hence $V = s^\kappa_\lambda V \cap s^\lambda_\kappa V$, and we conclude that $V \in F$, as desired.

We wish now to demonstrate the necessity of both (ii) and (iii). To this end, we show first that every set $V \in F$ is the unit element of a $Gws_\alpha$. (This portion of the proof is due to Andréka and Németi.) For each $f \in V$ let

$$Y_f = \{u \in U : f^0_u \in V\} .$$

We need the following facts about $Y_f$.

(1) If $f \in V$, then $f \in {}^\alpha Y_f$.

For, if $\kappa < \alpha$, then $f \in V \subseteq s^0_\kappa V$, so $f^0_{f\kappa} \in V$ and hence $f\kappa \in Y_f$.

(2) If $f \in V$ and $u \in Y_f$, then $Y_f = Y_{f(0/u)}$.

(3) If $f \in V$, $u \in Y_f$, and $\kappa < \alpha$, then $f^\kappa_u \in V$.

For we may assume that $\kappa \neq 0$. Then $f^0_u \in V \subseteq s^\kappa_0 V$, so $f^{0\kappa}_{uu} \in V$. Also, $f \in V \subseteq s^\kappa_0 V$, whence $f^\kappa_{f0} \in V$. Hence $f^\kappa_u \in s^0_\kappa V \cap s^\kappa_0 V = V$, as desired.

(4) If $f \in V$, $\kappa < \alpha$, and $u \in Y_f$, then $f^\kappa_u \in V$ and $Y_f \subseteq Y_{f(\kappa/u)}$.

Indeed, (4) clearly follows from (2) if $\kappa = 0$. Assume $\kappa \neq 0$, and assume the hypothesis of (4). Then $f^\kappa_u \in V$ by (3). Consider now any $v \in Y_f$. We have $f^0_u, f^0_v \in V \subseteq s^\kappa_0 V$, so that $f^{0\kappa}_{uu}, f^{0\kappa}_{vv} \in V$. Hence $f^{0\kappa}_{vu} \in s^\kappa_0 V \cap s^0_\kappa V = V$. Thus $v \in Y_{f(\kappa/u)}$, as desired. From (4) we get (5) and (6):

(5) ${}^\alpha Y_f^{(f)} \subseteq V$.

(6) If $f, g \in V$ and $f \in {}^\alpha Y_g^{(g)}$, then $Y_g \subseteq Y_f$.

(7) If $f, g \in V$ and ${}^\alpha Y_f^{(f)} \subseteq {}^\alpha Y_g^{(g)}$, then ${}^\alpha Y_f^{(f)} = {}^\alpha Y_g^{(g)}$.

In fact, $f \in {}^{\alpha}Y_f^{(f)}$, so that $Y_g \subseteq Y_f$ by (6), whence ${}^{\alpha}Y_f^{(f)} = {}^{\alpha}Y_g^{(g)}$.

(8) If $f,g \in V$ and ${}^{\alpha}Y_f^{(f)} \cap {}^{\alpha}Y_g^{(g)} \neq 0$, then ${}^{\alpha}Y_f^{(f)} = {}^{\alpha}Y_g^{(g)}$.

Let indeed $h \in {}^{\alpha}Y_f^{(f)} \cap {}^{\alpha}Y_g^{(g)}$. Then by (6), $Y_f \cup Y_g \subseteq Y_h$, so that ${}^{\alpha}Y_f^{(f)} \subseteq {}^{\alpha}Y_h^{(h)}$ and ${}^{\alpha}Y_g^{(g)} \subseteq {}^{\alpha}Y_h^{(h)}$. Therefore, by (7), ${}^{\alpha}Y_f^{(f)} = {}^{\alpha}Y_g^{(g)}$.

By (8), $V$ is the unit element of a $Gws_\alpha$, as desired in (iii). In case $\alpha = 2$, we have ${}^{\alpha}Y_f^{(f)} = {}^{\alpha}Y_f$ for all $f \in V$. Furthermore, if $f,g \in V$ then $f^0_{g1} \in s^0_1 V \cap s^1_0 V \subseteq V$, and consequently, $g1 \in Y_f$. Thus by (1), $Y_f \cap Y_g \neq 0$ and hence ${}^{\alpha}Y_f \cap {}^{\alpha}Y_g \neq 0$. Therefore ${}^{\alpha}Y_f = {}^{\alpha}Y_g$ by (8). It follows that $V$ is a Cartesian space, as desired in (ii). Finally, (iv) is clear since the condition $C_{(\alpha\sim 1)}V = 1$ is equivalent to $\bigcup_{p \in V} Rgp = U$.

<u>Theorem</u> I.2.2. Assume that $\alpha \geq 3$. Let $\mathfrak{A}$ be the full $Cs_\alpha$ with base a set $U$, and let $\mathfrak{B}$ be a $Crs_\alpha$ with unit set $V \subseteq {}^{\alpha}U$. Then the following conditions are equivalent:

(i) $\mathfrak{B}$ is a $Gws_\alpha$;

(ii) $s^\kappa_\lambda V \cap s^\lambda_\kappa V = V$ for all $\kappa, \lambda < \alpha$ (operating in $\mathfrak{A}$).

<u>Proof</u>: By I.2.1.

<u>Remark</u> I.2.3. For $\alpha \geq 3$ Theorem I.2.2 gives a simple elementary characterization of those unit elements of $Gws_\alpha$'s which are members of a $Cs_\alpha$. For $\alpha \leq 1$ the corresponding result is trivial, in view of I.1.8. On the other hand, for $\alpha = 2$, there is no elementary characterization of those elements of a $Cs_2$ which are unit elements of a $Gs_2$. In fact, let $U_0$ and $U_1$ be disjoint sets each having at least two elements. Let $\mathfrak{A}$ be the full $Cs_2$ with base $U_0 \cup U_1$, let $X = {}^2U_0 \cup {}^2U_1$, and let $\mathfrak{B} = \mathfrak{Sg}^{(\mathfrak{A})}X$. One easily checks that $\mathfrak{B}$ is atomic and actually possesses just

three atoms $D_{01}, X \sim D_{01}$ , and ${}^2(U_0 \cup U_1) \sim X$ and that its structure is completely determined by the fact that $C_0Y = C_1Y = 1$ for each atom $Y$ . It follows directly that there is an automorphism of $\mathfrak{B}$ which interchanges the atoms $X \sim D_{01}$ and ${}^2(U_0 \cup U_2) \sim X$ , and which therefore interchanges the elements $X$ and $D_{01} \cup {}^2(U_0 \cup U_1) \sim X$ . As $X$ is the unit element of a $Gs_2$ , this implies that $D_{01} \cup {}^2(U_0 \cup U_1) \sim X$ satisfies all first-order (and indeed higher-order) sentences that hold for all unit elements of $Gs_2$'s (in arbitrary $Cs_2$'s). As $D_{01} \cup ({}^2(U_0 \cup U_1) \sim X)$ is not itself such a unit element, no elementary (or even higher-order) characterization of such elements of $Cs_2$'s is possible. It is also of interest that if $\mathfrak{A}$ is a full $Cs_2$ with base $U$ , then an element $V \in A$ is a unit element of a $Gs_2$ iff $V$ is an equivalence relation with field included in $U$ .

**In [AN3] it is shown that for $\alpha \geq \omega$ there is no elementary characterization of the unit elements of $Gs_\alpha$'s which are members of a given $Cs_\alpha$.**

The other main results of this section are related to the following obvious theorem.

<u>Theorem</u> I.2.4. For any algebra $\mathfrak{A}$ similar to $CA_\alpha$'s the following three conditions are equivalent:

(i) $\mathfrak{A} \in Crs_\alpha$ ;

(ii) $\mathfrak{A} \subseteq \mathfrak{Rl}_V\mathfrak{B}$ for some full $Cs_\alpha\mathfrak{B}$ and some $V \in B$ ;

(iii) $\mathfrak{A} \subseteq \mathfrak{Rl}_V\mathfrak{B}$ for some $Cs_\alpha\mathfrak{B}$ and some $V \in B$.

The question naturally arises whether the condition $\mathfrak{A} \subseteq \mathfrak{Rl}_V\mathfrak{B}$ in (iii) can be replaced by $\mathfrak{A} = \mathfrak{Rl}_V\mathfrak{B}$ . For arbitrary $Crs_\alpha$'s the answer in general is negative. Here we want to carefully consider this question for our classes $Ws_\alpha$ , $Gs_\alpha$ , and $Gws_\alpha$ . We begin with the following simple result.

Theorem I.2.5. $Ws_\alpha \subseteq Rl Cs_\alpha$. More specifically, assuming $\mathfrak{A}$ is a $Ws_\alpha$ with base $U$ and unit set $V$, we have $\mathfrak{A} = \mathfrak{Rl}_V\mathfrak{B}$ for some $Cs_\alpha\ \mathfrak{B}$ with $V \in B$; in fact, if we let $\mathfrak{C}$ be the full $Cs_\alpha$ with base $U$, we can take $\mathfrak{Sg}^{(\mathfrak{C})}A$ for $\mathfrak{B}$.

Proof. Say $V = {}^\alpha U^{(p)}$. Obviously $\mathfrak{A} \subseteq \mathfrak{Rl}_V\mathfrak{B}$. For the converse first note that $\underset{\sim}{\Delta}^{(\mathfrak{C})}V = 0$, and hence $A \subseteq \{X \in C : X \cap V \in A\} \in Su\mathfrak{C}$. Hence $B \subseteq \{X \in C : X \cap V \in A\}$ so $\mathfrak{Rl}_V\mathfrak{B} \subseteq \mathfrak{A}$, as desired.

Remarks I.2.6. The last part of I.2.5 does not extend to $Gws_\alpha$'s with $\alpha \geq \omega$. In fact, let $U_0 = \omega$, $U_1 = \omega \sim 1$, $p = \langle 0 : \kappa < \alpha\rangle$, and $q = \langle 1 : \kappa < \alpha\rangle$. Thus ${}^\alpha U_0^{(p)} \cap {}^\alpha U_1^{(q)} = 0$. Let $V = {}^\alpha U_0^{(p)} \cup {}^\alpha U_1^{(q)}$ and let $\mathfrak{C}$ be the full $Cs_\alpha$ with base $\omega$; note that $U_0 \cup U_1 = \omega$, so that $V \in C$. We let $\mathfrak{A}$ be the minimal subalgebra of $\mathfrak{Rl}_V\mathfrak{C}$. Thus $\mathfrak{A}$ is a $Gws_\alpha$. It has characteristic $0$ (cf. 2.4.61), and hence $\mathfrak{A}$ is simply the Boolean subalgebra of $\mathfrak{Rl}_V\mathfrak{C}$ generated by its diagonal elements. On the other hand, if $\mathfrak{D}$ is any $Cs_\alpha$ with $V \in D$ and with base $\omega$, then in $\mathfrak{D}$

$$c_0(-V) \cap V = {}^\alpha U_1^{(q)},$$

so that ${}^\alpha U_1^{(q)}$ is in $\mathfrak{Rl}_V\mathfrak{D}$ but not in $A$. Hence $\mathfrak{A} \neq \mathfrak{Rl}_V\mathfrak{D}$.

The $Gws_\alpha$ just constructed is unusual, in that the subbase $U_1$ is included in the subbase $U_0$. Usually a $Gws_\alpha$ is normal, in the following sense. Let $\mathfrak{B}$ be a $Gws_\alpha$, $\alpha \geq \omega$, and suppose the unit element of $\mathfrak{B}$ is $\bigcup_{i\in I} {}^\alpha W_i^{(p_i)}$, with ${}^\alpha W_i^{(p_i)} \cap {}^\alpha W_i^{(p_j)} = 0$ whenever $i, j \in I$ and $i \neq j$. We call $\mathfrak{B}$ <u>normal</u> if $W_i = W_j$ or $W_i \cap W_j = 0$ for all $i, j$; <u>widely-distributed</u> if $W_i \cap W_j = 0$ whenever $i \neq j$; <u>compressed</u> if

$W_i = W_j$ for all $i,j$. Thus every $Gs_\alpha$ is a normal $Gws_\alpha$, and every $Cs_\alpha$ is a compressed $Gws_\alpha$. It follows from results which will be established in section 7 of this paper that every $Gws_\alpha$ is isomorphic to a widely-distributed $Gws_\alpha$. It is also clear that every compressed $Gws_\alpha\ \mathfrak{B}$ is a direct factor of a $Cs_\alpha$ which has the same base as $\mathfrak{B}$; so $\mathfrak{B}$ is a $Rl Cs_\alpha$ with that same base.

We do not know whether a normal $Gws_\alpha\ \mathfrak{A}$ with $\alpha \geq \omega$ can always be obtained by relativization from a $Cs_\alpha$ having the same base as $\mathfrak{A}$. We show below in I.2.9 that this is true in case $3 \leq \alpha < \omega$ and $\mathfrak{A}$ is a $Gs_\alpha$. On the other hand, we show in I.2.11 that in case $\alpha \geq \omega$ we have at least $Gws_\alpha \subseteq Rl Cs_\alpha$. (Thus, of course, the last part of I.2.5 holds for $Gws_\alpha$'s with $\alpha < \omega$, since we have then $Gws_\alpha = Gs_\alpha$. We also have $Gs_\alpha \subseteq Rl Cs_\alpha$ in case $\alpha \geq \omega$, since $Gs_\alpha \subseteq Gws_\alpha$.) I.2.9 is due to Henkin, while I.2.11 has been obtained jointly by Henkin and Monk.

We need two lemmas about arbitrary $CA_\alpha$'s. In connection with these lemmas recall that, by 2.2.10, $\mathfrak{Rl}_b \mathfrak{A}$ is a $CA_\alpha$ for every $\mathfrak{A} \in CA_\alpha$ and every $b \in A$ having the property that $s^\kappa_\lambda b \cdot s^\lambda_\kappa b = b$ for all $\kappa, \lambda < \alpha$.

<u>Lemma</u> I.2.7. Let $\mathfrak{A}$ be any $CA_\alpha$, let $b \in A$, and assume that $s^\kappa_\lambda b \cdot s^\lambda_\kappa b = b$ for all $\kappa,\lambda < \alpha$. Let $+', -', c'_\kappa$, etc. be the operations of the algebra $\mathfrak{Rl}_b \mathfrak{A}$. The following conditions (i)-(vi) hold then for any $\kappa,\lambda < \alpha$, any finite $\Gamma,\Delta \subseteq \alpha$, and any $x,y \leq b$:

(i) If $\Gamma \cup \Delta \subset \alpha$ and $\Gamma \cap \Delta = 0$, then $c_{(\Gamma)} b \cdot c_{(\Delta)} b = b$.

(ii) If $\Gamma \cup \{\kappa\} \subset \alpha$ then $b \cdot c_{(\Gamma \cup \{\kappa\})} x \leq c_\kappa (b \cdot c_{(\Gamma)} x)$.

(iii) If $\Gamma \subset \alpha$, then $b \cdot c_{(\Gamma)} x = c'_{(\Gamma)} x$.

(iv) $c_{(\Gamma)} c'_{(\Gamma)} x = c_{(\Gamma)} x$.

(v) If $\Gamma \cup \Delta \subset \alpha$, then $c_{(\Gamma)} x \cdot c_{(\Delta)} y = c_{(\Gamma \cap \Delta)} (c'_{(\Gamma)} x \cdot c'_{(\Delta)} y)$.

(vi) If $\Gamma \subset \alpha$, then $-c_{(\Gamma)}x = c_{(\Gamma)}(b \cdot -c'_{(\Gamma)}x) + -c_{(\Gamma)}b$.

Furthermore, assume that $\alpha < \omega$ and that $c_{(\alpha\sim 1)}b = 1$. For all $\kappa,\lambda < \alpha$ let $e_{\kappa\lambda} = c_{(\alpha\sim\{\kappa,\lambda\})}b$. Then the following conditions (vii)-(xii) hold.

(vii) $c_{(\alpha\sim\{\kappa\})}b = 1$

(viii) $c_{(\Gamma)}b = \prod_{\mu,\nu\in\alpha\sim\Gamma} e_{\mu\nu}$.

(ix) $c_\mu e_{\kappa\lambda} = e_{\kappa\lambda}$ whenever $\mu < \alpha$ and $\mu \neq \kappa,\lambda$.

(x) $d_{\kappa\lambda} \leq e_{\kappa\lambda}$.

(xi) $d_{\kappa\lambda} = c_{(\alpha\sim\{\kappa,\lambda\})}(b \cdot d_{\kappa\lambda})$.

(xii) Suppose that $\nu \in \omega \sim 1$, $\lambda \in {}^\nu\alpha$, $\lambda_0 \notin \Gamma \cup \{\lambda_1,\ldots,\lambda_{\nu-1}\}$, $\Gamma \cup \{\lambda_0\} \subset \alpha$, and $w = c_{(\Gamma)}x \cdot \sum_{\iota\in\nu\sim 1} e_{\lambda_0\lambda_\iota}$. Then

$$c_{\lambda_0}(c_{(\Gamma)}x \cdot \prod_{\iota\in\nu\sim 1} -e_{\lambda_0\lambda_\iota}) = c_{(\Gamma\cup\{\lambda_0\})}x \cdot -c_{\lambda_0}w .$$

Proof. (i). We prove (i) by induction on $|\Gamma \cup \Delta|$. It is obvious for $\Gamma = 0$ or $\Delta = 0$, so suppose $\Gamma \neq 0 \neq \Delta$. If $|\Gamma \cup \Delta| = 2$, then $|\Gamma| = 1 = |\Delta|$; say $\Gamma = \{\kappa\}$ and $\Delta = \{\lambda\}$. Then

$$b \leq c_\kappa b \cdot c_\lambda b \leq c_\kappa s^\kappa_\lambda b \cdot c_\lambda s^\lambda_\kappa b = s^\kappa_\lambda b \cdot s^\lambda_\kappa b = b ,$$

so (i) holds in this case. Now suppose that $|\Gamma \cup \Delta| > 2$ and that (i) holds for sets $\Gamma'$, $\Delta'$ with $|\Gamma' \cup \Delta'| < |\Gamma \cup \Delta|$. Say $|\Delta| > 1$. Choose $\kappa \in \Delta$ and $\lambda \in \alpha \sim (\Gamma \cup \Delta)$. Then

$$\begin{aligned} b &\leq c_{(\Gamma)}b \cdot c_{(\Delta)}b \leq c_{(\Gamma)}b \cdot c_{(\Gamma)}s^\kappa_\lambda b \cdot c_{(\Delta)}s^\kappa_\lambda b \\ &= c_{(\Gamma)}b \cdot s^\kappa_\lambda c_{(\Gamma)}b \cdot s^\kappa_\lambda c_{(\Delta\sim\{\kappa\})}b \\ &= c_{(\Gamma)}b \cdot s^\kappa_\lambda(c_{(\Gamma)}b \cdot c_{(\Delta\sim\{\kappa\})}b) \\ &= c_{(\Gamma)}b \cdot s^\kappa_\lambda b \leq c_{(\Gamma)}b \cdot c_\kappa b = b . \end{aligned}$$

Thus (i) has been established.

(ii). Assume $\Gamma \cup \{\kappa\} \subset \alpha$. We may assume that $\kappa \notin \Gamma$. Then

$$\begin{aligned} b \cdot c_{(\Gamma\cup\{\kappa\})}x \le c_\kappa b \cdot c_\kappa c_{(\Gamma)}x &= c_\kappa(c_\kappa b \cdot c_{(\Gamma)}x) \\ &= c_\kappa(c_\kappa b \cdot c_{(\Gamma)}b \cdot c_{(\Gamma)}x) \\ &= c_\kappa(b \cdot c_{(\Gamma)}x) \qquad \text{(by (i))}, \end{aligned}$$

and (ii) holds.

(iii). This follows from (ii) by an easy induction on $|\Gamma|$.

(iv). Obvious, by direct calculation.

(v). Assume that $\Gamma \cup \Delta \subset \alpha$. Then

$$\begin{aligned} c_{(\Gamma)}x \cdot c_{(\Delta)}y &= c_{(\Gamma)}x \cdot c_{(\Delta)}c'_{(\Delta)}y \qquad \text{(by (iv))} \\ &= c_{(\Gamma\cap\Delta)}(c_{(\Gamma)}x \cdot c_{(\Delta\sim\Gamma)}c'_{(\Delta)}y) \\ &= c_{(\Gamma\cap\Delta)}(c_{(\Gamma)}x \cdot c_{(\Delta\sim\Gamma)}c'_{(\Delta)}y \cdot b) \qquad \text{(by (i))} \\ &= c_{(\Gamma\cap\Delta)}(c'_{(\Gamma)}x \cdot c'_{(\Delta\sim\Gamma)}c'_{(\Delta)}y) \qquad \text{(by (iii))} \\ &= c_{(\Gamma\cap\Delta)}(c'_{(\Gamma)}x \cdot c'_{(\Delta)}y) . \end{aligned}$$

(vi) Assume $\Gamma \subset \alpha$. Then

$$\begin{aligned} c_{(\Gamma)}(b \cdot -c'_{(\Gamma)}x) &= c_{(\Gamma)}(b \cdot -(b \cdot c_{(\Gamma)}x)) \qquad \text{(by (iii))} \\ &= c_{(\Gamma)}(b \cdot -c_{(\Gamma)}x) \\ &= c_{(\Gamma)}b \cdot -c_{(\Gamma)}x , \end{aligned}$$

and (vi) follows, since $c_{(\Gamma)}x \le c_{(\Gamma)}b$.

Now we assume the additional premises for (vii)-(ix).

(vii) We may assume that $\kappa \neq 0$. Then

$$\begin{aligned} 1 = s^0_\kappa c_{(\alpha\sim 1)} b &= s^0_\kappa c_{(\alpha\sim\{0,\kappa\})} c_\kappa b \\ &= c_{(\alpha\sim\{0,\kappa\})} s^0_\kappa c_\kappa (s^0_\kappa b \cdot s^\kappa_0 b) \\ &= c_{(\alpha\sim\{\kappa\})} (d_{0\kappa} \cdot s^\kappa_0 b \cdot c_\kappa c_0 (d_{0\kappa} \cdot b)) \\ &= c_{(\alpha\sim\{\kappa\})} (d_{0\kappa} \cdot b \cdot c_\kappa c_0 (d_{0\kappa} \cdot b)) \\ &\le c_{(\alpha\sim\{\kappa\})} b \ . \end{aligned}$$

(viii). Since $c'_{(\Delta)} b = b$ for all finite $\Delta \subseteq \alpha$ , this is an immediate consequence of (v) and (vii) .

(ix). Obvious.

(x). We have, assuming $\kappa \neq \lambda$ ,

$$\begin{aligned} d_{\kappa\lambda} &= d_{\kappa\lambda} \cdot c_{(\alpha\sim\{\kappa\})} b & \text{(by (vii))} \\ &= d_{\kappa\lambda} \cdot c_{(\alpha\sim\{\kappa,\lambda\})} c_\lambda b \\ &\le c_{(\alpha\sim\{\kappa,\lambda\})} (d_{\kappa\lambda} \cdot s^\lambda_\kappa b) \\ &= c_{(\alpha\sim\{\kappa,\lambda\})} (d_{\kappa\lambda} \cdot b) \le e_{\kappa\lambda} \ . \end{aligned}$$

(xi). It follows immediately from (x) .

(xii). Assume the hypothesis of (xii) . Also let $z = c_{(\Gamma)} x \cdot \Pi_{\iota \in \nu\sim 1} - e_{\lambda_0 \lambda_\iota}$ . Then

$$c_{\lambda_0} z + c_{\lambda_0} w = c_{\lambda_0} (z + w) = c_{\lambda_0} c_{(\Gamma)} x \ .$$

Hence it suffices to show that $c_{\lambda_0} z \cdot c_{\lambda_0} w = 0$ , and for this it is enough to show that $z \cdot c_{\lambda_0} w = 0$ or, by the definition of $w$ , to take any $\iota \in \nu \sim 1$ and show that $z \cdot c_{\lambda_0} (c_{(\Gamma)} x \cdot e_{\lambda_0 \lambda_\iota}) = 0$ . And in fact

$$z \cdot c_{\lambda_0}(c_{(\Gamma)}x \cdot e_{\lambda_0\lambda_\iota}) \le c_{(\Gamma)}b \cdot -e_{\lambda_0\lambda_\iota} \cdot$$

$$c_{\lambda_0}(c_{(\Gamma)}b \cdot c_{(\alpha\sim\{\lambda_0,\lambda_\iota\})}b)$$

$$= c_{(\Gamma)}b \cdot -e_{\lambda_0\lambda_\iota} \cdot c_{((\Gamma\cup\{\lambda_0\})\sim\{\lambda_\iota\})}b \quad \text{(by (v))}$$

$$= c_{(\Gamma\sim\{\lambda_0,\lambda_\iota\})}b \cdot -e_{\lambda_0\lambda_\iota} = 0 \ .$$

This completes the proof of I.2.7 .

<u>Lemma</u> I.2.8. Let $3 \le \alpha < \omega$ , and let $\mathfrak{A}$ be a simple $CA_\alpha$ . Let $b \in A$ satisfy the following conditions:

(i) $s^\kappa_\lambda b \cdot s^\lambda_\kappa b = b$ for all $\kappa, \lambda < \alpha$ .

(ii) $c_{(\alpha\sim 1)}b = 1$ .

(iii) Setting

$$e_{\xi\zeta} = c_{(\alpha\sim\{\xi\zeta\})}b \quad \text{for all } \xi,\zeta < \alpha \ ,$$

we have, for every $x \le b$ , every $\nu \in \omega \sim 1$ , and every sequence $\lambda \in {}^\nu\alpha$ without repeating terms,

$$c_{\lambda_0}(c_{(\alpha\sim\{\lambda_0\})}x \cdot \Pi_{\iota<\nu-1} - e_{\lambda_0\lambda_\iota}) = z_\nu + \sum_{\kappa<\nu}(z_\kappa \cdot -z_{\kappa+1} \cdot w_\kappa) \ ,$$

where the sequences $z$ and $w$ are uniquely determined by the following stipulations:

$$z_\kappa = c_{(Rg\lambda)}(\Pi_{\iota<\kappa} c_{(\alpha\sim\{\lambda_\iota\})}x \cdot \Pi_{\iota<\theta<\kappa} - e_{\lambda_\iota\lambda_\theta}) \quad \text{for each } \kappa \le \nu \ ,$$

$$S_\kappa = \{\Delta : \Delta \subsetneq \nu - 1, \ |\Delta| = \kappa\} \quad \text{for each } \kappa < \nu \ ,$$

$$w_\kappa = -\sum_{\Delta\in S_\kappa}(\Pi_{\iota\in\Delta} c_{(\alpha\sim\{\lambda_\iota\})} c'_{(\alpha)}x \cdot \Pi_{\iota,\theta\in\Delta,\iota\ne\theta} - e_{\lambda_\iota\lambda_\theta}) \quad (\text{with } c'_{(\alpha)}$$

denoting the generalized cylindrification in $\mathfrak{Rl}_b\mathfrak{A}$ ) for each $\kappa < \nu$ .

Under this hypothesis we have $\mathfrak{B} = \mathfrak{Rl}_b \mathfrak{Sg}^{(\mathfrak{A})} B$ whenever $\mathfrak{B} \subseteq \mathfrak{Rl}_b \mathfrak{A}$ .

Proof. By the assumptions (i) and (ii) we have all the parts of I.2.7 available. Again, we denote operations of $\mathfrak{Rl}_b\mathfrak{A}$ by $-'$, $d'_{\kappa\lambda}$, $c'_\kappa$, etc. Now let $A^* = \{c_{(\Gamma)}x : x \in B \text{ and } \Gamma \subseteq \alpha\} \cup \{-e_{\kappa\lambda} : \kappa,\lambda < \alpha\}$ , and let $A^{**}$ be the set of all finite sums of finite products of elements of $A^*$ . Obviously $A^{**} \subseteq \mathfrak{Sg}^{(\mathfrak{A})}B$ . In view of I.2.7 (iii) it is now clearly enough to see how to prove the converse, and in fact to show that $B \subseteq A^{**} \in \mathsf{Su}\mathfrak{A}$ . Obviously $B \subseteq A^{**}$. For any $\kappa,\lambda < \alpha$ we have $d_{\kappa\lambda} \in A^* \subseteq A^{**}$ by I.2.7 (xi). Clearly $A^{**}$ is closed under $+$ . To show that $A^{**}$ is closed under $-$ , it suffice to show that $-y \in A^{**}$ for every $y \in A^*$ . If $y = -e_{\kappa\lambda}$ , then $-y = e_{\kappa\lambda} = c_{(\alpha\sim\{\kappa\lambda\})}b \in A^*$ . Suppose that $y = c_{(\Gamma)}x$ . If $\Gamma = \alpha$ , then $y = 1$ or $y = 0$ by simplicity, so $-y \in A^{**}$ . If $\Gamma \neq \alpha$ , then by I.2.7 (vi) we have

$$-y = -c_{(\Gamma)}x = c_{(\Gamma)}(b \cdot -c'_{(\Gamma)}x) + -c_{(\Gamma)}b ,$$

and obviously $c_{(\Gamma)}(b \cdot -c'_{(\Gamma)}x) \in A^* \subseteq A$ , while $-c_{(\Gamma)}b \in A^{**}$ by I.2.7 (viii) .

Finally, we must show that $A^{**}$ is closed under each $c_\lambda$ . (Here we give an algebraic version of an elimination of quantifiers.) From the definition of $A^{**}$ we see that it suffices to show that $c_\lambda y \in A^{**}$ for every $y$ which is a product of elements of $A^*$ . Say

$$y = \Pi_{\iota<\beta} c_{(\Gamma_\iota)} x_\iota \cdot \Pi_{\iota<\gamma} - e_{\kappa_\iota \mu_\iota}$$

with each $x_\iota \in B$ , where $\beta,\gamma < \omega$ . Clearly we may assume that $\lambda \notin \Gamma_\iota$ for each $\iota < \beta$ , and $\lambda \in \{\kappa_\iota, \mu_\iota\}$ for each $\iota < \gamma$ . Thus by I.2.7 (v) we may assume that $\beta \leq 1$ . Since $c_{(\alpha\sim\{\lambda\})}b = 1$ , by I.2.7 (vii) , we may assume that $\beta = 1$ . Thus we are concerned with y of the form

$$y = c_{(\Delta)}x \cdot \Pi_{\iota<\gamma} - e_{\lambda\nu_\iota}$$

where $x \in B$ , $\lambda \notin \Delta$ , and $\lambda \neq \nu_\iota$ for each $\iota < \gamma$ , and $\nu_\iota \neq \nu_\theta$ in case $\iota < \theta < \gamma$ . If $\Delta \cup \{\lambda\} \subset \alpha$ , then by I.2.7 (xii) we have

$$c_\lambda y = c_{(\Delta\cup\{\lambda\})}x \cdot -c_\lambda w ,$$

where $w = c_{(\Delta)}x \cdot \Sigma_{\iota<\gamma} e_{\lambda\nu_\iota}$ and

$$\begin{aligned} c_\lambda w &= c_\lambda(c_{(\Delta)}x \cdot \Sigma_{\iota<\gamma} e_{\lambda\nu_\iota}) \\ &= c_\lambda \Sigma_{\iota<\gamma}(c_{(\Delta)}x \cdot c_{(\alpha\sim\{\lambda,\nu_\iota\})}b) \\ &= c_\lambda \Sigma_{\iota<\gamma} c_{(\Delta\sim\{\nu_\iota\})} c'_{(\Delta)}x \text{ (by I.2.7(v))} \\ &= \Sigma_{\iota<\gamma} c_{((\Delta\cup\{\lambda\})\sim\{\nu_\iota\})} c'_{(\Delta)}x \in A^{**} , \end{aligned}$$

so that $c_\lambda y \in A^{**}$ . On the other hand, if $\Delta \cup \{\lambda\} = \alpha$ , then $c_\lambda y \in A^{**}$ by premise (iii) of our lemma, since under the notation used there each $z_\kappa$ is closed and hence is either 0 or 1 by the simplicity of $\mathfrak{A}$ while obviously each $w_\kappa$ is in $A^{**}$ .

This completes the proof.

Theorem I.2.9. In case $3 \leq \alpha < \omega$ we have $Gs_\alpha \subseteq \mathbf{Rl}Cs_\alpha$ . More specifically, assuming $\mathfrak{A}$ to be a $Gs_\alpha$ with base U and unit element V , we have $\mathfrak{A} = \mathfrak{Rl}_V\mathfrak{B}$ for some $Cs_\alpha$ $\mathfrak{B}$ with $V \in B$ ; in fact, if we let $\mathfrak{C}$ be the full $Cs_\alpha$ with base U , we can take $\mathfrak{Sg}^{(\mathfrak{C})}A$ for $\mathfrak{B}$ .

Proof. It suffices to verify the hypotheses of I.2.8 (with $\mathfrak{C}$ and V in place of $\mathfrak{A}$ and b ). It is easy to see that $\mathfrak{C}$ is simple; see also 2.3.14 or 2.3.15. Condition I.2.8 (i) follows from I.2.2, while I.2.8

(ii) is obvious. Assume the notation of I.2.8 (iii) (with $V$ in place of $b$ ). Say $V = \bigcup_{i\in I} {}^{\alpha}W_i$ where $W_i \cap W_j = 0$ for $i \neq j$. Thus $U = \bigcup_{i\in I} W_i$. For each $u \in U$ there is a unique $i \in I$ denoted by $in(u)$, such that $u \in W_i$. Thus clearly

(1) for every $\kappa,\lambda < \alpha$ and $f \in {}^{\alpha}U$, we have $f \in e_{\kappa\lambda}$ iff $in(f_\kappa) = in(f_\lambda)$.

Choosing any $x \subseteq V$, we then get

(2) for every $\kappa < \alpha$ and $f \in {}^{\alpha}U$, $f \in c_{(\alpha\sim\{\kappa\})}x$ iff $f_\kappa \in pj^*_\kappa x$.

(Recall from the Preliminaries that $pj^*_\lambda x = \{g_\lambda : g \in x\}$.) Hence for every $f \in {}^{\alpha}U$,

(3) $f \in c_{\lambda_0}(c_{(\alpha\sim\{\lambda_0\})}x \cdot \prod_{\iota\in\nu\sim 1} - e_{\lambda_0\lambda_\iota})$ iff there is a $u \in pj^*_{\lambda_0}x$ such that $in(u) \neq in(f_{\lambda_\iota})$ for all $\iota \in \nu \sim 1$.

On the other hand, obviously

(4) for each $\kappa \leq \nu$, $z_\kappa = {}^{\alpha}U$ or $z_\kappa = 0$; $z_\kappa = {}^{\alpha}U$ iff there are elements $w_0 \in pj^*_{\lambda_0}x, \ldots, w_{\kappa-1} \in pj^*_{\lambda_{\kappa-1}}x$ such that $in(w_\iota) \neq in(w_\theta)$ whenever $\iota \neq \theta$.

Since $in^* pj^*_\kappa x = in^* pj^*_\rho x$ for all $\kappa,\rho < \alpha$, we may use the fact that $x \subseteq V$ to rewrite (4) as

(5) for each $\kappa \leq \nu$, $z_\kappa = {}^{\alpha}U$ or $z_\kappa = 0$; $z_\kappa = {}^{\alpha}U$ iff $|\{in(u) : u \in pj^*_{\lambda_0}x\}| \geq \kappa$.

(6) $c'_{(\alpha)}x = \bigcup\{{}^{\alpha}W_i : i \in I, {}^{\alpha}W_i \cap x \neq 0\}$.

One can easily prove (6) by observing from I.2.7 (iii) that $c'_{(\alpha)}x = c_0(c_{(\alpha\sim 1)}x \cdot V) \cdot V$ and using (2) . From (6) we easily get

(7) if $\kappa < \nu$ , $\Delta \subseteq \nu \sim 1$ , $|\Delta| = \kappa$ and $f \in {}^{\alpha}U$ , then $f \in \Pi_{\iota\in\Delta} c_{(\alpha\sim\{\lambda_\iota\})} c'_{(\alpha)}x \cdot \Pi_{\iota,\theta\in\Delta,\iota\neq\theta} - e_{\lambda_\iota\lambda_\theta}$ iff $\mathrm{in}(f_{\lambda_\iota}) \neq \mathrm{in}(f_{\lambda_\theta})$ for all distinct $\iota,\theta \in \Delta$ , and $f_{\lambda_\iota} \in \bigcup\{W_i : i \in I, {}^{\alpha}W_i \cap x \neq 0\}$ for all $\iota \in \Delta$ .

Hence for any $f \in {}^{\alpha}U$ we have

(8) $f \in w_\kappa$ iff $|\{\mathrm{in}(f_{\lambda_1}),\ldots,\mathrm{in}(f_{\lambda_{\nu-1}})\} \cap \{i : {}^{\alpha}W_i \cap x \neq 0\}| < \kappa$ .

Now let us compare (3), (5), and (8) . If $z_\nu = {}^{\alpha}U$ , we see that $c_{\lambda_0}(c_{(\alpha\sim\{\lambda_0\})}x \cdot \Pi_{\iota\in\nu\sim 1} - e_{\lambda_0\lambda_\iota}) = {}^{\alpha}U$ also, and I.2.8 (iii) holds. Suppose $z_\nu = 0$ , and let $\kappa = |\{\mathrm{in}(u) : u \in pj^*_{\lambda_0} x\}|$ . By (5) we have $\kappa < \nu$ . Clearly $\{\mathrm{in}(u) : u \in pj^*_{\lambda_0} x\} = \{i : {}^{\alpha}W_i \cap x \neq 0\}$ . It follows that

$$\begin{aligned} z_\nu + \Sigma_{\mu<\nu} z_\mu \cdot {}^{-}z_{\mu+1} \cdot w_\mu &= w_\kappa \\ &= c_{\lambda_0}(c_{(\alpha\sim\{\lambda_0\})}x \cdot \Pi_{\iota\in\nu\sim 1} - e_{\lambda_0\lambda_\iota}) . \end{aligned}$$

This completes the proof.

<u>Remark</u> I.2.10. The preceding theorem can be extended to the case $\alpha = 2$ . To cover that case we would first establish an analog of Lemma I.2.8 having the same conclusion and same premise (iii) , but with premises (i) and (ii) replaced by the following:

(1) $b \leq s^0_1 b \cdot s^1_0 b$ ;

(2) $d_{01} \leq b$ ;

(3) $c_\kappa(c_\lambda x \cdot b) \cdot b \leq c_\lambda(c_\kappa x \cdot b)$ for any $\lambda,\kappa < 2$ and $x \leq b$ .

The proof of such a lemma for the case $\alpha = 2$ is merely a simplified version of the proof given for I.2.8. With such a lemma at hand, there is no difficulty in extending I.2.9 to cover the case $\alpha = 2$, by verifying that the above hypotheses (1), (2), and (3) hold under the conditions assumed in I.2.9.

For $\alpha = 2$, I.2.9 specializes to $Gs_2 \subseteq R\ell Cs_2$. In a later paper of this series we shall show that $CA_2 \subseteq IR\ell Cs_2$. In view of results from [HR] this suggests that perhaps $Crs_2 = R\ell Cs_2$, but this has been disproved by Andréka and Németi, see 2.3 of [AN3] (p.155).

<u>Theorem</u> I.2.11. In case $\alpha \geq \omega$ we have $Gws_\alpha \subseteq R\ell Cs_\alpha$.

<u>Proof</u>. Let $\mathfrak{A}$ be a $Gws_\alpha$ with unit set $V = \bigcup_{i \in I} {}^{\alpha}W_i^{(pi)}$, where ${}^{\alpha}W_i^{(pi)} \cap {}^{\alpha}W_j^{(pj)} = 0$ for distinct $i,j \in I$. Let $U = \bigcup_{i \in I} W_i$ be the base of $\mathfrak{A}$ and choose $U'$ such that $U' \sim U$ is infinite. Let $\mathfrak{C}$ be the full $Cs_\alpha$ with base $U'$, and let $\mathfrak{B} = \mathfrak{Sg}^{(\mathfrak{C})}A$. We claim that $\mathfrak{A} = \mathfrak{R}\ell_V\mathfrak{B}$. To prove this, we shall again follow an elimination of quantifiers argument, as in the proof of I.2.8. Let

$$C^* = \{C_{(\Gamma)}x : x \in A, \Gamma \text{ a finite subset of } \alpha\} \cup \{-C_{(\Gamma)}V : \Gamma \text{ a finite subset of } \alpha\} \cup \{D_{\kappa\lambda} : \kappa,\lambda < \alpha\} \cup \{-D_{\kappa\lambda} : \kappa,\lambda < \alpha\}.$$

Let $C^{**}$ be the set of all finite unions of finite intersections of members of $C^*$. Obviously $C^{**} \subseteq Sg^{(\mathfrak{C})}A$, and it suffices to prove the converse, i.e., that $A \subseteq C^{**} \in Su\mathfrak{C}$ . Obviously $A \subseteq C^{**}$ , $D_{\kappa\lambda} \in C^{**}$ for all $\kappa,\lambda < \alpha$ , and $C^{**}$ is closed under unions. To show that $C^{**}$ is closed under $-$ , it suffices to show that $-y \in C^{**}$ for each $y \in C^*$ . This is obvious for $y$ of the form $-C_{(\Gamma)}V, D_{\kappa\lambda}$ or $-D_{\kappa\lambda}$ , while $-y \in C^{**}$ for $y = C_{(\Gamma)}x$ by I.2.7 (vi) , which applies by I.2.1 .

So it remains to show that $C_\kappa y \in C^{**}$ whenever $\kappa < \alpha$ and $y \in C^{**}$ . We may assume that $y$ is merely a product of members of $C^*$ , say

$$\begin{aligned} y = {} & C_{(\Gamma_0)}x_0 \cap \cdots \cap C_{(\Gamma_{\beta-1})}x_{\beta-1} \\ & \cap -C_{(\Delta_0)}V \cap \cdots \cap -C_{(\Delta_{\gamma-1})}V \\ & \cap D_{\lambda_0\mu_0} \cap \cdots \cap D_{\lambda_{\delta-1}\mu_{\delta-1}} \\ & \cap -D_{\xi_0\rho_0} \cap \cdots \cap -D_{\xi_{\varepsilon-1}\rho_{\varepsilon-1}} \quad , \end{aligned}$$

where $\beta,\gamma,\delta,\varepsilon < \omega$ , $\Gamma_\iota$ and $\Delta_\theta$ are finite subsets of $\alpha$ for $\iota < \beta$ , $\theta < \gamma$ , and $\lambda_\iota, \mu_\iota, \xi_\theta, \rho_\theta < \alpha$ for $\iota < \delta$ , $\theta < \varepsilon$ . Further, we may assume that $\kappa \notin \Gamma_\iota$ for all $\iota < \beta$ , $\kappa \notin \Delta_\iota$ for all $\iota < \gamma$ , $\kappa \in \{\lambda_\iota, \mu_\iota\}$ for all $\iota < \delta$ , and $\kappa \in \{\xi_\theta, \rho_\theta\}$ for all $\theta < \varepsilon$ . Say $\kappa = \lambda_0 = \cdots = \lambda_{\delta-1} = \rho_0 = \cdots = \rho_{\varepsilon-1}$ .
Further, we may assume that the $\mu_\iota$ 's are distinct from each other and from $\kappa$ , and similarly for the $\delta_\iota$ 's . The following easily verified facts show that we can assume that $\delta = 0$ (here $\kappa \neq \nu$ ):

$$C_\kappa(X \cap Y \cap D_{\kappa\nu}) = C_\kappa(X \cap D_{\kappa\nu}) \cap C_\kappa(Y \cap D_{\kappa\nu}) \ ;$$

$$C_\kappa(C_{(\Gamma_\iota)}x_\iota \cap D_{\kappa\nu}) = C_{(\Gamma_\iota \cup \{\kappa\})}(x_\iota \cap D_{\kappa\nu}) \quad \text{if} \quad \nu \notin \Gamma_\iota \ ;$$

$$C_\kappa(C_{(\Gamma_\iota)}x_\iota \cap D_{\kappa\nu}) = C_{(\Gamma_\iota \sim \{\nu\})\cup\{\kappa\})}(D_{\kappa\nu} \cap C_\nu x_\iota \cap V) \quad \text{if} \quad \nu \in \Gamma_\iota \ ;$$

$$C_\kappa(-C_{(\Delta_\iota)}^V \cap D_{\kappa\nu}) = -C_{((\Delta_\iota\sim\{\nu\})\cup\{\kappa\})}(D_{\kappa\nu} \cap V) \in C^{**}$$

by closure under $-$ , proved above;

$$C_\kappa(D_{\kappa\mu} \cap D_{\kappa\nu}) = D_{\mu\nu} \ ;$$

$$C_\kappa(-D_{\kappa\mu} \cap D_{\kappa\nu} = -D_{\mu\nu} \ .$$

Also, by I.2.7 (v) we can assume that $\beta \leq 1$ . Thus we have reduced our considerations to $y$ of the form

$$C_{(\Gamma)}x \cap -C_{(\Delta_0)}^V \cap \cdots \cap -C_{(\Delta_{\gamma-1})}^V$$
$$\cap -D_{\kappa\rho_0} \cap \cdots \cap -D_{\kappa\rho_{\varepsilon-1}} \ ,$$

where $\kappa \notin \Gamma$ , $\kappa \notin \Delta_\iota$ for each $\iota < \gamma$ , $\kappa \neq \rho_\iota$ for each $\iota < \varepsilon$ , all the $\rho_\iota$'s are distinct, and where perhaps the factor $C_{(\Gamma)}x$ does not appear, and possibly $\gamma = 0$ or $\varepsilon = 0$ . Now we consider two cases.

Case 1. $C_{(\Gamma)}x$ is a factor of $y$ . We proceed by induction on $\varepsilon$ . First take $\varepsilon = 0$ . Clearly we may assume that $\gamma > 0$ . Then, as in the proof of I.2.7 (xii) we see that

$$C_\kappa y = C_{(\Gamma\cup\{\kappa\})}x \cdot -C_\kappa(C_{(\Gamma)}x \cap \bigcup_{\iota<\gamma} C_{(\Delta_\iota)}^V) \ ;$$

so by I.2.7 (v) and (vi) $C_\kappa y \in C^{**}$ . Now assume, inductively, that $\varepsilon > 0$ . For each $\iota < \varepsilon$ we have $C_{(\Gamma)}x \cap -C_{(\Gamma\sim\{\rho_\iota\})}^V \subseteq -D_{\kappa\rho_\iota}$ . Thus

(9) $$C_\kappa y = C_\kappa(C_{(\Gamma)}x \cap \bigcap_{\iota<\gamma} - C_{(\Delta_\iota)}^V \cap \bigcap_{\iota<\varepsilon} - C_{(\Gamma\sim\{\rho_\iota\})}^V)$$
$$\cup\ C_\kappa(C_{(\Gamma)}x \cap \bigcap_{\iota<\gamma} - C_{(\Delta_\iota)}^V$$
$$\cap \bigcap_{\iota<\varepsilon} - D_{\kappa\rho_\iota} \cap \bigcup_{\iota<\varepsilon} C_{(\Gamma\sim\{\rho_\iota\})}^V) \ .$$

Now we have $C_\kappa(C_{(\Gamma)}x \cap \bigcap_{\iota<\gamma} - C_{(\Delta_\iota)}^V \cap \bigcap_{\iota<\varepsilon} - C_{(\Gamma\sim\{\rho_\iota\})}^V) \in C^{**}$ by the case $\varepsilon = 0$ just disposed of. And for each $\mu < \varepsilon$ we have

$$C_\kappa(C_{(\Gamma)}x \cap \bigcap_{\iota<\gamma} -C_{(\Delta_\iota)}V \cap \bigcap_{\iota<\varepsilon} -D_{\kappa\rho_\iota} \cap C_{(\Gamma\sim\{\rho_\mu\})}V)$$

$$= C_\kappa(C_{(\Gamma\sim\{\rho_\mu\})}(C'_{(\Gamma)}x) \cap \bigcap_{\iota<\gamma} -C_{(\Delta_\iota)}V \cap \bigcap_{\iota<\varepsilon} -D_{\kappa\rho_\iota}) \quad \text{by I.2.7 (v)}$$

$$= C_\kappa(C_{(\Gamma\sim\{\rho_\mu\})}(C'_{(\Gamma)}x \cap -D_{\kappa\rho_\mu}) \cap \bigcap_{\iota<\gamma} -C_{(\Delta_\iota)}V \cap \bigcap_{\iota<\varepsilon,\iota\neq\mu} -D_{\kappa\rho_\iota}),$$

to which the induction hypothesis applies.

Case 2. $C_{(\Gamma)}x$ is not a factor of $y$. Then

(*) $$C_\kappa(\bigcap_{\iota<\gamma} -C_{(\Delta_\iota)}V \cap \bigcap_{\iota<\varepsilon} -D_{\kappa\rho_\iota}) = {}^\alpha U'.$$

In fact, let $u \in {}^\alpha U'$. Choose $s \in U' \sim (U \cup \{u\rho_\iota : \iota < \varepsilon\})$. Then clearly $u^\kappa_s \in \bigcap_{\iota<\gamma} -C_{(\Delta_\iota)}V \cap \bigcap_{\iota<\varepsilon} -D_{\kappa\rho_\iota})$, and (*) follows.

By (*) we clearly have $C_\kappa y \in C^{**}$ in this case also, so the proof of I.2.11 is complete.

Corollary I.2.12. For every $\alpha \geq 3$ we have $Gs_\alpha \subseteq Gws_\alpha \subseteq Rl Cs_\alpha$.

Remark I.2.13. For each $\alpha \geq 2$ there is a $Cs_\alpha \mathfrak{A}$ and a $V \in A$ such that $\mathfrak{Rl}_V\mathfrak{A}$ is not a $Gws_\alpha$, and in fact is not even isomorphic to a $Gws_\alpha$. Thus $Gs_\alpha \subset RlCs_\alpha$, $IGs_\alpha \subset IRlCs_\alpha$, and $IGws_\alpha \subset IRlCs_\alpha$. It is a trivial matter to construct such $\mathfrak{A}$ and $V$. For example, let $\mathfrak{A}$ be the full $Cs_\alpha$ with base 2, and let $V = \{f \in {}^\alpha 2 : f0 \neq f1\}$. Then $\mathfrak{Rl}_V\mathfrak{A}$ is not a $CA_\alpha$, since $D_{01} \cap V = 0$ (see 2.2.3 (ii)). So the above facts are trivial. It is of more interest to construct an example in which $\mathfrak{Rl}_V\mathfrak{A}$ is a $CA_\alpha$. (Since $IGs_\alpha$ is the class of all representable $CA_\alpha$'s, this will provide another example of a non-representable $CA_\alpha$; we gave one in 2.6.42. See 1.1.13 for the notion of a representable

$CA_\alpha$ ; it will be proved in I.6.3 (rather easily) that $IGs_\alpha$ is the class of all such.) The example which we shall now give is due to Henkin.

<u>Theorem</u> I.2.14. For each $\alpha \geq 2$ there is a $Cs_\alpha$ $\mathfrak{A}$ and a $V \in A$ such that $\mathfrak{Rl}_V\mathfrak{A} \in CA_\alpha \sim IGws_\alpha$ .

<u>Proof</u>. Let $W$ be a set of cardinality $|\alpha|$ with $\alpha \cap W = 0$ , and let $'$ be a one-to-one function mapping $\alpha$ onto $W$ . The base $U$ of our $Cs_\alpha$ $\mathfrak{A}$ is to be $\alpha \cup W$ . Let $G$ be the set of all permutations $f$ of $U$ such that for some permutation $\tau$ of $\alpha$ we have $\alpha \upharpoonright f = \tau$ , while $f\kappa' = (\tau\kappa)'$ for each $\kappa < \alpha$ . Clearly $G$ is the universe of a group of permutations of $U$ . Now for $g,h \in {}^\alpha U$ define $g \equiv h$ iff $f \circ g = h$ for some $f \in G$ . It is easily verified that $\equiv$ is an equivalence relation on $U$ . Let $A$ be the collection of all sets $X \subseteq {}^\alpha U$ which are unions of equivalence classes under $\equiv$ . It is easily verified that $A$ is a cylindric field of subsets of ${}^\alpha U$ . We let $\mathfrak{A}$ be the cylindric set algebra with universe $A$ . For each $u \in U$ let $gu = u$ if $u \in \alpha$ , and $gu = \kappa$ if $u = \kappa'$ , $\kappa < \alpha$ . We now set

$$V = \{f \in {}^\alpha U : g \circ f \text{ is not one-to-one}\} \cup (\langle 0',1,2,\ldots\rangle/\equiv) .$$

Clearly $V \in A$ . Now we want to show that $\mathfrak{Rl}_V\mathfrak{A}$ is a $CA_\alpha$ ; and to do this we shall apply 2.2.3. Since $D_{\kappa\lambda} \subseteq V$ for all distinct $\kappa,\lambda < \alpha$ , condition 2.2.3 (ii) is clear. Now suppose that $\kappa,\lambda < \alpha$ , $X \in A$ , $X \subseteq V$ , and $h \in C_\kappa(C_\lambda X \cap V) \cap V$ . We may assume that $\kappa \neq \lambda$ . Now our assumption on $h$ implies that there is an $a \in U$ with $h^\kappa_a \in C_\lambda X \cap V$ , and so there is a $b \in U$ with $h^{\kappa\lambda}_{ab} \in X$ . If $h^\lambda_b \in V$ , then

$h_b^\lambda \in C_\kappa X \cap V$ and hence $h \in C_\lambda(C_\kappa X \cap V)$, as desired. Hence suppose $h_b^\lambda \notin V$. Thus

(1) $g \circ h_b^\lambda$ is one-to-one.

(2) $h\lambda \neq b$.

For, otherwise $h_b^\lambda = h \in V$.

(3) $h\kappa \neq a$.

For, otherwise $h_b^\lambda = h_{ab}^{\kappa\lambda} \in V$. Now we consider several cases.

Case 1. $ga = gb$. Let $c = gh\kappa$ or $(gh\kappa)'$ according as $b \in \alpha$ or $b \in W$, and let $d = gh\kappa$ or $(gh\kappa)'$ according as $a \in \alpha$ or $a \in W$. Then $h_{cd}^{\lambda\kappa} \in X$ since $h_{ab}^{\kappa\lambda} \in X$ and $h_{cd}^{\lambda\kappa} \equiv h_{ab}^{\kappa\lambda}$, and $h_c^\lambda \in V$ since $gc = gh\kappa$, so $h \in C_\lambda(C_\kappa X \cap V)$.

Case 2. $gb \neq ga \neq gh\kappa$. Then by (1), $gb \neq gh\mu$ for all $\mu \neq \kappa,\lambda$ and $gb \neq ga$. And also $gh\kappa \neq gh\mu$ for all $\mu \neq \kappa,\lambda$ and $gh\kappa \neq ga$. Let $c = gh\kappa$ if $b \in \omega$, $c = (gh\kappa)'$ if $b \in W$. Then $h_{ca}^{\lambda\kappa} \in X$ since $h_{ba}^{\lambda\kappa} \in X$, and $h_c^\lambda \in V$ since $gc = gh\kappa$, so $h \in C_\lambda(C_\kappa X \cap V)$.

Case 3. $gb \neq ga = gh\kappa$, $0 \notin \{\kappa,\lambda\}$. Thus since $g \circ h_{ab}^{\kappa\lambda}$ is one-to-one by (1), we have $a,b \in \alpha$. Hence $h_{ab}^{\lambda\kappa} \in X$, and $h_a^\lambda \in V$ since $ga = gh\kappa$, so $h \in C_\lambda(C_\kappa X \cap V)$.

Case 4. $gb \neq ga = gh\kappa$, $\kappa = 0$. Again by (1), $a \in W$ and $b \in \alpha$. Hence $h_{(ga)b'}^{\lambda\ \kappa} \in X$ and $h_{ga}^{\lambda} \in V$, so $h \in C_\lambda(C_\kappa X \cap V)$.

Case 5. $gb \neq ga = gh\kappa$, $\lambda = 0$. Then $a \in \alpha$ and $b \in W$, so $h_{a'(gb)}^{\lambda\ \kappa} \in X$, $h_{a'}^{\lambda} \in V$, $h \in C_\lambda(C_\kappa X \cap V)$.

Thus $\mathfrak{Rl}_V\mathfrak{A}$ is a $CA_\alpha$. It is obvious that $V$ is not the unit set of a $Gws_\alpha$. To show that $\mathfrak{Rl}_V\mathfrak{A}$ is not even isomorphic to a $Gws_\alpha$, we shall use the equation of 2.6.42 :

(4) $$c_1(y \cdot c_0(c_1 y \cdot -y)) \cdot -c_0(c_1 y \cdot -d_{01}) = 0 \quad .$$

It is easy to check that this equation holds in every $Gws_\alpha$ . To say that (4) holds in $\mathfrak{Rl}_V\mathfrak{A}$ is to say that the following equation holds in $\mathfrak{A}$ itself, for any $X \subseteq V$ such that $X \in A$ :

(5) $$C_1(X \cap C_0(C_1 X \cap \sim X \cap V)) \cap V \cap -C_0(C_1 X \cap \sim D_{01} \cap V) = 0 \ .$$

Now let $X = \langle 0', 0, 2, 3, 4, \ldots \rangle/\equiv$ . Let $f = \langle 0', 0', 2, 3, 4, \ldots \rangle$ . Then $f^1_0 \in X$ , and $f^{1\,0}_{01'} = \langle 1', 0, 2, 3, 4, \ldots \rangle \in \sim X \cap V$ , and $f^{1\,0\,1}_{01'1} = \langle 1', 1, 2, 3, 4, \ldots \rangle \in X$ . Thus $f \in C_1(X \cap C_0(C_1 X \cap \sim X \cap V))$ . Hence if (5) holds in $\mathfrak{A}$ there is an $a \in U$ with $f^0_a \in \sim D_{01} \cap V$ , and a $b \in U$ with $f^{01}_{ab} \in X$ . From the form of $X$ it follows that $a = 0'$ and $b = 0$ , or $a = 1'$ and $b = 1$ . But the first possibility yields $f^0_a = f \in D_{01}$ , and the second possibility yields $f^0_a = \langle 1', 0', 2, 3, 4, \ldots \rangle \notin V$ . Thus (5) does not hold in $\mathfrak{A}$ . So $\mathfrak{Rl}_V\mathfrak{A}$ is not isomorphic to a $Gws_\alpha$ .

## 3. Change of base

Given a set algebra $\mathfrak{A}$ with base $U$ , and given some set $W$ , is $\mathfrak{A}$ isomorphic to a set algebra with base $W$? We first consider the case $|U| = |W|$ , where the answer is obviously yes.

Theorem I.3.1. (i) Let $\mathfrak{A}$ be a $Crs_\alpha$ with base $U$ and unit element $V$ . Suppose $f$ is a one-one function from $U$ onto a set $W$ . For any $X \in A$ let $FX = \{y \in {}^\alpha W : f^{-1} \circ y \in X\}$ . Then $F$ is an isomorphism from $\mathfrak{A}$ onto a $Crs_\alpha$ $\mathfrak{B}$ with base $W$ and unit element $FV$ .

(ii) If in (i) $\mathfrak{A}$ is a $Cs_\alpha$, then $\mathfrak{B}$ is a $Cs_\alpha$.

(iii) If in (i) $\mathfrak{A}$ is a $Ws_\alpha$ with unit element ${}^\alpha U^{(p)}$, then $\mathfrak{B}$ is a $Ws_\alpha$ with unit element ${}^\alpha W^{(f\circ p)}$.

(iv) If in (i) $\mathfrak{A}$ is a $Gs_\alpha$ with unit element $\bigcup_{i\in I}{}^\alpha S_i$, where $S_i \cap S_j = 0$ for $i \neq j$, then $\mathfrak{B}$ is a $Gs_\alpha$ with unit element $\bigcup_{i\in I}{}^\alpha f^*S_i$, where $f^*S_i \cap f^*S_j = 0$ for $i \neq j$.

(v) If in (i) $\mathfrak{A}$ is a $Gws_\alpha$ with unit element $\bigcup_{i\in I}{}^\alpha S_i^{(pi)}$, where ${}^\alpha S_i^{(pi)} \cap {}^\alpha S_i^{(pj)} = 0$ for $i \neq j$, then $\mathfrak{B}$ is a $Gws_\alpha$ with unit element $\bigcup_{i\in I}{}^\alpha(f^*S_i)^{(f\circ pi)}$, where ${}^\alpha(f^*S_i)^{(f\circ pi)} \cap {}^\alpha(f^*S_j)^{(f\circ pj)} = 0$ for $i \neq j$.

If we apply I.3.1 to the special case $U = W$, in some cases the function $F$ is an automorphism of the algebra $\mathfrak{A}$. This is always true, e.g., if $\mathfrak{A}$ is a full $Cs_\alpha$. In this way one can develop a general kind of Galois theory. We shall not go into this theory, which is rather extensive. See, e.g., Daigneault [D].

Remark I.3.2. The less trivial question concerning change of base arises when the two bases have different cardinalities. To begin our discussion of this case we shall show that for each $\alpha \geq 3$ there is a $Cs_\alpha$ with an infinite base not isomorphic to a $Cs_\alpha$ with a finite base. First suppose that $3 \leq \alpha < \omega$. Then, we claim, the following equation $\varepsilon$ holds identically in every $Cs_\alpha$ with a finite base, but fails in some finite $Cs_\alpha$ with an infinite base (this equation is due to Andréka and Németi, and replaces a longer one originally found for this purpose):

$$c_{(\alpha)}[-c_{(\alpha\sim 1)}x + c_{(\alpha\sim 2)}x \cdot d_{01} + c_{(\alpha\sim 2)}x \cdot s^0_1 s^1_2 c_{(\alpha\sim 2)}x \cdot -s^1_2 c_{(\alpha\sim 2)}x] = 1 .$$

To show this, let $\mathfrak{A}$ be a $Cs_\alpha$ with base $U$ in which $\varepsilon$ does not hold identically. Then it is easy to see that there is an $X \in A$ such that the following conditions hold:

(1) $C_{(\alpha\sim 1)}X = {}^\alpha U$ ;

(2) $C_{(\alpha\sim 2)}X \subseteq \sim D_{01}$ ;

(3) $C_{(\alpha\sim 2)}X \cap s^0_1 s^1_2 C_{(\alpha\sim 2)}X \subseteq s^1_2 C_{(\alpha\sim 2)}X$ .

Now let $R = \{2 1 u : u \in X\}$ . Then $R$ is a binary relation on $U$ satisfying the following conditions:

(4) for all $u \in U$ there is a $v \in U$ with $uRv$ ;

(5) for all $u \in U$ , not $(uRu)$ ;

(6) $R$ is transitive.

It follows that $U$ is infinite. Thus $\varepsilon$ holds in every $Cs_\alpha$ with a finite base.

Now we construct a finite $Cs_\alpha$ $\mathfrak{A}$ with an infinite base such that $\varepsilon$ fails to hold in $\mathfrak{A}$ . Let $\mathfrak{B} = \langle B,<\rangle$ , where $B$ is the set of rational numbers and $<$ is the usual ordering on $B$ . Let $\Lambda$ be a discourse language for $\mathfrak{B}$ , with a sequence $\langle v_\xi : \xi < \alpha\rangle$ of variables. Then $\{\tilde{\varphi}^{(\mathfrak{B})} : \varphi$ a formula of $\Lambda\}$ is the universe of a $Cs_\alpha$ $\mathfrak{A}$ with infinite base $B$ . From the usual decision procedure for sentences holding in $\mathfrak{B}$ we see that $\mathfrak{A}$ is finite. Letting $X = \tilde{\varphi}^{(\mathfrak{B})}$ , $\varphi$ the formula $v_0 < v_1$ , we see that $\varepsilon$ fails in $\mathfrak{A}$ .

For $\alpha \geq \omega$, let $\mathfrak{A}$ be any $Cs_\alpha$ with an infinite base. By Theorem I.3.3, $\mathfrak{A}$ is not isomorphic to a $Cs_\alpha$ with a finite base.

Finally, some remarks on the case $\alpha \leq 2$. Any $Crs_0$ has base $0$. If $\mathfrak{A}$ is a finite $Cs_1$, then $\mathfrak{A}$ is isomorphic to a $Cs_1$ with a finite base. In fact, say $U$ is the base of $\mathfrak{A}$. For each non-zero $a \in A$ choose $u_a \in U$ so that $\langle u_a \rangle \in a$. Let $U' = \{u_a : a \in A\}$, and for any $a \in A$ let $fa = a \cap {}^1U'$. Then $f$ is an isomorphism of $\mathfrak{A}$ onto a $Cs_1$ with finite base $U'$. In a later article in this series we shall show that any finite $Gs_2$ (resp. $Cs_2$) is isomorphic to a $Gs_2$ (resp. $Cs_2$) with a finite base.

<u>Theorem</u> I.3.3. Let $\mathfrak{A}$ and $\mathfrak{B}$ be $Gws_\alpha$'s, and assume that $\mathfrak{A} \succsim \mathfrak{B}$. If $|W| < \alpha \cap \omega$ for some subbase $W$ of $\mathfrak{B}$, then $|W| = |W'|$ for some subbase $W'$ of $\mathfrak{A}$.

<u>Proof</u>. Let $\kappa = |W|$. Then

$$(1) \qquad \bigcap_{\lambda<\mu<\kappa} \sim D_{\lambda\mu} \cap c_\kappa^\partial \bigcup_{\lambda<\kappa} D_{\lambda\kappa} \neq 0$$

in $\mathfrak{B}$. In fact, ${}^\alpha W^{(p)}$ is included in the unit element of $\mathfrak{B}$ for some $p \in {}^\alpha W$, and any $q \in {}^\alpha W^{(p)}$ such that $\kappa 1 q$ maps $\kappa$ one-one onto $W$ will be a member of the left side of (1). It follows that (1) also holds in $\mathfrak{A}$, and this gives the desired set $W'$.

<u>Remark</u> I.3.4. The implication in Theorem I.3.3 does not hold in the other direction. Namely, for each $\alpha \geq 2$ there are a $Gs_\alpha$ $\mathfrak{A}$, a $Cs_\alpha$ $\mathfrak{B}$ with $\mathfrak{A} \succ \mathfrak{B}$, and a subbase $W$ of $\mathfrak{A}$ with $|W| < \alpha \cap \omega$, such that the base of $\mathfrak{B}$ has cardinality $\neq |W|$. To construct these objects, let $U$ and $W$ be disjoint sets with $|U| > |W| = 1$. Let

$\mathfrak{A}$ be the full $Gs_\alpha$ with unit element ${}^\alpha U \cup {}^\alpha W$ , and $\mathfrak{B}$ the full $Cs_\alpha$ with unit element ${}^\alpha U$ . The mapping f from A onto B such that $fX = X \cap {}^\alpha U$ is a homomorphism by 2.2.12. Clearly the above properties hold.

We can strengthen a part of I.3.3 by use of the following notion of base-isomorphism.

Definition I.3.5. (i) With f as in Theorem I.3.1, we denote by $\tilde{f}$ the function F defined in I.3.1.

(ii) Let $\mathfrak{A}$ and $\mathfrak{B}$ be $Crs_\alpha$'s with bases U and W and unit elements V and Y respectively. We say that $\mathfrak{A}$ and $\mathfrak{B}$ are base-isomorphic if there is a one-one function f mapping U onto W such $\tilde{f}$ is an isomorphism from $\mathfrak{A}$ onto $\mathfrak{B}$ .

The following result is an algebraic version of the logical result according to which any two elementarily equivalent finite structures are isomorphic; it is due to Monk.

Theorem I.3.6. Let $\mathfrak{A}$ and $\mathfrak{B}$ be locally finite-dimensional regular $Cs_\alpha$'s , and assume that $\mathfrak{A} \cong \mathfrak{B}$ . If the base of either $\mathfrak{A}$ or $\mathfrak{B}$ has power $< \alpha \cap \omega$ , then $\mathfrak{A}$ and $\mathfrak{B}$ are base-isomorphic.

Proof. Assume the hypotheses, and suppose that $\mathfrak{A}$ and $\mathfrak{B}$ have bases U and W respectively, and that $|U| = \beta < \alpha \cap \omega$ . Let G be the given isomorphism from $\mathfrak{A}$ onto $\mathfrak{B}$ . Note that by I.3.3, $|W| = |U|$ . Now we need the following

Claim. For each regular $Cs_\alpha$ $\mathfrak{C}$ having a base T of power $< \alpha \cap \omega$

there is a function $s^{\mathfrak{C}}$ which assigns to each $\tau \in {}^{\alpha}\alpha$ a mapping $s^{\mathfrak{C}}_{\tau}$ of $C$ into $C$ such that:

(a) for any $x \in C$ and any $f \in {}^{\alpha}T$ we have $f \in s^{\mathfrak{C}}_{\tau}x$ iff $f \circ \tau \in x$ ;

(b) if $F$ is an isomorphism from $\mathfrak{C}$ onto any regular $Cs_{\alpha}\ \mathfrak{D}$ with base of power $< \alpha \cap \omega$ , then $Fs^{\mathfrak{C}}_{\tau}x = s^{\mathfrak{D}}_{\tau}Fx$ for all $x \in C$ and $\tau \in {}^{\alpha}\alpha$ .

This claim is of course related to our considerations in section 1.11. For $\alpha \geq \omega$ it follows easily by the methods of section 1.11. In fact, for any $x \in C$ let $\Delta x = \{\kappa_0, \ldots, \kappa_{\iota-1}\}$ , with $\kappa_0 < \ldots < \kappa_{\iota-1}$ , let $\tau \in {}^{\alpha}\alpha$ , and let $\lambda_0 < \ldots < \lambda_{\iota-1}$ be the first $\iota$ ordinals $< \alpha$ not in $\Delta x \cup \{\tau\kappa_0, \ldots, \tau\kappa_{\iota-1}\}$ . Then we set

$$s^{\mathfrak{C}}_{\tau}x = s^{\lambda_0}_{\tau\kappa_0} \ldots s^{\lambda_{\iota-1}}_{\tau\kappa_{\iota-1}} s^{\kappa_0}_{\lambda_0} \ldots s^{\kappa_{\iota-1}}_{\lambda_{\iota-1}} x .$$

Now note that $f \in s^{\mu}_{\nu}y$ iff $f^{\mu}_{f\nu} \in y$ for all $\mu, \nu < \alpha$ and all $y \in C$ . The claim now follows easily, using the regularity of $\mathfrak{C}$ in checking (a) .

To establish the claim in the case $\alpha < \omega$ it suffices to consider the case in which $\tau$ is a transposition $[\kappa/\lambda, \lambda/\kappa]$ with $\kappa \neq \lambda$ (see the preliminaries), since every transformation of $\alpha$ is the composition of a finite sequence of replacements and transpositions. In this case we let

$$s^{\mathfrak{C}}_{\tau}x = (x \cap D_{\kappa\lambda}) \cup \bigcup_{\mu\neq\kappa,\lambda}(s^{\mu}_{\lambda}s^{\lambda}_{\kappa}s^{\kappa}_{\mu}x \cap D_{\lambda\mu})$$

$$\cup \bigcup_{\mu\neq\kappa,\lambda}(s^{\mu}_{\kappa}s^{\kappa}_{\lambda}s^{\lambda}_{\mu}x \cap D_{\kappa\mu})$$

$$\cup \bigcup_{\mu,\nu\neq\kappa,\lambda;\mu\neq\nu}(s^{\mu}_{\lambda}s^{\lambda}_{\kappa}s^{\kappa}_{\mu}s^{\mu}_{\nu}x \cap D_{\mu\nu}) .$$

Then condition (b) of the claim is clear. For condition (a), let $x \in C$ and $f \in {}^{\alpha}T$ . First note:

(1) if $f \in D_{\kappa\lambda}$ , then $f = f \circ \tau$ ;

(2) if $f \in D_{\lambda\mu}$ with $\mu \neq \kappa,\lambda$ , then $f \circ \tau = f \circ [\mu/\lambda] \circ [\lambda/\kappa] \circ [\kappa/\mu]$ ;

(3) if $f \in D_{\kappa\mu}$ with $\mu \neq \kappa,\lambda$ , then $f \circ \tau = f \circ [\mu/\kappa] \circ [\kappa/\lambda] \circ [\lambda/\mu]$ ;

(4) if $f \in D_{\mu\nu}$ with $\mu,\nu \neq \kappa,\lambda$ and $\mu \neq \nu$ , then $f \circ \tau = f \circ [\mu/\lambda] \circ [\lambda/\kappa] \circ [\kappa/\mu] \circ [\mu/\nu]$ .

Now to verify (a), first suppose $f \in s^{\mathfrak{C}}_{\tau}x$ . If $f \in x \cap D_{\kappa\lambda}$ , then $f \circ \tau \in x$ by (1) ; the other possibilities are taken care of by (2)-(4) . Second, suppose $f \circ \tau \in x$ . Since $|T| < \alpha$ , we have $\bigcup_{\mu<\nu<\alpha} D_{\mu\nu} = {}^{\alpha}T$ , and so $f \in D_{\mu\nu}$ for some distinct $\mu,\nu < \alpha$ . Then (1)-(4) yield $f \in s^{\mathfrak{C}}_{\tau}x$ , as desired. We have now fully established the claim.

Now let $u$ be a one-to-one function mapping $U$ onto $\beta$ . To show that $\mathfrak{A}$ and $\mathfrak{B}$ are base-isomorphic we shall first take the case in which $\mathfrak{A}$ is finitely generated. Say $\mathfrak{A} = \mathrm{Sg}X$ , where $0 \neq X \subseteq A$ and $X$ is finite. For each $w$ mapping a subset $\Gamma$ of $\alpha$ into $U$ we set $w' = w \cup \langle u^{-1}0 : \kappa \in \alpha \sim \Gamma\rangle$ ; thus $w' \in {}^{\alpha}U$ . Now consider the following element of $A$ :

$$(5)\quad \bigcap\{s^{\mathfrak{A}}_{u\circ w'}x : x \in X,\ w \in {}^{\Delta x}U,\ w' \in x\}$$
$$\cap \bigcap\{\sim s^{\mathfrak{A}}_{u\circ w'}x : x \in X,\ w \in {}^{\Delta x}U,\ w' \notin x\}$$
$$\cap \bigcap_{\kappa<\lambda<\beta} \sim D_{\kappa\lambda} \cap c^{\partial}_{\beta} \bigcup_{\kappa<\beta} D_{\kappa\beta} .$$

(This element is an algebraic expression of a complete diagram of a finite structure.) Let $h = u^{-1} \cup \langle u^{-1}0 : \kappa \in \alpha\sim\beta\rangle$ . Then $h \circ u \circ w'$ $= w'$ for each $w$ in (5), so it follows from (a) of the claim that $h$ is a member of the element (5), which is thus shown to be non-zero. Applying $G$ to (5) and using (b) of the claim we conclude that the following element of $B$ is non-zero:

$$(6)\quad \bigcap\{s^{\mathfrak{B}}_{u\circ w'}Gx : x \in X,\ w \in {}^{\Delta x}U,\ w' \in x\}$$
$$\cap \bigcap\{\sim s^{\mathfrak{B}}_{u\circ w'}Gx : x \in X,\ w \in {}^{\Delta x}U,\ w' \notin x\}$$
$$\cap \bigcap_{\kappa<\lambda<\beta} \sim D_{\kappa\lambda} \cap c^{\partial}_{\beta} \bigcup_{\kappa<\beta} D_{\kappa\beta} .$$

Let $g \in {}^{\alpha}W$ be any member of (6) . It is easily checked that $f = g \circ u$ is a one-to-one function mapping $U$ onto $W$ . By I.3.1, $\tilde{f}$ is an isomorphism of $\mathfrak{A}$ into the full $Cs_\alpha$ with base $W$ . Thus by 0.2.14 (iii) it suffices now to show that $X \upharpoonright \tilde{f} = X \upharpoonright G$ (hence $\tilde{f} = G$ , as desired). So, let $x \in X$ , and suppose that $k \in \tilde{f}x$ . Thus $f^{-1} \circ k \in x$ . Let $w = \Delta x \upharpoonright (f^{-1} \circ k)$ . Since $x$ is regular, $w' \in x$ . Thus because $g$ is a member of (6) we infer using the claim that $g \circ u \circ w' \in Gx$ . Now $\Delta x \upharpoonright g \circ u \circ w' = \Delta x \upharpoonright k$ , so by the regularity of $Gx$ we see that $k \in Gx$ . Similarly, $k \notin Fx$ implies $k \notin Gx$ , so $Fx = Gx$ .

Having taken care of the finitely generated case, we turn to the general case. Let $\mathcal{F} = \{C : 0 \neq C \subseteq A,\ |C| < \omega\}$ . For each

$C \in \mathfrak{F}$ let $I_C = \{f : f$ is a one-to-one function mapping $U$ onto $W$, and $\tilde{f}$ is a base-isomorphism of $\mathfrak{S}\mathfrak{g}^{\mathfrak{A}}C$ onto a subalgebra of $\mathfrak{B}\}$. Thus the finitely generated case treated above shows that $I_C \neq 0$ for all $C \in \mathfrak{F}$. Since each set $I_C$ is finite, choose $C_0 \in \mathfrak{F}$ with $|I_{C_0}|$ minimum. For each $D \in \mathfrak{F}$ we have $I_{C_0 \cup D} \subseteq I_{C_0}$, and hence $I_{C_0 \cup D} = I_{C_0}$. Therefore any member of $I_{C_0}$ induces a base-isomorphism of $\mathfrak{A}$ onto $\mathfrak{B}$, as desired.

The next few results I.3.7-I.3.10, and Remark I.3.11, are due to Andréka and Németi, and address the question concerning possible improvements of I.3.6.

Lemma I.3.7. Let $\mathfrak{A}$ and $\mathfrak{B}$ be base-isomorphic $Crs_\alpha$'s via $F$. If $x \in A$ is regular, then so is $Fx$.

Lemma I.3.8. Let $\mathfrak{A}$ be the full $Cs_\alpha$ with base $U$. Let $x$ be a regular element of $\mathfrak{A}$ not in the minimal subalgebra of $\mathfrak{A}$, with $|\underset{\sim}{\Delta} x| < \omega$. Then there is a base-automorphism $F$ of $\mathfrak{A}$ such that $Fx \neq x$.

Proof. For each equivalence relation $E$ on $\underset{\sim}{\Delta} x$, let

$$m_E = \bigcap\{D_{\kappa\lambda} : \kappa E \lambda\} \cap \bigcap\{\sim D_{\kappa\lambda} : \kappa,\lambda \in \underset{\sim}{\Delta} x,\ \kappa \not\!E\ \lambda\} .$$

For each $s \in {}^\alpha U$, let $Es = \{\langle\kappa,\lambda\rangle : \kappa,\lambda \in \underset{\sim}{\Delta} x,\ s\kappa = s\lambda\}$. Then we put

$$y = \bigcup_{s \in x} m_{Es} .$$

Thus $y$ is in the minimal subalgebra of $\mathfrak{A}$. Clearly $x \subseteq y$, so there is a $t \in y \sim x$. Say $t \in m_{Es}$, $s \in x$. Then for all $\kappa,\lambda \in \underset{\sim}{\Delta} x$ we have $t\kappa = t\lambda$ iff $s\kappa = s\lambda$. Hence there is a permutation $f$ of

$U$ such that $fs\kappa = t\kappa$ for all $\kappa \in \Delta x$ . Thus $\tilde{f}$ is a base-automorphism of $\mathfrak{A}$ . Since $s \in x$ , we have $f \circ s \in \tilde{f}x$ . Now $\Delta x \upharpoonright (f \circ s) = \Delta x \upharpoonright t$ , $t \notin x$ , and $\tilde{f}x$ is regular (cf. I.3.7), so $f \circ s \notin x$ . Thus $x \neq \tilde{f}x$ , as desired.

<u>Lemma</u> I.3.9. Let $\alpha \geq \omega$ , let $\mathfrak{A} \in Cs_\alpha$ , and suppose that $x \in A$ , $x$ is regular, $|\Delta x| < \omega$ , and $x$ is not in the minimal subalgebra of $\mathfrak{A}$ . Then there is a homomorphism $f$ of $\mathfrak{A}$ onto a $Cs_\alpha$ $\mathfrak{B}$ such that $fx$ is not regular.

<u>Proof</u>. Let $U$ be the base of $\mathfrak{A}$ , and let $\mathfrak{C}$ be the full $Cs_\alpha$ with base $U$ . By Lemma I.3.8, let $F$ be a base-automorphism of $\mathfrak{C}$ such that $Fx \neq x$ . Let $q \in {}^\alpha U$ be arbitrary. For any $c \in C$ let $Gc = (c \cap {}^\alpha U^{(q)}) \cup (Fc \cap ({}^\alpha U \sim {}^\alpha U^{(q)}))$ . Because $\Delta({}^\alpha U^{(q)}) = 0$ , it is easy to verify that $G$ is an endomorphism of $\mathfrak{C}$ . Let $f = A \upharpoonright G$ and $\mathfrak{B} = f^*\mathfrak{A}$ . It remains only to check that $Gx$ is not regular. Since $Fx \neq x$ , say $s \in Fx \sim x$ . Let $z \in {}^\alpha U \sim {}^\alpha U^{(q)}$ be arbitrary (note that $|U| > 1$ since $\mathfrak{A} \neq \mathfrak{S}\mathfrak{g}^{\mathfrak{A}}0$). Now we set

$$z' = (\Delta x \upharpoonright s) \cup (\alpha \sim \Delta x) \upharpoonright z ,$$

$$q' = (\Delta x \upharpoonright s) \cup (\alpha \sim \Delta x) \upharpoonright q .$$

Thus $z' \in {}^\alpha U \sim {}^\alpha U^{(q)}$ and $q' \in {}^\alpha U^{(q)}$ . Furthermore, since $x$ and $Fx$ are regular (cf. I.3.7), we have $z', q' \in Fx \sim x$ . Hence $z' \in Gx$ and $q' \notin Gx$ . But $\Delta x \upharpoonright z' = \Delta x \upharpoonright q'$ and $\Delta Gx \subseteq \Delta x$ . Thus $Gx$ is not regular.

<u>Theorem</u> I.3.10. Let $\alpha \geq \omega$ . Then every non-minimal locally finite dimensional regular $Cs_\alpha$ $\mathfrak{A}$ is isomorphic to a non-regular $Cs_\alpha$ .

Proof. By I.5.2 (i) below, $\mathfrak{A}$ is simple (the proof is easy and direct). Hence the theorem follows from Lemma I.3.9.

Remark I.3.11. From Lemma I.3.7 and Theorem I.3.10 it follows that for each $\alpha \geq \omega$ there exist locally finite-dimensional isomorphic $Cs_\alpha$'s $\mathfrak{A}, \mathfrak{B}$ with finite bases, $\mathfrak{A}$ regular, which are not base-isomorphic. Thus regularity cannot be dropped, for $\mathfrak{A}$ or $\mathfrak{B}$, in Theorem I.3.6. Andréka and Németi have also proved the following:

(1) In I.3.6 we cannot replace $\mathfrak{A},\mathfrak{B} \in Lf_\alpha$ by $\mathfrak{A},\mathfrak{B} \in Dc_\alpha$; see [AN3], Prop. 3.5(iv).

(2) In I.3.6 we cannot replace $\mathfrak{A},\mathfrak{B} \in Cs_\alpha$ by $\mathfrak{A},\mathfrak{B} \in Gs_\alpha$.

(3) The condition that one of the bases is finite cannot be removed.

They also noted that for $\alpha \geq \omega$ $Cs_\alpha$ cannot be replaced by $Ws_\alpha$ (or $Gws_\alpha$); we give the simple example. Let $p = \langle 0 : \kappa < \alpha \rangle$ and $q = \langle 0 : \kappa \text{ even} < \alpha \rangle \cup \langle 1 : \kappa \text{ odd} < \alpha \rangle$. Let $\mathfrak{A}$ and $\mathfrak{B}$ be the minimal $Ws_\alpha$'s with unit elements ${}^\alpha 2^{(p)}$ and ${}^\alpha 2^{(q)}$ respectively, and let $\mathfrak{C}$ be the minimal $Cs_\alpha$ with base $2$. Since $\underset{\sim}{\Delta}({}^\alpha 2^{(p)}) = 0$, we have $\mathfrak{C} \succcurlyeq \mathfrak{A}$, and since $\mathfrak{C}$ is simple (cf. I.5.2), it follows that $\mathfrak{C} \cong \mathfrak{A}$. Similarly, $\mathfrak{C} \cong \mathfrak{B}$. Clearly $\mathfrak{A}$ and $\mathfrak{B}$ are not base-isomorphic.

Remark I.3.12. To complete the discussion of change of base when one base is finite, consider the case of two $Cs_\alpha$'s $\mathfrak{A}$ and $\mathfrak{B}$ with bases $U,W$ respectively, where $\alpha \leq |U| < \omega$. Then it is possible to have $\mathfrak{A} \cong \mathfrak{B}$ even though $|U| \neq |W|$. For example, let $\mathfrak{A}$ and $\mathfrak{B}$ be minimal $Cs_\alpha$'s with bases $U$ and $W$ respectively, subject only to the condition $\alpha \leq |U|, |W|$ ; then $\mathfrak{A} \cong \mathfrak{B}$ by 2.5.30. But also the following simple result shows that not all possibilities can be realized (also recall Remark I.3.2).

Theorem I.3.13. Assume that $\mathfrak{A}$ is the $Cs_\alpha$ with base U generated by $\{\{\langle u : \kappa < \alpha\rangle\} : u \in U\}$ . Then if $\mathfrak{B}$ is any $Crs_\alpha$ with base W such that $\mathfrak{A} \cong \mathfrak{B}$ , we have $|U| \leq |W|$ .

Proof. Let f be the given isomorphism from $\mathfrak{A}$ onto $\mathfrak{B}$ , and for each $u \in U$ let $x_u = \{\langle u : \kappa < \alpha\rangle\}$ . Then $\langle fx_u : u \in U\rangle$ is a system of pairwise disjoint elements of B . Furthermore, for any $u \in U$ and $\kappa, \lambda < \alpha$ we have $x_u \subseteq D_{\kappa\lambda}$ , and so $fx_u \leq d^{\mathfrak{B}}_{\kappa\lambda}$ . Hence for every $u \in U$ there is a $w \in W$ such that $\langle w : \kappa < \alpha\rangle \in fx_u$ . Hence $|U| \leq |W|$ .

Remark I.3.14. By Cor. 1.4 of Andréka, Németi [AN3], the algebra $\mathfrak{A}$ of I.3.13 is regular. The construction can be modified to give a $Ws_\alpha$ with the same conclusion as I.3.13. Namely, fix $u_0 \in U$ and let $p = \langle u_0 : \kappa < \alpha\rangle$ . Let $\mathfrak{A}$ be the $Ws_\alpha$ with unit element ${}^\alpha U^{(p)}$ generated by $\{p^0_u : u \in U\}$ .

Theorem I.3.13 and Remark I.3.14 show that in general the size of a base cannot be reduced. We now prove a theorem giving important special cases in which it <u>is</u> possible to decrease the base; the theorem is due to Andréka, Monk, and Németi. It is an algebraic version of the downward Löwenheim-Skolem theorem, and is proved as that theorem is proved in Tarski, Vaught [TV] . It generalizes Lemma 5 in Henkin, Monk [HM] . First we need a definition and a lemma.

<u>Definition</u> I.3.15. Let $\mathfrak{A}$ and $\mathfrak{B}$ be $Crs_\alpha$'s with unit elements $V_0$ and $V_1$ respectively. If the mapping $\langle X \cap V_0 : X \in B\rangle$ is an isomorphism of $\mathfrak{B}$ onto $\mathfrak{A}$, then we say that $\mathfrak{A}$ is <u>sub-isomorphic</u> to $\mathfrak{B}$, and $\mathfrak{B}$ is <u>ext-isomorphic</u> to $\mathfrak{A}$.

This definition gives algebraic versions of the notions of elementary substructures and elementary extensions. Note that if $\mathfrak{A}$ is sub-isomorphic to $\mathfrak{B}$ then the unit element and base of $\mathfrak{A}$ are contained in those of $\mathfrak{B}$.

<u>Lemma</u> I.3.16. Let $\mathfrak{A}$ and $\mathfrak{B}$ be $Crs_\alpha$'s with unit elements $V_0$ and $V_1$ respectively, and assume that $\mathfrak{A}$ is sub-isomorphic to $\mathfrak{B}$. Then if $X \in B$ is regular in $\mathfrak{B}$, so is $X \cap V_0$ in $\mathfrak{A}$.

<u>Remark</u> I.3.17. For each $\alpha \geq \omega$ there is a regular $Cs_\alpha$ sub-isomorphic to a non-regular $Cs_\alpha$ (thus I.3.16 holds only in the direction given); the example is due to Andréka and Németi. Let $\mathfrak{C}$, resp. $\mathfrak{D}$, be the full $Cs_\alpha$ with base $\omega+1$, resp. $\omega$. Set $p = \langle \omega : \kappa < \alpha\rangle$, $q = \langle 0 : \kappa < \alpha\rangle^0_\omega$ (thus $q0 = \omega$ and $q\kappa = 0$ for all $\kappa \neq 0$). Set

$$X = \{u \in {}^\alpha(\omega+1)^{(p)} : u0 = \omega\}$$
$$\cup \{u \in {}^\alpha(\omega+1) \sim {}^\alpha(\omega+1)^{(p)} : u0 = 0\} .$$

Let $\mathfrak{B} = \mathfrak{Sg}^{\mathfrak{C}}\{X\}$ and $\mathfrak{A} = \mathfrak{Sg}^{\mathfrak{D}}\{X \cap {}^\alpha\omega\}$. Now $\Delta^{\mathfrak{B}} X = 1$, $11p = 11q$, $p \in X$, but $q \notin X$. Hence $X$ is not regular in $\mathfrak{B}$. Clearly, however, $X \cap {}^\alpha\omega = \{u \in {}^\alpha\omega : u0 = 0\}$ is regular in $\mathfrak{A}$ and $\Delta^{\mathfrak{A}}(X \cap {}^\alpha\omega) = 1$; hence by I.4.1 below, $\mathfrak{A}$ is regular. Finally, $\mathfrak{A}$ is subisomorphic to $\mathfrak{B}$. This follows from the following three facts:

(1) If $\mathfrak{A}'$ and $\mathfrak{B}'$ are simple $CA_\alpha$'s generated by $\{x_0\}$ and $\{x_1\}$ respectively, with $\Delta x_0 = \Delta x_1 = 1$, and if for every $\kappa < \omega$ and $\varepsilon < 2$ we have

$$c_{(\kappa)}[\bar{d}(\kappa \times \kappa) \cdot \Pi_{\lambda<\kappa} s^0_\lambda x_\varepsilon] = 0 \quad \text{and}$$

$$c_{(\kappa)}[\bar{d}(\kappa \times \kappa) \cdot \Pi_{\lambda<\kappa} - s^0_\lambda x_\varepsilon] = 1 \quad ,$$

then there is an isomorphism $f$ of $\mathfrak{B}'$ onto $\mathfrak{A}'$ such that $fx_1 = x_0$.

(2) $\mathfrak{A}$ and $\mathfrak{B}$ satisfy (1), with $\mathfrak{A} = \mathfrak{A}'$, $\mathfrak{B} = \mathfrak{B}'$, $X = x_1$, $X \cap {}^\alpha\omega = x_0$.

(3) If $f$ is an isomorphism of $\mathfrak{B}$ onto $\mathfrak{A}$ such that $fX = X \cap {}^\alpha\omega$, then $f = \langle Y \cap {}^\alpha\omega : Y \in B\rangle$.

It is straightforward to check (2), except possibly that $\mathfrak{A}$ and $\mathfrak{B}$ are simple. $\mathfrak{A}$ is simple by I.5.2 below. That both $\mathfrak{A}$ and $\mathfrak{B}$ are simple is seen by 2.2.24. For (3), note that $f(s^0_\kappa X) = s^0_\kappa X \cap {}^\alpha\omega$ for all $\kappa < \alpha$, and hence that both $f$ and $\langle Y \cap {}^\alpha\omega : Y \in B\rangle$ are homomorphisms from $\mathfrak{Bl}\mathfrak{B}$ into $\mathfrak{Bl}\mathfrak{A}$ agreeing on

$$\{D^{\mathfrak{B}}_{\kappa\lambda} : \kappa,\lambda < \alpha\} \cup \{s^0_\kappa X : \kappa < \alpha\} \quad ,$$

which generates $\mathfrak{Bl}\mathfrak{B}$ by 2.2.24 and (2). Hence (3) holds. Finally, (1) is a special case of Theorem 17 of Monk [M1]. We sketch the proof of (1) for completeness. It follows from the following statement that there is an isomorphism $f$ of $\mathfrak{Bl}\mathfrak{B}'$ onto $\mathfrak{Bl}\mathfrak{A}'$ such that $fx_1 = x_0$ and $fd_{\kappa\lambda} = d_{\kappa\lambda}$ for all $\kappa,\lambda < \alpha$:

(4) For all finite $R,S \subseteq \alpha \times \alpha$ and all finite $\Gamma,\Delta \subseteq \alpha$,

$$\Pi_{\langle\kappa,\lambda\rangle\in R} d_{\kappa\lambda} \cdot \Pi_{\langle\kappa,\lambda\rangle\in S} - d_{\kappa\lambda} \cdot \Pi_{\kappa\in\Gamma} s^0_\kappa x_0 \cdot \Pi_{\kappa\in\Delta} s^0_\kappa(-x_0) = 0$$

iff
$$\Pi_{\langle\kappa,\lambda\rangle\in R} d_{\kappa\lambda} \cdot \Pi_{\langle\kappa,\lambda\rangle\in S} - d_{\kappa\lambda} \cdot \Pi_{\kappa\in\Gamma} s^0_\kappa x_1 \cdot \Pi_{\kappa\in\Delta} s^0_\kappa(-x_1) = 0 .$$

Since for all $y$ and $\kappa$, $y = 0$ iff $c_\kappa y = 0$, one sees that (4) is true by using the hypothesis of (1) and 2.2.22, proceeding by induction on $|FdR \cup FdS \cup \Gamma \cup \Delta|$. Given such an isomorphism $f$, that $f$ preserves $c_\kappa$ is also easily seen by 2.2.22.

<u>Theorem</u> I.3.18. Let $\mathfrak{A}$ be a $Crs_\alpha$ with unit element $V$ and base $U$. Let $\kappa$ be an infinite cardinal such that $|A| \leq \kappa \leq |U|$. Assume $S \subseteq U$ and $|S| \leq \kappa$. Then there is a $W$ with $S \subseteq W \subseteq U$ such that $|W| = \kappa$ and:

(i) Each of the following conditions a) - c) implies that $\mathfrak{A}$ is ext-isomorphic to a $Crs_\alpha$ with unit element $V \cap {}^\alpha W$:

a) $\mathfrak{A} \in Ws_\alpha$ ;

b) $\kappa = \kappa^{|\alpha|}$ ;

c) $\mathfrak{A}$ is a regular $Gs_\alpha$, and $\kappa = \sum_{\mu<\lambda} \kappa^\mu$, where $\lambda$ is the least infinite cardinal such that $|\Delta X| < \lambda$ for all $X \in A$.

(ii) If $\mathfrak{A}$ is a $Ws_\alpha$ with unit element ${}^\alpha U^{(p)}$, then $\mathfrak{A}$ is ext-isomorphic to a $Ws_\alpha$ with unit element ${}^\alpha W^{(p)}$.

(iii) If $\mathfrak{A}$ is a $Cs_\alpha$ and $\kappa = \kappa^{|\alpha|}$, then $\mathfrak{A}$ is ext-isomorphic to a $Cs_\alpha$ with base $W$.

(iv) If $\mathfrak{A}$ is a regular $Gs_\alpha$ (resp. $Cs_\alpha$) and (i) (c) holds, then $\mathfrak{A}$ is ext-isomorphic to a regular $Gs_\alpha$ (resp. $Cs_\alpha$) with base $W$.

(v) If $\mathfrak{A}$ is a $Gws_\alpha$ then $\mathfrak{A}$ is ext-isomorphic to a $Gws_\alpha$ with base $W$.

(vi) If $\alpha \leq \kappa$, then $\mathfrak{A}$ is ext-isomorphic to a $Crs_\alpha$ with base $W$.

Proof. We assume given well-orderings of $U$ and $V$. (i) (a) and (ii): Note that $|\alpha| \leq |A|$. There is a subset $T_0$ of $U$ such that $|T_0| = \kappa$, $S \cup Rgp \subseteq T_0$, and $X \cap {}^\alpha T_0 \neq 0$ whenever $0 \neq X \in A$. Now suppose that $0 < \beta < \kappa$ and $T_\gamma$ has been defined for all $\gamma < \beta$. Let $M = \bigcup_{\gamma<\beta} T_\gamma$ and let

$$T_\beta = M \cup \{a \in U : \text{there is an } X \in A, \text{ a } \mu < \alpha, \text{ and a}$$

$u \in {}^\alpha M^{(p)}$ such that $a$ is the first element of $U$ with the property that $u^\mu_a \in X\}$.

Let $W = T_\kappa = \bigcup_{\gamma<M} T_\gamma$. By induction it is easily seen that $|T_\beta| = \kappa$ for all $\beta \leq \kappa$; in particular, $|W| = \kappa$. The desired conclusion is easy to check. (i) (b) and (iii): We make the same construction, beginning with $T_0 \subseteq U$ such that $|T_0| = \kappa$, $S \subseteq T_0$, and $X \cap {}^\alpha T_0 \neq 0$ whenever $0 \neq X \in A$; to construct $T_\beta$ we replace ${}^\alpha M^{(p)}$ above by ${}^\alpha M \cap V$. The condition $\kappa^{|\alpha|} = \kappa$ is used to check that $|T_\beta| = \kappa$ for all $\beta \leq \kappa$. To check that $\langle X \cap {}^\alpha W \cap V : X \in A\rangle$ preserves $C_\mu$ it is enough to note that $\alpha < cf\kappa$ because $\kappa^{|\alpha|} = \kappa$, and hence any $p \in {}^\alpha W$ is in ${}^\alpha T_\beta$ for some $\beta < \kappa$. (i) (c) and (iv): If $\mathfrak{A}$ is discrete then the conclusion is clear. Now suppose $\mathfrak{A}$ is non-discrete. Then $|\alpha| \leq |A| \leq \kappa$. Furthermore, we may assume that $\alpha \geq 2$, since the case $\alpha \leq 1$ is treated by (i) (a) and (ii) above (cf. I.1.8). Now we proceed as in (i) (b), except that $T_\beta$ is defined as follows:

$$T_\beta = M \cup \{a \in U : \text{there is an } X \in A, \text{ a } \mu \in \underset{\sim}{\Delta} X, \text{ a}$$

$\nu \in \alpha \sim \{\mu\}$ and a $u \in {}^{((\underset{\sim}{\Delta}X \sim \{\mu\}) \cup \{\nu\})}M$ such that a is the first element of U with the property that $v \in X$ for some $v \in V$ with $((\underset{\sim}{\Delta}X \sim \{\mu\}) \cup \{\nu\}) \upharpoonright v = u$ and $v\mu = a\}$ .

The condition in (i) (c) is used to check that $|T_\beta| = \kappa$ for all $\beta \leq \kappa$ , and that $u \in {}^{\Gamma}W$ with $|\Gamma| < \lambda$ implies $u \in {}^{\Gamma}T_\beta$ for some $\beta < \kappa$ . To check that $h = \langle X \cap {}^{\alpha}W \cap V : X \in A \rangle$ preserves $C_\mu$ , we need to prove

$$(*) \qquad C_\mu^{[{}^{\alpha}W \cap V]}(X \cap {}^{\alpha}W \cap V) = C_\mu^{[V]}X \cap {}^{\alpha}W \cap V$$

for $X \in A$ . If $\mu \notin \underset{\sim}{\Delta}^{(\mathfrak{A})}X$ , then (*) is clear. Assume $\mu \in \underset{\sim}{\Delta}^{(\mathfrak{A})}X$ . Then the inclusion $\subseteq$ in (*) is clear. Suppose $p \in C_\mu^{[V]}X \cap {}^{\alpha}W \cap V$ . So $p \in V$ , and $p_a^\mu \in X$ for some a . Choose $\nu \in \alpha \sim \{\mu\}$ and let $u = ((\underset{\sim}{\Delta}X \sim \{\mu\}) \cup \{\nu\}) \upharpoonright p$ . Then $u \in {}^{(\underset{\sim}{\Delta}X \sim \{\mu\}) \cup \{\nu\}}(\bigcup_{\gamma < \beta} T_\gamma)$ for some $\beta < \kappa$ , so by the definition of $T_\beta$ , $v \in X$ for some v with $((\underset{\sim}{\Delta}X \sim \{\mu\}) \cup \{\nu\}) \upharpoonright v = u$ and $v\mu \in T_\beta \subseteq W$ . Then $v\nu = u\nu = p\nu$ , so it follows since $\mathfrak{A} \in Gs_\alpha$ and $\alpha \geq 2$ that $p_{v\mu}^\mu \in V$ ,while obviously $p_{v\mu}^\mu \in {}^{\alpha}W$ . Thus $\underset{\sim}{\Delta}X \upharpoonright v = \underset{\sim}{\Delta}X \upharpoonright p_{v\mu}^\mu$ , so by regularity of X , $p_{v\mu}^\mu \in X$ . Thus $p \in C_\mu^{[{}^{\alpha}W \cap V]}(X \cap {}^{\alpha}W \cap V)$ , as desired. For the regularity required in (iv) , see Lemma I.3.16. (v): Let the unit element of $\mathfrak{A}$ be

$$\bigcup_{i \in I} {}^{\alpha}T_i^{(pi)} ,$$

where ${}^{\alpha}T_i^{(pi)} \cap {}^{\alpha}T_j^{(pj)} = 0$ for $i \neq j$ . Choose $J \subseteq I$ with $|J| \leq \kappa$ such that $S \subseteq \bigcup_{j \in J} T_j$ and $X \cap \bigcup_{j \in J} {}^{\alpha}T_j^{(pj)} \neq 0$ for all non-zero $X \in A$ . Now take any $j \in J$ . The set $\{X \cap {}^{\alpha}T_j^{(pj)} : X \in A\}$ is the universe of a $Ws_\alpha$ , as is easily checked (since ${}^{\alpha}T_j^{(pj)}$ may not be a member of A , we cannot use the notation $\mathfrak{Rl}_y \mathfrak{A}$ , $y = {}^{\alpha}T_j^{(pj)}$ ) . We denote this $Ws_\alpha$ by $\mathfrak{B}_j$ . If $|T_j| < \kappa$ , we set $W_j = T_j$ . If $|T_j| \geq \kappa$ , by (ii), $\mathfrak{B}_j$ is ext-isomorphic to a $Ws_\alpha$ with unit element ${}^\alpha W_j^{(pj)}$ , where $\mathrm{rg}p_j \cup (S \cap T_j) \subseteq W_j \subseteq T_j$ and $|W_j| = \kappa$ . With $V' = \bigcup_{j\in J} {}^\alpha W_j^{(pj)}$ the desired conclusion is easily checked. (vi): Let $Z_0$ be a subset of $V$ such that $|Z_0| \leq \kappa$ , $S \subseteq \bigcup_{z\in Z_0} \mathrm{Rg}z$ , $|\bigcup_{z\in Z_0} \mathrm{Rg}z| = \kappa$ , and $X \cap Z_0 \neq 0$ whenever $0 \neq X \in A$. If $n \in \omega$ and $Z_n$ has been defined, let

$$Z_{n+1} = Z_n \cup \{z \in V : \text{there is a } q \in Z_n \text{, an } X \in A \text{, and a } \mu < \alpha \text{ such that } z \text{ is the least element of } V \text{ with } z \in X \text{ and } (\alpha\sim\{\mu\}) \upharpoonright z = (\alpha\sim\{\mu\}) \upharpoonright q\} .$$

Let $V' = \bigcup_{n\in\omega} Z_n$ and $W = \bigcup_{z\in V'} \mathrm{Rg}z$ . It is easily checked that $\mathfrak{A}$ is ext-isomorphic to a $Crs_\alpha$ with unit element $V'$ and base $W$ satisfying the desired conditions.

<u>Remark</u> I.3.19. The conditions in I.3.18 are necessary for the truth of the theorem. First, Andréka and Németi have noticed that for each $\alpha \geq \omega$ and each infinite $\kappa$ with $\kappa^{|\alpha|} \neq \kappa$ there is a locally-finite dimensional $Cs_\alpha$ $\mathfrak{A}$ of power $|\alpha|$ , with base of power $\kappa^{|\alpha|}$ , such that $\mathfrak{A}$ is not ext-isomorphic to any $Cs_\alpha$ with base of power $\kappa$ . This shows that $Ws_\alpha$ cannot be replaced by $Cs_\alpha$ in (i) a), the condition $\kappa = \kappa^{|\alpha|}$ cannot be weakened in (i) b) or (iii), regularity cannot be dropped for $Cs_\alpha$'s in (iv), and "$Gws_\alpha$" cannot be replaced by "$Cs_\alpha$" in (v). To construct this algebra, let $U = {}^\alpha\kappa$ . Let

$$R = \{\langle f,g\rangle : f,g \in {}^\alpha U \text{ and } |\{\kappa < \alpha : f\kappa \neq g\kappa\}| < \omega\} .$$

Let $F$ be a function from ${}^{\alpha}U$ into $U$ such that for all $f,g \in {}^{\alpha}U$ we have $Ff = Fg$ iff $\langle f,g\rangle \in R$. Let $\tau = \langle \kappa + 1 : \kappa < \omega\rangle \cup \langle \kappa : \kappa \in \alpha \sim \omega\rangle$. Set $X = \{q \in {}^{\alpha}U : q_0 = F(q \circ \tau)\}$. Clearly $c_0^{[{}^{\alpha}U]}X = {}^{\alpha}U$ and $\Delta X = \{0\}$. Let $\mathfrak{B}$ be the full $Cs_\alpha$ with base $U$ and let $\mathfrak{A} = \mathfrak{Sg}^{(\mathfrak{B})}\{X\}$. Thus $\mathfrak{A}$ is locally finite-dimensional, $|A| = \alpha$, and the base of $\mathfrak{A}$ has power $\kappa^{|\alpha|}$. Suppose $W \subseteq U$ and $|W| = \kappa$. To show that $\mathfrak{A}$ is not ext-isomorphic to any $Cs_\alpha$ with base $W$ it suffices to show that $c_0^{[{}^{\alpha}W]}(X \cap {}^{\alpha}W) \neq {}^{\alpha}W$. Note that each equivalence class under $R \cap ({}^{\alpha}W \times {}^{\alpha}W)$ has cardinality $|\alpha| \cup \kappa < \kappa^{|\alpha|}$, and so there are $\kappa^{|\alpha|}$ equivalence classes altogether. Hence we can choose $f \in {}^{\alpha}W$ such that $Ff \notin W$. Let $g = \{\langle 0,f_0\rangle\} \cup \langle f_{\kappa-1} : \kappa \in \omega \sim 1\rangle \cup (\alpha \sim \omega) 1 f$. Then $g \circ \tau = f$ and hence $F(g \circ \tau) \notin W$. Therefore $g \in {}^{\alpha}W$ but $g \notin c_0^{[{}^{\alpha}W]}(X \cap {}^{\alpha}W)$, as desired.

Second, Andréka and Németi have constructed for each $\alpha \geq \omega$ a regular dimension-complemented $Cs_\alpha$ of power $|\alpha|$ with base of power $|\alpha|^+$ which is not ext-isomorphic to any $Cs_\alpha$ with base of power $|\alpha|$. Thus the condition $\kappa = \sum_{\mu<\lambda} \kappa^{\mu}$ cannot be dropped in (i) c). Third, Theorem I.3.13 and Remark I.3.14 show that the condition $|A| \leq \kappa$ is needed. Fourth, the hypothesis that $\kappa$ is infinite is essential by Remark I.3.2. Finally, Andréka and Németi have shown that in (i) (c) one cannot replace "$Gs_\alpha$" by "$Gws_\alpha$". We describe their interesting example: for each $\alpha \geq \omega$ we construct an $\mathfrak{A} \in Gws_\alpha^{reg} \cap Lf_\alpha$ with a base $U$ such that $|A| \leq |\alpha| \leq |U|$ and having the property that for all $W \subseteq U$, if $|W| = |\alpha|$ then $\langle a \cap {}^{\alpha}W : a \in A\rangle$ is not an isomorphism. This provides the desired counterexample, with $\kappa = |\alpha|$. Let $\beta = |{}^{\alpha}2|$. It is easy to define $p \in {}^{\beta}({}^{\alpha}\omega)$ such that for all $\gamma,\delta$, if $\gamma < \delta < \beta$ then

$p_\gamma \notin {}^\alpha\omega^{(p_\delta)}$ and such that for every infinite $\Gamma \subseteq \omega$ and every $\gamma < \beta$ there is a $\delta < \beta$ with $\gamma \le \delta$ and $p_\delta \in {}^\alpha\Gamma$ . For each $\gamma < \beta$ let $U_\gamma = \omega \cup (\beta \sim \gamma)$ , set $V = \bigcup_{\gamma<\beta} {}^\alpha U_\gamma^{(p_\gamma)^\delta}$ , and let $X = \{f \in V : f0 \notin \omega\}$ . Let $\mathfrak{A}$ be the $Gws_\alpha$ with unit element $V$ generated by $\{X\}$ . We claim that $\mathfrak{A}$ is the desired algebra. Note that $\mathfrak{A}$ has base $\beta$ , and that $U_\gamma \supseteq U_\delta$ whenever $\gamma \le \delta < \beta$ . Since $\Delta X = 1$ , by 2.1.15 (i) we have $\mathfrak{A} \in Lf_\alpha$ . Now we show that if $W \subseteq U$ and $|W| = |\alpha|$ , then $\mathfrak{A}$ is not ext-isomorphic to any $Crs_\alpha$ with unit element $V \cap {}^\alpha W$ . Let $Z = V \cap {}^\alpha W$ , $\kappa = |W \cap \omega| + 1$ . We now consider two cases.

Case 1. $\kappa \in \omega$ . Now

$$c_{(\kappa)}^{[V]}(\bar{d}(\kappa \times \kappa) \cdot \Pi_{\iota<\kappa} - s_\iota^0 X) = V$$

while

$$c_{(\kappa)}^{[Z]}(\bar{d}(\kappa \times \kappa) \cdot \Pi_{\iota<\kappa} - s_\iota^0 X) = 0 ,$$

so clearly $\mathfrak{A}$ is not ext-isomorphic to a $Crs_\alpha$ with unit element $Z$ .

Case 2. $\kappa \ge \omega$ . Since $|W| = |\alpha| < cf|{}^\alpha 2| = cf\beta$ , there is a $\gamma < \beta$ such that for all $\delta \in W$ , $\delta < \gamma$ . By our choice of $p$ there then exists a $\delta$ , $\gamma \le \delta < \beta$ , such that $p_\delta \in {}^\alpha(W \cap \omega)$ . Thus $p_\delta \in c_0^{[V]} X \cap Z$ . But we now show that $p_\delta \notin c_0^{[Z]}(X \cap Z)$ , and hence again $\mathfrak{A}$ is not ext-isomorphic to a $Crs_\alpha$ with unit element $Z$ . In fact, otherwise there is an $\varepsilon < \beta$ such that $(p_\delta)_\varepsilon^0 \in X \cap Z$ . Thus $\varepsilon \ge \omega$ by the definition of $X$ , and hence $\varepsilon \ge \delta$ since $(p_\delta)_\varepsilon^0 \in {}^\alpha U_\delta^{(p_\delta)}$ (which is true since $(p_\delta)_\varepsilon^0 \in Z \subseteq V$ , using our assumptions on $p$ ). But then $\varepsilon \notin W$ , by our choice of $\delta$ , contradicting the fact that $(p_\delta)_\varepsilon^0 \in Z \subseteq {}^\alpha W$ .

It remains to show that $\mathfrak{A}$ is regular. For this purpose we need the following fact about regularity. It is a part of Lemma 1.3.4 of Andréka, Németi [AN 3], but we include its short proof for completeness.

(*) Let $\mathfrak{A}$ be a $Gws_\alpha$ with unit element $V$, $X \in A$, and $\Gamma$ a finite subset of $\alpha$. Suppose that for all $f,g$, if $f \in X$, $g \in V$ and $(\Delta X \cup \Gamma) \upharpoonright f = (\Delta X \cup \Gamma) \upharpoonright g$, then $g \in X$. Then $X$ is regular.

To prove (*), assume its hypothesis, and suppose that $f \in X$, $g \in V$, and $(\Delta X \cup 1) \upharpoonright f = (\Delta X \cup 1) \upharpoonright g$; we are to show that $g \in X$. Let $\Theta = \Gamma \sim (\Delta X \cup 1)$, and for each $k \in V$ let $k' = (\alpha \sim \Theta) \upharpoonright k \cup (\Theta \times \{f0\})$. Since $f0 = g0$ and $\Theta$ is finite, we have $f', g' \in V$. Since $\{\kappa : f\kappa \neq f'\kappa\}$ is a finite subset of $\alpha \sim \Delta X$, we have $f' \in X$. Clearly $(\Delta X \cup \Gamma) \upharpoonright f' = (\Delta X \cup \Gamma) \upharpoonright g'$, so by the hypothesis of (*), $g' \in X$. Finally, $\{\kappa : g\kappa \neq g'\kappa\}$ is a finite subset of $\alpha \sim \Delta X$, so $g \in X$, as desired.

Now we prove that $\mathfrak{A}$ is regular. Let $Y \in A$. Since $\Delta X = 1$, we can apply 2.2.24. Note that for any $\kappa \in \omega$ we have

$$c_{(\kappa)}[\bar{d}(\kappa \times \kappa) \cap \bigcap_{\iota<\kappa} s^0_\iota Z] = V ,$$

for $Z = X$ or $Z = V \sim X$. Thus by 2.2.24 we can write

$$Y = \bigcup_{\gamma\in\Gamma}[\bigcap_{\langle\kappa,\lambda\rangle\in R\gamma} D_{\kappa\lambda} \cap \bigcap_{\langle\kappa,\lambda\rangle\in S\delta} \sim D_{\kappa\lambda} \cap \bigcap_{\delta\in\Theta\gamma} s^0_{\mu\gamma\delta} X$$
$$\cap \bigcap_{\delta\in\Xi\gamma} s^0_{\nu\gamma\delta}(V \sim X)] ,$$

where $|\Gamma| < \omega$, $R\gamma, S\gamma \subseteq \alpha \times \alpha$ and $|R\gamma|, |S\gamma| < \omega$ for $\gamma \in \Gamma$, $|\Theta\gamma|, |\Xi\gamma| < \omega$ for $\gamma \in \Gamma$, and $\mu\gamma\delta, \nu\gamma\varepsilon \in \alpha$ for $\gamma \in \Gamma$, $\delta \in \Theta\gamma$, $\varepsilon \in \Xi\gamma$. Let

$$\Omega = \bigcup_{\gamma\in\Gamma}(FdR\gamma \cup FdS\gamma \cup \{\mu\gamma\delta : \delta \in \Theta\gamma\} \cup \{\nu\gamma\delta : \delta \in \Xi\gamma\}) .$$

Note that $\Delta Y \subseteq \Omega$ . Thus to prove that $Y$ is regular it suffices by (*) to suppose that $f \in Y$ , $g \in V$ , and $\Omega 1 f = \Omega 1 g$ , and show that $g \in Y$ . Since $f \in Y$ , choose $\gamma \in \Gamma$ so that

$$f \in \bigcap_{\langle\kappa,\lambda\rangle\in R\gamma} D_{\kappa\lambda} \cap \bigcap_{\langle\kappa,\lambda\rangle\in S\gamma} \sim D_{\kappa\lambda}$$
$$\cap \bigcap_{\delta\in\Theta\gamma} s^0_{\mu\gamma\delta}X \cap \bigcap_{\delta\in\Xi\gamma} s^0_{\nu\gamma\delta}(V\sim X) .$$

Since $s^0_\rho X = \{h \in V : h\rho \notin \omega\}$ , for any $\rho < \alpha$ it is now easy to see that $g \in Y$ .

We shall consider the question of increasing bases in section I.7, since we need ultraproducts to establish these results; see I.7.19 - I.7.30. We wish to conclude this section by considering a question related to the changing base question: when is a $Ws_\alpha$ with unit element ${}^\alpha W^{(p)}$ isomorphic to one with unit element ${}^\alpha W^{(q)}$? The following theorem is a generalization of Lemma 6 of [HM] due to Andréka and Németi:

<u>Theorem</u> I.3.20. Let $\mathfrak{A}$ (resp. $\mathfrak{A}'$) and $\mathfrak{B}$ (resp. $\mathfrak{B}'$) be (the full) $Ws_\alpha$'s with unit elements $V_0$ and $V_1$ , and bases $U_0$ and $U_1$ , respectively. Consider the following conditions:

(i) $\mathfrak{A}$ and $\mathfrak{B}$ are base-isomorphic;

(ii) there exist $p' \in V_0$ and $q' \in V_1$ such that $p'|p'^{-1} = q'|q'^{-1}$ and $|U_0 \sim Rgp'| = |U_1 \sim Rgq'|$ ;

(iii) $\mathfrak{A}' \cong \mathfrak{B}'$ ;

(iv) $\mathfrak{A}'$ is base-isomorphic to $\mathfrak{B}'$ .

Then (i) $\Rightarrow$ (ii), while (ii) $\Leftrightarrow$ (iii) $\Leftrightarrow$ (iv) .

<u>Proof</u>. Say $V_0 = {}^{\alpha}U_0^{(p)}$ and $V_1 = {}^{\alpha}U_1^{(q)}$ . (i) $\Rightarrow$ (iv): trivial. (iv) $\Rightarrow$ (iii): trivial. (iii) $\Rightarrow$ (ii): Let $f$ be an isomorphism from $\mathfrak{A}'$ onto $\mathfrak{B}'$ . Choose $q' \in V_1$ so that $f\{p\} = \{q'\}$ . If $p\kappa = p\lambda$ , then $\{p\} \subseteq D_{\kappa\lambda}$ , so $\{q'\} \subseteq D_{\kappa\lambda}$ and $q'\kappa = q'\lambda$ . By symmetry, $p|p^{-1} = q'|q'^{-1}$ . Also, it is easy to check that

$$|U \sim Rgp| = |\{d \in A' : d \text{ is an atom} \leq$$

$$C_0\{p\} \cap \bigcap_{0<\kappa<\alpha} \sim D_{0\kappa} \sim \{p\}\}| =$$

$$|\{d \in B' : d \text{ is an atom} \leq C_0\{q'\} \cap$$

$$\bigcap_{0<\kappa<\alpha} \sim D_{0\kappa} \sim \{q'\}\}| = |U \sim Rgq'| .$$

Thus (ii) holds. (ii) $\Rightarrow$ (iv): this is proved in [HM] , but we sketch the proof here. Let $f = \{\langle p'\kappa, q'\kappa\rangle : \kappa < \alpha\}$ . Then $f$ is a one-to-one function from a subset of $U_0$ onto a subset of $U_1$ , and it can be extended to a one-to-one function $f'$ from $U_0$ onto $U_1$ . By Theorem I.3.1, $\tilde{f}'$ is an isomorphism from $\mathfrak{A}'$ onto the full $Ws_\alpha\, \mathfrak{B}''$ with unit element ${}^{\alpha}U_1^{(f' \circ p)}$ . If $\Gamma$ and $\Delta$ are finite sets such that $(\alpha \sim \Gamma) 1 p = (\alpha \sim \Gamma) 1 p'$ and $(\alpha \sim \Delta) 1 q = (\alpha \sim \Delta) 1 q'$ , clearly $(\alpha \sim (\Gamma \cup \Delta)) 1 (f' \circ p) = (\alpha \sim (\Gamma \cup \Delta)) 1 q$ . Thus ${}^{\alpha}U_1^{(f' \circ p)} = {}^{\alpha}U_1^{(q)}$ , as desired.

<u>Remark</u> I.3.21. It is easy to see that in I.3.20, (ii) does not imply (i) in general. The condition of base-isomorphism in (i) cannot be replaced by isomorphism. This follows from the following theorem of Andréka and Németi.

<u>Theorem</u> I.3.22. If $\mathfrak{A}$ is a locally finite-dimensional $Ws_\alpha$ with base $U$ and $q \in {}^\alpha U$, then $\mathfrak{A}$ is isomorphic to a $Ws_\alpha$ with unit element ${}^\alpha U^{(q)}$.

<u>Proof</u>. Let $\mathfrak{A}$ have unit element ${}^\alpha U^{(p)}$. For any $X \in A$ let

$$fX = \{u \in {}^\alpha U^{(q)} : \text{there is a } v \in X \text{ with } \Delta X \upharpoonright u = \Delta X \upharpoonright v\}.$$

Using the regularity of $\mathfrak{A}$ (I.1.16), it is easy to see that $f$ is an isomorphism from $\mathfrak{Bl}\mathfrak{A}$ into $\mathfrak{Bl}\mathfrak{B}$, $\mathfrak{B}$ the full $Ws_\alpha$ with unit element ${}^\alpha U^{(q)}$. Now let $\kappa < \alpha$ and $X \in A$. If $u \in fC_\kappa X$, choose $v \in C_\kappa X$ so that $\Delta C_\kappa X \upharpoonright u = \Delta C_\kappa X \upharpoonright v$. Define $w \in {}^\alpha U$ be setting for any $\lambda < \alpha$

$$w\lambda = \begin{cases} u\lambda & \lambda \in \Delta X, \\ v\lambda & \lambda \notin \Delta X. \end{cases}$$

Since $\Delta X$ is finite, $w \in {}^\alpha U^{(p)}$. Now $\Delta C_\kappa X \subseteq \Delta X$, so $\Delta C_\kappa X \upharpoonright v = \Delta C_\kappa X \upharpoonright w$. Hence by regularity $w \in C_\kappa X$. Choose $a \in U$ such that $w^\kappa_a \in X$. Now $\Delta X \upharpoonright w^\kappa_a = \Delta X \upharpoonright u^\kappa_a$, so $u^\kappa_a \in fX$. Thus $u \in C_\kappa fX$. The converse is straightforward.

Some results related to I.3.20 and I.3.22 are given in I.7.27-I.7.30.

## 4. Subalgebras

Our various classes of set algebras are clearly closed under the formation of subalgebras, and we shall not formulate a theorem to this effect. The following theorem gives an important method for forming regular set algebras. The proof is due to Andréka and Németi.

Theorem I.4.1. If $\mathfrak{A}$ is a $Cs_\alpha$ generated by a set of regular elements with finite dimension sets, then $\mathfrak{A}$ is regular.

Proof. We shall use (*) from I.3.19. Let $B$ be the set of all finite dimensional regular elements of $\mathfrak{A}$ ; it suffices to show that $B \in Su\,\mathfrak{A}$ . Clearly $D_{\kappa\lambda} \in B$ for all $\kappa,\lambda < \alpha$ , and clearly $B$ is closed under $-$ , since $\underset{\sim}{\Delta}X = \underset{\sim}{\Delta}(-X)$ .

Now let $X,Y \in B$ ; we show that $X \cap Y \in B$ . In fact, we shall verify (*) with $\Gamma = \underset{\sim}{\Delta}X \cup \underset{\sim}{\Delta}Y \supseteq \underset{\sim}{\Delta}(X \cap Y)$ . Suppose $\Gamma 1 f = \Gamma 1 g$ , $f \in X \cap Y$ and $g \in {}^\alpha U$ . Then $\underset{\sim}{\Delta}X 1 f = \underset{\sim}{\Delta}X 1 g$ , so $g \in X$ since $X$ is regular. Similarly, $g \in Y$ , as desired.

Finally, suppose $X \in B$ and $\kappa < \alpha$ ; we show that $C_\kappa X \in B$ . To this end we verify (*) with $\Gamma = \underset{\sim}{\Delta} X \supseteq \underset{\sim}{\Delta}C_\kappa X$ . So, suppose $\underset{\sim}{\Delta}X 1 f = \underset{\sim}{\Delta}X 1 g$ , $f \in C_\kappa X$ , and $g \in {}^\alpha U$ . Then for some $u \in U$ we have $f^\kappa_u \in X$ . Thus $\underset{\sim}{\Delta}X 1 f^\kappa_u = \underset{\sim}{\Delta}X 1 g^\kappa_u$ and $g^\kappa_u \in {}^\alpha U$ , so by the regularity of $X$ , $g^\kappa_u \in X$ . Thus $g \in C_\kappa X$ , as desired.

Remarks. I.4.2. As mentioned in the introduction to this paper, regular cylindric set algebras arise naturally from relational structures and the notion of satisfaction in an associated first-order language. Using I.4.1 we can express this construction of regular $Cs_\alpha$'s without recourse to an auxiliary language. Namely, let $\mathfrak{A} = \langle A, R_i, \mathcal{O}_j\rangle_{i\in I, j\in J}$ be a relational structure, and let $\alpha \geq \omega$ . Let $\rho_i$ be the rank of $R_i$ and $\sigma_j$ the rank of $\mathcal{O}_j$ for each $i \in I$ and $j \in J$ . Set

$$X = \{\{x \in {}^\alpha A : \rho_i 1 x \in R_i\} : i \in I\}$$
$$\cup \{\{x \in {}^\alpha A : \mathcal{O}_j\langle x_\kappa : \kappa < \sigma_j\rangle = x_{\sigma_j}\} : j \in J\} .$$

Clearly each member of $X$ is regular and finite dimensional. Hence by I.4.1, the $Cs_\alpha$ of subsets of ${}^\alpha A$ generated by $X$ is regular. This is the same $Cs_\alpha$ described in the introduction to this paper in terms of a language for $\mathfrak{A}$ .

Conversely, given any $\mathfrak{B} \in Cs_\alpha^{reg} \cap Lf_\alpha$, there is a relational structure $\mathfrak{A}$ such that $\mathfrak{B}$ is obtained from $\mathfrak{A}$ in the way just described. We shall prove this, which is rather easy, in a later paper where we discuss this correspondence in detail.

The assumption that the dimension sets are finite in I.4.1 is essential, and cannot even be replaced by the assumption that $\mathfrak{A}$ is dimension complemented, or by the weaker assumption that the regular elements mentioned have dimension sets with infinite complements. To see this, let $\alpha \geq \omega$ , let $\mathfrak{B}$ be the full $Cs_\alpha$ with base $\omega$ , and let $X = \{ x \in {}^\alpha\omega : \text{for every odd } \kappa < \alpha, x\kappa \leq x0\}$ . Clearly $\Delta X = \{\kappa < \alpha : \kappa \text{ is odd}\} \cup 1$ , and hence $\mathfrak{A} = \mathfrak{Sg}^{(\mathfrak{B})}\{X\} \in Dc_\alpha$ . Furthermore, $X$ is clearly regular. But $c_0 X = \{x \in {}^\alpha\omega : \text{there is a } \lambda < \omega \text{ such that } x\kappa \leq \lambda \text{ for every odd } \kappa < \alpha\}$ , so $\Delta c_0 X = 0$ while $0 \neq c_0 X \neq 1$ , so $c_0 X$ is not regular.

Andréka and Németi have shown that "$Cs_\alpha$" cannot be replaced by "$Gws_\alpha$" in I.4.1. They also established the following interesting facts about $Gws_\alpha$'s (where $\mathfrak{Mn}\mathfrak{A}$ is the minimal subalgebra of $\mathfrak{A}$ ):

(1) For any $Gws_\alpha\,\mathfrak{A}$ , $\alpha \geq \omega$ , $\mathfrak{Mn}\mathfrak{A}$ is regular iff for every two subbases $Y$ and $W$ of $\mathfrak{A}$ , $|Y| = |W| < \omega$ or $|Y|, |W| \geq \omega$ ;

(2) There is a $Gws_\alpha\,\mathfrak{A}$ , $\alpha \geq \omega$ , having elements $X,Y$ such that $\Delta X = \Delta Y = 1$ and both $\mathfrak{Sg}\{X\}$ and $\mathfrak{Sg}\{Y\}$ are regular but $\mathfrak{Sg}\{X,Y\}$ is not;

(3) There is a $Gws_\alpha$ $\mathfrak{A}$ , $\alpha \geq \omega$ , such that $\mathfrak{Mn}\mathfrak{A}$ is the largest regular subalgebra of $\mathfrak{A}$ and there is an element $X \in A$ such that $\Delta X = 1$ , $X$ is regular, and $C_0X$ is not regular.

(4) For every $Gws_\alpha$ $\mathfrak{A}$ the following two conditions are equivalent:

(a) $\mathfrak{A}$ is normal (see I.2.6);

(b) if $\mathfrak{B}$ is the full $Gws_\alpha$ such that $\mathfrak{A} \subseteq \mathfrak{B}$ , then every subset $X$ of $\mathfrak{B}$ consisting of regular finite-dimensional elements is such that $\mathfrak{Sg}^{(\mathfrak{B})}X$ is regular.

We also mention the following useful and obvious property of regular Cs's:

Theorem I.4.3. If $\mathfrak{A}$ is a regular $Cs_\alpha$ , then $Zd\mathfrak{A} = \{0,1\}$ .

For the rest of this section we consider the problem of the number of generators of set algebras, in particular, conditions under which a set algebra has a single generator. This question was considered in 2.1.11, 2.3.22, 2.3.23, and 2.6.25. In particular, following 2.1.11 the following result was stated, the proof being easily obtained from the proof of 2.1.11:

(*) If $2 \leq \alpha < \omega$ and $\kappa < \omega$ , then the full $Cs_\alpha$ with base $\kappa$ is generated by a single element.

By generalizing the proof of 2.1.11 further we obtain the following generalization of (*) , due to Monk.

Theorem I.4.4. Let $\alpha < \omega$ , and let $\mathfrak{A}$ be the full $Gs_\alpha$ with unit element $V = \bigcup_{i\in I} {}^\alpha U_i$ , where $U_i \cap U_j = 0$ for distinct $i,j \in I$ ,

$2 \cdot |I| \leq \alpha$ , and $1 < |U_i| < \omega$ for all $i \in I$ . Then $\mathfrak{A}$ is generated by a single element.

<u>Proof</u>. The theorem is trivial if $\alpha \leq 1$ , so assume that $\alpha \geq 2$ . We may assume that $I = \beta < \omega$ with $2\beta \leq \alpha$ . For each $\kappa < \beta$ let $t_\kappa$ be a one-one function mapping $U_\kappa$ onto some $\mu_\kappa < \omega$ . Our single generator is

$$X = \{u \in V : \text{if } \kappa \text{ is the element of } \beta \text{ such that } u \in {}^\alpha U_\kappa, \text{ then } t_\kappa u_{2\kappa} < t_\kappa u_{2\kappa+1}\} .$$

Now for each $\kappa < \beta$ we define a sequence $Y_\kappa \in {}^\omega A$ by recursion:

$$Y_{\kappa 0} = C_{2\kappa+1}(\sim X) \cap C_{2\kappa} C_{2\kappa+1} X ,$$

while for $0 \leq \lambda < \omega$ we set

$$Y_{\kappa,\lambda+1} = Y_{\kappa\lambda} \cap C_{2\kappa}(C_{2\kappa+1}(Y_{\kappa\lambda} \sim X) \cap X) .$$

Then it is easily seen by induction on $\lambda$ that

$$Y_{\kappa\lambda} = \{u \in {}^\alpha U_\kappa : \lambda \leq t_\kappa u_{2\kappa+1}\}$$

for all $\lambda < \omega$ . Hence if $\kappa < \beta$ and $v \in U_\kappa$ we have

$$\{u \in {}^\alpha U_\kappa : u_{2\kappa+1} = v\} = Y_{\kappa, t_\kappa v} \sim Y_{\kappa, t_\kappa v+1} .$$

Thus for $\kappa < \beta$ and $u \in {}^\alpha U_\kappa$ we have

$$\{u\} = \bigcap_{\nu<\alpha} s_\nu^{2\kappa+1}(Y_{\kappa, t_\kappa u_\nu} \sim Y_{\kappa, t_\kappa u_\nu+1}) .$$

Thus $X$ generates $\mathfrak{A}$ , since $A$ is finite.

Remark I.4.5. Note that the assumption $\alpha < \omega$ in I.4.4 is inessential if we replace "generated" by "completely generated". Theorem I.4.4 has been generalized by Andréka and Németi, who showed in [AN4] that if for each $\alpha$ with $2 \leq \alpha < \omega$ we let

$f\alpha$ = the smallest $\beta$ such that there is a system $\langle U_\gamma : \gamma < \beta \rangle$ of pairwise disjoint sets with $\omega > |U_\gamma| > 1$ for all $\gamma < \beta$ such that the full $Gs_\alpha$ with unit element $\bigcup_{\gamma<\beta} {}^\alpha U_\gamma$ is not generated by a single element,

then $f$ is given by the simple arithmetic formula

$$f\alpha = \frac{1}{2}(2^{2^\alpha} - 2^{2^{\alpha-1}}) + 1 .$$

Also note that in Theorem I.4.4 one may assume that $|U_i| = 0$ is possible or even replace the assumption that $1 < |U_i| < \omega$ for all $i \in I$ by the condition: $|U_i| < \omega$ for all $i \in I$, and $|U_i| = 1$ for at most one $i \in I$. I.4.4 even remains true if we delete the assumption that $U_i \cap U_j = 0$ for distinct $i,j \in I$, since the assumption that $\mathfrak{A}$ is a $Gs_\alpha$ implies the existence of pairwise disjoint non-empty $W_j$, $j \in J$, such that $V = \bigcup_{j\in J} {}^\alpha W_j$, and an easy argument shows that $|J| \leq |I|$.

Closely related to (*) above is the following theorem of Stephen Comer.

Theorem I.4.6. Assume that $\alpha < \omega$, and that $\mathfrak{A}$ is a $Cs_\alpha$ with a base $U$ such that $|U| \leq \alpha$. Then $\mathfrak{A}$ can be generated by a single element.

Proof. We may assume that $\alpha \geq 2$ and $U \neq 0$. For each $x \in {}^{\alpha}U$ let $x' = \bigcap\{X : x \in X \in A\}$. Thus $x' \in A$, since $A$ is finite. We shall show that if $x$ maps $\alpha$ onto $U$ then $x'$ generates $\mathfrak{A}$. To prove this we need four preliminary results.

(1) If $x,y \in {}^{\alpha}U$, $\kappa,\lambda < \alpha$, $\kappa \neq \lambda$, and $x \circ [\kappa/\lambda] = y$, then $y' = c_{\kappa}x' \cap d_{\kappa\lambda}$, so $y' \in Sg\{x'\}$.

For, $x \in x'$, and $(\alpha\sim\{\kappa\}) \upharpoonright x = (\alpha\sim\{\kappa\}) \upharpoonright y$, so $y \in c_{\kappa}x' \cap d_{\kappa\lambda}$. Hence $y' \subseteq c_{\kappa}x' \cap d_{\kappa\lambda}$. Now let $X \in A$ be arbitrary such that $y \in X$. Clearly $x \in c_{\kappa}(X \cap d_{\kappa\lambda})$, so $x' \subseteq c_{\kappa}(X \cap d_{\kappa\lambda})$. Thus

$$c_{\kappa}x' \cap d_{\kappa\lambda} \subseteq c_{\kappa}(X \cap d_{\kappa\lambda}) \cap d_{\kappa\lambda} = X \cap d_{\kappa\lambda} \subseteq X .$$

Hence $c_{\kappa}x' \cap d_{\kappa\lambda} \subseteq y'$, and (1) is established.

(2) If $x,y \in {}^{\alpha}U$, $\kappa,\lambda,\mu$ are distinct ordinals $< \alpha$, $x \circ [\kappa/\lambda,\lambda/\kappa] = y$, and $x_{\kappa} = x_{\mu}$, then $y' \in Sg\{x'\}$.

For, under the hypotheses of (2) we have $y = x \circ [\kappa/\lambda] \circ [\lambda/\mu]$, so (2) follows from (1).

(3) If $x,y \in {}^{\alpha}U$, $\kappa,\lambda,\mu,\nu$ are distinct ordinals $< \alpha$, $x \circ [\kappa/\lambda,\lambda/\kappa] = y$ and $x\mu = x\nu$, then $y' = s^{\mu}_{\lambda}s^{\lambda}_{\kappa}s^{\kappa}_{\mu}c_{\mu}x' \cap d_{\mu\nu}$.

In fact, $y = x \circ [\mu/\kappa] \circ [\kappa/\lambda] \circ [\lambda/\mu] \circ [\mu/\nu]$, so (3) follows easily from (1).

(4) If $x,y \in {}^{\alpha}U$, $Rgy \subseteq Rgx$, and $|Rgx| < \alpha$, then $y' \in Sg\{x'\}$.

For, write $y = x \circ \sigma$ with $\sigma \in {}^{\alpha}\alpha$; express $\sigma$ as a product of transpositions and replacements, and use (1)-(3).

Now let $x$ map $\alpha$ onto $U$; we show $Sg\{x'\} = A$. If $|U| < \alpha$, then by (4) $y' \in Sg\{x'\}$ for each $y \in {}^{\alpha}U$, and for any $X \in A$ we have $X = \bigcup_{y \in X} y' \in Sg\{x'\}$. Assume that $|U| = \alpha$. For each $\kappa < \alpha$ let ${}_{\kappa}y = x \circ [\kappa/0]$ ; thus $({}_{\kappa}y)' \in Sg\{x'\}$ by (1). Also let $z = x \circ [0/1]$; again $z' \in Sg\{x'\}$ by (1). Now if $w \in {}^{\alpha}U$ and $|Rgw| < \alpha$, then either $Rgw \subseteq Rg_{\kappa}y$ for some $\kappa \in \alpha \sim 1$, and hence by (4) $w' \in Sg\{({}_{\kappa}y)'\} \subseteq Sg\{x'\}$, or $Rgw \subseteq Rgz$ and $w' \in Sg\{z'\} \subseteq Sg\{x'\}$. Hence it suffices to take $w \in {}^{\alpha}U$ with $|Rgw| = \alpha$ and show that $w' \in Sg\{x'\}$. Let $t = w \circ [0/1]$. Then $t' \in Sg\{x'\}$ by the situation just discussed. Thus the proof is finished as soon as we prove

(5) $w' = C_0 t' \cap \bigcap_{0<\kappa<\alpha} \sim D_{0\kappa}$ .

Since $w \in C_0 t' \cap \bigcap_{0<\kappa<\alpha} \sim D_{0\kappa}$, the inclusion $\subseteq$ is clear. Now suppose $w \in X \in A$. Then $t \in C_0(X \cap \bigcap_{\kappa<\lambda<\alpha} \sim D_{\kappa\lambda})$, so

$$C_0 t' \cap \bigcap_{0<\kappa<\alpha} \sim D_{0\kappa} \subseteq C_0(X \cap \bigcap_{\kappa<\lambda<\alpha} \sim D_{\kappa\lambda}) \cap \bigcap_{0<\kappa<\alpha} \sim D_{0\kappa} \subseteq X ,$$

so $\supseteq$ in (5) follows.

<u>Remark</u> I.4.7. In contrast to I.4.6 we now show that for $1 \leq \alpha, \beta < \omega$ there is a $Cs_{\alpha}$ $\mathfrak{A}$ with base $U$ such that $|U| = \beta \cdot \alpha$ and $\mathfrak{A}$ cannot be generated by fewer than $\log_2 \beta$ elements. The example is due to Henkin, with some simplifications by Andréka and Németi. Let $\mathfrak{C}$ be the full $Cs_{\alpha}$ with base $\alpha \cdot \beta$. For each $\iota < \beta$ let

$$X_{\iota} = \{q \in {}^{\alpha}(\alpha \cdot \beta) : q \text{ is one-to-one and } \textstyle\sum_{\nu<\alpha} q_{\nu} \equiv \iota \pmod{\beta}\} .$$

Clearly $\langle X_{\iota} : \iota < \beta \rangle$ is a partition of $\overline{d}(\alpha \times \alpha)$ into pairwise disjoint non-empty sets, and $C_\kappa X_\iota = C_\kappa \bar{d}(\alpha \times \alpha)$ for all $\iota < \beta$ and $\kappa < \alpha$. Let $Y = \{X_\iota : \iota < \beta\}$ and let $\mathfrak{A} = \mathfrak{Sg}^{(\mathfrak{C})}Y$. We shall show that $\mathfrak{A}$ cannot be generated by less than $\log_2\beta$ elements. First let $\mathfrak{M}$ be the minimal subalgebra of $\mathfrak{A}$, and let $\mathfrak{B} = \mathfrak{Sg}^{\mathfrak{Bl}\mathfrak{A}}(Y \cup M)$. Note that $At\mathfrak{B} = Y \cup \{x : x \in AtM, x \neq \bar{d}(\alpha \times \alpha)\}$. Now we claim

(1) for all $b \in B$ and all $\kappa < \alpha$, $C_\kappa b \in M$.

In fact, since $B$ is finite it is enough to prove (1) for an atom $b$, and by virtue of the mentioned description of $At\mathfrak{B}$ this is clear, since $C_\kappa X_\iota = C_\kappa \bar{d}(\alpha \times \alpha)$ for all $\iota < \beta$.

Because of (1) we have, first of all, $B = A$.

Now suppose that $G \subseteq A$ and $\mathfrak{A} = \mathfrak{Sg}^{(\mathfrak{A})}G$; we show that $\beta \leq 2^{|G|}$. Let $\mathfrak{E} = \mathfrak{Sg}^{(\mathfrak{B})}G$. Then by (1) $\mathfrak{A} = \mathfrak{Sg}^{(\mathfrak{B})}(G \cup M) = \mathfrak{Sg}^{(\mathfrak{B})}(E \cup M)$. Since $X_\iota \subseteq \bar{d}(\alpha \times \alpha) \in At\mathfrak{M}$ for all $\iota < \beta$, it follows that $Y \subseteq \{z \cap \bar{d}(\alpha \times \alpha) : z \in At\mathfrak{E}\}$, and so $\beta \leq |At\mathfrak{E}| \leq 2^{|G|}$, as desired.

It is perhaps interesting that the algebra just constructed <u>can</u> be generated by $\gamma$ elements, where $\gamma$ is the least integer $\geq \log_2\beta$. In fact, let $f \in {}^\beta({}^\gamma 2)$ be one-to-one, with $f0 = \langle 1 : \nu < \gamma \rangle$. For each $\nu < \gamma$, let

$$Z_\nu = \bigcup\{X_\iota : \iota < \beta, f_\iota \nu = 1\} .$$

Then, as is easily checked, $X_0 = \bigcap_{\nu<\gamma} Z_\nu$, $\bar{d}(\alpha \times \alpha) = C_0X_0 \cap C_1X_0$, and for each $\iota < \beta$ with $\iota \neq 0$,

$$X_\iota = (\bar{d}(\alpha \times \alpha) \sim \bigcup\{Z_\nu : f_\iota\nu = 0\}) \cap \bigcap\{Z_\nu : f_\iota\nu = 1\} .$$

Thus $\{Z_\nu : \nu < \gamma\}$ generates $\mathfrak{A}$, as desired.

Remarks I.4.8. G. Bergman independently obtained examples of $Cs_2$'s which require a large number of generators. In connection with I.4.6 and I.4.7 it is natural to define the following function. For $2 \leq \alpha,\beta < \omega$ let

$q(\alpha,\beta)$ = the smallest $\gamma < \omega$ such that every $Cs_\alpha$ with base $\beta$ can be generated by $\gamma$ elements.

Thus the example in I.4.7 shows that $q(\alpha,\alpha\cdot\beta) \geq \log_2\beta$ , and Theorem I.4.6 says that $q(\alpha,\beta) = 1$ for $\beta \leq \alpha$ . Andréka and Németi have established many properties of this function $q$ . For example, $q(\alpha,\alpha+1) = 1$ , $q(2,\beta)$ = least integer $\geq \log_2(\beta - 1)$ for $\beta > 2$ , and

$$q(\alpha,\alpha\cdot\beta) = \text{least integer} \geq \log_2(\beta + \frac{\beta}{\alpha - 1}) ,$$

generalizing I.4.6 and I.4.7.

Note that these results on the function $q$ are relevant to our discussion in I.3 of change of base. For example I.4.6 implies that if a $Cs_\alpha$ cannot be generated by a single element, then it is not even isomorphic to a $Cs_\alpha$ with base of power $\leq \alpha$ .

J. Larson has shown that for $2 \leq \alpha < \omega$ there are $2^{\aleph_0}$ isomorphism types of one-generated $Cs_\alpha$'s. P. Erdös, V. Faber and J. Larson [EFL] show that there is a countable $Cs_2$ not embeddable in any finitely generated $Cs_2$ .

In I.7.10 and I.7.11 we show that $IGs_\alpha$ and $IGws_\alpha$ (for arbitrary $\alpha$ ), $ICs_\alpha$ and $IWs_\alpha$ (for $\alpha < \omega$ ) , and $ICs_\alpha^{reg} \cap Lf_\alpha$ are closed under directed unions. Andréka and Németi have shown that $ICs_\alpha$ and $ICs_\alpha^{reg}$ are not closed under directed unions for $\alpha \geq \omega$ . It remains open whether $IWs_\alpha$ is closed under directed unions for $\alpha \geq \omega$ ; the proof of I.7.11 may be relevant to this problem.

We also should mention that H. Andréka and I. Németi have solved Problem 2.3 of [HMT] by showing that for each $\alpha > 0$ there is a simple finitely generated $Cs_\alpha$ not generated by a single element; see [AN2].

## 5. Homomorphisms

The following result about CA's in general will be useful in what follows.

Theorem I.5.1. Let $\mathfrak{A}$ be a $CA_\alpha$ and $I \in Il\mathfrak{A}$ . Then $Sg^{(\mathfrak{A})}I = \{x \oplus d : x \in I \text{ and } d \in Sg^{\mathfrak{A}}\{0\}\}$ .

Proof. This is clear since $\mathfrak{Sg}^{(\mathfrak{A})}I/I$ is a minimal $CA_\alpha$ .

Turning now to set algebras, we begin with a result concerning simple algebras.

Theorem I.5.2. (i) Any regular locally finite-dimensional $Cs_\alpha$ with non-empty base is simple.

(ii) Any locally finite-dimensional $Ws_\alpha$ is simple.

(iii) For $\alpha < \omega$ any $Cs_\alpha$ with non-empty base is simple.

Proof. Trivial, using 2.3.14.

Corollary I.5.3. Let $\mathfrak{A} \in Cs_\alpha \cup Ws_\alpha$ with base $U$ , $|U| > 0$ . Then the minimal subalgebra of $\mathfrak{A}$ is simple. The characteristic of $\mathfrak{A}$ is $|U|$ if $|U| < \alpha \cap \omega$ , and it is $0$ if $|U| \geq \alpha \cap \omega$ .

Corollary I.5.4. Let $\mathfrak{A}$ be a $Gws_\alpha$ with subbases $\langle U_i : i \in I\rangle$, each $U_i \neq 0$, $I \neq 0$ and $\alpha \geq 1$. Then the minimal subalgebra of $\mathfrak{A}$ is simple iff either (1) for all $i,j \in I$ we have $|U_i| = |U_j| < \alpha \cap \omega$, in which case $\mathfrak{A}$ has characteristic $|U_i|$ (for any $i \in I$), or (2) for all $i \in I$ we have $|U_i| \geq \alpha \cap \omega$, in which case $\mathfrak{A}$ has characteristic $0$.

Remarks I.5.5. Theorem I.5.2 (iii) does not extend to $Gs_\alpha$'s and $Gws_\alpha$'s. In fact, if $1 \leq \alpha$, $\mathfrak{A}$ is a full $Gs_\alpha$ with unit element $\bigcup_{i\in I} {}^\alpha U_i$, where $U_i \cap U_j = 0$ for $i \neq j$, and if $|I| > 1$ and each subbase $U_i$ is non-empty, then $\mathfrak{A}$ is not simple. For, choose $i_0 \in I$, and let $fX = X \cap {}^\alpha U_{i_0}$ for all $X \in A$. Then it is easy to verify that $f$ is a homomorphism of $\mathfrak{A}$ into a $Cs_\alpha$ $\mathfrak{B}$ with base $U_{i_0}$ (and hence that $|B| > 1$), but $f$ is not one-to-one. So $\mathfrak{A}$ is not simple. A similar construction works for $Gws_\alpha$'s. This phenomenon is more fully explained by the fact that $Gs_\alpha$'s and $Gws_\alpha$'s are often subdirectly decomposable; see I.6.3, I.6.4.

For $\alpha \geq \omega$, I.5.2 (i) does not extend to arbitrary locally finite-dimensional $Cs_\alpha$'s. For, let $\mathfrak{B}$ be the full $Cs_\alpha$ with base $2$, let $p = \langle 0 : \kappa < \alpha\rangle$, and let $\mathfrak{A} = \mathfrak{Sg}^{(\mathfrak{B})}\{{}^\alpha 2^{(p)}\}$. Now $\underset{\sim}{\Delta}({}^\alpha 2^{(p)}) = 0$, so $\mathfrak{A} \in Lf_\alpha$ by 2.15 (i), while $\mathfrak{A}$ fails to be simple by 2.3.14. Also, for $\alpha \geq \omega$ I.5.2 (ii) does not extend to arbitrary $Ws_\alpha$'s; in fact, one cannot even replace "locally finite dimensional" by "dimension complemented". In fact, let $\alpha \geq \omega$, and choose $\Gamma \subseteq \alpha$ with $\Gamma$ and $\alpha \sim \Gamma$ infinite. Let $p = \langle 0 : \kappa < \alpha\rangle$ and let $\mathfrak{B}$ be the full $Ws_\alpha$ with unit element ${}^\alpha 2^{(p)}$. Let $X = \{f \in {}^\alpha 2^{(p)} : \Gamma 1 f \subseteq p\}$, and let $\mathfrak{A} = \mathfrak{Sg}^{(\mathfrak{B})}\{X\}$ Then clearly $\mathfrak{A} \in Ws_\alpha \cap Dc_\alpha$, while by 2.3.14 $\mathfrak{A}$ is not simple, since $c_{(\Delta)}X \neq {}^{\alpha}2^{(p)}$ for every finite $\Delta \subseteq \alpha$ .

Andréka and Németi have shown that I.5.2 (ii) does not extend to regular dimension-complemented $Cs_\alpha$'s for $\alpha \geq \omega$ . They also noted that I.5.3 does not extend to $Gs_\alpha$'s or $Gws_\alpha$'s for $\alpha \geq 2$ . In fact, let $U_0 = \{0\}$ and $U_1 = \{1,2\}$ , and put $p = \langle 0 : \kappa < \alpha \rangle$ , $q = \langle 1 : \kappa < \alpha \rangle$ . Let $\mathfrak{A}$ be any $Gs_\alpha$ (resp. $Gws_\alpha$ ) with unit element ${}^{\alpha}U_0 \cup {}^{\alpha}U_1$ (resp. ${}^{\alpha}U_0^{(p)} \cup {}^{\alpha}U_1^{(q)}$ ); in particular, $\mathfrak{A}$ could be minimal. Then $\mathfrak{A}$ is not simple. To see this, observe that some $Cs_\alpha$ $\mathfrak{B}$ with base $U_0$ is a homomorphic image of $\mathfrak{A}$ , but $\mathfrak{A}$ is not isomorphic to $\mathfrak{B}$ since $\overline{d}(|U_1| \times |U_1|) \neq 0$ in $\mathfrak{A}$ but it is $0$ in $\mathfrak{B}$ . Finally, concerning I.5.4 it is worth remarking that for any $Gws_\alpha$ $\mathfrak{A}$ with subbases $\langle U_i : i \in I \rangle$ we have $\nabla\mathfrak{A} = \{|U_i| : 0 < |U_i| < \alpha \cap \omega\}$ (see 2.5.25).

Remarks I.5.6. We now discuss possible closure under homomorphisms of our various classes of set algebras, where not all natural questions have been answered. The situation is summarized in Figure I.5.7, which we now discuss; cf. Figure I.1.4 .

(1) It will be shown in I.7.15 that $HGs_\alpha = IGs_\alpha = HGws_\alpha$ for all $\alpha$ .

(2) For $0 < \alpha < \omega$ we have $IWs_\alpha \cup \{\underset{\sim}{1}\} = ICs_\alpha^{reg}$ (where $\underset{\sim}{1}$ is the one-element $Cs_\alpha$ ) , and hence $IWs_\alpha \cup \{\underset{\sim}{1}\} = ICs_\alpha^{reg} = ICs_\alpha = HCs_\alpha = HWs_\alpha = HCs_\alpha^{reg}$ .

For the remainder of these remarks assume that $\alpha \geq \omega$ .

(3) Andréka and Németi have shown that there is an $\mathfrak{A} \in Ws_\alpha$ with infinite base such that $H\mathfrak{A} \not\subseteq IWs_\alpha \cup \{\underset{\sim}{1}\}$ .

(4) In I.7.21 we show that any homomorphic image of a $Cs_\alpha$ or a $Ws_\alpha$ with an infinite base is isomorphic to a $Cs_\alpha$ . This is not

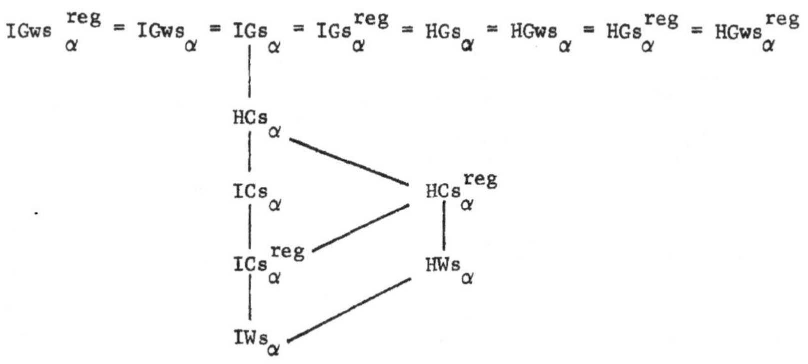


Figure I.5.7

true if we replace "$Cs_\alpha$" by "$Cs_\alpha^{reg}$", as is shown by the following example of Andréka and Németi. Let U be the set of finite subsets of $\alpha$, and let H be a one-one function from $\alpha$ onto U. For each $\kappa < \alpha$ let $u_\kappa = \alpha \times \{H\kappa\}$ and let $y_\kappa = C_{(H\kappa)} u_\kappa$. Set $x = \bigcup_{\kappa<\alpha} y_\kappa$. With $\mathfrak{C}$ the full $Cs_\alpha$ with base U we set $\mathfrak{A} = \mathfrak{Sg}^{(\mathfrak{C})}\{x\}$. We now claim that $\mathfrak{A} \in Cs_\alpha^{reg}$ but $H\mathfrak{A} \not\subseteq ICs_\alpha^{reg}$. First, to show that $\mathfrak{A}$ is regular, let $y \in A$. We may assume that $|\alpha \sim \Delta y| \geq \omega$. We show then that $y \in Sg^{(\mathfrak{A})}\{0\}$ (and hence $y$ is regular). Since $\mathfrak{A} = \mathfrak{Sg}^{(\mathfrak{C})}\{x\}$ $= \mathfrak{Sg}^{(\mathfrak{C})} Ig^{(\mathfrak{A})}\{x\}$, there is a $d \in Sg^{(\mathfrak{A})}\{0\}$ such that $y \oplus d \in Ig^{(\mathfrak{A})}\{x\}$ (using I.5.1). Let $z = y \oplus d$; we show that $z = 0$ (hence $y = d$, as desired). Suppose $z \neq 0$. Since $z \in Sg\{y\}$, we have $|\Delta z \sim \Delta y| < \omega$. Moreover, $z \in Ig\{x\}$ implies that $z \leq C_{(\Gamma)} x$ for some finite $\Gamma \subseteq \alpha$. Now fix $q \in z$. Then $q \in C_{(\Gamma)} y_\kappa$ for some $\kappa < \alpha$. Thus $[\alpha \sim (H\kappa \cup \Gamma)] \upharpoonright q \subseteq u_\kappa$. Since $|\alpha \sim \Delta y| \geq \omega$ and $|\Delta z \sim \Delta y| < \omega$, we can choose $\lambda \in \alpha \sim (\Delta z \cup H\kappa \cup \Gamma)$. Fix $\mu \in \alpha \sim \{\kappa\}$. Then $q^\lambda_{H\mu} \in z$ but $q^\lambda_{H\mu} \notin c_{(\Gamma)} x$, as is easily checked, contradicting $z \subseteq c_{(\Gamma)} x$.

Next we construct $\mathfrak{B} \in \mathbf{H}\mathfrak{A}$ with $|Zd\mathfrak{B}| > 2$ (hence $\mathfrak{B} \notin ICs_\alpha^{reg}$, by I.4.3). Let $I = Ig^{(\mathfrak{A})}\{c_\kappa x \cdot -x : \kappa < \alpha\}$, and let $\mathfrak{B} = \mathfrak{A}/I$. Clearly $x/I \in Zd\mathfrak{B}$, so we only need to show that $x \notin I$ and $-x \notin I$. Now if $y \in I$ then there exist finite subsets $\Gamma, \Delta$ of $\alpha$ such that

$$(*) \quad y \subseteq c_{(\Gamma)}(\bigcup_{\delta\in\Delta}(\bigcup\{c_\delta y_\lambda : \lambda < \alpha,\ \delta \notin H\lambda\} \cdot -x)) .$$

Suppose $x \in I$, with $\Gamma$ and $\Delta$ as in (*) where $y = x$. Choose $\lambda < \alpha$ so that $\Delta = H\lambda$. Then $u_\lambda \in x$, but $u_\lambda$ is not in the right side of (*), a contradiction. Suppose $-x \in I$, with $\Gamma$ and $\Delta$ as in (*) where $y = -x$. Then $H \in -x$, but $H$ is not in the right side of (*), a contradiction.

(5) In contrast to (3) and (4) we now show that $HCs_\alpha \not\subseteq ICs_\alpha$, $HWs_\alpha \not\subseteq ICs_\alpha$, and $HWs_\alpha \not\subseteq IWs_\alpha \cup \{\underset{\sim}{1}\}$, where $\underset{\sim}{1}$ is the one-element $Cs_\alpha$. Of course our example will have a finite base. The example is due to Monk; in Demaree [De] there is an error in the construction of such an example. Let $2 \leq \kappa < \omega$. Our construction is based on the following obvious fact:

(*) If $\mathfrak{A}$ is an atomless $Cs_\alpha$ or $Ws_\alpha$ of characteristic $> 0$, then $\mathfrak{A}$ has no element $\subseteq \bigcap_{\kappa,\lambda<\alpha} D_{\kappa\lambda}$ except 0.

Let $\mathfrak{A}$ be the full $Cs_\alpha$ (resp. $Ws_\alpha$) with base $\kappa$ (resp. unit element ${}^\alpha\kappa^{(p)}$, where $p = \kappa \times \{0\}$. Let $I = \{X : X \in A,\ |X| < \omega\}$. Clearly $I$ is a proper ideal of $\mathfrak{A}$. We claim that $\mathfrak{A}/I$ is not isomorphic to a $Cs_\alpha$ or $Ws_\alpha$. Suppose to the contrary that $\mathfrak{A}/I$ is isomorphic to a $Cs_\alpha$ or $Ws_\alpha$ $\mathfrak{B}$. By I.5.3, $\mathfrak{A}$ and $\mathfrak{B}$ have isomorphic minimal subalgebras $\mathfrak{A}'$, $\mathfrak{B}'$ respectively. The two formulas

$$\bar{d}(\kappa \times \kappa) \neq 0 , \quad \bar{d}((\kappa + 1) \times (\kappa + 1)) = 0$$

hold in $\mathfrak{A}'$ and hence in $\mathfrak{B}'$. Hence the base of $\mathfrak{B}$ has power $\kappa$. Noting that $\mathfrak{A}/I$ is atomless, we can obtain a contradiction to (*) by exhibiting an element $X$ of $A$ such that $0 \neq X/I \leq d_{\lambda\mu}^{(\mathfrak{A}/I)}$ for all $\lambda, \mu < \alpha$. Set $X = \{x \in {}^{\alpha}\kappa : x\lambda \neq 0$ for exactly one $\lambda < \alpha\}$. It is easily checked that $X$ satisfies the above conditions.

(6) Andréka and Németi have modified the above construction to show that $\mathsf{HCs}_{\alpha}^{reg} \not\subseteq \mathsf{ICs}_{\alpha}$.

(7) Andréka and Németi have shown that $\mathsf{ICs}_{\alpha}^{reg} \not\subseteq \mathsf{HWs}_{\alpha}$.

(8) It is clear that $\mathsf{Gs}_{\alpha} \not\subseteq \mathsf{HCs}_{\alpha}$, since the minimal subalgebra of any $\mathfrak{A} \in \mathsf{HCs}_{\alpha}$ is simple or of power 1 by I.5.3, while there are clearly $\mathfrak{A} \in \mathsf{Gs}_{\alpha}$ without this property.

(9) The inclusion $\mathsf{IWs}_{\alpha} \subseteq \mathsf{ICs}_{\alpha}^{reg}$ will be established in I.7.13. It implies that $\mathsf{IWs}_{\alpha} \subseteq \mathsf{HCs}_{\alpha}$, and this inclusion is easy to establish. In fact, clearly $\mathsf{Ws}_{\alpha} \subseteq \mathsf{SHCs}_{\alpha}$ since the full $\mathsf{Ws}_{\alpha}$ with unit element ${}^{\alpha}U^{(p)}$ is a homomorphic image of the full $\mathsf{Cs}_{\alpha}$ with base $U$. Hence $\mathsf{IWs}_{\alpha} \subseteq \mathsf{SHCs}_{\alpha} \subseteq \mathsf{HSCs}_{\alpha} = \mathsf{HCs}_{\alpha}$.

(10) It remains open whether $\mathsf{ICs}_{\alpha} \subseteq \mathsf{HCs}_{\alpha}^{reg}$ or $\mathsf{HCs}_{\alpha} = \mathsf{HCs}_{\alpha}^{reg}$.

(11) Andréka and Németi have shown that $\mathsf{H}(\mathsf{Cs}_{\alpha} \cap \mathsf{Lf}_{\alpha}) \not\subseteq \mathsf{ICs}_{\alpha}$. By I.5.1, the inclusion holds trivially if "$\mathsf{Cs}_{\alpha}$" is replaced by "$\mathsf{Cs}_{\alpha}^{reg}$" or "$\mathsf{Ws}_{\alpha}$".

(12) From the definition of characteristic we know that if $\mathfrak{A}$ has characteristic $\kappa$, $\mathfrak{A} \succcurlyeq \mathfrak{B}$, and $|B| > 1$, then $\mathfrak{B}$ has characteristic $\kappa$. The meaning of characteristic for set algebras, described in I.5.3 and I.5.4, is further elucidated by a result of Andréka and Németi according to which for each cardinal $\kappa \geq 2$ there is a $\mathsf{Cs}_{\alpha}^{reg}$ $\mathfrak{A}$

with base of power $\kappa$ , having a homomorphic image $\mathfrak{B}$ such that every $Gws_\alpha$ $\mathfrak{C}$ isomorphic to $\mathfrak{B}$ has base of power $>\kappa$ .

(13) Recall from I.3.9 that for every $\mathfrak{A} \in Cs_\alpha^{reg}$ , if there is in $\mathfrak{A}$ an element $x$ , not in the minimal algebra of $\mathfrak{A}$ , with $|\Delta x| < \omega$ , then $\mathbf{H}\mathfrak{A} \cap Cs_\alpha \not\subseteq Cs_\alpha^{reg}$ .

(14) In contrast to (13) , Andréka and Németi have constructed a $Cs_\alpha^{reg}$ $\mathfrak{A}$ such that $\mathbf{H}\mathfrak{A} \subseteq \mathbf{I}Cs_\alpha^{reg}$ , $\mathbf{H}\mathfrak{A} \cap Cs_\alpha \subseteq Cs_\alpha^{reg}$ , and $\mathfrak{A}$ is not simple. The construction is simple: let $\mathfrak{B}$ be the full $Cs_\alpha$ with base $\omega$ , and let $\mathfrak{A} = \mathfrak{Sg}^{(\mathfrak{B})}\{\{x\} : x \in {}^\alpha\omega\}$ . Note that for any $y \in \{\{x\} : x \in {}^\alpha\omega\}$ we have

(*) for every finite $\Gamma \subseteq \alpha$ and every $\kappa \in \alpha \sim \Gamma$ we have $c_\kappa^\partial c_{(\Gamma)} y = 0$ .

Now if $f \in Ho(\mathfrak{A},\mathfrak{C})$ and $\mathfrak{C} \in Cs_\alpha$ , then each $y \in \{f\{x\} : x \in {}^\alpha\omega\}$ satisfies (*) , and so by Theorem 1.3 of [AN3] we have $\mathfrak{C} \in Cs_\alpha^{reg}$ . Thus $\mathbf{H}\mathfrak{A} \cap Cs_\alpha \subseteq Cs_\alpha^{reg}$ . By (4) above, it follows that $\mathbf{H}\mathfrak{A} \subseteq \mathbf{I}Cs_\alpha^{reg}$ . Finally, by 2.3.14 it is clear that $\mathfrak{A}$ is not simple.

(15) Generalizing the construction given in (4) above, Andréka and Németi have shown that for any $\kappa \geq 2$ there is an $\mathfrak{A} \in Cs_\alpha^{reg}$ with base $\kappa$ and some $\mathfrak{B} \in \mathbf{H}\mathfrak{A}$ with $|Zd\mathfrak{B}| > 2$ .

(16) Contrasting to (15) , Andréka and Németi have shown that for any $\kappa \geq 2$ there is an $\mathfrak{A} \in Cs_\alpha^{reg}$ with base $\kappa$ such that $\mathfrak{A}$ is not simple, but $|Zd\mathfrak{B}| \leq 2$ for all $\mathfrak{B} \in \mathbf{H}\mathfrak{A}$ .

(17) Figure I.5.7 simplifies considerably if we restrict ourselves to set algebras with bases and subbases infinite and to $\alpha \geq \omega$ . Let us denote by ${}_\infty Cs_\alpha$, ${}_\infty Gs_\alpha$ , etc. the corresponding classes. Then we

obtain Figure I.5.8; see (3), (4), (9) . Here $= ?$ means that we do not know whether equality holds in the two indicated cases.

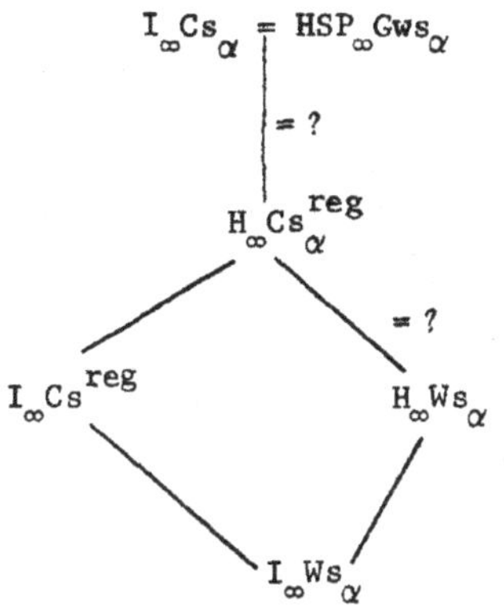


$(\alpha \geq \omega)$

Figure I.5.8

## 6. Products

In terms of products, we can express a simple relationship between $Gs_\alpha$'s and $Cs_\alpha$'s , and between $Gws_\alpha$'s and $Ws_\alpha$'s . We first express this relationship more generally for $Crs_\alpha$'s . For this purpose it is convenient to introduce the following special notation.

Definition I.6.1. Let $\mathfrak{A}$ be a $Crs_\alpha$ with unit element $V$ and let $W \subseteq V$ . Then $rl_W^{\mathfrak{A}}$ is the function with domain $A$ such that for any $a \in A$ , $rl_W^{\mathfrak{A}} a = W \cap a$ .

Theorem I.6.2. Let $\mathfrak{B}$ be a full $Crs_\alpha$ with unit element $\bigcup_{i \in I} V_i$ , where $V_i \cap V_j = 0$ for $i,j \in I$ and $i \neq j$ , and $\Delta^{\mathfrak{B}} V_i = 0$ for all $i \in I$ . Assume that $\mathfrak{B} \in CA_\alpha$ . For each $i \in I$ let $\mathfrak{A}_i$ be the full $Crs_\alpha$ with unit element $V_i$. Then $\mathfrak{B} \cong P_{i\in I}\,\mathfrak{A}_i$. In fact, there is a unique $f \in Is(\mathfrak{B}, P_{i\in I}\mathfrak{A}_i)$ such that $r\ell^{\mathfrak{B}}_{V_i} = pj_i \circ f$ for each $i \in I$.

Proof. Clearly there is a unique $f$ mapping $B$ into $P_{i\in I}A_i$ and satisfying the final condition. By 2.3.26, $r\ell^{\mathfrak{B}}_{V_i} \in Ho(\mathfrak{B},\mathfrak{A}_i)$ for each $i \in I$, so $f \in Hom(\mathfrak{B}, P_{i\in I}\mathfrak{A}_i)$ by 0.3.6 (ii). Clearly $f$ is one-to-one and onto.

The assumption $\mathfrak{B} \in CA_\alpha$ is not actually needed in I.6.2.

Corollary I.6.3. For $\alpha \geq 2$ we have $IGs_\alpha = SPCs_\alpha$ and $IGs^{reg}_\alpha = SPGs^{reg}_\alpha$.

Proof. First suppose that $\mathfrak{C} \in Gs_\alpha$; say $\mathfrak{C}$ has unit element $\bigcup_{i\in I}{}^\alpha U_i$, where $U_i \cap U_j = 0$ for distinct $i,j \in I$, and $U_i \neq 0$ for all $i \in I$. Let $V_i = {}^\alpha U_i$ for each $i \in I$, and let $\mathfrak{B}$, $\mathfrak{A}$, $f$ be as in Theorem I.6.2; clearly $\underset{\approx}{\Delta}^{\mathfrak{B}} V_i = 0$ for all $i \in I$ since $\alpha \geq 2$ and $\mathfrak{B} \in CA_\alpha$. Clearly $C \restriction f$ is an isomorphism of $\mathfrak{C}$ onto a subdirect product of $Cs_\alpha$'s, as desired.

Second, suppose $\mathfrak{C} \subseteq_d P_{i\in I}\mathfrak{D}_i$, each $\mathfrak{D}_i$ a $Cs_\alpha$ with base $U_i \neq 0$. We may assume that $U_i \cap U_j = 0$ for distinct $i,j \in I$. Let $V_i = {}^\alpha U_i$ for each $i \in I$, and again let $\mathfrak{B}$, $\mathfrak{A}$ and $f$ be as in Theorem I.6.2. Clearly $C \restriction f^{-1}$ is an isomorphism of $\mathfrak{C}$ onto a $Gs_\alpha$.

The second part of I.6.3 is handled by I.1.15.

In an entirely analogous way we obtain

<u>Corollary</u> I.6.4. For $\alpha \geq 2$ we have $\mathrm{IGws}_\alpha = \mathrm{SPWs}_\alpha$ .

<u>Corollary</u> I.6.5. Let $\mathfrak{A} \in \mathrm{IGs}_\alpha$ , $|A| > 1$ , $\alpha < \omega$ . Then the following conditions are equivalent:

(i) $\mathfrak{A} \in \mathrm{ICs}_\alpha$ ;

(ii) $\mathfrak{A}$ is simple.

<u>Proof</u>. By I.6.3 and I.5.2 (iii) (treating $\alpha < 2$ separately).

<u>Corollary</u> I.6.6. For $\alpha \geq 2$ we have $\mathrm{PGs}_\alpha = \mathrm{IGs}_\alpha$ , $\mathrm{PGs}_\alpha^{reg} = \mathrm{IGs}_\alpha^{reg}$ , $\mathrm{PGws}_\alpha = \mathrm{IGws}_\alpha$ , and $\mathrm{PGws}_\alpha^{reg} = \mathrm{IGws}_\alpha^{reg}$ .

<u>Remark</u> I.6.7. None of I.6.3, I.6.4, I.6.6 extend to $\alpha \leq 1$ .

<u>Remarks</u> I.6.8. Closure properties under H, S, P of our classes of set algebras are summarized in Figure I.6.9 for $\alpha \geq 2$ ; see also Figures I.1.4, I.5.7, and I.5.8. We now discuss this figure.

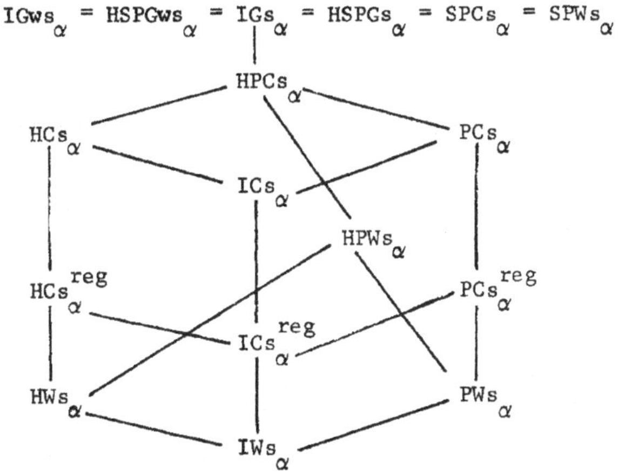


Figure I.6.9

$(\alpha \geq 2)$

(1) For $\alpha \le 1$ the diagram is different; then the classes are just five in number, increasing under inclusion:

(a) $IWs_\alpha = \{\mathfrak{A} \in CA_\alpha : \mathfrak{A}$ is simple$\}$ ;

(b) $ICs_\alpha = \{\mathfrak{A} \in CA_\alpha : \mathfrak{A}$ is simple or $|A| = 1\}$ ;

(c) $PCs_\alpha = \{\mathfrak{A} \in CA_\alpha : \mathfrak{A}$ is a product of simple $CA_\alpha$'s$\}$ ;

(d) $HPCs_\alpha$ ;

(e) $HSPCs_\alpha = CA_\alpha = SPWs_\alpha$ .

(2) The example $\mathfrak{A}$ in I.5.6 (5) also shows that $HWs_\alpha \not\subseteq PCs_\alpha$ in general, for $\alpha \ge \omega$ . In fact, (*) continues to hold for all $\mathfrak{A} \in PCs_\alpha$ .

(3) Andréka and Németi have shown that $Cs_\alpha^{reg} \not\subseteq PWs_\alpha$ and $Cs_\alpha \not\subseteq PCs_\alpha^{reg}$ for $\alpha \ge \omega$ .

(4) To show that $PWs_\alpha \not\subseteq HCs_\alpha$ for $\alpha \ge \omega$ , let $\mathfrak{A}$ and $\mathfrak{B}$ be $Ws_\alpha$'s with bases $\kappa,\lambda$ respectively, where $1 < \kappa < \lambda < \omega$ . Then $\mathfrak{A} \times \mathfrak{B} \notin HCs_\alpha$ , since each non-trivial homomorphic image of a $Cs_\alpha$ has a well-defined characteristic, while $\mathfrak{A} \times \mathfrak{B}$ does not (cf. 2.4.61 for the definition of characteristic).

(5) For any $\alpha$ we have $SPCs_\alpha \not\subseteq HPCs_\alpha$ (for $\alpha \ge \omega$ this was shown by

Andréka and Németi). For $\alpha = 0$ this is well-known ($\mathrm{SPCs}_\alpha = \mathrm{BA}$, while if $\mathfrak{A} \in \mathrm{HPCs}_\alpha$ and $|A| \geq \omega$, then $|A|^\omega = |A|$ by S. Koppelberg [Ko]). For $0 < \alpha < \omega$ let $\mathfrak{A}$ be the full $\mathrm{Cs}_\alpha$ with base $\omega$, and let $\mathfrak{B}$ be the subalgebra of ${}^\omega\mathfrak{A}$ generated by $\omega$ zero-dimensional elements. Suppose $\mathfrak{B} \in \mathrm{HPCs}_\alpha$; say $\mathrm{P}_{i\in I}\mathfrak{C}_i \succcurlyeq \mathfrak{B}$, each $\mathfrak{C}_i$ a non-trivial $\mathrm{Cs}_\alpha$. By the above result of Koppelberg, since $|\mathrm{Zd}\mathfrak{B}| = \omega$ we must have $|I| < \omega$. Since each $\mathfrak{C}_i$ is simple by I.5.2 (iii), it follows that $\mathfrak{B} \cong \mathrm{P}_{i\in J}\mathfrak{C}_j$ for some $J \subseteq I$. But $\mathrm{P}_{i\in J}\mathfrak{C}_j$ has only finitely many 0- dimensional elements, a contradiction.

Now suppose $\alpha \geq \omega$. For each $\kappa \in \omega \sim 2$ let $\mathfrak{A}_\kappa$ be the minimal $\mathrm{Cs}_\alpha$ with base $\kappa$, let $\mathfrak{B} = \mathrm{P}_{\kappa\in\omega\sim 2}\mathfrak{A}_\kappa$, and let $\mathfrak{M}$ be the minimal subalgebra of $\mathfrak{B}$. Then $\mathfrak{M} \notin \mathrm{HPCs}_\alpha$. For, assume that $h \in \mathrm{Ho}(\mathrm{P}_{i\in I}\mathfrak{C}_i, \mathfrak{M})$, where $\mathfrak{C}_i \in \mathrm{Cs}_\alpha$ for all $i \in I$. Let $\mathfrak{D} = \mathrm{P}_{i\in I}\mathfrak{C}_i$. For each $\kappa \in \omega \sim 2$ let $\sigma_\kappa$ be the term

$$c_{(\kappa)}\overline{d}(\kappa \times \kappa) \cdot - c_{(\kappa+1)}\overline{d}((\kappa + 1) \times (\kappa + 1)) .$$

Thus $\sigma_\kappa = 1$ holds in a $\mathrm{Cs}_\alpha$ $\mathfrak{C}$ iff $\mathfrak{C}$ has base of cardinality $\kappa$. It follows that $\sigma_\kappa^{(\mathfrak{D})} \neq 0$ for all $\kappa \in \omega \sim 2$, so for all $\kappa \in \omega \sim 2$ there is an $i \in I$ such that $\mathfrak{C}_i$ has base of power $\kappa$. Now define $f \in \mathrm{P}_{i\in I}C_i$ by defining $fi = 1^{(\mathfrak{C}_i)}$ if the base of $\mathfrak{C}_i$ has $2\kappa$ elements for some $\kappa \in \omega \sim 1$, $fi = 0^{(\mathfrak{C}_i)}$ otherwise. Thus $\sigma_\kappa^{(\mathfrak{D})} \leq f$ if $\kappa \in \omega \sim 2$ is even, and $\sigma_\kappa^{(\mathfrak{D})} \cdot f = 0$ if $\kappa \in \omega \sim 2$ is odd. Hence $\sigma_\kappa^{(\mathfrak{M})} \leq hf$ and $\sigma_\kappa^{(\mathfrak{M})} \cdot hf = 0$ in these two cases. But $\Delta^{(\mathfrak{D})}f = 0$, so $\Delta^{(\mathfrak{M})}hf = 0$, so by 2.1.17 (iii), $hf \in \mathrm{Sg}^{(\mathfrak{Bl}\mathfrak{M})}\{\sigma_\kappa^{(\mathfrak{M})} : \kappa \in \omega \sim 2\}$, which is clearly impossible.

(6) From (4) it follows, of course, that $\mathrm{PCs}_\alpha \not\subseteq \mathrm{ICs}_\alpha$ even for

$\alpha \geq \omega$ . But we show in I.7.21 that a product of $Cs_\alpha$'s with infinite bases is isomorphic to a $Cs_\alpha$ for $\alpha \geq \omega$ .

(7) Andréka and Németi have noted that $Cs_\alpha^{reg} \not\subseteq SPDc_\alpha$ and $Ws_\alpha \not\subseteq SPDc_\alpha$ for all $\alpha > 0$ . In fact, let $\mathfrak{A}$ be the full $Cs_\alpha$ with base $\kappa \geq 2$ , and let $\mathfrak{B}$ be the subalgebra of $\mathfrak{A}$ generated by the atoms of $\mathfrak{A}$ . Then by Cor. 1.4 of [AN3] , $\mathfrak{B} \in Cs_\alpha^{reg}$ . But the statement

(*) for all $x$ , if $c_\lambda^\partial x = 0$ for all $\lambda < \alpha$ then $x = 0$

holds in every $Dc_\alpha$ , and hence in every $\mathfrak{C} \in SPDc_\alpha$ . Since every atom $x$ of $\mathfrak{B}$ falsifies (*) , we have $\mathfrak{B} \notin SPDc_\alpha$ . The case of $Ws_\alpha$'s is similar: any full $Ws_\alpha$ has an atom.

(8) In connection with (7) we should mention the following general fact about $Lf_\alpha$'s and $Dc_\alpha$'s not found in [HMT]: $SPDc_\alpha \not\subseteq SPLf_\alpha$ for $\alpha \geq \omega$ . In fact, write $\alpha = \Gamma \cup \Delta$ with $\Gamma \cap \Delta = 0$ and $|\Gamma|, |\Delta| \geq \omega$ . Then the statement

(**) for all $x$ , if $c_\lambda^\partial x = 0$ for all $\lambda \in \Gamma$ then $x = 0$

holds in every $Lf_\alpha$ , hence in every $SPLf_\alpha$ , but fails in some $Dc_\alpha$ .

(9) From I.6.13 it follows that $PWs_\alpha \not\subseteq IWs_\alpha$ .

(10) Among the questions about Figure I.6.9 which are open the most important seems to be whether $ICs_\alpha^{reg} \subseteq HPWs_\alpha$ .

(11) If we restrict ourselves to $\alpha \geq \omega$ and to set algebras with bases and subbases infinite, I.6.9 simplifies as in Figure I.6.10, where we use the notation of I.5.7 (17) .

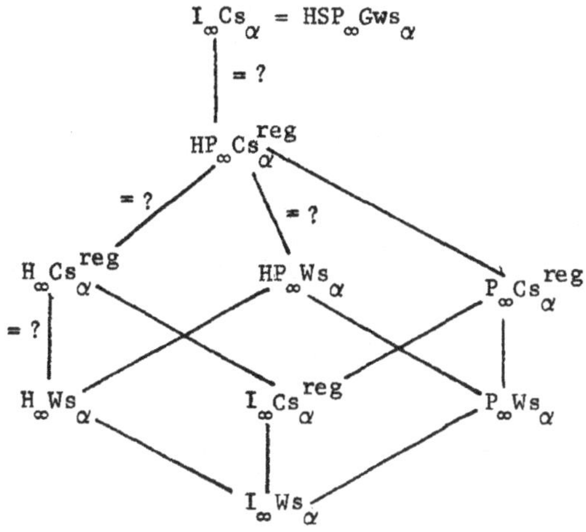


$(\alpha \geq \omega)$

Figure I.6.10

Again $=?$ means that equality of the classes in question is not known. Some of the theorems needed to check this figure are in [AN3]6.2.

Now we discuss direct indecomposability, subdirect indecomposability, and weak subdirect indecomposability. We give some simple results about these notions and then we discuss some examples and problems.

<u>Theorem</u> I.6.11. Every full $Ws_\alpha$ is subdirectly indecomposable.

<u>Proof</u>. Let $\mathfrak{A}$ be a full $Ws_\alpha$, with unit element ${}^\alpha U^{(p)}$. Given $0 \neq y \in A$, choose $f \in y$. Then there is a finite $\Gamma \subseteq \alpha$ such that $(\alpha \sim \Gamma) 1 f = (\alpha \sim \Gamma) 1 p$. Thus $\{p\} \subseteq c_{(\Gamma)} y$. So $\mathfrak{A}$ is subdirectly indecomposable by 2.4.44.

<u>Corollary</u> I.6.12. Any subdirectly indecomposable $Cs_\alpha$ is isomorphic to a $Ws_\alpha$.

Proof. By I.1.11 and I.6.4.

Corollary I.6.13. Every $Ws_\alpha$ is weakly subdirectly indecomposable.

Proof. By 0.3.58 (ii), 2.4.47 (i), and I.6.11.

Corollary I.6.14. Let $\mathfrak{A} \in IGws_\alpha$. Then the following two conditions are equivalent:

(i) $\mathfrak{A} \in IWs_\alpha$ ;

(ii) $\mathfrak{A} \subseteq \mathfrak{B}$ for some subdirectly indecomposable $\mathfrak{B} \in IGws_\alpha$ .

Proof. (i) implies (ii) by I.6.11. (ii) implies (i) by I.6.2.

Corollary I.6.15. Any regular $Cs_\alpha$ with non-empty base is directly indecomposable.

Proof. By I.4.3 and 2.4.14.

Remarks I.6.16. Throughout these remarks let $\alpha \geq \omega$ .

(1) Examples (I) and (II) in 2.4.50 are $Ws_\alpha$'s which are respectively subdirectly indecomposable but not simple, and weakly subdirectly indecomposable but not subdirectly indecomposable.

(2) To supplement our discussion of homomorphisms we shall now show that for any $\kappa \geq 2$ there is a $Ws_\alpha$ with base $\kappa$ having a homomorphic image not isomorphic to a $Ws_\alpha$ . The first such example was due to Monk; the present simpler example is due to Andréka and Németi. Let $p = \langle 0 : \kappa < \alpha \rangle$ and let $\mathfrak{A}$ be the full $Ws_\alpha$ with unit element ${}^\alpha\kappa^{(p)}$ . Let $x = \{f \in V : \omega 1 f \subseteq p$ or the greatest $\kappa < \omega$ such that $f\kappa \neq 0$ is even$\}$ . Let

$$I = Ig^{(\mathfrak{A})}\{c_{(\Gamma)}x \cdot -x : \Gamma \subseteq \omega,\ |\Gamma| < \omega\} .$$

We claim that $|Zd(\mathfrak{A}/I)| > 2$ , hence $\mathfrak{A}/I \notin Ws_{\alpha}$ by I.6.13 and 0.3.58. In fact, clearly $x/I \in Zd(\mathfrak{A}/I)$ , so it suffices to show that $x,-x \notin I$ . If $\Gamma \subseteq \omega$ and $\Gamma \subseteq \kappa \in \omega$ , then $c_{(\Gamma)}x \cdot -x \subseteq \{f \in V : (\omega\sim\kappa) 1 f = (\omega\sim\kappa) 1 p\}$ . Thus for every $y \in I$ there is a $\kappa < \omega$ such that for all $f \in y$ we have $(\omega\sim\kappa) 1 f = (\omega\sim\kappa) 1 p$ . Hence, $x,-x \notin I$ .

(3) In contrast to the situation for homomorphic images (see I.5.6(5)), we show in I.7.17 that any direct factor of a $Cs_{\alpha}$ is isomorphic to a $Cs_{\alpha}$ . Note that a $CA_{\alpha}$ $\mathfrak{A}$ is a direct factor of some $Cs_{\alpha}$ iff $\mathfrak{A}$ is isomorphic to a compressed $Gws_{\alpha}$ (cf. I.2.6).

(4) The full $Cs_{\alpha}$ with base of any cardinality $\geq 2$ is directly decomposable.

(5) A $Cs_{\alpha}^{reg}$ which is directly indecomposable (by I.6.15) but not weakly subdirectly indecomposable can be obtained by modifying Example (III) of 2.4.50. Namely, let $p = \langle 0 : \kappa < \alpha\rangle$ , $q = \langle 1 : \kappa < \alpha\rangle$ , and let $\mathfrak{A}$ be the $Cs_{\alpha}$ of subsets of ${}^{\alpha}2$ generated by $\{\{p\},\{q\}\}$ . By Cor 1.4 of [AN3], $\mathfrak{A}$ is regular, and it is clearly not weakly subdirectly indecomposable.

(6) Example (I) in 2.4.50 can be similarly modified to yield a $Cs_{\alpha}^{reg}$ $\mathfrak{A}$ which is subdirectly indecomposable but not simple: $\mathfrak{A}$ is the $Cs_{\alpha}$ of subsets of ${}^{\alpha}2$ generated by $\{\{p\}\}$ , with $p$ as above. Clearly $\mathfrak{A}$ is not simple, while $\mathfrak{A} \in Cs_{\alpha}^{reg}$ by Cor. 1.4 of [AN3] . Now let $0 \neq x \in A$ . By I.5.1 write $x = y \oplus d$ ,where $y \in Ig^{(\mathfrak{A})}\{\{p\}\}$ and $d \in Sg^{(\mathfrak{A})}\{0\}$ . Clearly then there is a finite $\Gamma \subseteq \alpha$ such that $\{p\} \subseteq c_{(\Gamma)}x$ . Hence $\mathfrak{A}$ is subdirectly indecomposable.

(7) Example (II) in 2.4.50 can be modified to yield a $Cs_\alpha$ $\mathfrak{A}$ which is weakly subdirectly indecomposable but not subdirectly indecomposable. This was noted by Andréka and Németi, who also constructed a $Cs_\alpha^{reg}$ with these properties. To construct such a $Cs_\alpha$, let $\mathfrak{A}$ be the full $Cs_\alpha$ with base 2. Let $\langle \Gamma_\kappa : \kappa < \omega \rangle$ be a system of pairwise disjoint infinite subsets of $\alpha$. Let $p = \langle 0 : \kappa < \alpha \rangle$, and for each $\kappa < \alpha$ let

$$X_\kappa = \{f \in {}^{\alpha}2^{(p)} : \Gamma_\kappa 1 f = \Gamma_\kappa 1 p\} .$$

Let $\mathfrak{B} = \mathfrak{Sg}^{(\mathfrak{A})}\{X_\kappa : \kappa < \omega\}$. Let $I = Ig^{(\mathfrak{B})}\{X_\kappa : \kappa < \omega\}$. Thus by I.5.1 we have

(*) $$B = \{x \oplus d : x \in I \text{ and } d \in Sg^{(\mathfrak{B})}\{0\}\} .$$

Now $\mathfrak{B}$ is not subdirectly indecomposable, by the same argument as in 2.4.50 (II). Now suppose that $0 \neq y \in B$; by (*) write $y = x \oplus d$ with $x \in I$ and $d \in Sg^{(\mathfrak{B})}\{0\}$. Then we can choose $\kappa \in \omega$ and a finite $\Omega \subseteq \alpha$ such that $x \subseteq \bigcup_{\lambda<\kappa} c_{(\Omega)}X_\lambda$. We shall show that $p \in c_{(\Theta)}y$ for some finite $\Theta \subseteq \alpha$. Since this is then true for every non-zero $y \in B$, it then easily follows from 2.4.46 that $\mathfrak{B}$ is weakly subdirectly indecomposable. Since $y \neq 0$, we have two possibilities.

<u>Case</u> 1. $x \cdot -d \neq 0$. For any $f \in x \cdot -d$ we have $f \in {}^{\alpha}2^{(p)}$, and hence $p \in c_{(\Theta)}(x \cdot -d)$ for some finite $\Theta \subseteq \alpha$.

<u>Case</u> 2. $d \cdot -x \neq 0$. Fix $f \in d \cdot -x$. For all $\lambda < \kappa$ choose $\mu\lambda \in \Gamma_\lambda \sim (\Omega \cup \underset{\sim}{\Delta} d)$. Let $\Phi = \{\mu\lambda : \lambda < \kappa\}$, $h = [(\underset{\sim}{\Delta} d) 1 f] \cup [(\alpha \sim \underset{\sim}{\Delta} d) 1 p]$, $k = [(\alpha \sim \Phi) 1 h] \cup \langle 1 : \nu \in \Phi \rangle$. Then $h \in d$ since $d$ is regular, $k \in d$ similarly, $k \notin x$, but $k \in {}^{\alpha}2^{(p)}$. Hence the desired conclusion follows

as in Case 1.

(8) Andréka and Németi have shown that the converse of Theorem I.6.15 fails: there is a directly indecomposable $Cs_\alpha$ not isomorphic to a $Cs_\alpha^{reg}$. Let $\kappa \geq 2$ be arbitrary, and let $\mathfrak{B}$ be the full $Cs_\alpha$ with base $\kappa$. Choose $\Gamma \subseteq \alpha$ with $0 \in \Gamma$, and both $\Gamma, \alpha \sim \Gamma$ infinite. Let $p = \langle 0 : \kappa < \alpha \rangle$. Choose $q_0, q_1 \in {}^\alpha\kappa$ so that $\Gamma 1 q_0 = \Gamma 1 q_1 = \Gamma 1 p$ and $q_0 \notin {}^\alpha\kappa^{(q1)}$. For each $\iota < 2$ let $x_\iota = \{f \in {}^\alpha\kappa^{(q\iota)} : \Gamma 1 f = \Gamma 1 p\}$. Let $\mathfrak{A} = \mathfrak{S}\mathfrak{g}^{(\mathfrak{C})}\{y, x_0, x_1\}$. We claim that $\mathfrak{A}$ is the desired algebra. To check that $\mathfrak{A}$ is directly indecomposable, we need an auxiliary construction. Let $\mathfrak{B} = \mathfrak{S}\mathfrak{g}^{(\mathfrak{A})}\{y\}$. Thus $\mathfrak{B}$ is a regular $Cs_\alpha$, by Theorem I.4.1. Now let

$$C = \{a \in A : a \text{ or } -a \text{ is } \leq c_{(\Theta)}x_\iota \text{ for some } \iota < 2 \text{ and some finite } \Theta \subseteq \alpha\} .$$

Note that $C$ is closed under $-$. Let $D = \{\prod_{\iota<\kappa} a_\iota : \kappa < \omega,\ a \in {}^\kappa(C \cup B)\}$ and $E = \{\sum_{\iota<\lambda} a_\iota : \lambda < \omega,\ a \in {}^\lambda D\}$. Now we claim:

(*) $A = E$.

To prove (*), since $y, x_0, x_1 \in E$, it suffices to show that $E \in \mathrm{Su}\mathfrak{A}$. Clearly $E$ is closed under $+$, and $d_{\kappa\lambda} \in E$ for all $\kappa, \lambda < \alpha$. Since $C \cup B$ is closed under $-$, so is $E$. To show that $E$ is closed under $c_\kappa$, where $\kappa < \alpha$, it suffices to prove that $c_\kappa Z \in E$ for all $Z \in D$. Thus let $\lambda < \omega$ and $a \in {}^\lambda(C \cup B)$. If $a_\nu \leq c_{(\Theta)}x_\iota$ for some finite $\Theta \subseteq \alpha$, some $\iota < 2$, and some $\nu < \lambda$, then

$$c_\kappa \prod_{\mu<\lambda} a_\mu \leq c_\kappa a_\nu \leq c_{(\Theta \cup \{\kappa\})}x_\iota ,$$

and so $c_\kappa \prod_{\mu<\lambda} a_\mu \in C \subseteq E$ . Thus we may assume that

$$\prod_{\mu<\lambda} a_\mu = \prod_{\mu<\varphi} d_\mu \cdot b ,$$

where $b \in B$ , $\varphi < \omega$ , and for each $\mu < \varphi$ there is a finite $\Theta_\mu \subseteq \alpha$ and an $\iota_\mu < 2$ such that $-d_\mu \leq c_{(\Theta\mu)} x_{\iota\mu}$ . Let $z = \prod_{\mu<\lambda} a_\mu$ . Then

$$(c^\partial_\kappa \prod_{\mu<\varphi} d_\mu) \cdot c_\kappa z = (c^\partial_\kappa \prod_{\mu<\varphi} d_\mu) \cdot c_\kappa b = (\prod_{\mu<\varphi} c^\partial_\kappa d_\mu) \cdot c_\kappa b ,$$

and for each $\mu < \varphi$ we have

$$c_\kappa(-d_\mu) \leq c_{(\Theta\mu\cup\{\kappa\})} x_{\iota\mu}$$

and hence $-c^\partial_\kappa d_\mu \in C$ . Thus we conclude

(**) $\quad (c^\partial_\kappa \prod_{\mu<\varphi} d_\mu) \cdot c_\kappa z \in E$ .

On the other hand,

$$-c^\partial_\kappa \prod_{\mu<\varphi} d_\mu \cdot c_\kappa z = \sum_{\mu<\varphi} [c_\kappa(-d_\mu) \cdot c_\kappa z] ,$$

and for each $\mu < \varphi$ we have

$$c_\kappa(-d_\mu) \cdot c_\kappa z \leq c_{(\Theta\mu\cup\{\kappa\})} x_{\iota\mu} .$$

Thus $-c^\partial_\kappa \prod_{\mu<\varphi} d_\mu \cdot c_\kappa z \in E$ , so by (**) , $c_\kappa z \in E$ . Hence we have proved (*) . Now let $z \in Zd\mathfrak{A}$ . We shall show that $z \in B$, hence $z \in \{0,1\}$ by I.6.15. By (*) we can write

$$z = \sum_{\nu<\kappa} d_\nu \cdot b_\nu + e + g ,$$

where $\kappa < \omega$ , $b \in {}^\kappa B$ , $-d_\nu \leq c_{(\Theta\nu)} x_{\iota\nu}$ for each $\iota < \kappa$ , where $\Theta_\nu$ is a finite subset of $\alpha$ and $\iota\nu < 2$ , $e \leq c_{(\Omega)} x_0$ and $g \leq x_{(\Xi)} x_1$ , where $\Omega$ and $\Xi$ are finite subsets of $\alpha$. Choose distinct

$$\lambda,\mu \in \Gamma \sim (\bigcup_{\nu<\kappa}\Delta b_\nu \cup \bigcup_{\nu<\kappa}\Theta_\nu \cup \Omega \cup \Xi) .$$

Note that $c_\mu - c_{(\Theta\nu)}x_{\iota\nu} = 1$ for each $\nu < \kappa$. Hence

$$z = c_\mu z = \sum_{\nu<\kappa} b_\nu + c_\mu e + c_\mu g .$$

Furthermore, $c^\partial_\lambda c_\mu c_{(\Omega)}x_0 = 0 = c^\partial_\lambda c_\mu c_{(\Xi)}x_1$, so

$$z = c^\partial_\lambda c_\mu z = \sum_{\nu<\kappa} b_\nu \in B ,$$

as desired.

It remains to show that $\mathfrak{A} \notin \mathrm{ICs}_\alpha^{reg}$. Suppose $F \in \mathrm{Is}(\mathfrak{A},\mathfrak{C})$, where $\mathfrak{C}$ is a $\mathrm{Cs}_\alpha$ with base $U$. Let $y' = Fy$ and $x'_\iota = Fx_\iota$ for $\iota < 2$. Since $c_0 y = 1$ and $y \cdot s^0_1 y \cdot -d_{01} = 0$, similar equations hold for $y'$, and so there is a $u \in U$ such that $f0 = u$ for all $f \in y'$. Also, $x_0 \leq d_{0\kappa}$ for all $\kappa \in \Gamma$, and $x_0 \leq y$; it follows that $f\kappa = u$ for all $\kappa \in \Gamma$, for all $f \in x'_0$. Similarly for $x'_1$. Since $\Delta x'_0 = \Delta x'_1 = \Gamma$ and $x'_0 \cdot x'_1 = 0$, $\mathfrak{C}$ cannot be regular.

(9) It follows from I.7.13 that any subdirectly indecomposable $\mathrm{Cs}_\alpha$ is isomorphic to a regular $\mathrm{Cs}_\alpha$.

(10) The following problems about these notions remain open:

(a) Is every weakly subdirectly indecomposable $\mathrm{Cs}_\alpha$ isomorphic to a regular $\mathrm{Cs}_\alpha$ ?

(b) Is every weakly subdirectly indecomposable $\mathrm{Gws}_\alpha$ (or $\mathrm{Cs}_\alpha^{reg}$) isomorphic to a $\mathrm{Ws}_\alpha$ ?

## 7. Ultraproducts

We shall begin with a general lemma (due to Monk) about ultraproducts of $Crs_\alpha$'s . To formulate this lemma it is convenient to introduce some special notation.

Definition I.7.1. Let $F$ be an ultrafilter on a set $I$ , $U = \langle U_i : i \in I \rangle$ a system of sets, and $\alpha$ an ordinal.

(i) By an $(F,U,\alpha)$ - choice function we mean a function $c$ mapping $\alpha \times (P_{i\in I}U_i/\overline{F})$ into $P_{i\in I}U_i$ such that for all $\kappa < \alpha$ and all $y \in P_{i\in I}U_i/\overline{F}$ we have $c(\kappa,y) \in y$ .

(ii) If $c$ is an $(F,U,\alpha)$ - choice function, then we define $c^+$ mapping ${}^\alpha(P_{i\in I}U_i/\overline{F})$ into $P_{i\in I}{}^\alpha U_i$ by setting, for all $q \in {}^\alpha(P_{i\in I}U_i/\overline{F})$ and all $i \in I$

$$(c^+q)_i = \langle c(\kappa,q_\kappa)_i : \kappa < \alpha \rangle \ .$$

(iii) Let $A = \langle A_i : i \in I \rangle$ be a system of sets such that $A_i \subseteq Sb({}^\alpha U_i)$ for all $i \in I$ , and let $c$ be an $(F,U,\alpha)$ - choice function. Then there is a unique function $Rep(F,U,\alpha,A,c)$ (usually abbreviated by omitting one or more of the five arguments) mapping $P_{i\in I}A_i/\overline{F}$ into $Sb^\alpha(P_{i\in I}U_i/\overline{F})$ such that, for any $a \in P_{i\in I}A_i$ ,

$$Rep(a/\overline{F}) = \{q \in {}^\alpha(P_{i\in I}U_i/\overline{F}) : \{i \in I : (c^+q)_i \in a_i\} \in F\} \ .$$

Lemma I.7.2. Let $F$ be an ultrafilter on a set $I$ , $U = \langle U_i : i \in I \rangle$ a system of non-empty sets, and $\alpha$ an ordinal. Let $c$ be an $(F,U,\alpha)$ - choice function. Further, let $\mathfrak{A} \in {}^I Crs_\alpha$ , where each $\mathfrak{A}_i$ has base $U_i$

and unit element $V_i$ , and set $V = \langle V_i : i \in I\rangle$ .

Then $\mathrm{Rep}(c)$ is a homomorphism from $P_{i\in I}\mathfrak{A}_i/\overline{F}$ into a $\mathrm{Crs}_\alpha$ . Furthermore, for every $0 \neq a/\overline{F} \in P_{i\in I}A_i/\overline{F}$ , if $Z \in F$ , $s \in P_{i\in I}V_i$ , $s_i \in a_i$ for all $i \in Z$ , and $w = \langle\langle s_i\kappa : i \in I\rangle : \kappa < \alpha\rangle$ it follows that if $c'$ is any $(F,U,\alpha)$ - choice function such that $c'(\kappa, w\kappa/\overline{F}) = w\kappa$ for all $\kappa < \alpha$ , then $\mathrm{Rep}(c')(a/\overline{F}) \neq 0$.

<u>Proof</u>. Let $f = \mathrm{Rep}(c)$ , $X = P_{i\in I}U_i/\overline{F}$ , $T = f(V/\overline{F})$ . Clearly $f$ preserves $+$ . Now let $\kappa,\lambda < \alpha$ ; we show that $f$ preserves $d_{\kappa\lambda}$ . Note that $fd_{\kappa\lambda} \subseteq T$ since $f$ preserves $+$ . Now let $q \in T$ . Then $\{i \in I : (c^+q)_i \in V_i\} \in F$ , and so

$$q \in fd_{\kappa\lambda} \text{ iff } \{i \in I : (c^+q)_i \in D_{\kappa\lambda}^{[V_i]}\} \in F \text{ iff } \{i \in I : (c^+q)_i\kappa = (c^+q)_i\lambda\} \in F \text{ iff } \{i \in I : c(\kappa,q\kappa)_i = c(\kappa,q\lambda)_i\} \in F \text{ iff } q_\kappa = q_\lambda \text{ iff } q \in D_{\kappa\lambda}^{[T]} .$$

Now let $a \in P_{i\in I}A_i$ . We show that $f$ preserves $-$ . Clearly $f(-a/\overline{F}) \subseteq T \sim f(a/\overline{F})$ . Now let $q \in T \sim f(a/\overline{F})$ . Thus $\{i \in I : (c^+q)_i \in V_i\} \in F$ and $\{i \in I : (c^+q)_i \in a_i\} \notin F$ , i.e., $\{i \in I : (c^+q)_i \in V_i \sim a_i\} \in F$ . Therefore $q \in f(-a/\overline{F})$ .

Next, let also $\kappa < \alpha$ ; we show that $f$ preserves $c_\kappa$ . First suppose that $q \in f(c_\kappa a/\overline{F})$ . Let $M = \{i \in I : (c^+q)_i \in C_\kappa^{[V_i]}a_i\}$ ; thus $M \in F$ . Then there is an $s \in P_{i\in I}U_i$ such that $[(c^+q)_i]^\kappa_{si} \in a_i$ for all $i \in M$ . We show that $q^\kappa_{s/\overline{F}} \in f(a/\overline{F})$ . Let $Z = \{i \in I : si = c(\kappa,s/\overline{F})_i\}$ . Then $Z \in F$ since $c(\kappa,s/\overline{F}) \in s/\overline{F}$ . For any $i \in Z \cap M$ we have

$$(c^+q^\kappa_{s/\bar F})_i = \langle c(\lambda,(q^\kappa_{s/\bar F})\lambda)_i : \lambda < \alpha\rangle$$
$$= [(c^+q)_i]^\kappa_{c(\kappa,s/\bar F)i} = [(c^+q)_i]^\kappa_{si} \in a_i \ .$$

Thus, indeed, $q^\kappa_{s/\bar F} \in f(a/\bar F)$ . Hence $q \in C^{[T]}_\kappa f(a/\bar F)$ . Second, suppose that $q \in C^{[T]}_\kappa f(a/\bar F)$ . Thus $q \in T$ and $(\alpha \sim \{\kappa\}) 1 q = (\alpha \sim \{\kappa\}) 1 p$ for some $p \in f(a/\bar F)$ . Let $M = \{i \in I : (c^+p)_i \in a_i\}$ ; thus $M \in F$ . Also since $q \in T$ the set $Z = \{i \in I : (c^+q)_i \in V_i\}$ is in $F$ . Now let $i \in M \cap Z$ . Then $(c^+q)_i \in V_i$ and $(\alpha \sim \{\kappa\}) 1 (c^+q)_i \subseteq (c^+p)_i \in a_i$ , proving that $(c^+q)_i \in C^{[V_i]}_\kappa a_i$ . Thus $q \in f(c_\kappa a/\bar F)$ , since $M \cap Z \in F$ .

We have now verified that f is a homomorphism. For the second part of the conclusion of the lemma, assume its additional hypotheses. Let $q = \langle w\kappa/\bar F : \kappa < \alpha\rangle$ and $f' = \mathrm{Rep}(c')$ . We show that $q \in f(a/\bar F)$ (hence $f(a/\bar F) \neq 0$ , as desired). In fact, for any $i \in Z$ we have

$$(c^+q)_i = \langle c(\kappa,q\kappa)_i : \kappa < \alpha\rangle = \langle c(\kappa,w\kappa/\bar F)_i : \kappa < \alpha\rangle$$
$$= \langle (w\kappa)_i : \kappa < \alpha\rangle = s_i \in a_i \ .$$

So $q \in f(a/\bar F)$ , as desired .

Next, we derive some specialized versions of Lemma I.7.2 which are more easily applicable; they are due to Andréka and Németi.

<u>Lemma</u> I.7.3. Assume the hypotheses of Lemma I.7.2. Also suppose that F is $|\alpha|^+$ - complete (which holds, e.g., if $\alpha < \omega$). Then

(i) for any $(F,U,\alpha)$ - choice function $c'$ we have $\mathrm{Rep}(c) = \mathrm{Rep}(c')$ ;

(ii) $\mathrm{Rep}(c)$ is an isomorphism;

(iii) For any $a \in P_{i\in I}A_i$ we have

$$(\mathrm{Rep}(c))(a/\overline{F}) = \{q \in {}^{\alpha}(P_{i\in I}U_i/\overline{F}) : \text{ there is } w \in {}^{\alpha}(P_{i\in I}U_i) \text{ such that } q\kappa = w\kappa/\overline{F} \text{ for all } \kappa < \alpha \text{ and } \{i \in I : pj_i \circ w \in a_i\} \in F\} .$$

<u>Proof</u>. It suffices to prove (iii), since the fact that $c$ is arbitrary then implies (i), and thus (ii) follows from Lemma I.7.2. So, let $a \in P_{i\in I}A_i$ and $q \in {}^{\alpha}(P_{i\in I}U_i/\overline{F})$ . We want to show that the following two conditions are equivalent:

(1) $\{i \in I : (c^+q)_i \in a_i\} \in F$

(2) there is a $w \in {}^{\alpha}(P_{i\in I}U_i)$ such that $q\kappa = w\kappa/\overline{F}$ for all $\kappa < \alpha$ and $\{i \in I : pj_i \circ w \in a_i\} \in F$ .

If (1) holds, we let $w = \langle c(\kappa, q\kappa) : \kappa < \alpha\rangle$ ; then (2) is clear. Now suppose that (2) holds, and let $H = \{i \in I : pj_i \circ w \in a_i\}$ . Then for any $\kappa < \alpha$ , both $w\kappa$ and $c(\kappa, q\kappa)$ are members of $q\kappa$ , and hence the set $Z_\kappa = \{i \in I : (w\kappa)_i = c(\kappa, q\kappa)_i\}$ is in $F$ . Therefore the set $Y = H \cap \bigcap_{\kappa<\alpha} Z_\kappa$ is also in $F$ . For any $i \in Y$ we have

$$(c^+q)_i = \langle c(\kappa, q\kappa)_i : \kappa < \alpha\rangle = \langle (w\kappa)_i : \kappa < \alpha\rangle = pj_i \circ \kappa \in a_i ,$$

and hence (1) holds.

<u>Lemma</u> I.7.4. Assume the hypotheses of Lemma I.7.2. Let $\mathfrak{B} = \mathrm{Rep}(c)^*(P_{i\in I}\mathfrak{A}_i/\overline{F})$ . Then:

(i) $\mathfrak{A} \in {}^{I}K$ implies $\mathfrak{B} \in K$ for $K = \mathrm{Gws}_\alpha$ or $K = \mathrm{Cs}_\alpha$ ; in the latter case, $\mathfrak{B}$ has base $P_{i\in I}U_i/\overline{F}$ .

(ii) If in addition $F$ is $|\alpha|^+$-complete, then $\mathfrak{A} \in {}^{I}K$ implies $\mathfrak{B} \in K$ for $K = \mathrm{Gs}_\alpha$ or $K = \mathrm{Ws}_\alpha$ .

Proof. Let $f = Rep(c)$ . First suppose $\mathfrak{A} \in {}^I Cs_\alpha$ . Thus $V_i = {}^\alpha U_i$ for each $i \in I$ . We need to prove that $f(V/\overline{F}) = {}^\alpha X$ , where $X = P_{i\in I} U_i/\overline{F}$ . Let $q \in {}^\alpha X$ . Then $(c^+q)_i \in {}^\alpha U_i = V_i$ for all $i \in I$ , so $q \in f(V/\overline{F})$ . The converse being obvious, we thus have $\mathfrak{B} \in Cs_\alpha$ with base $X$ .

Next, suppose that $\mathfrak{A} \in {}^I Gws_\alpha$ . Then for each $i \in I$ we can write

$$V_i = \bigcup\{{}^\alpha Y_{ij}^{(pij)} : j \in J_i\} ,$$

where ${}^\alpha Y_{ij}^{(pij)} \cap {}^\alpha Y_{ik}^{(pik)} = 0$ whenever $j \neq k$ . Let $S_{ij} = {}^\alpha Y_{ij}^{(pij)}$ . Now for each $j \in P_{i\in I} J_i$ we set

$$W_j = \{q \in {}^\alpha X : \{i \in I : (c^+q)_i \in S_{i,ji}\} \in F\},$$

$$Q_j = P_{i\in I}\, Y_{i,ji}/\overline{F}^{(U)} .$$

Now we claim

(1) $f(V/\overline{F}) = \bigcup\{W_j : j \in P_{i\in I} J_i\}$ .

For, let $q \in f(V/\overline{F})$ . Let $M = \{i \in I : (c^+q)_i \in V_i\}$ , so that $M \in F$ . Choose $j \in P_{i\in I} J_i$ so that $(c^+q)_i \in S_{i,ji}$ for all $i \in M$ . Thus $q \in W_j$ , as desired. Clearly each $W_i \subseteq f(V/\overline{F})$ , so (1) holds.

(2) If $j,k \in P_{i\in I} J_i$ and $j/\overline{F} \neq k/\overline{F}$ , then $W_j \cap W_k = 0$ .

For assume the hypothesis of (2) and let $q \in W_j$ . Let $Z = \{i \in I : (c^+q)_i \in S_{i,ji}\}$ and let $H = \{i \in I : (c^+q)_i \in S_{i,ki}\}$ . Then $Z \cap H \subseteq \{i \in I : ji = ki\} \notin F$ , while $Z \in F$ , so $H \notin F$ . Thus $q \notin W_k$ , as desired.

(3) For any $j \in P_{i\in I}J_i$ we have $W_j = \bigcup_{q\in W_j} {}^{\alpha}Q_j^{(q)}$ .

For, first let $q \in W_j$ . Let $\kappa < \alpha$ . Now $\{i \in I : (c^+q)_i \in S_{i,ji}\} \in F$ and hence $\{i \in I : c(\kappa,q\kappa)_i \in Y_{i,ji}\} \in F$ , and further, since $c(\kappa,q\kappa) \in q\kappa$ , $q\kappa \in Q_j$ . Thus $\subseteq$ in (3) holds. For the other direction, it suffices to take $q \in W_j$ , $\kappa < \alpha$ , $u \in Q_j$ , and show that $q_u^\kappa \in W_j$ . Let $M = \{i \in I : (c^+q)_i \in S_{i,ji}\}$ and $Z = \{i \in I : c(\kappa,u)_i \in Y_{i,ji}\}$ . Then $M \in F$ since $q \in W_j$ and $Z \in F$ since $u \in Q_j$ and $c(\kappa,u) \in u$ . So $M \cap Z \in F$ . Let $i \in M \cap Z$ . Then

$$
\begin{aligned}
(c^+q_u^\kappa)_i &= \langle c(\lambda,q_u^\kappa\lambda)_i : \lambda < \alpha\rangle \\
&= \langle c(\lambda,q\lambda)_i : \lambda < \alpha\rangle^\kappa_{c(\kappa,u)i} \\
&= [(c^+q)_i]^\kappa_{c(\kappa,u)i} \in S_{i,ji} .
\end{aligned}
$$

Thus $q_u^\kappa \in W_j$ , as desired.

Now (1), (2), (3) immediately yield that $\mathfrak{B} \in Gws_\alpha$ , upon noting that $W_j = W_k$ if $j/\overline{F} = k/\overline{F}$ .

Now we turn to the proof of (ii) . So, assume that $F$ is $|\alpha|^+$-complete. Let $\mathfrak{A} \in {}^IGs_\alpha$ . Since $Gs_\alpha \subseteq Gws_\alpha$ , we can assume the above notation, where in addition for each $i \in I$ and $j,k \in J_i$ we have $Y_{ij} = Y_{ik}$ or $Y_{ij} \cap Y_{ik} = 0$ (that is, in the terminology of I.2.6, $\mathfrak{A}_i$ is a normal $Gws_\alpha$). We claim

(4) If $j,k \in P_{i\in I}J_i$ and $j/\overline{F} \neq k/\overline{F}$ , then $Q_j \cap Q_k = 0$ .

In fact, assume $y \in P_{i\in I}Y_{i,ji}$ . Let $H = \{i \in I : y_i \in Y_{i,ki}\}$ . Clearly $H \subseteq \{i \in I : ji = ki\} \notin F$ , so $H \notin F$ and hence $y \notin Q_k$ . So (4) holds.

By (1), (2), (3) and (4) it now suffices to show

(5) For any $j \in P_{i\in I}J_i$ we have ${}^{\alpha}Q_j \subseteq f(V/\overline{F})$ .

To prove (5), let $q \in {}^{\alpha}Q_j$ . Say $y \in {}^{\alpha}(P_{i\in I}Y_{i,ji})$ and $q\kappa = y\kappa/\overline{F}$ for all $\kappa < \alpha$ . Then for all $i \in I$ we have $pj_i \circ y \in {}^{\alpha}Y_{i,ji} \subseteq V_i$ . Hence by Lemma I.7.3 we conclude that $q \in f(V/\overline{F})$ .

It remains to check (ii) for $K = Ws_\alpha$ . So, we suppose that $\mathfrak{A} \in {}^{I}Ws_\alpha$ . Since $Ws_\alpha \subseteq Gws_\alpha$ , we still have the above situation, with each $J_i$ a singleton $\{t_i\}$ ; we write $Y_i$ for $Y_{i,ti}$ , $p_i$ for $p_{i,ti}$ , $W$ and $Q$ for $W_j$ and $Q_j$ where $j$ is the unique member of $P_{i\in I}J_i$ . Thus by (1) $f(V/\overline{F}) = W$ , and by (3) , $W = \bigcup_{q\in W}{}^{\alpha}Q^{(q)}$ . Let $r = \langle\langle p_i\kappa : i \in I\rangle/\overline{F} : \kappa < \alpha\rangle$ . Thus $r \in {}^{\alpha}X$ . Now we claim:

(6) For any $q \in {}^{\alpha}X$ , $q \in W$ iff $\{\kappa < \alpha : q\kappa \neq r\kappa\}$ is finite.

(Hence $W = {}^{\alpha}Q^{(r)}$ , as desired.) To prove (6) , let $q \in {}^{\alpha}X$ . Choose $y \in {}^{\alpha}(P_{i\in I}Y_i)$ so that $q\kappa = y_\kappa/\overline{F}$ for all $\kappa < \alpha$ .

Let $G = \{\kappa < \alpha : q\kappa \neq r\kappa\}$ . For each $\kappa < \alpha$ let $Z_\kappa = \{i \in I : y_\kappa i = p_i\kappa\}$ . Thus $Z_\kappa \in F$ iff $\kappa \notin G$ . Hence the set

$$Z = \bigcap\{Z_\kappa : \kappa \in \alpha \sim G\} \cap \bigcap\{I \sim Z_\kappa : \kappa \in G\}$$

is in $F$ (by $|\alpha|^+$-completeness). Now for all $i \in Z$ we have $\{\kappa : y_\kappa i = p_i\kappa\} = \alpha \sim G$ . Let $M = \{i \in I : c(\kappa,q\kappa)_i = y_\kappa i\}$ . Thus $M \in F$ also.

Now if $q \in W$ , then the set $N = \{i \in I : (c^+q)_i \in {}^{\alpha}Y_i^{(pi)}\}$ is in $F$ . For $i \in M \cap N \cap Z$ we have, for all $\kappa < \alpha$ , $(c^+q)_i\kappa = c(\kappa,q\kappa)_i = y_\kappa i$ and hence (by the definition of $N$), $G = \{\kappa : y_\kappa i \neq p_i\kappa\}$ is finite.

On the other hand, if $G$ is finite then for any $i \in M \cap Z$ and any $\kappa \in \alpha \sim G$ we have $(c^+q)_i\kappa = c(\kappa, q\kappa)_i = y_\kappa i = p_i\kappa$ and so, since $M \cap Z \in F$, $\{i \in I : (c^+q)_i \in {}^\alpha Y_i^{(pi)}\} \in F$ and $q \in W$. This completes the proof.

Lemma I.7.5. Assume the hypotheses of Lemma I.7.2. Let $K = Ws_\alpha$ or $Gs_\alpha$. Then for every non-zero $a \in P_{i\in I}A_i/\overline{F}$ there is an $(F,U,\alpha)$-choice function $c'$ such that $Rep(c')a \neq 0$ and $Rep(c')^*(P_{i\in I}\mathfrak{A}_i/\overline{F}) \in K$.

Proof. If $F$ is $|\alpha|^+$-complete, the desired conclusion follows from Lemma I.7.3 and Lemma I.7.4 (ii). So we assume henceforth that $F$ is not $|\alpha|^+$-complete. In particular, $\alpha \geq \omega$. We let $s$ and $w$ be as in the last part of the proof of Lemma I.7.2. Let $X = P_{i\in I}U_i/\overline{F}$.

Assume now $\mathfrak{A} \in {}^I Ws_\alpha$. Let $N = \{i \in I : |U_i| > 1\}$. Then it is easily seen that there is an $(F,U,\alpha)$-choice function $c'$ satisfying the following two conditions:

(1) $c'(\kappa, w\kappa/\overline{F}) = w\kappa$ for all $\kappa < \alpha$;

(2) $c'(\kappa, y)_i \neq w_\kappa i$ whenever $\kappa < \alpha$, $y \in X$, $y \neq w_\kappa/\overline{F}$, and $i \in N$.

Again let $f = Rep(c')$. By Lemma I.7.2, $fa \neq 0$. Now let $q = \langle w_\kappa/\overline{F} : \kappa < \alpha \rangle$. We shall show that $f(V/\overline{F}) = {}^\alpha X^{(q)}$. Note that $q \in {}^\alpha X$.

If $N \notin F$, then $|X| = 1$ and hence $0 \neq f(V/\overline{F}) \subseteq {}^\alpha X = \{q\} = {}^\alpha X^{(q)}$, so $f(V/\overline{F}) = {}^\alpha X^{(q)}$.

Assume that $N \in F$. Let $p \in {}^\alpha X$, $i \in N$, and $\kappa < \alpha$. Then $(c'^+p)_i\kappa = s_i\kappa$ iff $y/\overline{F} = w_\kappa/\overline{F}$, and hence $(c'^+p)_i \in V_i$ iff $p \in {}^\alpha X^{(q)}$. Since $N \in F$ it follows that $p \in f(V/\overline{F})$ iff $p \in {}^\alpha X^{(q)}$, as desired.

Next, assume $\mathfrak{A} \in {}^I Gs_\alpha$ . Since $F$ is not $|\alpha|^+$-complete, there is an $h \in {}^I\alpha$ such that $[I/(h|h^{-1})] \cap F = 0$ (see, e.g., Chang-Keisler [CK], p. 180). Since $\mathfrak{A} \in {}^I Gs_\alpha$ , for each $i \in I$ there is a subbase $Y_i$ of $V_i$ such that $s_i \in {}^\alpha Y_i$ . Now there clearly is an $(F,U,\alpha)$-choice function $c'$ satisfying the following three conditions:

(3) $c'(hi,y/\overline{F})_i \in Y_i$ for all $i \in I$ and all $y \in X$ ;

(4) $c'(\kappa,y/\overline{F}) \in P_{i\in I}Y_i$ for all $\kappa < \alpha$ and all $y \in P_{i\in I}Y_i$ ;

(5) $c'(\kappa,w\kappa/\overline{F}) = w\kappa$ for all $\kappa < \alpha$ .

Let $W = P_{i\in I}Y_i/\overline{F}^{(U)}$ . Again let $f = Rep(c')$ . By Lemma I.7.2, $fa \neq 0$ . Now we shall show that $f(V/\overline{F}) = {}^\alpha W$ . By (4) we have ${}^\alpha W \subseteq f(V/\overline{F})$ . Now let $q \in {}^\alpha X$ such that $q_\kappa \notin W$ for some $\kappa < \alpha$ . Thus the set $H = \{i \in I : c(\kappa,q_\kappa)_i \notin Y_i\} \in F$ . If $i \in H$ , then by (3) we have $c(hi,q_{hi})_i \in Y_i$ while $c(\kappa,q_\kappa)_i \notin Y_i$ ; hence $(c^+q)_i \notin V_i$ . Since $H \in F$ , it follows that $q \notin f(V/\overline{F})$ . This completes the proof.

Our final version of I.7.2 concerns regularity.

<u>Lemma</u> I.7.6. Assume the hypotheses of Lemma I.7.2. Let $f = Rep(c)$ . Also suppose $a \in P_{i\in I}A_i$ and for each $i \in I$ , $a_i$ is regular in $\mathfrak{A}_i$ and $\underset{\sim}{\Delta} a_i \subseteq \underset{\sim}{\Delta} f(a/\overline{F})$ . Then $f(a/\overline{F})$ is regular.

<u>Proof</u>. We assume all the hypotheses. Let $\Gamma = 1 \cup \Delta f(a/\overline{F})$ , and assume that $p \in f(a/\overline{F})$ , $q \in f(V/\overline{F})$ , and $\Gamma 1 p = \Gamma 1 q$ . We want to show that $q \in f(a/\overline{F})$ . Let

$$H = \{i \in I : (c^+p)_i \in a_i \text{ and } (c^+q)_i \in V_i\} .$$

Thus $H \in F$. By the definition of $c^+$, for all $i \in I$ we have $\Gamma 1 (c^+p)_i = \Gamma 1 (c^+q)_i$. Thus for any $i \in H$, the regularity of $a_i$ implies that $(c^+q)_i \in a_i$. Since $H \in F$, it follows that $q \in f(a/\overline{F})$.

Remarks I.7.7. The above lemmas are algebraic forms of the Łoś lemma for ultraproducts. The exact relationships between them and Łoś's lemma will be discussed in a later article.

Andréka and Németi have shown that various hypotheses in I.7.2-I.7.6 are essential, and that these lemmas do not generalize to arbitrary reduced products.

Now we shall use the above lemmas to prove various closure properties of our classes of set algebras.

Theorem I.7.8. $\mathrm{UpK} = \mathrm{IK}$ for $K \in \{Gws_\alpha, Gs_\alpha\}$.

Proof. By Lemmas I.7.2, I.7.4 (i), and I.7.5, each member of UpK is a subdirect product of memebers of K. If $\alpha \neq 1$, then $\mathrm{IK} = \mathrm{SPK}$; if $\alpha = 1$, then by I.7.3 and I.7.4 we have $\mathrm{UpK} = \mathrm{IK}$; the proof is complete.

Theorem I.7.9. $\mathrm{UpCs}_\alpha = \mathrm{ICs}_\alpha$ for $\alpha < \omega$.

Proof. By Lemmas I.7.3 and I.7.4 (ii).

Theorem I.7.10. $\bigcup L \in K$ whenever $L$ is a non-empty subset of $K$ directed by $\subseteq$, for $K \in \{IGws_\alpha, IGs_\alpha\} \cup \{ICs_\alpha : \alpha < \omega\} \cup \{I(Ws_\alpha \cap Lf_\alpha) : \alpha \text{ an ordinal}\}$.

Proof. This is immediate from I.7.8 and I.7.9, since $\mathrm{SK} = K$, for all choices of $K$ except the last. Let $K = Ws_\alpha \cap Lf_\alpha$, and let

$L$ be as indicated. Then $\bigcup L \in Gws_\alpha$ by what was already proved, while $\bigcup L$ is simple by I.5.2 (ii), 2.4.43, and 0.3.51. Hence $\bigcup L \in IWs_\alpha$ by I.6.4.

The following generalization of a part of Theorem I.7.10 is due to Andréka and Németi [AN1]:

Theorem I.7.11. If $0 \neq L \subseteq ICs_\alpha^{reg} \cap Lf_\alpha$ is a set directed by $\subseteq$, then $\bigcup L \in ICs_\alpha^{reg} \cap Lf_\alpha$.

Proof. By Theorem I.7.10 we may assume that $\alpha \geq \omega$. For each $\mathfrak{B} \in L$ let $f_\mathfrak{B}$ be an isomorphism onto a regular $Cs_\alpha$ $\mathfrak{A}_\mathfrak{B}$ with base $U_\mathfrak{B}$; further, let $M_\mathfrak{B} = \{\mathfrak{C} \in L : \mathfrak{B} \subseteq \mathfrak{C}\}$. Let $F$ be an ultrafilter on $L$ such that $M_\mathfrak{B} \in F$ for each $\mathfrak{B} \in L$. There is an isomorphism $g$ of $P_{\mathfrak{B}\in L}\mathfrak{B}/\overline{F}$ onto $P_{\mathfrak{B}\in L}\mathfrak{A}_\mathfrak{B}/\overline{F}$ such that $g(b/\overline{F}) = \langle f_\mathfrak{B} b_\mathfrak{B} : \mathfrak{B} \in L\rangle/\overline{F}$ for all $b \in P_{\mathfrak{B}\in L}B$. Let $c$ be an $(F,U,\alpha)$-choice function, and let $f = Rep(c)$. By Lemmas I.7.2 and I.7.4 (i), $f$ is a homomorphism from $P_{\mathfrak{B}\in L}\mathfrak{A}_\mathfrak{B}/\overline{F}$ onto a $Cs_\alpha$ with base $P_{\mathfrak{B}\in L}U_\mathfrak{B}/\overline{F}$. Now for each $b \in \bigcup_{\mathfrak{B}\in L}B$ let

$$h'b = \langle b : b \in B, \mathfrak{B} \in L\rangle \cup \langle 0^{(\mathfrak{B})} : b \notin B, \mathfrak{B} \in L\rangle,$$

and let $hb = h'b/\overline{F}$. Then $h$ is an isomorphism of $\bigcup L$ into $P_{\mathfrak{B}\in L}\mathfrak{B}/\overline{F}$ (see the proof of 0.3.71). Now let $\mathfrak{C} = (f \circ g \circ h)^* \bigcup L$. We claim that $f \circ g \circ h$ is an isomorphism, and $\mathfrak{C} \in Cs_\alpha^{reg} \cap Lf_\alpha$.

First note that each member of $L$ is simple, by I.5.2 (i); hence $\bigcup L$ is simple, by 2.3.16 (ii). Thus $f \circ g \circ h$ is an isomorphism. By 2.1.13, $\mathfrak{C} \in Lf_\alpha$. To show that $\mathfrak{C}$ is regular we apply I.7.6. Let $b \in \bigcup_{\mathfrak{B}\in L}B$. Then for each $\mathfrak{B} \in L$, $f_\mathfrak{B}(h'b)_\mathfrak{B}$ is regular in $\mathfrak{B}$, and

$$\underset{\sim}{\Delta} f_{\mathfrak{B}}(h'b)_{\mathfrak{B}} = \underset{\sim}{\Delta}(h'b)_{\mathfrak{B}} \subseteq \underset{\sim}{\Delta} b \subseteq \underset{\sim}{\Delta}(fghb)$$

since $f \circ g \circ h$ is an isomorphism. Therefore by I.7.6, $fghb$ is regular, as desired.

The following lemma is due to Andréka and Németi.

Lemma I.7.12. Let $\mathfrak{A}$ be a $Crs_\alpha$ with base $U$ and unit element $V$, and let $F$ be an ultrafilter on a set $I$. Let $c$ be an $(F, \langle U : i \in I\rangle, \alpha)$ - choice function, and let $f = Rep(F, \langle U : i \in I\rangle, \alpha, \langle \mathfrak{A} : i \in I\rangle, c)$. Define $\delta \in {}^A({}^I A/\overline{F})$ and $\varepsilon \in {}^U({}^I U/\overline{F})$ by

$$\delta = \langle \langle a : i \in I\rangle/\overline{F} : a \in A\rangle ,$$

$$\varepsilon = \langle \langle u : i \in I\rangle/\overline{F} : u \in U\rangle .$$

We assume also that for every $r \in V$ there is a $Z \in F$ such that, for all $\kappa < \alpha$, $c(\kappa, \varepsilon(r\kappa)) \supseteq \langle r\kappa : i \in Z\rangle$. Finally, let $g = f \circ \delta$ and $\mathfrak{B} = g^* \mathfrak{A}$. Then

(i) $g$ is an isomorphism from $\mathfrak{A}$ onto $\mathfrak{B}$;

(ii) $\tilde{\varepsilon}^* \mathfrak{A}$ is sub-isomorphic to $\mathfrak{B}$ (recall the definition of $\tilde{\varepsilon}$ from I.3.5, and subisomorphism from I.3.15);

(iii) $\tilde{\varepsilon} = rl^{\mathfrak{B}}_{\tilde{\varepsilon}V} \circ g$ (recall the definition of $rl$ from I.6.1).

Proof. Clearly $g$ is a homomorphism from $\mathfrak{A}$ onto $\mathfrak{B}$. By I.3.1, $\tilde{\varepsilon}$ is an isomorphism from $\mathfrak{A}$ onto $\tilde{\varepsilon}^* \mathfrak{A}$, so if we establish (iii), then (i) and (ii) will follow. To establish (iii), let $a \in A$; we want to show

(1) $ga \cap \tilde{\varepsilon}V = \{\varepsilon \circ s : s \in a\}$ .

First suppose $q \in ga \cap {}^{\sim}_{\varepsilon}V$ . Since $q \in {}^{\sim}_{\varepsilon}V$ , there is an $s \in V$ such that $q = \varepsilon \circ s$ . By the hypothesis of the lemma choose $Z \in F$ such that $c(\kappa, \varepsilon(s\kappa)) \supseteq \langle s\kappa : i \in Z\rangle$ for all $\kappa < \alpha$ . Let $H = \{i \in I : (c^{+}q)_i \in a\}$ ; thus $H \in F$ . Since $H \cap Z \in F$ , we can choose $i \in H \cap Z$ . Then

$$s = \langle s\kappa : \kappa < \alpha\rangle = \langle c(\kappa, \varepsilon(s\kappa))_i : \kappa < \alpha\rangle$$
$$= \langle c(\kappa, q\kappa)_i : \kappa < \alpha\rangle = (c^{+}q)_i \in a .$$

Thus $q = \varepsilon \circ s$ implies that $q$ is in the right side of (1).

Second suppose $q = \varepsilon \circ s$ with $s \in a$ . Since $a \subseteq V$ , we have $q \in {}^{\sim}_{\varepsilon}V$ . Again by the hypothesis of the lemma let $Z \in F$ be such that $c(\kappa, \varepsilon s\kappa) \supseteq \langle s\kappa : i \in Z\rangle$ for all $\kappa < \alpha$ . Then $(c^{+}q)_i = s \in a$ for all $i \in Z$ , so $q \in ga \cap {}^{\sim}_{\varepsilon}V$ .

We now use this lemma to establish that $Ws_\alpha \subseteq ICs_\alpha^{reg}$ , also due to Andréka and Németi. It generalizes the fact that $Ws_\alpha \subseteq ICs_\alpha$ , established in Henkin, Monk [HM] .

<u>Theorem</u> I.7.13. $Ws_\alpha \subseteq ICs_\alpha^{reg}$ .

<u>Proof</u>. Let $\mathfrak{A}$ be a $Ws_\alpha$ with unit element $V = {}^{\alpha}U^{(p)}$ . Let $|I| \geq |\alpha| \cup \omega$ (in a later proof we shall choose $I$ in a special way; for now we could take $I = \alpha$). Let $F$ be a $|I|$ - regular ultrafilter over $I$ (for the notion of $\kappa$ - regular ultrafilter see Chang, Keisler [CK], p. 201). Then, as is easily seen, there is a function $h \in {}^{I}\{\Gamma \subseteq \alpha : |\Gamma| < \omega\}$ such that $\{i \in I : \kappa \in hi\} \in F$ for all $\kappa < \alpha$ . Now let $\delta$ and $\varepsilon$ be as in I.7.12, and set $X = {}^{I}U/\overline{F}$ . Let $c$ be an $(F, \langle U : i \in I\rangle, \alpha)$ - choice function satisfying the following condition:

(1) For all $\kappa < \alpha$, $i \in I$, and all $y \in X$, if $\kappa \notin hi$ then $c(\kappa,y)_i = p\kappa$ ; if $\kappa \in hi$ and $y = \varepsilon u$ with $u \in U$ then $c(\kappa,y)_i = u$ .

Let $f = Rep(c)$ . We shall show that $f \circ \delta$ is the desired isomorphism. By I.7.2, $f \circ \delta$ is a homomorphism onto a $Crs_\alpha$ $\mathfrak{B}$ . Now we show that $f\delta V = {}^\alpha X$ , so that $\mathfrak{B}$ is a $Cs_\alpha$ . Since $f\delta V \subseteq {}^\alpha X$ trivially, we show the other inclusion. Let $q \in {}^\alpha X$ . It suffices to show that $(c^+q)_i \in V$ for <u>all</u> $i \in I$ . So, let $i \in I$ . Note that $(c^+q)_i \in {}^\alpha U$ . If $\kappa \notin hi$ , then by (1) we have $(c^+q)_i\kappa = c(\kappa,q\kappa)_i = p\kappa$ . Since $hi$ is finite, it follows that $(c^+q)_i \in {}^\alpha U^{(p)}$ , as desired.

To show that $f \circ \delta$ is an isomorphism we apply I.7.12. We have only one hypothesis of I.7.12 left to check. Let $r \in V$ . Then there is a finite $\Gamma \subseteq \alpha$ such that $(\alpha \sim \Gamma) \upharpoonright r = (\alpha \sim \Gamma) \upharpoonright p$ . Let $Z = \{i \in I : \Gamma \subseteq hi\}$ . By the choice of $h$ we have $Z \in F$ . Let $\kappa < \alpha$ and $i \in Z$ ; we show that $c(\kappa,\varepsilon r\kappa)_i = r\kappa$ (as desired). If $\kappa \in \Gamma$ , then $\kappa \in hi$ and so $c(\kappa,\varepsilon r\kappa)_i = r\kappa$ by (1) . If $\kappa \notin \Gamma$ then $r\kappa = p\kappa$ and $c(\kappa,\varepsilon r\kappa)_i = r\kappa$ by either clause of (1) . Thus the hypotheses of I.7.12 hold, and hence $f \circ \delta$ is an isomorphism.

It remains to show that $\mathfrak{B}$ is regular; but since $f \circ \delta$ is an isomorphism, this is immediate from I.7.6 and I.1.16.

<u>Theorem</u> I.7.14. $IGs_\alpha = IGs_\alpha^{reg} = IGws_\alpha^{reg} = IGws_\alpha \subseteq SPCs_\alpha^{reg}$ .

<u>Proof</u>. Using I.1.10, I.1.11, I.1.18, I.1.19, I.6.4, I.6.6, and I.7.13 we have

$$IGws_\alpha \subseteq SPWs_\alpha \subseteq SPCs_\alpha^{reg} \subseteq IGs_\alpha^{reg} \subseteq IK \subseteq IGws_\alpha \subseteq SPCs_\alpha^{reg} ,$$

where $K = Gs_\alpha$ or $Gws_\alpha^{reg}$ . The theorem follows.

The following result, $HGws_\alpha \subseteq IGs_\alpha$ , was known to the authors for a long time using representation theory. The present direct proof is due to Andréka and Németi.

<u>Theorem</u> I.7.15. $HGws_\alpha \subseteq IGs_\alpha$ .

<u>Proof</u>. The case $\alpha < 2$ is trivial, so assume $\alpha \geq 2$ . Let $\mathfrak{A} \in Gws_\alpha$ , and let $L$ be an ideal of $\mathfrak{A}$ . We want to show that $\mathfrak{A}/L \in IGs_\alpha$ .

<u>Case</u> 1. $\alpha < \omega$ . For each $z \in A$ let $kz = -c_{(\alpha)}z$ . Note that $kz$ is zero-dimensional and $\mathfrak{Rl}_{kz}\mathfrak{A} \in Gs_\alpha$ . Let $\mathfrak{B} = \langle \mathfrak{Rl}_{kz}\mathfrak{A} : z \in L \rangle$ . Let $F$ be an ultrafilter over $L$ such that $\{v \in L : v \geq z\} \in F$ for all $z \in L$ . Now define $h$ mapping $A$ into $P_{z \in L}B_z/\overline{F}$ by setting, for any $a \in A$ ,

$$ha = \langle a \cdot kz : z \in L \rangle / \overline{F} .$$

Now $\langle a \cdot kz : a \in A \rangle \in Ho(\mathfrak{A}, \mathfrak{B}_z)$ for each $z \in L$ by 2.3.26, so $\langle \langle a \cdot kz : z \in L \rangle : a \in A \rangle \in Hom(\mathfrak{A}, P_{z \in L}\mathfrak{B}_z)$ by 0.3.6(ii). Hence by 0.3.61 we infer that $h \in Hom(\mathfrak{A}, P_{z \in L}\mathfrak{B}_z/\overline{F})$ . Now we claim that $h^*\mathfrak{A} \cong \mathfrak{A}/L$ ; to show this it suffices to show that $h^{-1\star}0 = L$ . If $a \in L$ , then $a \cdot -c_{(\alpha)}v = 0$ for all $v \geq a$ , so $\{v \in L : v \geq a\} \subseteq \{v \in L : a \cdot kv = 0\}$ ; since $\{v \in L : v \geq a\} \in F$ , it follows that $ha = 0$ . On the other hand, if $a \in A \sim L$ and $z \in L$ then $a \nleq c_{(\alpha)}z$ and hence $a \cdot kz \neq 0$ ; thus $ha \neq 0$ , as desired.

Thus $h^*\mathfrak{A} \cong \mathfrak{A}/L$ , so $\mathfrak{A}/L \in SUpGs_\alpha$ . By I.7.8 , $\mathfrak{A}/L \in IGs_\alpha$ .

*Case* 2. $\alpha \geq \omega$ . Using I.6.3, it is enough to take any $a \in A \sim L$ and find a homomorphism $h$ of $\mathfrak{A}$ onto a $Cs_\alpha$ such that $ha \neq 0$ and $h^*L = \{0\}$ . Let

$$I = L \times \{\Gamma \subseteq \alpha : |\Gamma| < \omega\} ,$$

and let $F$ be an ultrafilter on $I$ such that $\{\langle v,\Delta\rangle \in I : v \geq z, \Delta \supseteq \Gamma\} \in F$ for all $\langle z,\Gamma\rangle \in I$ . Let $\mathfrak{A}$ have base $U$ and unit element

$$V = \bigcup\{{}^\alpha Y_j^{(p_j)} : j \in J\} ,$$

where ${}^\alpha Y_i^{(p_i)} \cap {}^\alpha Y_j^{(p_j)} = 0$ for distinct $i,j \in J$ . Let $X = {}^I U/\overline{F}$ . Since $a \notin L$ , there is a function $r \in {}^I V$ such that $r(z,\Gamma) \in a \cdot -c_{(\Gamma)}z$ for all $\langle z,\Gamma\rangle \in I$ . Then there is a function $j \in {}^I J$ such that $r_i \in {}^\alpha Y_{ji}^{(p_{ji})}$ for all $i \in I$ . Next we let

$$Q = \{k/\overline{F} : k \in P_{i\in I}Y_{ji}\} ,$$
$$s = \langle\langle r_i\kappa : i \in I\rangle : \kappa < \alpha\rangle .$$

Let $c$ be an $(F,\langle U : i \in I\rangle,\alpha)$ - choice function such that the following conditions hold.

(1) If $\kappa < \alpha$ and $y \in Q$ , then $c(\kappa,y) \in P_{i\in I}Y_{ji}$ .

(2) If $\kappa < \alpha$ , then $c(\kappa,s\kappa/\overline{F}) = s\kappa$ .

(3) If $\kappa < \alpha$ , $y \in X$ , $(z,\Gamma) \in I$ , and $\kappa \notin \Gamma$ , then $c(\kappa,y)_{z\Gamma} = r(z,\Gamma)_\kappa$ .

Let $f = Rep(F,\langle U : i \in I\rangle,\alpha,\langle A : i \in I\rangle,c)$ . Also let $\delta$ be as in I.7.12 . Then we claim that $f \circ \delta$ is the desired homomorphism. First we note that by the second part of I.7.2 and by (2), we have $f\delta a \neq 0$. Next we show that $(f \circ \delta)^* L = \{0\}$. Let $z \in L$. Set $Z = \{\langle v,\Gamma\rangle \in I : v \geq z\}$; thus $Z \in F$. Let $q \in {}^{\alpha}X$. We show, for any $i \in Z$, that $(c^+q)_i \notin z$ (thus $q \notin f\delta z$, as desired). Let $i \in Z$, say $i = \langle v,\Gamma\rangle$, where $v \geq z$. By (3) we have $c(\kappa,q\kappa)_i = r_i\kappa$ for all $\kappa \in \alpha \sim \Gamma$; thus $(\alpha \sim \Gamma)1(c^+q)_i = (\alpha \sim \Gamma)1r_i$. Since $r_i \notin c_{(\Gamma)}v$ it follows that $(c^+q)_i \notin c_{(\Gamma)}v$ and, since $v \geq z$, $(c^+q)_i \notin z$.

It remains only to show that $(f \circ \delta)^*\mathfrak{A}$ is a $Cs_\alpha$; we show that $f\delta V = {}^{\alpha}Q$.

(4) For any $q \in {}^{\alpha}X$ and any $i \in I$, $(c^+q)_i \in V$ iff $(c^+q)_i \in {}^{\alpha}Y_{ji}$.

For, if $q \in {}^{\alpha}X$ and $i \in I$, then with $i = (z,\Gamma)$ we have by (3) $(\alpha \sim \Gamma)1(c^+q)_i = (\alpha \sim \Gamma)1r_i$. An easy argument yields (4).

Now suppose $q \in {}^{\alpha}Q$. By (1), $(c^+q)_i \in {}^{\alpha}Y_{ji}$ for all $i \in I$, so by (4) $(c^+q)_i \in V$ for all $i \in I$ and hence $q \in f\delta V$. On the other hand, let $q \in f\delta V$. Then the set $Z = \{i \in I : (c^+q)_i \in V\}$ is in $F$. By (4), $(c^+q)_i \in {}^{\alpha}Y_{ji}$ for each $i \in Z$, i.e., $c(\kappa,q\kappa)_i \in Y_{ji}$ for all $\kappa < \alpha$ and all $i \in Z$. Thus $q \in {}^{\alpha}Q$. This completes the proof.

Theorem I.7.16. For $\alpha > 1$ we have $IGs_\alpha = HSPGs_\alpha = HSPGws_\alpha = HSPWs_\alpha = HSPCs_\alpha = HSPCs_\alpha^{reg}$.

Proof. $HSPGws_\alpha = HGws_\alpha = IGs_\alpha \subseteq IGws_\alpha$ using I.7.15; the other parts of the theorem are easily established by using I.7.13.

The following result is due to Monk.

<u>Theorem</u> I.7.17. Any direct factor of a $Cs_\alpha$ is isomorphic to a $Cs_\alpha$ .

<u>Proof</u>. We may assume that $\alpha \geq \omega$ . We use the notion of compressed $Gws_\alpha$ in I.2.6. In fact, we first prove the following independently interesting result noticed by Andréka and Németi.

(1) $ICs_\alpha = \{\mathfrak{A} : \mathfrak{A}$ is isomorphic to a compressed $Gws_\alpha\}$ ; in fact, any compressed $Gws_\alpha$ is subisomorphic to a $Cs_\alpha$ .

The theorem follows immediately from (1), using 2.4.8, since any zero-dimensional element of a compressed $Gws_\alpha$ is clearly the unit element of a compressed $Gws_\alpha$ . Now let $\mathfrak{A}$ be a compressed $Gws_\alpha$ . Say the unit element of $\mathfrak{A}$ is $V$ and its base is $U$ . Let $F$ be a $|\alpha|$ - regular ultrafilter on $\alpha$ . Let $\varepsilon$ and $\delta$ be as in I.7.12. Let $X = {}^\alpha U/\overline{F}$ . Let $c$ be an $(F, \langle U : \lambda < \alpha\rangle, \alpha)$ - choice function such that

(2) for all $\kappa < \alpha$ and all $u \in U$ , $c(\kappa, \varepsilon u) = \langle u : \lambda < \alpha\rangle$ .

Let $f = Rep(F, \langle U : \lambda < \alpha\rangle, \alpha, \langle A : \lambda < \alpha\rangle, c)$ . Then the hypotheses of I.7.12 are met, and hence $f \circ \delta$ is an isomorphism. Now by I.7.4(i), $(f \circ \delta)^*\mathfrak{A} \in Gws_\alpha$ . Actually if we examine the proof of I.7.4(i) we see that $Q_j = Q_k$ for all $j,k \in P_{i\in I}J_i$ in its notation, with $I = \alpha$ , and hence $(f \circ \delta)^*\mathfrak{A}$ is a compressed $Gws_\alpha$ too. Also recall from that proof that the base of $(f \circ \delta)^*\mathfrak{A}$ is $X$ . Let $W = {}^\alpha X \sim f\delta V$ . If $W = 0$ we are finished, so assume that $W \neq 0$ . Note that $W$ itself is the unit element of some $Gws_\alpha$ . Now we claim

(3) $\mathfrak{A} \succcurlyeq \mathfrak{B}$ for some $Gws_\alpha$ with unit element $W$.

In fact, applying 2.3.26(ii) to the full $Gws_\alpha$ with unit element $V$ and then restricting to $\mathfrak{A}$ we find that $\mathfrak{A} \succcurlyeq \mathfrak{C}$ for some $Ws_\alpha$ $\mathfrak{C}$ with base $U$. By I.7.13, $\mathfrak{C}$ is isomorphic to a $Cs_\alpha$ $\mathfrak{D}$; looking at the proof of I.7.13 we see that we may assume that the base of $\mathfrak{D}$ is $X$. Since $W \subseteq {}^\alpha X$ is zero-dimensional, we get a $Gws_\alpha$ $\mathfrak{B}$ with unit element $W$ such that $\mathfrak{D} \succcurlyeq \mathfrak{B}$, as desired in (3); let $g$ be a homomorphism from $\mathfrak{A}$ onto $\mathfrak{B}$.

By 0.3.6(ii), $\mathfrak{A}$ is isomorphic to a subalgebra of $\mathfrak{B} \times (f \circ \delta)^*\mathfrak{A}$, and by I.6.2 $\mathfrak{B} \times (f \circ \delta)^*\mathfrak{A}$ is isomorphic to a $Cs_\alpha$ $\mathfrak{A}'$; in fact, the function $h = \langle ga \cup f\delta a : a \in A \rangle$ is an isomorphism from $\mathfrak{A}$ into $\mathfrak{A}'$. It is easily checked that $ha \cap \tilde{\varepsilon}V = \tilde{\varepsilon}a$ for all $a \in A$, so (1) holds.

We shall establish a few more results about closure properties and relationships between our classes of set algebras in I.7.28-I.7.30, after discussing again change of base. First we make some remarks about the results already established in this section.

Remarks I.7.18. Many of the results above cannot be improved in the obvious ways; we now make several specific arguments to this effect.

(1) If $\alpha \geq \omega$ and $\mathfrak{A}$ is a $Cs_\alpha$ with base $U$ such that $2 \leq |U| < \omega$, then there is a set $I$ and an ultrafilter $F$ over $I$ such that ${}^I\mathfrak{A}/\overline{F} \notin ICs_\alpha$. This was first noticed by Monk [M2]. It does not extend to $|U| \geq \omega$, by I.7.22. In fact, let $I = 2^{2^{|\alpha|}}$, and choose an ultrafilter $F$ over $I$ such that $|{}^IA/\overline{F}| = 2^{2^{2^{|\alpha|}}}$

(see, e.g., Chang, Keisler [CK] p. 202). Now $\mathfrak{A}$ has characteristic $|U|$, and hence so does ${}^I\mathfrak{A}/\overline{F}$. But any $Cs_\alpha$ of characteristic $|U|$ has base of cardinality $|U|$, and hence has at most $2^{2^{|\alpha|}}$ elements. Thus ${}^I\mathfrak{A}/\overline{F} \notin ICs_\alpha$. This same remark and proof apply to $Ws_\alpha$'s.

(2) For any $\alpha \geq \omega$, any non-discrete $\mathfrak{A} \in Ws_\alpha$, and any non-prinicpal ultrafilter $F$ on $\omega$ we have ${}^\omega\mathfrak{A}/\overline{F} \notin IWs_\alpha$. For, let $x = \langle d_{\kappa,\kappa+1} : \kappa < \omega \rangle$. Then $0 < x/\overline{F} < 1$ and $\Delta(x/F) = 0$. By I.6.13, ${}^\omega\mathfrak{A}/\overline{F} \notin IWs_\alpha$.

Now we again discuss change of base (see section I.3), using ultraproducts. First we prove a sharper form of part of I.7.13, due to Andréka and Németi.

Theorem I.7.19. Assume $\alpha \geq 2$. Let $\mathfrak{A}$ be a $Ws_\alpha$ with infinite base $U$. Let $\gamma$ be a cardinal such that $|A| \cdot |U| \leq \gamma \leq \sum_{\mu<\lambda} \gamma^\mu$, where $\lambda$ is the least infinite cardinal such that $|\Delta x| < \lambda$ for all $x \in A$. Then $\mathfrak{A}$ is sub-isomorphic to a $Cs_\alpha^{reg}$ with base of power $\gamma$.

Proof. Let $\mathfrak{A}$ have unit element ${}^\alpha U^{(p)}$. Let $I = \max(\alpha,\gamma)$, and let $F$ be a $|I|$-regular ultrafilter on $I$. Introducing the notation in the proof of I.7.13, we see from that proof that $f \circ \delta$ is an isomorphism of $\mathfrak{A}$ onto a $Cs_\alpha^{reg}$ $\mathfrak{B}$, and $\tilde{\varepsilon}^* \mathfrak{A}$ is sub-isomorphic to $\mathfrak{B}$. Also note that $\mathfrak{B}$ has base $X$. By Proposition 4.3.7 of [CK] we have $|X| = |{}^I U| > \gamma$. Hence we may apply I.3.18(iv) to get a subset $W$ of $X$ such that $\varepsilon^* U \subseteq W$, $|W| = \gamma$, and such that $\mathfrak{B}$ is ext-isomorphic to a $Cs_\alpha^{reg}$ $\mathfrak{C}$ with base $W$. Thus $\tilde{\varepsilon}^* \mathfrak{A}$ is sub-isomorphic to $\mathfrak{C}$. Choose a set $W' \supseteq U$ and a one-one function $\varepsilon'$ from $W'$ onto $W$ such that $\varepsilon \subseteq \varepsilon'$. Let $\mathfrak{D} = (\varepsilon'^{-1})^{\sim *} \mathfrak{C}$ (cf. I.3.5).

Thus $\mathfrak{D} \in Cs_\alpha^{reg}$ by I.3.1 and I.3.7, and it is easily checked that $\mathfrak{A}$ is sub-isomorphic to $\mathfrak{D}$.

Using this theorem we can prove one of the basic results about set algebras, due to Henkin and Monk. First we formally give a definition used in I.5.6.

<u>Definition I.7.20</u>. For $K \in \{Cs_\alpha, Gs_\alpha, Gws_\alpha, Ws_\alpha\}$ we denote by ${}_\infty K$ the class of all $\mathfrak{A} \in K$ having all subbases infinite.

<u>Theorem</u> I.7.21. For $\alpha \geq \omega$ we have ${}_\infty Gws_\alpha \subseteq I{}_\infty Cs_\alpha$.

<u>Proof</u>. Let $\mathfrak{A}$ be a ${}_\infty Gws_\alpha$, say with unit element $\bigcup_{i\in I} V_i$, where $V_i = {}^\alpha U_i^{(p_i)}$ and $U_i$ is infinite for all $i \in I$, and $V_i \cap V_j = 0$ for all $a \in A \sim \{0\}$. For each $a \in A$ let $\mathfrak{B}_a$ be the full $Ws_\alpha$ with unit element $V_{ia}$, and set $h_a = rl^{\mathfrak{A}}_{V_{ia}}$ (recall I.6.1). Thus $h_a \in Hom(\mathfrak{A}, \mathfrak{B}_a)$ for all $a \in A$, and $h_a a \neq 0$ if $a \neq 0$. Let

$$\kappa = \bigcup_{i\in I} |U_i| \cup |A| \cup |\alpha| .$$

Now let $\langle W_a : a \in A\rangle$ be such that $2^\kappa = \bigcup_{a\in A} W_a$, $|W_a| = 2^\kappa$ for each $a \in A$, and $W_a \cap W_{a'} = 0$ for distinct $a, a' \in A$. By I.7.19, $\mathfrak{B}_a$ is isomorphic to a $Cs_\alpha$ $\mathfrak{C}_a$ with base $W_a$, for each $a \in A$; let $j_a \in Is(\mathfrak{B}_a, \mathfrak{C}_a)$ for each $a \in A$. Choose $z \in {}^A({}^\alpha(2^\kappa))$ such that $z_a \in j_a h_a a$ for each $a \in A \sim \{0\}$. For each $a \in A \sim \{0\}$ let $X_a = {}^\alpha(2^\kappa)^{(za)}$, and let $X_0 = {}^\alpha(2^\kappa) \sim \bigcup\{X_a : a \in A \sim \{0\}\}$. For every $a \in A$ let $\mathfrak{D}_a$ be the full $Crs_\alpha$ with unit element $X_a$ and set $g_a = rl_{X_a} \circ j_a \circ h_a$. Thus $g_a \in Hom(\mathfrak{A}, \mathfrak{D}_a)$ for all $a \in A$, and $g_a a \neq 0$ for $a \neq 0$. Hence $\mathfrak{A} \cong | \subseteq \mathsf{P}_{a\in A} \mathfrak{D}_a$. By I.6.2 we have $\mathsf{P}_{a\in A} \mathfrak{D}_a \in ICs_\alpha$, as desired.

Remarks I.7.22. By I.7.16 and I.7.21, $I_{\infty}Cs_{\alpha}$ is an algebraically closed class for $\alpha \geq \omega$. Thus for many purposes, cylindric set algebras with infinite bases are the simplest CA's with which to isomorphically represent abstract CA's. Andréka and Németi have improved I.7.21 by showing that any ${}_{\infty}Gws_{\alpha}$ is sub-isomorphic to a $Cs_{\alpha}$.

Before turning to results concerning change of base by increasing the cardinality of the base, algebraic analogs of the upward Löwenheim-Skolem-Tarski theorem, we give an example supplementing those of section I.3. The example, due to Andréka and Németi, shows that if $1 < \alpha < \omega$ and $\kappa < \omega$, then there is a $Cs_{\alpha}$ $\mathfrak{A}$ with base $\kappa$ such that if $\mathfrak{A} \cong \mathfrak{B} \in Cs_{\alpha}$ then the base of $\mathfrak{B}$ has power $\kappa$ (whether or not $\kappa < \alpha$); because of this example we assume in most of our theorems that the base is infinite. For each $\lambda < \kappa$ let $a_{\lambda} = \{\langle \lambda : \mu < \alpha \rangle\}$. Let $\mathfrak{A}$ be the full $Cs_{\alpha}$ with base $\kappa$. Thus $a_0, \ldots, a_{\kappa-1}$ are the distinct atoms $\leq d_{\alpha}^{\mathfrak{A}}$. Now suppose $\mathfrak{A} \cong \mathfrak{B} \in Cs_{\alpha}$, say $f \in Is(\mathfrak{A},\mathfrak{B})$. Then $fa_0, \ldots, fa_{\kappa-1}$ are distinct atoms $\leq d_{\alpha}^{\mathfrak{B}}$, so the base $U$ of $\mathfrak{B}$ has at least $\kappa$ elements. Now suppose $|U| > \kappa$. Since $d_{\alpha}^{\mathfrak{B}} = \sum_{\lambda<\kappa} fa_{\lambda}$, it follows that $|fa_{\lambda}| > 1$ for some $\lambda < \kappa$. Now $(c_0 a_{\lambda} \sim a_{\lambda}) \cap c_{(\alpha\sim 1)} a_{\lambda} = 0$, so $(c_0 fa_{\lambda} \sim fa_{\lambda}) \cap c_{(\alpha\sim 1)} fa_{\lambda} = 0$, contradiction.

The following general lemma and theorem about increasing a base are due to Andréka and Németi; their parts dealing with $Ws_{\alpha}$ and $Cs_{\alpha}^{reg}$ are due to Henkin and Monk.

Lemma I.7.23. Let $\mathfrak{A}$ be a $Gws_{\alpha}$ with base $U$ and unit element $V$. Let $F$ be a $|\alpha|$-regular ultrafilter on some set $I$. Then there is an $(F,\langle U : i \in I\rangle,\alpha)$-choice function $c$ such that, letting

$f = Rep(F, \langle U : i \in I\rangle, \alpha, \langle A : i \in I\rangle, c)$ and letting $\delta$ and $\varepsilon$ be as in I.7.12, we have:

(i) $f \circ \delta$ is an isomorphism from $\mathfrak{A}$ onto $(f \circ \delta)^* \mathfrak{A}$ ;

(ii) $\tilde{\varepsilon}^* \mathfrak{A}$ is sub-isomorphic to $(f \circ \delta)^* \mathfrak{A}$ ;

(iii) $\tilde{\varepsilon} = r\ell_{\tilde{\varepsilon}V} \circ f \circ \delta$ ;

(iv) $\tilde{\varepsilon}V = {}^{\alpha}(\varepsilon^* U) \cap f\delta V$ ;

(v) the base of $(f \circ \delta)^* \mathfrak{A}$ is ${}^{I}U/\overline{F}$ ;

(vi) $\mathfrak{A} \in K$ implies $(f \circ \delta)^* \mathfrak{A} \in K$ for all $K \in \{Ws_\alpha, Cs_\alpha, Gs_\alpha, Gws_\alpha, Gws_\alpha^{reg}\}$ .

<u>Proof</u>. Assume the hypotheses. Say $V = \bigcup_{j \in J} {}^{\alpha}Y_j^{(pj)}$ , where $0 = {}^{\alpha}Y_j^{(pj)} \cap {}^{\alpha}Y_k^{(pk)}$ for distinct $j,k \in J$ . Let $Y = \langle Y_j : j \in J\rangle$ and $X = {}^{I}U/\overline{F}$ . Let $R = \{\langle j,k\rangle \in {}^{2}J : Y_j = Y_k\}$ . Thus $R$ is an equivalence relation on $J$ . Let $K$ be a subset of $J$ having exactly one element in common with each equivalence class under $R$ . Let $L \subseteq {}^{I}K$ have exactly one element in common with each equivalence class under $\overline{F}(\langle K : i \in I\rangle)$ , with $\langle j : i \in I\rangle \in L$ for each $j \in K$ . Now we define functions $w \in {}^{X}({}^{I}U)$ and $v \in {}^{X}({}^{I}J)$ . Let $y \in X$ . Choose $k_y = k \in {}^{I}K$ such that $y \cap P_{i \in I} Y_{ki} \neq 0$ , and $k$ constant if $y = \varepsilon u$ for some $u \in U$ . Let $vy$ be the unique element of $L \cap k/\overline{F}$ . Thus $y \cap P_{i \in I} Y_{(vy)i} \neq 0$ , so we can pick $wy \in y \cap P_{i \in I} Y_{(vy)i}$ with $wy = \langle u : i \in I\rangle$ if $y = \varepsilon u$ . Now $w$ and $v$ have the following properties:

(1) $wy \in y$ for all $y \in X$ ;

(2) $w\varepsilon u = \langle u : i \in I\rangle$ for all $u \in U$ ;

(3) for all $y \in X$ and $i \in I$ we have $(wy)i \in Y_{(vy)i}$ ;

(4) if $k,q \in Rgv$ and $\{i \in I : Y_{ki} = Y_{qi}\} \in F$ , then $k = q$ .

Since $F$ is $|\alpha|$ - regular, choose $h \in {}^{I}\{\Gamma \subseteq \alpha : |\Gamma| < \omega\}$ such that $\{i \in I : \kappa \in hi\} \in F$ for all $\kappa < \alpha$ . Now let $c$ be an $(F,\langle U : i \in I\rangle,\alpha)$ - choice function such that for all $\kappa < \alpha$ , $y \in X$ , and $i \in I$ ,

$$c(\kappa,y)_i = \begin{cases} p_{(vy)i}\kappa & \text{if } \kappa \notin hi \text{ and } y \notin \varepsilon^{*}U , \\ (wy)i & \text{otherwise.} \end{cases}$$

Note that $c(\kappa,y) \in P_{i\in I}Y_{(vy)i}$ for all $\kappa < \alpha$ and $y \in X$ . Let $f,\delta,\varepsilon$ be as in the statement of the lemma. By I.7.12, (i), (ii) and (iii) hold. For (iv), use (2) and the definition of c. To prove (v), let Z be the base of $(f \circ \delta)^{*}\mathfrak{A}$ ; we are to show that $Z = X$ , and it is obvious that $Z \subseteq X$ . Suppose $y \in \varepsilon^{*}U$ ; say $y = \varepsilon u$ with $u \in U$ ; in fact say $u \in Y_j$ with $j \in J$ . Now let $q = \langle \varepsilon p_j\kappa : \kappa < \alpha\rangle^{0}_{y}$ . Thus $q \in {}^{\alpha}X$ , and for any $i \in I$ we have $(c^{+}q)_i = (p_j)^{0}_{u} \in {}^{\alpha}Y_j{}^{(p_j)} \subseteq V$ using (2) and the definition of $c$ . It follows that $q \in f\delta V$ , and hence $y \in Z$ . Now let $y \in X \sim \varepsilon^{*}U$ . Let $q = \langle y : \kappa < \alpha\rangle$ . Then for any $i \in I$ ,

$$\begin{aligned}(c^{+}q)_i &= \langle c(\kappa,y)_i : \kappa < \alpha\rangle \\ &= (\alpha \sim hi)1 p_{(vy)i} \cup \langle (wy)i : \kappa \in hi\rangle \\ &\in {}^{\alpha}Y_{(vy)i}{}^{(p(vy)i)} \subseteq V .\end{aligned}$$

Hence again $q \in f\delta V$ and $y \in Z$ . So (v) holds.

Now we turn to the parts of (vi) . The cases $K = Gws_\alpha$ and $K = Cs_\alpha$ are taken care of by I.7.4. If $\mathfrak{A}$ is regular, then so is $(f \circ \delta)^{*}\mathfrak{A}$ by I.7.6, since $f \circ \delta$ is an isomorphism. Next suppose

$K = Gs_\alpha$ . Thus for all $j,k \in J$ we have $Y_j = Y_k$ or $Y_j \cap Y_k = 0$ . For each $r \in Rgv$ we set $Q_r = P_{i\in I}Y_{ri}/\overline{F}^U$ . Now

(5) if $r,r' \in Rgv$ and $r \neq r'$ , then $Q_r \cap Q_{r'} = 0$ .

For, suppose $y \in P_{i\in I}Y_{ri}$ , $z \in P_{i\in I}Y_{r'i}$ , and $\{i \in I : y_i = z_i\} \in F$ . Then $\{i \in I : Y_{ri} = Y_{r'i}\} \in F$ , so by (4) $r = r'$ .

(6) $f\delta V \subseteq \bigcup_{r\in Rgv} {}^\alpha Q_r$ .

For, let $q \in f\delta V$ . Thus the set $M = \{i \in I : (c^+q)_i \in V\}$ is in $F$ . We claim that $q \in {}^\alpha Q_{vq0}$ . For each $i \in M$ choose $ji \in J$ so that $(c^+q)_i \in {}^\alpha Y_{ji}$ . Now for any $i \in M$ and $\kappa < \alpha$ we have $(c^+q)_i\kappa \in Y_{ji}$ and also $(c^+q)_i\kappa = c(\kappa,q\kappa)_i \in Y_{(vq\kappa)i}$ by the note following the definition of $c$ , so $Y_{ji} = Y_{(vq\kappa)i}$ . So for any $i \in M$ and $\kappa < \alpha$ we have $c(\kappa,q\kappa)_i \in Y_{(vq\kappa)i} = Y_{ji} = Y_{(vq0)i}$ . Thus $q\kappa \in Q_{vq0}$ for any $\kappa < \alpha$ , as desired.

(7) If $r \in Rgv$ , $\kappa < \alpha$ , and $y \in Q_r$ , then $c(\kappa,y) \in P_{i\in I}Y_{ri}$ .

For, choose $z \in y \cap P_{i\in I}Y_{ri}$ . By the remark after the definition of $c$ we also have $c(\kappa,y) \in y \cap P_{i\in I}Y_{(vy)i}$ . Thus the set $M = \{i \in I : z_i = c(\kappa,y)_i\}$ is in $F$ . Since $M \subseteq \{i \in I : Y_{ri} = Y_{(vy)i}\}$ , we infer from (4) that $vy = r$ , so (7) follows.

(8) For all $r \in Rgv$ we have ${}^\alpha Q_r \subseteq f\delta V$ .

In fact, let $q \in {}^\alpha Q_r$ . By (7) we have, for all $\kappa < \alpha$ , $c(\kappa,q\kappa) \in P_{i\in I}Y_{ri}$ . Hence for all $i \in I$ , $(c^+q)_i = \langle c(\kappa,q\kappa)_i : \kappa < \alpha\rangle \in {}^\alpha Y_{ri} \subseteq V$ . Hence $q \in f\delta V$ .

By (5), (6), (8) we have $(f\circ\delta)^*\mathfrak{A} \in Gs_\alpha$ .

In the case $K = Ws_\alpha$ we redefine $c$ . Let $\mathfrak{A} \in Ws_\alpha$ , say $V = {}^\alpha U^{(p)}$ . We may assume that $|U| > 1$ . For each $u \in U$ and $\kappa < \alpha$ let $c(\kappa,\varepsilon u) = \langle u : i \in I\rangle$ . Now suppose $y \in X \sim \varepsilon^*U$ and $\kappa < \alpha$ . Choose $k_y \in y$ and let $Z_y = \{i \in I : k_y i \neq p\kappa\}$ . Thus $Z_y \in F$ since $y \neq \varepsilon p\kappa$ . Let $c(\kappa,y) = Z_y 1 k_y \cup \langle u_y : i \in \alpha \sim Z\rangle$ , where $u_y \in U \sim \{p\kappa\}$ . Thus $c$ is an $(F,\langle U : i \in I\rangle,\alpha)$-choice function, with the additional property

(9) for all $\kappa < \alpha$ and all $y \in X \sim \{\varepsilon p\kappa\}$ we have $p\kappa \notin Rg\, c(\kappa,y)$ .

Now the hypotheses of I.7.12 clearly hold, so (i) - (iv) are true. To establish (v) and $(f\circ\delta)^*\mathfrak{A} \in Ws_\alpha$ it suffices to show that $f\delta V = {}^\alpha X^{(\varepsilon\circ p)}$ . First suppose $q \in {}^\alpha X^{(\varepsilon\circ p)}$ . Let $i \in I$ . Then $(c^+q)_i \in {}^\alpha U$ , obviously. Let $\Gamma = \{\kappa < \alpha : q\kappa \neq \varepsilon p\kappa\}$ . Then $\Gamma$ is finite, and for $\kappa \in \alpha \sim \Gamma$ we have $(c^+q)_i\kappa = c(\kappa,q\kappa)_i = c(\kappa,\varepsilon p\kappa)_i = p\kappa$ . Thus $(c^+q)_i \in {}^\alpha U^{(p)} = V$ . Hence $q \in f\delta V$ .

Conversely, suppose $q \in {}^\alpha X \sim {}^\alpha X^{(\varepsilon\circ p)}$ ; we show that $q \notin f\delta V$ . Let $\Gamma = \{\kappa < \alpha : q\kappa \neq \varepsilon p\kappa\}$ ; thus $\Gamma$ is infinite. For any $\kappa \in \Gamma$ and $i \in I$ we have $c(\kappa,q\kappa)_i \neq p\kappa$ by (9) . Therefore $(c^+q)_i \notin {}^\alpha U^{(p)} = V$ for all $i \in I$ , and consequently $q \notin f\delta V$ .

This completes the proof of I.7.23.

This lemma immediately gives

<u>Theorem</u> I.7.24. Let $\mathfrak{A}$ be a $Gws_\alpha$ with an infinite base, and let $\kappa$ be any cardinal. Then $\mathfrak{A}$ is sub-isomorphic to a $Gws_\alpha$ $\mathfrak{B}$ with base of cardinality $\geq \kappa$ , such that $1^{\mathfrak{A}} = 1^{\mathfrak{B}} \cap {}^\alpha W$ for some $W$ .

Moreover, if $\mathfrak{A} \in K \in \{Ws_\alpha, Cs_\alpha, Gs_\alpha, Gws_\alpha, Gws_\alpha^{reg}, Cs_\alpha^{reg}, Gs_\alpha^{reg}\}$ then also $\mathfrak{B} \in K$.

Using I.3.18 we easily obtain the following more specific result.

Theorem I.7.25. Let $\mathfrak{A}$ be a $Gws_\alpha$ with an infinite base $U$, and suppose that $|A| \cup |U| \leq \kappa$. Then $\mathfrak{A}$ is sub-isomorphic to a $Gws_\alpha$ $\mathfrak{B}$ with base of cardinality $\kappa$, such that $1^{\mathfrak{A}} = 1^{\mathfrak{B}} \cap {}^\alpha W$ for some $W$. Moreover:

(i) if $\mathfrak{A} \in K \in \{Ws_\alpha, Gws_\alpha, Gs_\alpha, Gws_\alpha^{reg}, Gs_\alpha^{reg}\}$, then $\mathfrak{B} \in K$;

(ii) if $\mathfrak{A} \in Cs_\alpha$ (resp. $\mathfrak{A} \in Cs_\alpha^{reg}$) and $\kappa = \kappa^{|\alpha|}$, then $\mathfrak{B} \in Cs_\alpha$ (resp. $\mathfrak{B} \in Cs_\alpha^{reg}$).

Proof. The parts concerning $Ws_\alpha$, $Gws_\alpha$, $Gws_\alpha^{reg}$, $Cs_\alpha$ and $Cs_\alpha^{reg}$ are immediate from I.3.16, I.3.18 and I.7.24. For the other parts dealing with $Gs_\alpha$ and $Gs_\alpha^{reg}$ we use a direct construction, not involving ultraproducts (which actually works for some other classes). Suppose $\mathfrak{A}$ is a $Gws_\alpha$, say with base $U$ and unit element $V = \bigcup_{j\in J} {}^\alpha Y_j^{(pj)}$, where ${}^\alpha Y_j^{(pj)} \cap {}^\alpha Y_k^{(pk)} = 0$ for distinct $j,k \in J$. Let $\langle Z_\beta : \beta < \kappa\rangle$ be a system of pairwise disjoint sets, with $|Z_\beta| = |U|$ for all $\beta < \kappa$, and let $f_\beta$ be a one-to-one function mapping $U$ onto $Z_\beta$ for each $\beta < \kappa$; further, we assume that $Z_0 = U$ and $f_0 = U \upharpoonright Id$. Then for distinct $\langle\beta,j\rangle, \langle\gamma,k\rangle \in \kappa \times J$ we have ${}^\alpha(f_\beta^* Y_j)^{(f\beta\circ pj)} \cap {}^\alpha(f_\gamma^* Y_k)^{(f\gamma\circ pk)} = 0$. Let $\mathfrak{B}$ be the full $Gws_\alpha$ with unit element $\bigcup_{\beta<\kappa}\bigcup_{j\in J} {}^\alpha(f_\beta^* Y_j)^{(f\beta\circ pj)}$. For each $x \in A$ let $gx = \bigcup_{\beta<\kappa}\{y \in {}^\alpha Z_\beta : f_\beta^{-1} \circ y \in x\}$. It is easily checked that $g$ is an isomorphism of $\mathfrak{A}$ onto a subalgebra $\mathfrak{C}$ of $\mathfrak{B}$, and in fact $g^{-1}c = V \cap c$ for each $c \in C$, i.e., $\mathfrak{A}$ is sub-isomorphic to $\mathfrak{C}$. Moreover, the base of $\mathfrak{C}$ is of power $\kappa$. In case $\mathfrak{A} \in Gs_\alpha$, clearly $\mathfrak{C} \in Gs_\alpha$. Finally, suppose $\mathfrak{A} \in Gws_\alpha^{reg}$. Suppose $x \in A$, $y \in gx$, $z \in gV$, and $(\Delta gx \cup 1)1y = (\Delta gx \cup 1)1z$. Say $y \in {}^\alpha Z_\beta$ and $f_\beta^{-1} \circ y \in X$, while $z \in {}^\alpha Z_\gamma$ and $f_\gamma^{-1} \circ z \in V$. Since $y0 = z0$ we have $\beta = \gamma$. Thus since $\Delta x = \Delta gx$, because $g$ is an isomorphism, we get $f_\beta^{-1} \circ z \in x$ by regularity of $x$ and hence $z \in gx$.

For $Gws_\alpha$'s in general it seems to be more interesting to increase the size of various subbases rather than to merely increase the size of the base; that is the purpose of our next theorem.

<u>Theorem</u> I.7.26. Let $\mathfrak{A}$ be a $Gws_\alpha$ with base $U$ and unit element $V = \bigcup_{j\in J} {}^\alpha Y_j^{(pj)}$, where ${}^\alpha Y_j^{(pj)} \cap {}^\alpha Y_k^{(pk)} = 0$ for distinct $j,k \in J$. Let $\kappa$ be a cardinal-number-valued function with domain $J$ such that $\kappa_j \geq (|A| \cap 2^{|\alpha| \cup |Y_j|}) \cup \omega$ and $\kappa_j \geq |Y_j| \geq \omega$ whenever $j \in J$ and $\kappa_j \neq |Y_j|$.

Then $\mathfrak{A}$ is sub-isomorphic to a $Gws_\alpha$ $\mathfrak{B}$ with unit element $\bigcup_{j\in J} {}^\alpha W_j^{(pj)}$, with ${}^\alpha W_j^{(pj)} \cap {}^\alpha W_k^{(pk)} = 0$ for distinct $j,k \in J$, where $Y_j = W_j$ for all $j \in J$ for which $\kappa_j = |Y_j|$, and $W_j \supseteq Y_j$ with $|W_j| = \kappa_j$ for all $j \in J$ such that $\kappa_j > |Y_j|$. Furthermore, $V = {}^\alpha U \cap 1^{\mathfrak{B}}$.

<u>Proof</u>. By I.6.2 we have $\mathfrak{A} \cong | \subseteq P_{j\in J} \mathfrak{B}_j$, where $\mathfrak{B}_j$ is a $Ws_\alpha$ with unit element ${}^\alpha Y_j^{(pj)}$ for each $j \in J$; in fact, the isomorphism $h$ is given by $(ha)_j = a \cap {}^\alpha Y_j^{(pj)}$ for all $a \in A$ and $j \in J$. Note that $|Y_j| \geq \omega$ and

$$|B_j| \leq |A| \cap 2^{|\alpha| \cup |Y_j|} \leq \kappa_j$$

for each $j \in J$ for which $|Y_j| \neq \kappa_j$. Hence by I.7.25, $\mathfrak{B}_j$ is sub-isomorphic to a $Ws_\alpha \mathfrak{C}_j$ with unit element ${}^\alpha W_j^{(pj)}$ such that $\mathfrak{B}_j = \mathfrak{C}_j$ and hence $Y_j = W_j$ if $\kappa_j = |Y_j|$, while $|W_j| = \kappa_j$ for all $j \in J$. We may assume that $W_j \cap W_k = Y_j \cap Y_k$ and hence ${}^\alpha W_j^{(pj)} \cap {}^\alpha W_k^{(pk)} = 0$ for distinct $j,k \in J$. For each $j \in J$ let $f_j = \langle c \cap {}^\alpha Y_j^{(pj)} : c \in C_j \rangle$; thus $f_j \in Is(\mathfrak{C}_j, \mathfrak{B}_j)$. Finally, let for each $a \in A$

$$ga = \bigcup_{j \in J} f_j^{-1}(ha)_j \ ;$$

then $g$ is an isomorphism from $\mathfrak{A}$ onto a $Gws_\alpha \mathfrak{D}$ with unit element $\bigcup_{j \in J} {}^\alpha W_j^{(pj)}$ (as is easily checked), and $g^{-1}d = V \cap d$ for all $d \in D$, as desired.

We now return to a question discussed in I.3.20-I.3.22: changing the function $p$ in the unit element ${}^\alpha U^{(p)}$ of a $Ws_\alpha$. The following theorem of Andréka and Németi strengthens results of Henkin and Monk.

<u>Theorem</u> I.7.27. Suppose $\alpha \geq \omega$. Let $\mathfrak{A}$ be a $Ws_\alpha$ with unit element ${}^\alpha U^{(p)}$, and let $q \in {}^\alpha U$. Then $\mathfrak{A}$ is homomorphic to a $Ws_\alpha$ with unit element ${}^\alpha Y^{(q)}$ with $U \subseteq Y$, where if $|U| < \omega$ or $|U \sim Rgq| = |U| \geq |A|$ then one may take $Y = U$.

<u>Proof</u>. We may assume that $|U| > 1$. Let $I = \{\Gamma \subseteq \alpha : |\Gamma| < \omega\}$, and let $F$ be an ultrafilter on $I$ such that $\{\Gamma \in I : \Delta \subseteq \Gamma\} \in F$ for all $\Delta \in I$. Let $X = {}^I U/\overline{F}$. Choose $k \in {}^\alpha U$ so that $k\kappa \neq p\kappa$ for all $\kappa < \alpha$. Let $\varepsilon$ be as in I.7.12. Then there is an $(F, \langle U : i \in I \rangle, \alpha)$-choice function $c$ such that for all $y \in X$, $\Gamma \in I$, and $\kappa \in \alpha \sim \Gamma$,

$$c(\kappa,y)_\Gamma = \begin{cases} p\kappa & \text{if } y = \varepsilon q\kappa , \\ k\kappa & \text{otherwise} . \end{cases}$$

Now let $f = \text{Rep}(F,\langle U : i \in I\rangle,\alpha,\langle A : i \in I\rangle,c)$ . Thus by I.7.2, $f$ is a homomorphism from ${}^I\mathfrak{A}/\overline{F}$ onto some $\text{Crs}_\alpha\, \mathfrak{B}$ . Since the function $\delta$ of I.7.12 is an isomorphism of $\mathfrak{A}$ into ${}^I\mathfrak{A}/\overline{F}$ , $f \circ \delta \in \text{Hom}(\mathfrak{A},\mathfrak{B})$ . Now if $h \in {}^\alpha X$ and $\Gamma \in I$ , then $\{\kappa \in \alpha \sim \Gamma : (c^+h)_\Gamma \kappa \neq p\kappa\} = \{\kappa \in \alpha \sim \Gamma : c(\kappa,h\kappa)_\Gamma \neq p\kappa\} = \{\kappa \in \alpha \sim \Gamma : h\kappa \neq \varepsilon q\kappa\}$ , so $(c^+h)_\Gamma \in {}^\alpha U^{(p)}$ iff $h \in {}^\alpha X^{(\varepsilon \circ q)}$ . Thus $\mathfrak{B}$ is a $\text{Ws}_\alpha$ with unit element ${}^\alpha X^{(\varepsilon\circ q)}$ . Choose $Y \supseteq U$ together with a function $\ell$ mapping $Y$ one-to-one onto $X$ such that $\varepsilon \subseteq \ell$ . By I.3.1, $\ell^{-1}$ induces an isomorphism $t$ from $\mathfrak{B}$ onto a $\text{Ws}_\alpha$ with unit element ${}^\alpha Y^{(q)}$ . Thus $t \circ f \circ \delta$ is the desired homomorphism.

If $|U| < \omega$ , then $|U| = |X|$ and hence $Y = U$ . Assume now $|U \sim Rgq| = |U| \geq |A|$ . By I.3.18(ii), $t^*f^*\delta^*\mathfrak{A}$ is ext-isomorphic to a $\text{Ws}_\alpha$ with unit element ${}^\alpha W^{(q)}$ for some $W$ with $|U| = |W|$ , $U \subseteq W$ . Thus there is a one-to-one function $s$ mapping $W$ onto $U$ such that $s \circ q = q$ . By I.3.1, $s$ induces an isomorphism $u$ from $t^*f^*\delta^*\mathfrak{A}$ onto a $\text{Ws}_\alpha$ with unit element ${}^\alpha U^{(q)}$ , and hence $u \circ t \circ f \circ \delta$ is the desired homomorphism for the last part of the theorem.

Our next theorem, due to Henkin and Monk, is related to I.7.21. Given a $\text{Gws}_\alpha$ with all subbases finite, it is not always isomorphic to a $\text{Cs}_\alpha$ ; see I.6.8(5). It is natural, however, to try to reduce the number of subbases.

<u>Theorem</u> I.7.28. Let $\kappa < \omega \leq \alpha$ . Suppose that $\mathfrak{A}$ is a $\text{Gws}_\alpha$ with unit element $V$ such that every subbase of $\mathfrak{A}$ is of power $\kappa$ .

Let $\lambda$ be the least cardinal such that for each equivalence relation $R$ on $\alpha$ having all equivalence classes infinite the following inequality holds:

$$|\{q \in V : q|q^{-1} = R\}| \leq \lambda \cdot |\{q \in {}^{\alpha}\kappa : q|q^{-1} = R\}| .$$

Then $\mathfrak{A}$ is isomorphic to a $Gs_\alpha$ with $\lambda$ subbases.

Proof. Let $\mathfrak{A}$ have unit element $V = \bigcup_{i\in I} {}^{\alpha}U_i^{(pi)}$, where ${}^{\alpha}U_i^{(pi)} \cap {}^{\alpha}U_j^{(pj)} = 0$ for distinct $i,j \in I$. Now set $\mathfrak{R} = \{q|q^{-1} : q \in {}^{\alpha}\kappa\}$ and $\mathfrak{R}^{\infty} = \{R \in \mathfrak{R} : \text{all equivalence classes of } R \text{ are infinite}\}$. Now we define a relation $\equiv$ on $\mathfrak{R}$ by setting $R \equiv R'$ iff there exist $q,q' \in {}^{\alpha}\kappa$ such that $R = q|q^{-1}$, $R' = q'|q'^{-1}$, and $|\{\xi : \xi < \alpha \text{ and } q\xi \neq q'\xi\}| < \omega$.

(1) $\equiv$ is an equivalence relation on $\mathfrak{R}$.

Indeed, only the transitivity of $\equiv$ is questionable. Suppose $R \equiv R' \equiv R''$, say $q,q',r',r'' \in {}^{\alpha}\kappa$ and $R = q|q^{-1}$, $R' = (q'|q'^{-1}) = (r'|r'^{-1})$, $R'' = r''|r''^{-1}$ and $q \in {}^{\alpha}\kappa^{(q')}$, $r'' \in {}^{\alpha}\kappa^{(r')}$. Then there is a one-to-one function $k$ from $\kappa$ onto $\kappa$ such that $k \circ q' = r'$, since $\kappa < \omega$. Thus $R = (k \circ q)|(k \circ q)^{-1}$ and $k \circ q \in {}^{\alpha}\kappa^{(r'')}$, so $R \equiv R''$. Similarly we have

(2) if $R \equiv R'$, $|U| = \kappa$, $q \in {}^{\alpha}U$, and $R = q|q^{-1}$, then there is a $q' \in {}^{\alpha}U^{(q)}$ such that $R' = q'|q'^{-1}$.

The following statement is clear.

(3) If $R \in \mathfrak{R}^{\infty}$, $|U| = \kappa$, and $q \in {}^{\alpha}U$, then there is at most one

$q' \in {}^{\alpha}U(q)$ such that $q'|q'^{-1} = R$ .

(4) For every $R \in \mathfrak{R}$ there is an $R' \in \mathfrak{R}^{\infty}$ such that $R \equiv R'$ .

For, say $R = q|q^{-1}$ , where $q \in {}^{\alpha}\kappa$ . Let $\Gamma = \bigcup\{\beta/R : \beta < \alpha$ and $|\beta/R| < \omega\}$ . Since $\kappa < \omega$ we have $|\alpha/R| < \omega$ and hence $|\Gamma| < \omega$ . Choose $\beta \in \alpha \sim \Gamma$ , and let $q' = (\alpha \sim \Gamma)1q \cup \langle q\beta : \beta \in \Gamma\rangle$ ; $q'$ is as desired.

Now by (1) - (4) there is a function $R \in {}^{I}\mathfrak{R}^{\infty}$ such that $p_i|p_i^{-1} = R_i$ for all $i \in I$ , and $R_i = R_j$ whenever $p_i|p_i^{-1} = p_j|p_j^{-1}$ .

Now let $\langle Y_\beta : \beta < \lambda\rangle$ be a system of pairwise disjoint sets, each of power $\kappa$ . Set $W = \bigcup_{\beta<\lambda} {}^{\alpha}Y_\beta$ . We define $q \sim q'$ iff there is a $\beta < \lambda$ such that $q \in {}^{\alpha}Y_\beta$ and $q' \in {}^{\alpha}Y_\beta^{(q)}$ . Thus $\sim$ is an equivalence relation on $W$ , and the $\sim$-classes are weak spaces. Let $K = W/\sim$ . Next, for each $i \in I$ let $L_i = \{S \in K :$ there is a $q \in S$ with $q|q^{-1} = R_i\}$ . Then by the choice of $R$ ,

(5) if $i,j \in I$ and $R_i \neq R_j$ , then $L_i \cap L_j = 0$;

(6) if $i \in I$ , then $|i/(R|R^{-1})| \leq |L_i|$ .

In fact, using finally the hypothesis of the theorem, and (3),

$$\begin{aligned} |i/(R|R^{-1})| &= |\{q \in V : q|q^{-1} = R_i\}| \\ &\leq |\{q \in W : q|q^{-1} = R_i\}| \\ &= |L_i| \ . \end{aligned}$$

By (6) we can choose a one-to-one $S \in {}^{I}K$ such that $S_i \in L_i$ for all $i \in I$ . Then

(7) for all $i \in I$ there is a $q \in S_i$ such that $q|q^{-1} = p_i|p_i^{-1}$ .

This is true since $p_i|p_i^{-1} = R_i$ and $S_i \in L_i$ . Now let $m$ map $K$ onto $I$ such that $mSi = i$ for all $i \in I$ .

By I.6.2 we have $\mathfrak{A} \cong| \subseteq P_{i\in I}\mathfrak{B}_i$ , where $\mathfrak{B}_i$ is a $Ws_\alpha$ with unit element ${}^{\alpha}U_i^{(pi)}$ for each $i \in I$ . For each $T \in K$ let $\mathfrak{C}_T$ be the full $Ws_\alpha$ with unit element $T$ . By I.6.2 $P_{T\in K}\mathfrak{C}_T$ is isomorphic to a $Gs_\alpha$ with unit element $W$ , which has $\lambda$ subbases. Thus to prove the theorem it suffices to show that $P_{i\in I}\mathfrak{B}_i \cong| \subseteq P_{T\in K}\mathfrak{C}_T$ . First we note:

(8) for every $T \in K$ there is a homomorphism from $\mathfrak{B}_{mT}$ into $\mathfrak{C}_T$ .

For, say $T = {}^{\alpha}Y_\beta^{(q)}$ . Let $b \in {}^{UmT}Y_\beta$ be one-to-one and onto, and set $r = b^{-1} \circ q$ . By I.7.27 let $h$ be a homomorphism from $\mathfrak{B}_{mT}$ onto a $Ws_\alpha$ $\mathfrak{B}'$ with unit element ${}^{\alpha}U_{mT}^{(r)}$ . Then $\tilde{b} \circ h$ is as desired in (8) , where $\tilde{b}$ is defined in I.3.5.

(9) For every $i \in I$ there is an isomorphism from $\mathfrak{B}_i$ into $\mathfrak{C}_{Si}$ .

For, by (7) choose $k \in S_i$ such that $k|k^{-1} = p_i|p_i^{-1}$ . Thus $W_{Si} = {}^{\alpha}Y_\beta^{(k)}$ for some $\beta < \lambda$ . There is a one-to-one function $b$ from $U_i$ onto $Y_\beta$ such that $b \circ p_i = k$ . Thus $\tilde{b}$ is the desired isomorphism, by I.3.5 and I.3.1 .

By (8) and (9) there is a function $h$ with domain $K$ such that $h_T \in Hom(\mathfrak{B}_{mT},\mathfrak{C}_T)$ for any $T \in K$ and $h_{Si} \in Ism(\mathfrak{B}_i,\mathfrak{C}_{Si})$ for each $i \in I$ . It follows easily that

$$g = \langle\langle h_T x_{mT} : T \in K\rangle : x \in P_{i\in I} B_i\rangle$$

is the desired isomorphism from $P_{i\in I}\mathfrak{B}_i$ into $P_{T\in K}\mathfrak{C}_T$ .

Corollary I.7.29. Let $\kappa < \omega \leq \alpha$ . Suppose that $\mathfrak{A}$ is a $Gws_\alpha$ with unit element $V = \bigcup_{i\in I} {}^{\alpha}U_i^{(pi)}$ , where ${}^{\alpha}U_i^{(pi)} \cap {}^{\alpha}U_j^{(pj)} = 0$ for distinct $i,j \in I$ . Assume that $|U_i| = \kappa$ for all $i \in I$ , and $|I| \leq \kappa$ . Then $\mathfrak{A}$ is isomorphic to a $Cs_\alpha$ .

Proof. We may assume that $\kappa > 1$ . Let R be an equivalence relation on $\alpha$ having every equivalence class infinite. Then $|\{q \in V : q|q^{-1} = R\}| \leq |I|$ , since if $q|q^{-1} = R = q'|q'^{-1}$ with $q,q' \in V$ , and $q \neq q'$ , then $\{\beta < \alpha : q\beta \neq q'\beta\}$ is infinite and so $q \in {}^{\alpha}U_i^{(pi)}$ , $q' \in {}^{\alpha}U_j^{(pj)}$ for distinct $i,j \in I$ . Thus by I.7.28 it suffices to check

(*) if R is an equivalence relation on $\alpha$ and $|\alpha/R| \leq \kappa$ , then the set $\{{}^{\alpha}\kappa^{(q)} : q|q^{-1} = R\}$ has at least $\kappa$ elements.

To prove (*) , first choose $q \in {}^{\alpha}\kappa$ such that $q|q^{-1} = R$ . Let f be the permutation of $\kappa$ such that $f\lambda = \lambda + 1$ for all $\lambda < \kappa - 1$ , while $f(\kappa - 1) = 0$ . Then for all distinct $\lambda,\mu < \kappa$ we clearly have $f^\lambda \circ q \notin {}^{\alpha}\kappa^{(f^\mu \circ q)}$ , while $(f^\lambda \circ q)|(f^\lambda \circ q)^{-1} = R$ , so (*) follows.

Remarks I.7.30. (These remarks are due to Andréka and Németi.)
(a) The special hypotheses in I.7.19 are necessary. Namely, if $\omega \leq |U| \leq \alpha$ then there is a $Ws_\alpha$ $\mathfrak{A}$ with base U such that $|A| \leq \alpha$ and any $\mathfrak{B} \in Cs_\alpha \cap H\mathfrak{A}$ either has only one element or else has base of power $> \alpha$ . To construct such a $Ws_\alpha$ , let $|U| = \kappa$ . Let $w \in {}^{U}\kappa$ be one-to-one and onto. Let $p = \langle 0 : \mu < \alpha \rangle$ , and let $\mathfrak{C}$ be the full $Ws_\alpha$ with unit element $V = {}^{\alpha}U^{(p)}$ . We now construct $a \in {}^{\kappa}C$ . For every $\lambda < \kappa$ let $k_\lambda$ be a one-to-one function mapping

$\kappa$ into $\{v \in U : \lambda < wv\}$ . Then for each $\lambda < \kappa$ let

$$a_\lambda = \{q \in V : (k\ \max\{wq_\mu : 0 < \mu < \alpha\})\lambda = q_0\} .$$

Now if $0 < \mu < \alpha$ , $\lambda < \kappa$ , and $q \in a_\lambda$ , then

$$wq_\mu \le \max\{wq_\nu : 0 < \nu < \alpha\} < w(k\ \max\{wq_\nu : 0 < \alpha\})\lambda = wq_0 ,$$

so $q_\mu \ne q_0$ . Thus

(1) if $0 < \mu < \alpha$ and $\lambda < \kappa$ , then $a_\lambda \subseteq -d_{0\mu}$ .

We also clearly have

(2) $c_0 a_\lambda = 1$ for all $\lambda < \kappa$ ;

(3) $a_\lambda \cap a_\mu = 0$ for distinct $\lambda, \mu < \kappa$ .

Let $\mathfrak{A} = \mathfrak{S}\mathfrak{g}^{(\mathfrak{C})}\{a_\lambda : \lambda < \kappa\}$ . Clearly $|A| \le \alpha$ . Now we claim

(4) if $\mathfrak{B} \in Ws_\alpha \cap H\mathfrak{A}$ with unit element ${}^{\alpha}Y^{(q)}$ , then $|Y \sim Rgq| \ge \kappa$ .

For, suppose that $f$ is a homomorphism from $\mathfrak{A}$ onto $\mathfrak{B}$ . Thus the conditions (1) - (3) hold with $\langle a_\lambda : \lambda < \kappa\rangle$ replaced by $\langle fa_\lambda : \lambda < \kappa\rangle$ . For each $\lambda < \kappa$ choose $y_\lambda \in Y$ so that $q^0_{y\lambda} \in fa_\lambda$ , using (2) for $fa_\lambda$ . Thus $y$ is one-to-one by (3) for $fa_\lambda$ and $fa_\mu$ . Furthermore, $y\lambda \notin Rgq$ for $0 < \lambda < \kappa$ by (1) for $fa_\lambda$ , if $y\lambda \ne q0$ . So (4) holds.

Now suppose $h \in Hom(\mathfrak{A},\mathfrak{C})$ for some $Cs_\alpha$ $\mathfrak{C}$ with base $W \ne 0$ , and assume that $|W| \le \alpha$ . Let $q$ map $\alpha$ onto $W$ , and set $V = {}^{\alpha}W^{(q)}$ . Then $r\ell_V^{\mathfrak{C}} \circ h \in Hom(\mathfrak{A},\mathfrak{B})$ , where $\mathfrak{B}$ is a $Ws_\alpha$ with unit element $V$ , contradicting (4) .

(b) In I.7.27 one cannot replace "homomorphic" by "isomorphic". In fact, let $\mathfrak{A}$ be the $Ws_\alpha$ with unit element ${}^\alpha U^{(p)}$ generated by $\{p\}$, where $U = \alpha + \alpha$ and $p = \langle 0 : \kappa < \alpha \rangle$, and let $\mathfrak{B}$ be the full $Ws_\alpha$ with unit element ${}^\alpha U^{(q)}$, where $q = \langle \kappa : \kappa < \alpha \rangle$. Suppose $f \in Ism(\mathfrak{A},\mathfrak{B})$. Since $\{p\} \subseteq d_{\kappa\lambda}$ for all $\kappa,\lambda < \alpha$, there is an $r \in f\{p\}$ with $r \in d_{\kappa\lambda}$ for all $\kappa,\lambda < \alpha$. Since $r \in {}^\alpha U^{(q)}$, this is impossible.

(c) By the argument for (a)(4) above, the condition $|U \sim Rgq| = |U|$ in I.7.27 is necessary.

(d) It is not known if the condition $|U| \geq |A|$ in I.7.27 is needed.

(e) The hypothesis on $\lambda$ in I.7.28 is in a sense best possible. Namely, suppose that $\mathfrak{A}$ is a $Gws_\alpha$ with unit element $V$ such that every subbase of $\mathfrak{A}$ is of power $\kappa$, and that $\mathfrak{A}$ is isomorphic to a $Gs_\alpha$ $\mathfrak{B}$ with $\lambda$ subbases; assume in addition that $\{q\} \in A$ for all $q \in V$; we show that the indicated inequality holds. Let $h \in Is(\mathfrak{A},\mathfrak{B})$, and let $W$ be the unit element of $\mathfrak{B}$. Let $R$ be an equivalence relation on $\alpha$ with all equivalence classes infinite. If $q \in V$ and $q|q^{-1} = R$, then $\langle \kappa,\lambda \rangle \in R$ implies that $\{q\} \subseteq d_{\kappa\lambda}$ and hence $p\kappa = p\lambda$ for each $p \in h\{q\}$; and similarly $p\kappa \neq p\lambda$ whenever $\langle \kappa,\lambda \rangle \notin R$ and $p \in h\{q\}$. So $q \in V$ and $q|q^{-1} = R$ imply that $p|p^{-1} = R$ for all $p \in h\{q\}$. Thus $\langle h\{q\} : q \in V, R = q|q^{-1} \rangle$ is a system of pairwise disjoint non-empty subsets of $\{q \in W : q|q^{-1} = R\}$. Hence

$$|\{q \in V : q|q^{-1} = R\} \leq |\{q \in W : q|q^{-1} = R\}|$$
$$= \lambda \cdot |\{q \in {}^\alpha\kappa : q|q^{-1} = R\}| .$$

(f) The assumption $|I| \leq \kappa$ in I.7.29 cannot be improved, by (e) upon considering $R = \alpha \times \alpha$.

(g) Andréka and Németi have proved the following algebraic version of the various logical theorems to the effect that elementarily equivalent structures have isomorphic elementary extensions: Let $\mathfrak{A}, \mathfrak{B} \in {}_\infty Cs_\alpha$ and $\mathfrak{A} \equiv \mathfrak{B}$. Then $\mathfrak{A}$ and $\mathfrak{B}$ are sub-isomorphic to $Cs_\alpha$'s $\mathfrak{A}'$ and $\mathfrak{B}'$ respectively such that $\mathfrak{A}'$ and $\mathfrak{B}'$ are base-isomorphic.

## 8. Reducts

We restrict ourselves in this section to the most basic results about reducts. A more detailed study is found in Andréka, Németi [AN3] to which we also refer for the statement of various open questions.

Lemma I.8.1. Let $\mathfrak{A}$ be a $Crs_\beta$ with base $U$ and unit element $V$. Let $\alpha$ be an ordinal and let $\rho \in {}^\alpha\beta$ be one-to-one. Fix $x \in X \in A$. For each $y \in {}^\alpha U$ set

$$y^+ = ((\beta \sim Rg\rho) 1 x) \cup (y \circ \rho^{-1}) ;$$

thus $y^+ \in {}^\beta U$. For all $Y \in A$ let $fY = \{y \in {}^\alpha U : y^+ \in Y\}$.

Then $f$ is a homomorphism of $\mathfrak{Rd}^{(\rho)}\mathfrak{A}$ into a $Crs_\alpha$, and $fX \neq 0$.

Proof. Let $W = fV$. Clearly $f$ preserves $+$ and $-$. Since $(x \circ \rho)^+ = x$, we have $x \circ \rho \in fX$, i.e., $fX \neq 0$. It is routine to check that $f$ preserves $d_{\kappa\lambda}$ for $\kappa, \lambda < \alpha$. Now suppose that $Y \in A$,

$\kappa < \alpha$ , and $y \in W$ . For brevity set $\mathfrak{B} = \mathfrak{Rd}^{(\rho)}\mathfrak{A}$ . To prove that $fc_\kappa^{(\mathfrak{B})}Y \subseteq C_\kappa^{[W]}fY$ , let $y \in fc_\kappa^{(\mathfrak{B})}Y$ . Thus $y \in fC_{\rho\kappa}^{[V]}Y$ , i.e., $y^+ \in C_{\rho\kappa}^{[V]}Y$ . Thus $y^+ \in V$ and $(y^+)_u^{\rho\kappa} \in Y$ for some $u \in U$ . It is easily checked that $(y_u^\kappa)^+ = (y^+)_u^{\rho\kappa}$ ; hence $(y_u^\kappa)^+ \in Y$ , so $y_u^\kappa \in fY$ and so $y \in C_\kappa^{[W]}fY$ . The other inclusion $\supseteq$ is established similarly.

<u>Theorem</u> I.8.2. If $\alpha$ and $\beta$ are ordinals with $\alpha \geq 2$ and $\rho \in {}^\alpha\beta$ is one-to-one, then $Rd_\alpha^{(\rho)}Gws_\beta \subseteq IGws_\alpha$ , and $Rd_\alpha^{(\rho)}\,{}_\infty Gws_\beta \subseteq I_\infty Gws_\alpha$ .

<u>Proof</u>. First we take any $\mathfrak{A} \in Ws_\beta$ and show that $\mathfrak{Rd}_\alpha^{(\rho)}\mathfrak{A} \in IGws_\alpha$ . To this end, by 2.4.39 and I.6.4 it suffices to take any non-zero $X \in A$ and find a homomorphism $f$ of $\mathfrak{Rd}_\alpha^{(\rho)}\mathfrak{A}$ into some $Ws_\alpha$ such that $fX \neq 0$ . Say $\mathfrak{A}$ has unit element ${}^\beta U^{(p)}$ , and $x \in X$ . For each $y \in {}^\alpha U$ define $y^+$ as in I.8.1; then define $f$ as there also. Applying I.8.1, we see that $f$ is a homomorphism of $\mathfrak{Rd}^{(\rho)}\mathfrak{A}$ into a $Crs_\alpha \mathfrak{B}$ , and $fX \neq 0$ . Now it is easily checked that $f({}^\beta U^{(p)}) = \{y \in {}^\alpha U : |\Gamma_y| < \omega\}$ , where $\Gamma_y = \{\kappa \in Rg\rho : y\rho^{-1}\kappa \neq p\kappa\}$ for all $y \in {}^\alpha U$ . Clearly $|\Gamma_y| < \omega$ iff $|\rho^{-1*}\Gamma_y| < \omega$ , and $\rho^{-1*}\Gamma_y = \{\kappa < \alpha : y\kappa \neq p\rho\kappa\}$ . Thus $f({}^\beta U^{(p)}) = {}^\alpha U^{(p\circ\rho)}$ , so $\mathfrak{B}$ is a $Ws_\alpha$ , as desired.

The theorem itself now follows easily from 0.5.13(iv) and I.6.4, the final statement being clear from the above.

<u>Remarks</u> I.8.3. Under the hypothesis of I.8.2 and using I.8.2 we also have $Rd_\alpha^{(\rho)}Gs_\beta \subseteq IGs_\alpha$ and $Rd_\alpha^{(\rho)}Gs_\beta^{reg} \subseteq IGs_\alpha^{reg}$ by I.7.14, and $Rd_\alpha^{(\rho)}\,{}_\infty Cs_\beta \subseteq I_\infty Cs_\alpha$ if $\alpha \geq \omega$ by I.7.21 . But Andréka and Németi have shown that if $Rg\rho \neq \beta$ then $Rd_\alpha^\rho Cs_\beta \not\subseteq ICs_\alpha$ and $Rd_\alpha^{(\rho)}Ws_\beta \not\subseteq IWs_\alpha$ , generalizing examples of Monk.

<u>Theorem</u> I.8.4. Let $\alpha$ and $\beta$ be ordinals and let $\rho \in {}^{\alpha}\beta$ be one-to-one and onto. Then $\mathbf{I}\mathfrak{Rd}_{\alpha}^{(\rho)} \mathsf{K}_{\beta} = \mathbf{I}\mathsf{K}_{\alpha}$ for $\mathsf{K} \in \{\mathsf{Crs}, \mathsf{Gws}, \mathsf{Gs}, \mathsf{Gws}^{reg}, \mathsf{Cs}, \mathsf{Ws}\}$ .

<u>Proof.</u> The arguments being very easy, we restrict ourselves to two representative cases, $\mathsf{K} = \mathsf{Crs}$ and $\mathsf{K} = \mathsf{Gws}^{reg}$ . First suppose $\mathfrak{A}$ is a $\mathsf{Crs}_{\beta}$ , say with base $U$ and unit element $V$ . Now the function $y^{+}$ in I.8.1 does not depend on any element $x$ ; we have simply $y^{+} = y \circ \rho^{-1}$ . The hypotheses of I.8.1 hold, so the function $f$ defined there is a homomorphism of $\mathfrak{Rd}^{(\rho)}\mathfrak{A}$ into a $\mathsf{Crs}_{\alpha}$ $\mathfrak{B}$ with unit element $fV$ , and $fX \neq 0$ for every non-zero $X \in A$ , i.e., $f$ is one-to-one. Thus $\mathfrak{Rd}^{(\rho)}\mathfrak{A} \in \mathbf{I}\mathsf{Crs}_{\alpha}$ , as desired.

Now suppose that $\mathfrak{A} \in \mathsf{Gws}_{\beta}^{reg}$ . It is easy to check that $\mathfrak{B} \in \mathsf{Gws}_{\alpha}$ . Assume that $Y \in A$ , $y \in fY$ , $z \in W$ , and $((\Delta^{(\mathfrak{B})}fY) \cup 1)\restriction y = ((\Delta^{(\mathfrak{B})}fY) \cup 1)\restriction z$ ; we want to show that $z \in fY$ . Now let $\mathfrak{C} = \mathfrak{Rd}^{(\rho)}\mathfrak{A}$ . For any $\kappa < \alpha$ we have $c_{\kappa}^{(\mathfrak{B})}fY = fY$ iff $c_{\kappa}^{(\mathfrak{C})}Y = Y$ iff $c_{\rho\kappa}^{(\mathfrak{A})}Y = Y$ . Thus $\Delta^{(\mathfrak{B})}fY = \rho^{-1*}\Delta^{(\mathfrak{A})}Y$ . First suppose that $0 \in \Delta^{(\mathfrak{A})}Y$ . Then $\rho^{-1}0 \in \Delta^{(\mathfrak{B})}fY$ , and so $y\rho^{-1}0 = z\rho^{-1}0$ , i.e., $y^{+}0 = z^{+}0$ . Hence $(\Delta^{(\mathfrak{A})}Y \cup 1)\restriction y^{+} = (\Delta^{(\mathfrak{A})}Y \cup 1)\restriction z^{+}$ . Clearly $y^{+} \in Y$ and $z^{+} \in V$ , so $z^{+} \in Y$ by the assumed regularity, so $z \in fY$ as desired. Second, assume that $0 \notin \Delta^{(\mathfrak{A})}Y$ . Now $y^{+} \in Y$ , so $(y^{+})_{y0}^{0} \in V$ (using $\mathfrak{A} \in \mathsf{Gws}_{\beta}$) , and hence $(y^{+})_{y0}^{0} \in Y$ . Clearly also $(z^{+})_{z0}^{0} \in V$ , and $(\Delta^{(\mathfrak{A})}Y \cup 1)\restriction(y^{+})_{y0}^{0} = (\Delta^{(\mathfrak{A})}Y \cup 1)\restriction(z^{+})_{y0}^{0}$ , so $(z^{+})_{y0}^{0} \in Y$ . Hence $z^{+} \in Y$ and $z \in fY$ , as desired.

We have shown $\mathbf{I}\mathfrak{Rd}_{\alpha}^{(\rho)}\mathsf{K}_{\beta} \subseteq \mathbf{I}\mathsf{K}_{\alpha}$ in our two representative cases. For the other inclusion, it suffices to note that $\rho^{-1} \in {}^{\beta}\alpha$ is one-to-one and onto, $\mathfrak{Rd}_{\beta}^{(\rho^{-1})}\mathsf{K}_{\alpha} \subseteq \mathbf{I}\mathsf{K}_{\beta}$ by what was already shown, and clearly

$Rd^{(p)}Rd^{(p^{-1})}K_\alpha = K_\alpha$ .

Now we turn to neat embeddings, for which we also require a technical lemma.

Lemma I.8.5. Let $\mathfrak{A}$ be a $Crs_\alpha$ with base $U$ and unit element $V$. Assume that $\alpha \le \beta$ and $W \subseteq {}^\beta U$. We also assume the following conditions:

(i) $V = \{x : x = \alpha 1 y \text{ for some } y \in W\}$ ;

(ii) for all $x \in W$, $\kappa < \alpha$, and $u \in U$, if $\alpha 1 x^\kappa_u \in V$ then $x^\kappa_u \in W$.

For any $X \in A$ let $fX = \{x \in W : \alpha 1 x \in X\}$ .

Then there is a $Crs_\beta$ $\mathfrak{B}$ with base $U$ and unit element $W$ such that $f \in Ism(\mathfrak{A}, \mathfrak{R}\mathfrak{d}_\alpha \mathfrak{B})$ and $c^{(\mathfrak{B})}_\kappa fX = fX$ for all $X \in A$ and $\kappa \in \beta \sim \alpha$ .

Proof. Let $\mathfrak{B}$ be the full $Crs_\beta$ with unit element $W$. Clearly $f$ preserves $+$. Now let $x \in W$ and $X \in A$. By (i) we have $\alpha 1 x \in V$. Hence $x \in f(V \sim X)$ iff $\alpha 1 x \in V \sim X$ iff $\alpha 1 x \notin X$ iff $x \in W \sim fX$. Hence $f$ preserves $-$. If $0 \neq X \in A$, choose $x \in X$. Then by (i) there is a $y \in W$ such that $x \subseteq y$. Thus $y \in fX$. This shows that $f$ is one-to-one. Clearly $f$ preserves $d_{\kappa\lambda}$ for $\kappa, \lambda < \alpha$ (again using (i)). Now suppose that $X \in A$, $\kappa < \alpha$, and $x \in C^{[V]}_\kappa X$ ; we want to show that $x \in C^{[W]}_\kappa fX$. By the definition of $f$ we have $x \in W$ and $\alpha 1 x \in C^{[V]}_\kappa X$. Hence $\alpha 1 x \in V$ and $(\alpha 1 x)^\kappa_u \in X$ for some $u \in U$. Since $(\alpha 1 x)^\kappa_u \in V$, our assumption (ii) yields that $x^\kappa_u \in W$. Hence $x \in C^{[W]}_\kappa fX$, as desired. The converse is similar.

Finally, suppose that $X \in A$, $\kappa \in \beta \sim \alpha$, and $x \in C^{[W]}_\kappa fX$. Thus $x \in W$ and $x^\kappa_u \in fX$ for some $u \in U$. Hence $x^\kappa_u \in W$ and

$\alpha 1 x_u^\kappa \in X$ . Since $\kappa \geq \alpha$ , this means that $\alpha 1 x \in X$ , and hence $x \in fX$ , as desired.

From this lemma it is easy to prove

Theorem I.8.6. Assume that $\alpha \leq \beta$ and $K \in \{Ws, Cs, Gws, Gs\}$ . Then $K_\alpha \subseteq ISNr_\alpha K_\beta$ .

Corollary I.8.7. If $2 \leq \alpha \leq \beta$ then $IGws_\alpha = SNr_\alpha IGws_\beta = SNr_\alpha IGs_\beta = IGs_\alpha$ .

Proof. By I.7.14, I.8.2, and I.8.6.

Remark I.8.8. It follows from 2.6.48 and I.8.6 that for any ordinal $\alpha$ we have $Cs_\alpha \cup Ws_\alpha \cup Gs_\alpha \cup Gws_\alpha \subseteq SNr_\alpha Dc_{\alpha+\omega}$ . A major result of the representation theory of $CA_\alpha$'s , to appear in a later paper, is that if $\alpha \geq 2$ then $SNr_\alpha Dc_{\alpha+\omega} = IGs_\alpha = IGws_\alpha$ .

## 9. Problems

We begin by indicating the status of the problems listed in [ HMT ] as of January 1981. In Problem 0.6 one should assume that $\alpha$ is less than the first uncountable measurable cardinal (see Chang, Keisler [CK]). Under this corrected formulation, the consistency of a positive answer relative to the consistency of ZFC plus certain other axioms has been shown by Magidor [Ma] and Laver [ L ] . Problem 1.2 has been solved affirmatively by B. Sobociński [ S ] . Andréka and Németi solved

Problem 2.3 affirmatively; see [AN2] and [N]. Problem 2.4 has been solved affirmatively by J. Ketonen [K] for Boolean algebras, and hence for discrete CA's. Problem 2.8 was solved affirmatively by D. Myers [My] and Problem 2.9 negatively by W. Hanf [H]. 2.11 was solved negatively (except for $\alpha < 2$) by Andréka and Németi; see [AN2] and [N]. Problem 2.12 was solved negatively by R. Maddux [Md].

Now we shall list some problems left open concerning set algebras.

Problem 1. Let $\alpha \geq \omega$. Given a normal $Gws_\alpha$ $\mathfrak{B}$ with base U, is there a $Cs_\alpha$ $\mathfrak{A}$ with same base U such that $\mathfrak{B} \in R\ell\{\mathfrak{A}\}$ ? (Cf. I.2.6-I.2.13).

Problem 2. Let q be the function defined in I.4.8. For every $\alpha \in \omega \sim 2$ let $q^+\alpha$ be the largest $\beta \in \omega$ such that $q(\alpha,\beta) = 1$. Give a simple arithmetic description of q, or at least of $q^+$.

Problem 3. Is $IWs_\alpha$ closed under directed unions for $\alpha \geq \omega$ ? (Cf. I.4.8 and I.7.11.)

Problem 4. Is $ICs_\alpha \subseteq HCs_\alpha^{reg}$ or $HCs_\alpha = HCs_\alpha^{reg}$ ? (Cf. I.5.6.)

Problem 5. Does $I_\infty Cs_\alpha = H_\infty Ws_\alpha$ ? (Cf. I.5.6(17) and I.5.8.)

Problem 6. Is $ICs_\alpha^{reg} \subseteq HPWs_\alpha$ or $ICs_\alpha \subseteq HPWs_\alpha$ ? (Cf. I.6.8.)

Problem 7. Is $H_\infty Ws_\alpha = HP_\infty Ws_\alpha$ ?

Problem 8. Is $HP_\infty Ws_\alpha = I_\infty Cs_\alpha$ ?

For these two questions cf. I.6.8 and I.6.10.

Problem 9. Is every weakly subdirectly indecomposable $Cs_\alpha$ isomorphic to a regular $Cs_\alpha$ ?

Problem 10. Is every weakly subdirectly indecomposable $Gws_\alpha$ (or $Cs_\alpha^{reg}$) isomorphic to a $Ws_\alpha$ ?

For these two questions cf. I.6.16.

Problem 11. Is the condition $|U| \geq |A|$ in II.7.27 needed? (Cf. here also I.7.30.)

## REFERENCES

[AN1] Andréka, H. and Németi, I., A simple, purely algebraic proof of the completeness of some first order logics, Alg. Univ. 5(1975), 8-15.

[AN2] Andréka, H. and Németi, I., On problems in cylindric algebra theory, Abstracts Amer. Math. Soc. 1(1980), 588.

[AN3] Andréka, H. and Németi, I., On cylindric-relativized set algebras, this vol.,

[AN4] Andréka, H. and Németi, I., Finite cylindric algebras generated by a single element, Finite algebra and multivalved logic (Proc. Coll. Szeged), eds. B. Csákány, I. Rosenberg, Colloq. Math. Soc. J. Bolyai vol. 28, North-Holland, to appear.

[CK] Chang, C.C. and Keisler, H.J., Model theory (second edition), North-Holland 1978, xii + 554 pp.

[D] Daigneault, A., On automorphisms of polyadic algebras, Trans. Amer. Math. Soc. 112(1964), 84-130.

[De] Demaree, D., Studies in algebraic logic, Doctoral Dissertation, Univ. of Calif., Berkeley 1970, 96pp.

[EFL] Erdös, P., Faber, V. and Larson, J., Sets of natural numbers of positive density and cylindric set algebras of dimension 2, to appear, Alg. Univ.

[H] Hanf, W., The Boolean algebra of logic, Bull. Amer. Math. Soc. 8(1975), 587-589.

[HM] Henkin, L. and Monk, J.D., Cylindric set algebras and related structures, Proc. of the Tarski Symposium, Proc. Symp. Pure Math. 25(1974), Amer. Math. Soc., 105-121.

[HMT] Henkin, L., Monk, J.D., and Tarski, A., Cylindric Algebras, Part I, North-Holland (1971), 508pp.

[HR] Henkin, L. and Resek, D., Relativization of cylindric algebras, Fund. Math. 82(1975), 363-383.

[HT] Henkin, L. and Tarski, A., Cylindric algebras, Lattice theory, Proc. Symp. pure math. 2(1961), Amer. Math. Soc., 83-113.

[K] Ketonen, J., The structure of countable Boolean algebras, Ann. Math. 108(1978), 41-89.

[Ko] Koppelberg, S., Homomorphic images of $\sigma$ - complete Boolean algebras, Proc. Amer. Math. Soc. 51(1975), 171-175.

[L] Laver, R., Saturated ideals and nonregular ultrafilters, to appear.

[Ma] Magidor, M., On the existence of nonregular ultrafilters and the cardinality of ultrapowers, Trans. Amer. Math. Soc.249 (1979),97-111 .

[M1] Monk, J.D., Singulary cylindric and polyadic equality algebras, Trans. Amer. Math. Soc. 112(1964), 185-205.

[M2] Monk, J.D., Model-theoretic methods and results in the theory of cylindric algebras, The Theory of Models, Proc. 1963 Symp., North- Holland, 238-250.

[Md] Maddux, R., Relation algebras and neat embeddings of cylindric algebras, Notices Amer. Math. Soc. 24(1977), A-298.

[My] Myers, D., Cylindric algebras of first-order languages, Trans. Amer. Math. Soc. 216(1976), 189-202.

[N] Németi, I., Connections between cylindric algebras and initial algebra semantics of CF languages, Mathematical logic in computer science, eds. B Dömölki, T. Gergely, Colloq. Math. Soc. J. Bolyai, vol. 26 North-Holland (1981), 561-606 .

[S] Sobociński, B., Solution to the problem concerning the Boolean bases for cylindric algebras, Notre Dame J. Formal Logic 13(1972), 529-545.

[TV] Tarski, A. and Vaught, R.L., Arithmetical extensions of relational systems, Compos. Math. 13(1957), 81-102.

# On cylindric-relativized set algebras

by H. Andréka and I. Németi

This work is based on the book [HMT] and the paper [HMTI]. The abstract theory of cylindric algebras (CA-s) is extensively developed in the book [HMT]. Most of the motivating examples for the abstract theory of CA-s are cylindric-relativized set algebras (Crs-s). The present work is devoted to the study of Crs-s, more precisely to certain distinguished classes of Crs-s introduced in [HMTI]. Such a distinguished class is $Gs^{reg}$. The role played by $Gs^{reg}$ in CA-theory is similar to the role played by Boolean set algebras in Boolean algebra theory. For example, the fundamental link between model theory and CA-theory is $Gs^{reg}$, see the introduction of [HMTI]. It was proved in [N] that the class connecting classical finitary model theory to CA-theory is exactly $Gs^{reg} \cap Lf$, in a sense at least. Recently much attention was given to the meta-structure consisting of all first order theories and all interpretations between them. It was proved in [G] that this structure can be represented isomorphically by $Gs^{reg}$ achieving considerable insight and simplification this way. Following these motivations we shall give special attention to $Gs^{reg}$.

We shall use the notations introduced in [HMT] without recalling them. The present paper is a continuation of [HMTI] and is organized parallel to [HMTI]. We refer to [HMTI] for an introductory discussion of the contents of the individual sections; we have practically the same section-titles as [HMTI]. The items in [HMTI] are numbered by three figures like I.2.2. The first figure is always I and therefore we omit it, e.g. the reference [HMTI]2.2 in this paper means item I.2.2 of [HMTI]. We refer to items of the present paper by strings of figures, e.g. 0.5.1 refers to item 0.5.1 of this paper, moreover this item is found in section 0, and it is a subitem of item 0.5. In general, the figures when read from left to right correspond to the subdivisions in which the item referred to is found.

We shall be glad to send full proofs of statements claimed but not proved (or not proved in detail) in the present work, whenever requested.

Acknowledgement. We are most grateful to Professor J.D. Monk for guiding us in this work as well as in our research concerning algebraic logic in general.

## 0. Basic concepts and notations

We use the notations and definitions of [HMTI] and [HMT] without recalling them. Especially we use [HMTI]1.1 where the classes $Cs_\alpha$, $Ws_\alpha$, $Gs_\alpha$, $Gws_\alpha$, $Crs_\alpha$, $Cs_\alpha^{reg}$, $Gs_\alpha^{reg}$, $Gws_\alpha^{reg}$, $Crs_\alpha^{reg}$ of cylindric-relativized set algebras were introduced. All these algebras are normal $Bo_\alpha$-s.

Notations: Let $\mathfrak{A}$ be an algebra similar to $CA_\alpha$-s. Then $1^{\mathfrak{A}}$ exists since $1$ is a constant symbol of $CA_\alpha$-s. We define $Mn(\mathfrak{A}) \overset{d}{=} Sg^{(\mathfrak{A})}\{1^{\mathfrak{A}}\}$ and $\mathfrak{Mn}(\mathfrak{A}) \overset{d}{=} \mathfrak{Sg}^{(\mathfrak{A})}\{1^{\mathfrak{A}}\}$.

Let $V$ be a $Crs_\alpha$-unit. Then $\mathfrak{Sb}\,V$ denotes the full $Crs_\alpha$ with unit $V$. This notation is ambiguous if $V=0$ but we hope context will help.

Let $x \subseteq V \subseteq {}^\alpha U$. Then $\Delta^{[V]}x \overset{d}{=} \{i\in\alpha : c_i^{[V]}x \neq x\}$ and $\Delta^{(U)}x \overset{d}{=} \Delta^{[{}^\alpha U]}x$.

Let $H$ be any set. Then $Sb_\omega H$ denotes the set of all finite subsets of $H$ and $G \subseteq_\omega H$ denotes that $G \in Sb_\omega H$.

As a generalization of the notation $f^x_u$ introduced in [HMTI], the following notation will be very useful. Let $f,k$ be two functions and let $H$ be a set. Then $f[H/k] \overset{d}{=} (Do f \sim H)1f \cup H1k$.

The notations $f : A \to B$, $f : A \twoheadrightarrow B$, $f : A \rightarrowtail B$, and $f : A \rightarrowtail\!\!\!\!\!\twoheadrightarrow B$ mean that $A1f$ is a function mapping $A$ into (onto, one-one into, one-one onto respectively) $B$. In accordance with [HMT],

$f \in Is(\mathfrak{A},\mathfrak{B})$ means that $A1f \in Is(\mathfrak{A},\mathfrak{B})$ and similarly for Hom etc.

We shall use the notations $\Delta$, Zd, etc. introduced for $CA_\alpha$-s in [HMT] to $Crs_\alpha$-s as well, despite of the fact that a $Crs_\alpha$ need not be a $CA_\alpha$. E.g. let $\mathfrak{A} \in Crs_\alpha$, and let $x \in A$. Then $\Delta x \overset{d}{=} \Delta^{(\mathfrak{A})} x \overset{d}{=} \{i \in \alpha : c_i^{\mathfrak{A}} x \neq x\}$, $Zd\mathfrak{A} \overset{d}{=} \{x \in A : \Delta^{(\mathfrak{A})} x = 0\}$ and $\mathfrak{Zd}\mathfrak{A} \overset{d}{=} \langle Zd\mathfrak{A}, +, \cdot, -, 0, 1 \rangle$, cf. [HMT] 1.6.1 and 1.6.18.

By [HMT] 2.2.3 we have that the axioms $(C_0)-(C_3)$, $(C_5)$ and $(C_7)$ of [HMT] 1.1.1 are valid in $Crs_\alpha$. Therefore [HMT] 1.2.1-1.2.12 are true for $Crs_\alpha$-s, although they are stated for $CA_\alpha$-s only, because in their proofs the only axioms used are $(C_0)-(C_3)$ (as it is explicitly noted on p.177 of [HMT]). Also [HMT] 1.6.2, 1.6.5-1.6.7 are true for $Crs_\alpha$-s, since their proofs use only $(C_0)-(C_3)$ and 1.2.1-1.2.12 of [HMT]. Because of the above, in the proofs we shall apply [HMT] 1.2.1-1.2.12 and 1.6.2, 1.6.5-1.6.7 to $Crs_\alpha$-s. By 1.6.2, 1.6.5-1.6.7 we have that $\mathfrak{Zd}\mathfrak{A} \in BA$ for every $\mathfrak{A} \in Crs_\alpha$.

Let $\mathfrak{A},\mathfrak{B} \in Crs_\alpha$. Then $A=B$ implies $\mathfrak{A}=\mathfrak{B}$. Therefore we let $Zd\,A \overset{d}{=} Zd\mathfrak{A}$ for $\mathfrak{A} \in Crs_\alpha$. In general, notions applicable to $Cr_\alpha$-s will be applied to cylindric-relativized fields of sets. The above argument holds for Boolean set algebras too. Particularly for any Boolean set algebra $\mathfrak{B}$ we let $At\,B \overset{d}{=} At\mathfrak{B}$.

<u>Definition 0.1.</u> Let $U$ be any set and let $V \subseteq {}^\alpha U$. By a <u>subunit of $V$</u> we understand an atom of the Boolean field Zd Sb $V$ of sets. <u>Subu(V)</u> denotes the set of all subunits of $V$. I.e. $Subu(V) = At\ Zd\ Sb\ V$. We define <u>base(V)</u> $\overset{d}{=} \bigcup\{Rgp : p \in V\}$. We say that $Y$ is a <u>subbase of $V$</u> iff $Y = base(W)$ for some $W \in Subu(V)$. <u>Subb(V)</u> denotes the set of all subbases of $V$.

Let $\mathfrak{A} \in Crs_\alpha$. Then $base(\mathfrak{A}) \overset{d}{=} base(1^{\mathfrak{A}})$, $Subu(\mathfrak{A}) = Subu(1^{\mathfrak{A}})$ and $Subb(\mathfrak{A}) = Subb(1^{\mathfrak{A}})$. $W$ is said to be a <u>subunit of $\mathfrak{A}$</u> iff $W$ is a subunit of $1^{\mathfrak{A}}$, and $Y$ is said to be a <u>subbase of $\mathfrak{A}$</u> iff $Y$ is a subbase of $1^{\mathfrak{A}}$.

The above definition of subbase agrees with [HMTI] 1.1 (vii). Note that a subbase might be empty iff $\alpha=0$ .

Notation: Let $K\subseteq Crs_\alpha$ and $\varkappa$ be a cardinal. Then

$${}_\varkappa K \stackrel{d}{=} \{\mathfrak{A}\in K : (\forall U\in Subb(\mathfrak{A}))|U|=\varkappa\} \quad \text{and}$$

$${}_\infty K \stackrel{d}{=} \{\mathfrak{A}\in K : (\forall U\in Subb(\mathfrak{A}))|U|\geq\omega\} .$$

The above notation agrees with [HMTI] 5.6(17) and [HMTI] 7.20. Note that the one-element $Crs_\alpha$ is in ${}_\varkappa Crs_\alpha \cap {}_\infty Crs_\alpha$ for all $\varkappa$ since it has no subbases.

In this connection we recall the following: Let $0<\varkappa<\omega\leq\alpha$ . Then

$$\mathsf{I}\,{}_\varkappa Gs_\alpha = \{\mathfrak{A}\in CA_\alpha : \mathfrak{A}\models (c_{(\varkappa)}\bar{d}(\varkappa\times\varkappa)-c_{(\varkappa+1)}\bar{d}((\varkappa+1)\times(\varkappa+1))=1)\} .$$

That is $\mathsf{I}\,{}_\varkappa Gs_\alpha$ is a variety definable by a finite scheme of equations consisting of $(C_0)$-$(C_7)$ and the above one. This is immediate by [HMT] 2.6.54 and [HMTI] 8.8.

Lemma 0.2. Let $\mathfrak{A}\in Crs_\alpha$ . Then (i)-(iii) below hold.

(i) $1^{\mathfrak{A}}$ is the disjoint union of all subunits of $\mathfrak{A}$ .

(ii) Let $W\in Subu(\mathfrak{A})$. Then $W\subseteq {}^\alpha Y^{(p)}$ for some $Y\in Subb(\mathfrak{A})$ and for some $p\in 1^{\mathfrak{A}}$ .

(iii) $\mathfrak{A}\in Gws_\alpha$ iff every subunit of $\mathfrak{A}$ is a weak space. Moreover, let $\alpha\geq 2$ and let $V = \cup\{{}^\alpha Y_i^{(pi)} : i\in I\}$ be such that $(\forall i,j\in I)[i\neq j \Rightarrow {}^\alpha Y_i^{(pi)}\cap{}^\alpha Y_j^{(pj)} = 0]$ . Then $Subu(V) = \{{}^\alpha Y_i^{(pi)} : i\in I\}$ .

The proof of Lemma 0.2 depends on the following lemma.

Lemma 0.2.1. Let $\mathfrak{A}$ be a complete $\mathbf{s}Cr_\alpha$ . Let $zd \stackrel{d}{=} \langle \Sigma\{c_{i0}\dots c_{in}x : n\in\omega,\ i\in{}^{n+1}\alpha\} : x\in A \rangle$ . Then (i)-(ii) below hold.

(i) $zd : A \to Zd\mathfrak{A}$ , $zd : At\mathfrak{A} \to At\ Zd\,\mathfrak{A}$, and $(\forall x\in A)\ zd(x) =$

$= \Pi\{y \in Zd\mathfrak{A} : x \leq y\}$, i.e. zd(x) is the "zero-dimensional closure" of x .

(ii) If $\mathfrak{A}$ is atomic then $1^{\mathfrak{A}} = \sum At\, \mathfrak{Z}\mathfrak{d}\,\mathfrak{A}$.

<u>Proof.</u> Let $\mathfrak{A}$ be a complete $\mathbf{s}Cr_\alpha$ and let the function zd be defined as in the statement of 0.2.1.

<u>Proof of 0.2.1(i):</u> Using [HMT] 1.2.6(i) it is clear that $zd : A \to Zd\mathfrak{A}$ . Next we show $zd(x) = \Pi\{y \in Zd\mathfrak{A} : x \leq y\}$ for any $x \in A$ . Let $y \in Zd\mathfrak{A}$ and suppose $x \leq y$ . Then $c_{i0} \dots c_{in} x \leq y$ for every $n \in \omega$ and $i \in {}^{n+1}\alpha$ by [HMT] 1.2.7 and therefore $zd(x) \leq y$ (by $(C_0)$) . By $x \leq zd(x) \in Zd\mathfrak{A}$ then $zd(x) = \Pi\{y \in Zd\mathfrak{A} : x \leq y\}$ . Suppose $x \in At\mathfrak{A}$ . We show that $zd(x) \in At\mathfrak{A}$ . Let $x \leq zd(x)$, $y \in Zd\mathfrak{A}$. Then $zd(x)-y \in Zd\mathfrak{A}$, by [HMT] 1.6.6 and 1.6.7. If $x \leq zd(x)-y$ then $zd(x) \leq zd(x)-y$, and therefore $y=0$ by $y \leq zd(x)$. Suppose $x \nleq zd(x)-y$ . Then $x \nleq -y$ by $x \leq zd(x)$ and therefore $x \leq y$ since $x \in At\mathfrak{A}$. Then $zd(x) \leq y$ by $x \leq y \in Zd\mathfrak{A}$ and by $zd(x) = \Pi\{y \in Zd\mathfrak{A} : x \leq y\}$ . Therefore $y = zd(x)$ by $y \leq zd(x)$. We have seen that $zd(x) \in At\, \mathfrak{Z}\mathfrak{d}\,\mathfrak{A}$. Therefore $zd : At\mathfrak{A} \to At\, \mathfrak{Z}\mathfrak{d}\,\mathfrak{A}$.

<u>Proof of 0.2.1(ii):</u> Suppose $\mathfrak{A}$ is atomic and complete. Then $1^{\mathfrak{A}} = \sum At\mathfrak{A}$ follows from the theory of Boolean algebras. $\sum At\mathfrak{A} \leq \sum At\, \mathfrak{Z}\mathfrak{d}\,\mathfrak{A}$ by $zd : At\mathfrak{A} \to At\mathfrak{Z}\mathfrak{d}\mathfrak{A}$ and by $(\forall x \in A) x \leq zd(x)$. Then $1^{\mathfrak{A}} = \sum At\mathfrak{A} \leq \sum At\, \mathfrak{Z}\mathfrak{d}\,\mathfrak{A}$ proves that $1^{\mathfrak{A}} = \sum At\, \mathfrak{Z}\mathfrak{d}\,\mathfrak{A}$.

<u>QED (Lemma 0.2.1.)</u>

Now we turn to the proof of Lemma 0.2. Let $\mathfrak{A} \in Crs_\alpha$ and let $\mathfrak{L} = \mathfrak{Sb}\, 1^{\mathfrak{A}}$ .

<u>Proof of 0.2(i):</u> $\mathfrak{L}$ is a complete and atomic $\mathbf{s}Cr_\alpha$ , hence $1^{\mathfrak{A}} = 1^{\mathfrak{L}} = \sum At\, Zd\mathfrak{L} = \sum Subu(\mathfrak{A})$ by 0.2.1.

<u>Proof of 0.2(ii)</u>: Let $W \in Subu(\mathfrak{A})$ and let $p \in W$ be arbitrary. We have $W = \sum\{c_{i0} \dots c_{in}\{p\} : n \in \omega,\ i \in {}^{n+1}\alpha\}$ by 0.2.1(i), since $p \in W \in At\, Zd\, Sb\, 1^{\mathfrak{A}}$ . Therefore $W \subseteq {}^{\alpha}base(W)^{(p)}$.

Proof of 0.2(iii): If every subunit of $\mathfrak{A}$ is a weak space then $\mathfrak{A}\in Gws_\alpha$ by 0.2(i) and by the definition of $Gws_\alpha$. If $\alpha\leq 1$ then the other direction holds too, since every nonempty subset of a $Gws_\alpha$-unit is an $\alpha$-dimensional weak space then. Suppose $\alpha\geq 2$. Let $V = \cup\{{}^\alpha Y_i^{(pi)} : i\in I\}$ be such that $(\forall i, k\in I)[i\neq k \Rightarrow$ $\Rightarrow {}^\alpha Y_i^{(pi)}\cap{}^\alpha Y_k^{(pk)}=0]$. Then by [HMTI] 1.12 we have At Zd Sb V = $= \{{}^\alpha Y_i^{(pi)} : i\in I\}$.

QED(Lemma 0.2.)

Proposition 0.3. Let V be a $Crs_\alpha$-unit. Then statements (i)-(iv) below are equivalent.

(i) V is a $Gws_\alpha$-unit.

(ii) The full $Crs_\alpha$ with unit V is a $CA_\alpha$.

(iii) The $Crs_\alpha$ with unit V and generated by $\{\{q\} : q\in V\}$ is a $CA_\alpha$.

(iv) Every $Crs_\alpha$ with unit V is a $CA_\alpha$.

Proof. Since $Gws_\alpha \subseteq CA_\alpha$ by [HMTI] 1.9(i), it is clear that (i) $\Rightarrow$ (iv) $\Rightarrow$ (ii) $\Rightarrow$ (iii). So, it is enough to prove that (iii) implies (i). We shall need the following lemma.

Lemma 0.3.1. Let $\mathfrak{A}\in CA_\alpha$, $b\in A$ and assume $CA_\alpha \cap S\mathfrak{Rl}_b\mathfrak{A} \neq 0$. Then $b \leq s^i_j b$ holds in $\mathfrak{A}$ for all $i,j\in\alpha$.

Proof. Assume the hypotheses of Lemma 0.3.1. Then the minimal subalgebra of $\mathfrak{Rl}_b\mathfrak{A}$ is a $CA_\alpha$ and hence $\mathfrak{Rl}_b\mathfrak{A} \models c_i d_{ij}=1$. Then $b\cdot c_i(d_{ij}\cdot b)=b$ holds in $\mathfrak{A}$, by the definition of $\mathfrak{Rl}_b\mathfrak{A}$. I.e. $b \leq s^i_j b$ holds in $\mathfrak{A}$.

QED(Lemma 0.3.1.)

Now we return to the proof of Prop.0.3. Let $V$ be a $Crs_\alpha$ -unit and let $\mathfrak{A}$ be the $Crs_\alpha$ with unit $V$ and generated by $\{\{q\} : q\in V\}$. Assume 0.3(iii), i.e. assume $\mathfrak{A}\in CA_\alpha$. Let $U = base(V)$ and let $\mathfrak{L} = \mathfrak{Sb}^\alpha U$. Then $\mathfrak{A}\in \mathbf{S}\mathfrak{R}\mathfrak{l}_V\mathfrak{L}$, and hence $V \le s^i_j V$ holds in $\mathfrak{L}$ by 0.3.1. In particular, $(\forall q\in V)(\forall i,j\in\alpha)\ \ q^i_{q(j)} \in V$ by $V \le s^i_j V = c_i(d_{ij}\cdot V)$. Now we show that $\mathfrak{A}\in Gws_\alpha$.

<u>Case 1</u> Assume $\alpha\ge 3$. By [HMTI] 2.2 and by $V \le s^i_j V\cdot s^j_i V$, it is enough to prove that $s^i_j V\cdot s^j_i V \le V$ holds in $\mathfrak{L}$, for all $i,j\in\alpha$, $i\ne j$. Let $i,j\in\alpha$, $i\ne j$. Let $q\in s^i_j V\cdot s^j_i V =$ $= c^{(U)}_i(D^{(U)}_{ij}\cap V)\cap c^{(U)}_j(D^{(U)}_{ij}\cap V)$. Then there are $f,h\in D^{(U)}_{ij}\cap V$ such that $q = f^i_{q(i)} = h^j_{q(j)}$. Let $k \in \alpha\sim\{i,j\}$. Let $u \overset{d}{=} q(k)$. Then $u = h(k) = f(k) = q(k)$, by $k \notin \{i,j\}$. By $V \le s^i_k V$ and $\{f,h\} \subseteq V$ we have $f^i_u = f^i_{f(k)} \in V$ and $h^i_u = h^i_{h(k)} \in V$. Note that $\{f\}\in A$ since $\mathfrak{A}$ is generated by $\{\{p\} : p\in V\}$, and $f\in V$. Then $h \in$ $\in c^{\mathfrak{A}}_i c^{\mathfrak{A}}_j c^{\mathfrak{A}}_i\{f\}$, since $h^i_u = (f^i_u)^j_{h(j)}$ and $\{h, f^i_u, h^i_u\} \subseteq V$. Since $\mathfrak{A}\in CA_\alpha$ by our assumption, we have by $(C_4)$ that $h \in c^{\mathfrak{A}}_j c^{\mathfrak{A}}_i\{f\}$. Then $(\exists p \in c_i\{f\})h = p^j_{h(j)}$. Then $p(i)= h(i)= q(i)$ by $i\ne j$ and $q=h^j_{q(j)}$. By $p \in c_i\{f\}$ we have $p=f^i_{p(i)}$ and therefore $q\in V$ by $p\in V$ and $p = f^i_{p(i)} = f^i_{q(i)} = q$. We have proved $s^i_j V\cdot s^j_i V \le V$ for all $i,j\in\alpha$.

<u>Case 2</u> Assume $\alpha\le 2$. If $\alpha\le 1$ then $Crs_\alpha = Gws_\alpha$ by [HMTI] 1.8 and therefore we are done. Let $\alpha=2$. We define $u \equiv v$ iff $u,v\in U$ and $\langle u,v\rangle\in V$. It suffices to show that $\equiv$ is an equivalence relation on $U$. Suppose $\langle u,v\rangle\in V$; we prove that $\langle v,u\rangle\in V$. In fact, $\langle u,u\rangle\in V$ by $V \subseteq s^1_0 V$ (Lemma 0.3.1), and similarly $\langle v,v\rangle\in V$. Clearly $\langle u,u\rangle\in c^{\mathfrak{A}}_1 c^{\mathfrak{A}}_0\{\langle v,v\rangle\}$, so $\langle u,u\rangle\in c^{\mathfrak{A}}_0 c^{\mathfrak{A}}_1\{\langle v,v\rangle\}$, and hence $\langle v,u\rangle\in V$. Now suppose $\langle u,v\rangle, \langle v,w\rangle\in V$; we show that $\langle u,w\rangle\in V$. We have $\langle w,v\rangle\in V$ by what was just proved, so $\langle u,v\rangle\in$ $\in c^{\mathfrak{A}}_0 c^{\mathfrak{A}}_1\{\langle w,w\rangle\}$ ($\langle w,w\rangle\in V$ by Lemma 0.3.1 again). Hence $\langle u,v\rangle\in$ $\in c^{\mathfrak{A}}_1 c^{\mathfrak{A}}_0\{\langle w,w\rangle\}$ and therefore $\langle u,w\rangle\in V$. By these we have proved that $V$ is a $Gs_2$ -unit.

QED(Proposition 0.3.)

Remark 0.4. In Proposition 0.3, condition (iii) cannot be replaced with the condition that some $Crs_\alpha$ with unit V be a $CA_\alpha$, since by [HMTI] 2.14 we have $\mathsf{I}Gws_\alpha \subset \mathsf{I}Crs_\alpha \cap CA_\alpha$.

Problem 0.4.1. Let $V \subseteq {}^\alpha U$. What are the sufficient and necessary conditions on V for $\mathfrak{Nr}(\mathfrak{Sb}\, V) \in CA_\alpha$ ?

Note that by [HMT] 2.6.57 we have $\mathfrak{Nr}(\mathfrak{Sb} V) \in CA_\alpha$ iff $\mathfrak{Nr}(\mathfrak{Sb}\, V) \in \in \mathsf{I}Gs_\alpha$. Similarly, by Prop. 0.3, $\mathfrak{Sb}\, V \in CA_\alpha$ iff $\mathfrak{Sb}\, V \in \mathsf{I}Gs_\alpha$.

By the above we have the following discussion of [HMT] 2.2.10 (stating $(\forall \mathfrak{A}\in CA_\alpha)(\forall b\in A)[(\forall i, j\in\alpha) b=s^i_j b\cdot s^j_i b \Rightarrow \mathfrak{Rl}_b\mathfrak{A}\in CA_\alpha])$. The condition $b \le s^i_j b\cdot s^j_i b$ is necessary for $\mathfrak{Rl}_b\mathfrak{A}\in CA_\alpha$, but the condition $b \ge s^i_j b\cdot s^j_i b$ is not necessary in general by [HMTI] 2.1, 2.14), though it cannot be omitted completely, as Prop.0.3 shows. E.g. $b \ge s^i_j b\cdot s^j_i b$ is necessary if $\mathfrak{A}$ is a $Crs_\alpha$ and $\{\{q\} : q\in b\} \subseteq A$.

Definition 0.5. Let $\mathfrak{A}\in Crs_\alpha$. Then $\mathfrak{A}$ is said to be normal iff $Subb(\mathfrak{A})$ is a partition of $base(\mathfrak{A})$, i.e. $\mathfrak{A}$ is normal iff $(\forall Y, Z\in Subb(\mathfrak{A}))[Y=Z$ or $Y\cap Z=0]$. $\mathfrak{A}$ is said to be compressed iff $|Subb(\mathfrak{A})|\le 1$. $\mathfrak{A}$ is said to be widely distributed iff $(\forall W, V\in Subu(\mathfrak{A}))[W=V$ or $base(W)\cap base(V)=0]$.

Let $K \subseteq Crs_\alpha$. Then $K^{norm}$, $K^{comp}$, and $K^{wd}$ denote the classes of all normal, compressed and widely distributed members of K, respectively. E.g. $Gws_\alpha^{comp} = \{\mathfrak{A}\in Gws_\alpha : \mathfrak{A}$ is compressed$\}$.

The above notions were introduced in [HMTI] 2.6.

In [HMTI] 2.1, an elementary characterization is given for those

$Gws_\alpha$ -units which are members of a $Cs_\alpha$, for $\alpha\geq 3$. It is shown in [HMTI] 2.3 that for $\alpha=2$ there is no abstract characterization of $Gws_\alpha$ -units as members of a given $Cs_\alpha$. At the end of [HMTI] 2.3, Theorem 0.6 below is quoted. Theorem 0.6 says that there is no abstract characterization of those $Gs_\alpha$ -units which are members of a given $Cs_\alpha$, for $\alpha\geq\omega$.

Theorem 0.6. (noncharacterizability of $Gs_\alpha$ -units) Let $\alpha\geq\omega$ and let $U$ be an arbitrary set. Let $X\subseteq {}^\alpha U$ be any nondiscrete $Gs_\alpha$ -unit, i.e. let $X \nsubseteq D_{01}^{(U)}$. Then there is an automorphism $t \in Is(\mathfrak{Sb}\,{}^\alpha U, \mathfrak{Sb}\,{}^\alpha U)$ such that $t(X)$ is not a $Gws_\alpha^{norm}$ -unit (hence $t(X)$ is not a $Gs_\alpha$ -unit).

We need the following lemma.

Lemma 0.6.1. Let $\alpha\geq\omega$, and let $\mathfrak{C} = \mathfrak{Sb}\,{}^\alpha U$. Let $f : U \rightarrowtail\!\!\!\twoheadrightarrow U$ be a permutation of $U$. Let $p \in {}^\alpha U$. Then there is $t \in Is(\mathfrak{C},\mathfrak{C})$ such that $t\{q\}=\{f\circ q\}$ for every $q\in {}^\alpha U^{(p)}$ and $t\{q\}=\{q\}$ for every $q\in {}^\alpha U\sim({}^\alpha U^{(p)}\cup {}^\alpha U^{(f\circ p)})$.

Proof. Let everything be as in the hypotheses of 0.6.1. Let $\mathfrak{Z} \overset{d}{=} \mathfrak{Zd}\,\mathfrak{C}$ and $\mathfrak{P} \overset{d}{=} P\langle \mathfrak{Rl}_d\mathfrak{C} : d \in At\ Z\rangle$. Let $r \overset{d}{=} \langle\langle x\cap d : d \in At\ Z\rangle : x\in C\rangle$. Then $r \in Is(\mathfrak{C},\mathfrak{P})$ by [HMTI] 6.2 and by [HMT] 0.3.6 (ii), since $1^{\mathfrak{C}} = \sum At\ Z$. We have $r^{-1} = \langle \cup Rgk : k\in P\rangle$, since $(\forall d, b\in At\ Z) b\cap d=0$. Let $a \overset{d}{=} {}^\alpha U^{(p)}$ and $b \overset{d}{=} {}^\alpha U^{(f\circ p)}$. Then $a,b \in At\ Z$. Let $z : At\ Z \rightarrowtail\!\!\!\twoheadrightarrow At\ Z$ be the permutation of $At\ Z$ defined as $z \overset{d}{=} (At\ Z\sim\{a,b\})1Id \cup\{\langle a,b\rangle, \langle b,a\rangle\}$. For every $d \in At\ Z$ let

$$h_d \overset{d}{=} \begin{cases} \tilde{f} & \text{if } d=a \\ \tilde{f}^{-1} & \text{if } d=b \text{ and } d\neq a \\ Id & \text{if } d\notin\{a,b\} \end{cases} .$$

(For the notation $\tilde{f}$ see [HMTI] 3.5(i).) Let $g \stackrel{d}{=} \langle\langle h_d(k_d) : : d \in At\ Z \rangle : k \in P \rangle$. Now $g \in Is(\mathfrak{P}, P\langle \mathfrak{Rl}_{z(d)}\mathfrak{L} : d \in At\ Z \rangle)$ by [HMT] 0.3.6 (iii), since for every $d \in At\ Z$ we have $h_d \in Is(\mathfrak{Rl}_d\mathfrak{L}, \mathfrak{Rl}_{z(d)}\mathfrak{L})$, by [HMTI] 3.1. By [HMTI] 6.2 (and by [HMT] 0.3.6 (ii)) again we have $r^{-1} \in Is(P\langle \mathfrak{Rl}_{z(d)}\mathfrak{L} : d \in At\ Z \rangle, \mathfrak{L})$, since $r^{-1} = = \langle \cup Rgk : k \in P\langle Rl_d\mathfrak{L} : d \in At\ Z \rangle\rangle = \langle \cup Rgk : k \in P\langle Rl_{z(d)}\mathfrak{L} : : d \in At\ Z \rangle\rangle$. Let $t \stackrel{d}{=} r^{-1} \circ g \circ r$. Then $t \in Is(\mathfrak{L},\mathfrak{L})$. Let $x \in C$ be arbitrary. Then by the definitions of $r$ and $g$ we have $t(x) = = r^{-1}gr(x) = r^{-1}g\langle x \cap d : d \in At\ Z \rangle = r^{-1}\langle h_d(x \cap d) : d \in At\ Z \rangle = = \cup\{h_d(x \cap d) : d \in At\ Z\} = \tilde{f}(x \cap a) \cup \widetilde{f^{-1}}(x \cap (b \sim a)) \cup x \cap ({}^{\alpha}U \sim (a \cup b))$. Therefore if $x \leq a$ then $t(x) = \tilde{f}(x) = \{f \circ q : q \in x\}$ and if $x \leq ({}^{\alpha}U \sim (a \cup b))$ then $t(x) = x$.

<u>QED(Lemma 0.6.1.)</u>

Now we turn to the proof of Theorem 0.6. Let $\alpha \geq \omega$ and let $U$ be any set. Let $\mathfrak{L} \stackrel{d}{=} \mathfrak{Sb}\,{}^{\alpha}U$. Let $X \subset {}^{\alpha}U$ be any nondiscrete $Gs_{\alpha}$ -unit. Then there is a subbase $Y$ of $X$ such that $|Y|>1$ and $Y \neq U$ (and ${}^{\alpha}Y \subseteq X \subseteq {}^{\alpha}Y \cup {}^{\alpha}(U \sim Y)$). Let $m,n \in Y$ and $w \in U \sim Y$ be such that $m \neq n$. Such $m,n,w$ exist by $|Y|>1$ and by $Y \neq U$. Let $f \stackrel{d}{=} \stackrel{d}{=} (U \sim \{m,w\}) 1 Id \cup \{\langle m,w \rangle, \langle w,m \rangle\}$. Then $f$ is a permutation of $U$. Let $p \stackrel{d}{=} \langle m : i \in \alpha \rangle$. By Lemma 0.6.1 there is $t \in Is(\mathfrak{L},\mathfrak{L})$ such that $(\forall q \in {}^{\alpha}U^{(p)})\, t\{q\} = \{f \circ q\}$ and $(\forall q \in {}^{\alpha}U \sim ({}^{\alpha}U^{(p)} \cup \cup {}^{\alpha}U^{(f \circ p)}))\, t\{q\} = \{q\}$. We shall show that $t(X)$ is not a $Gws_{\alpha}^{norm}$ -unit. Let $\bar{n} \stackrel{d}{=} \langle n : i \in \alpha \rangle$ and $\bar{w} \stackrel{d}{=} \langle w : i \in \alpha \rangle$. Then $t\{\bar{n}^{0}_{w}\} = \{\bar{n}^{0}_{w}\} \nsubseteq X$ hence $\bar{n}^{0}_{w} \notin t(X)$. But $\bar{w}^{0}_{n} = f \circ (p^{0}_{n}) \in t(X)$. Then $t(X)$ is not normal since the ranges of $\bar{w}^{0}_{n}$ and $\bar{n}^{0}_{w}$ are not disjoint and one of them is in $t(X)$ while the other is not.

<u>QED(Theorem 0.6.)</u>

Proposition 0.7 below says that neither the $Ws_{\alpha}$ -units nor the $Gws_{\alpha}^{wd}$ -units have any abstract characterizations as members of $Cs_{\alpha}$-s.

Moreover, there is a $Cs_\alpha^{reg}$ $\mathfrak{A}$ such that those members of A which are $Ws_\alpha$-units (or $Gws_\alpha^{wd}$ -units) have no abstract characterization even if $\mathfrak{A}$ is fixed, i.e. even relative to $\mathfrak{A}$. Indeed, let $\alpha\geq\omega$ and $\varkappa>2$. Let $\mathfrak{A}$, X and t be as in Prop.0.7. Then $\mathfrak{Sb}X \in$ $\in Ws_\alpha$ and $\mathfrak{Sb}(tX) \notin Gws_\alpha^{norm}$. Let $K \in \{Ws_\alpha, Gws_\alpha^{wd}, Gws_\alpha^{norm}, Gws_\alpha^{comp}\}$. By $Ws_\alpha \subseteq K \subseteq Gws_\alpha^{norm}$ then $\mathfrak{Sb}X \in K$ but $\mathfrak{Sb}t(X) \notin K$. Thus the K -units have no abstract characterization in $\mathfrak{A}$.

<u>Proposition 0.7.</u> (noncharacterizability of $Ws_\alpha$ -units) Let $\alpha\geq\omega$ and $\varkappa>1$. Then there are $\mathfrak{A}\in{}_\varkappa Cs_\alpha$, $t \in Is(\mathfrak{A},\mathfrak{A})$ and a $Ws_\alpha$ -unit $X\in A$ such that $t(X)$ is not a $Ws_\alpha$ -unit. If $\varkappa>2$ then $\mathfrak{A}$ is regular and $t(X)$ is not a $Gws_\alpha^{norm}$ -unit either.

<u>Proof.</u> Let $\alpha\geq\omega$, $\varkappa>1$ and $\bar{s} \overset{d}{=} \langle s : i\in\alpha \rangle$ for any set s. Let $X \overset{d}{=} {}^\alpha\varkappa^{(\bar{0})}$. Then X is a $Ws_\alpha$ -unit. Let $\mathfrak{L} = \mathfrak{Sb}\,{}^\alpha\varkappa$ and $\mathfrak{A} = \mathfrak{Sg}^{(\mathfrak{L})}\{X\}$. Let $\mathfrak{B} = rl_X{}^*\mathfrak{A}$ and $\mathfrak{N} \overset{d}{=} rl(-X)^*\mathfrak{A}$. Then $\mathfrak{B},\mathfrak{N}\in Mn_\alpha$ and by [HMT] 2.5.25 $\mathfrak{B}\cong\mathfrak{N}$ since $\nabla\mathfrak{B} = \nabla\mathfrak{N}$ is finite. Since $X\in Zd\,A$, $\mathfrak{A} \cong \mathfrak{B} \times \mathfrak{N}$. Clearly $f = \langle\langle z\cap X, z\cap -X \rangle : z\in A \rangle \in Is(\mathfrak{A}, \mathfrak{B}\times\mathfrak{N})$ and $f(X)=\langle 1,0 \rangle$. By [HMT] 0.3.6, there is $h \in Is(\mathfrak{B}\times\mathfrak{N}, \mathfrak{B}\times\mathfrak{N})$ with $h\langle 1,0 \rangle = \langle 0,1 \rangle$ and $h\circ h \subseteq Id$. Then $k \overset{d}{=} f^{-1}\circ h\circ f\in Is(\mathfrak{A},\mathfrak{A})$ and $k(X)=-X$. Observing that $-X={}^\alpha\varkappa\sim{}^\alpha\varkappa^{(\bar{0})}$ is not a $Ws_\alpha$ -unit completes the proof of the first statement.

To prove the second statement assume $\varkappa>2$. Let $U \overset{d}{=} \varkappa$. For every $n<3$ define $V_n \overset{d}{=} {}^\alpha U^{(\bar{n})}$ and $x_0 \overset{d}{=} {}^\alpha\{0,1\}^{(\bar{0})}$, $x_1 \overset{d}{=} {}^\alpha\{1,2\}^{(\bar{1})}$, $x_2 \overset{d}{=} {}^\alpha\{2,0\}^{(\bar{2})}$. Let $\mathfrak{A} \overset{d}{=} \mathfrak{Sg}^{(\mathfrak{Sb}\,{}^\alpha U)}\{x_0, x_1\cup x_2\}$. Now $x_0$ is a $Ws_\alpha$ -unit and $x_1\cup x_2$ is not a $Gws_\alpha^{norm}$ -unit. We show that $t(x_0)=x_1\cup x_2$ for some $t \in Is(\mathfrak{A},\mathfrak{A})$. For every $n<3$ let $\mathfrak{B}_n \overset{d}{=} \mathfrak{Sg}^{(\mathfrak{Sb}V_n)}\{x_n\}$ and let $\mathfrak{L} \overset{d}{=} \mathfrak{Sg}^{(\mathfrak{Sb}(V_1\cup V_2))}\{x_1\cup x_2\}$. For every $i<j<3$ there is a base-isomorphism $b_{ij}\in Is(\mathfrak{Sb}V_i, \mathfrak{Sb}V_j)$ such that $b_{ij}(x_i)=x_j$. By [HMT] 0.3.6(ii) and [HMTI] 6.2 we then get $g'\in Ism(\mathfrak{Sb}V_0, \mathfrak{Sb}(V_1\cup V_2))$ such that $g'(x_0)=x_1\cup x_2$. Hence we obtain

$g \in Is(\mathfrak{B}_0, \mathfrak{L})$ such that $g(x_0) = x_1 \cup x_2$. Let $Q \overset{d}{=} \cup\{V_n : n<3\}$. Now we show that $Q$, $\mathfrak{A}$ and $G \overset{d}{=} \{x_0, x_1 \cup x_2\}$ satisfy the conditions of Prop. 4.7. Clearly, $\Delta^{(U)}Q = 0$. It is easy to check that every element of $G$ is $Q$-wsmall in the sense of Def.4.5 (actually, $G \subseteq Sm^{\mathfrak{A}}$, see Def.1.2) and that $G$ satisfies condition (i). Condition (ii) is satisfied since $\cup G \subseteq Q$. Therefore $\mathfrak{A}$ is regular by 4.7(I) since every element $g$ of $G$ is regular by $\Delta g = \alpha$. Also, $rl_Q \in Is\,\mathfrak{A}$ by 4.7(II) since $\mathfrak{M}\mathfrak{w}(\mathfrak{A})$ is simple by [HMTI] 5.3. Let $\mathfrak{R} \overset{d}{=} rl_Q {}^* \mathfrak{A}$. Then $R = Sg\{x_0, x_1 \cup x_2\}$. Let $\mathfrak{N} \overset{d}{=} \mathfrak{Sg}^{(\mathfrak{B}_0 \times \mathfrak{L})}\{\langle x_0, 0\rangle, \langle 0, x_1 \cup x_2\rangle\}$. By [HMTI] 6.2 there is $h \in Is(\mathfrak{R}, \mathfrak{N})$ such that $h(x_0) = \langle x_0, 0\rangle$ and $h(x_1 \cup x_2) = \langle 0, x_1 \cup x_2\rangle$. Let $f \overset{d}{=} \langle\langle g^{-1}z, gy\rangle : \langle y,z\rangle \in B_0 \times C\rangle$. By $g \in Is(\mathfrak{B}_0, \mathfrak{L})$ and by general algebra then $f \in Is(\mathfrak{B}_0 \times \mathfrak{L}, \mathfrak{B}_0 \times \mathfrak{L})$. By $f(\langle x_0, 0\rangle) = \langle 0, x_1 \cup x_2\rangle$ and $f(\langle 0, x_1 \cup x_2\rangle) = \langle x_0, 0\rangle$ then $f \in Is(\mathfrak{N}, \mathfrak{N})$. Let $t \overset{d}{=} rl_Q{}^{-1} \circ h^{-1} \circ f \circ h \circ rl_Q$. Now $t \in Is(\mathfrak{A}, \mathfrak{A})$ and $t(x_0) = x_1 \cup x_2$.

<u>QED(Proposition 0.7.)</u>

<u>Remark:</u> All the conditions are needed in Prop.0.7 because of the following. Let $\mathfrak{A} \in {}_\varkappa Cs_\alpha$, $t \in Is(\mathfrak{A}, \mathfrak{A})$ and $X \in A$. If $\varkappa = 1$ then $X$ is a $Ws_\alpha$-unit iff $X = 1$ iff $t(X)$ is a $Ws_\alpha$-unit. If $\varkappa = 0$ then $X$ is not a $Ws_\alpha$-unit. Let $\varkappa = 2$, $base(\mathfrak{A}) = 2$ and $X \in A$ be a $Ws_\alpha$-unit. Assume $t(X)$ is not a $Gws_\alpha^{norm}$-unit. Then $Subb(t(X)) = \{\{u\}, 2\}$ for some $u \in 2$. Thus $\Delta(t(X)) = \alpha$ and $t(X) \nsubseteq$ $\nsubseteq d_{ij}$ for every $i<j<\alpha$. By $t \in Is(\mathfrak{A}, \mathfrak{A})$ then $\Delta(X) = \alpha$ which implies $base(X) \neq \varkappa$, hence $|base(X)| = 1$. Then $X \subseteq d_{ij}$ for every $i, j \in \alpha$; a contradiction.

<u>Remark 0.8.</u> By Theorem 0.6, Prop.0.7 and by [HMTI] 2.1 we have a complete description of abstract characterizability of the distinguished types of units defined in [HMTI] 1.1. Recall that in [HMTI] seven kinds of units were introduced: $Cs_\alpha$, $Gs_\alpha$, $Ws_\alpha$, $Gws_\alpha^{comp}$,

$Gws_\alpha^{wd}$, $Gws_\alpha^{norm}$, $Gws_\alpha$-units. Of these 7 kinds of units exactly one has abstract characterization, and this one is the class of $Gws_\alpha$-units. By [HMTI] 2.1 exactly those members x of $Cs_\alpha$-s are $Gws_\alpha$-units for which $(\forall i,j\in\alpha)s_j^i x\cdot s_i^j x=x$. This is an abstract characterization of $Gws_\alpha$-units as members of $Cs_\alpha$-s. We call this characterization abstract because the property $s_j^i x\cdot s_i^j x=x$ of x is preserved under isomorphisms. By Theorem 0.6 and Prop.0.7 none of the remaining 6 types of units has any kind of abstract characterizations. Moreover, any one of these classes of units is even destroyed by automorphisms. Thus they cannot be characterized even if the $Cs_\alpha$ in which they are characterized as its members is fixed. (Theorem 0.6 shows this for $Cs_\alpha$, $Gs_\alpha$, $Gws_\alpha^{comp}$ and $Gws_\alpha^{norm}$-units and Prop. 0.7 shows this for $Ws_\alpha$, $Gws_\alpha^{comp}$, $Gws_\alpha^{wd}$ and $Gws_\alpha^{norm}$-units.)

The $Cs_\alpha$-units and the $Gs_\alpha$-units have an even stronger negative property not shared by any one of the remaining 5 kinds of units. By Theorem 0.6, in any full $Cs_\alpha$ $\mathfrak{C}$ for any nondiscrete $Gs_\alpha$-unit $x\in C\sim\{1^{\mathfrak{C}}\}$ there is an automorphism taking x to a $Gws_\alpha$-unit which is not a $Gs_\alpha$-unit; hence the same holds for $Cs_\alpha$-units. Next we show that $Ws_\alpha$-units do have an abstract characterization as members of full $Cs_\alpha$-s. Let $\mathfrak{C}$ be any full $Cs_\alpha$ and let $x\in C$. Then x is a $Ws_\alpha$-unit iff $[(\exists y\in At\ Zd\ C)x\leq y$ and $(\forall i,j\in\alpha)$ $s_j^i x\cdot s_i^j x=x]$ since by [HMTI] 2.1 x is a $Gws_\alpha$-unit iff $(\forall i,j\in\alpha)$ $s_j^i x\cdot s_i^j x=x$ and since $At\ Zd\ C=\{{}^\alpha base(\mathfrak{C})^{(q)} : q\in 1^{\mathfrak{C}}\}$. Define the set of formulas

$$\Phi(x) \overset{d}{=} \{(\exists y)[\bigwedge_{i<\alpha} c_i y=y \ \wedge\ \forall z(\bigwedge_{i<\alpha} c_i z=z \rightarrow (z=0\vee z\geq y)) \wedge x\leq y] \wedge$$

$$\wedge\ s_j^i x\cdot s_i^j x=x : i,j\in\alpha\} .$$

By the above, for any full $Cs_\alpha$ $\mathfrak{C}$, an element $a\in C$ is a $Ws_\alpha$-unit iff $\mathfrak{C}\models\Phi[a]$. This is an abstract characterization of $Ws_\alpha$-units in full $Cs_\alpha$-s. However, this characterization is not elementary since the set $\Phi(x)$ of formulas contains infinitary formulas. Since $Ws_\alpha \subseteq Gws_\alpha^{comp} \cap Gws_\alpha^{wd} \cap Gws_\alpha^{norm}$ we have that Theorem 0.6 does not extend to any of the remaining 5 kinds of units (from $Cs_\alpha$ - and $Gs_\alpha$ -units) in its full strength.

Proposition 0.9. Let $\alpha \geq 2$ .

(i) $\mathbf{I}Crs_\alpha$ is a variety, i.e. $\mathbf{I}Crs_\alpha = \mathbf{HSP}Crs_\alpha$.

(ii) $\mathbf{I}Crs_\alpha^{reg}$ is a quasiequational class, i.e. $\mathbf{I}Crs_\alpha^{reg} = \mathbf{SPUp}Crs_\alpha^{reg}$ .

(iii) $\mathbf{I}Crs_\alpha^{reg} \neq \mathbf{H}Crs_\alpha^{reg}$ if $\alpha > 3$ .

(iv) $\mathbf{I}Crs_\alpha \neq \mathbf{I}Crs_\alpha^{reg}$ iff $\alpha > 2$ iff $Crs_\alpha \neq Crs_\alpha^{reg}$ .

The detailed proof can be found in the preprint [AN6]. A brief outline of proof appeared in [N]. To save space we omit the proof.

Lemma 0.10. Let $\mathfrak{A} \in \mathbf{S}Cr_\alpha$ . Let $x,y \in A$ . Then $\Delta(x \oplus y) \supseteq (\Delta x) \oplus (\Delta y)$.

Proof. Let $i \in (\Delta x) \oplus (\Delta y)$ . We may suppose $i \in \Delta y \sim \Delta x$ , since $\oplus$ is symmetric. Then $c_i x = x$ and $c_i y > y$ , i.e. $c_i y \cdot -y \neq 0$ . Therefore either $x \cdot c_i y \cdot -y \neq 0$ or $(-x) \cdot (c_i y \cdot -y) \neq 0$ .

Case 1 Assume $x \cdot c_i y \cdot -y \neq 0$ . Let $z \overset{d}{=} x \cdot c_i y \cdot -y$ . $z \leq x \cdot c_i y = c_i(c_i x \cdot y) = c_i(x \cdot y) \leq c_i(-(x \oplus y))$ , by $x = c_i x$ and by $(C_3)$. $z \leq x \cdot -y \leq x \oplus y$ . Therefore $0 \neq z \leq (x \oplus y) \cdot c_i(-(x \oplus y))$ . This implies $c_i(x \oplus y) \neq x \oplus y$ , since $c_i(x \oplus y) = x \oplus y$ implies $c_i(x \oplus y) \cdot c_i(-(x \oplus y)) = c_i((x \oplus y) \cdot -(x \oplus y)) = c_i 0 = 0$ by $(C_0)$, $(C_3)$ and $(C_1)$ .

Case 2 Assume $-x \cdot c_i y \cdot -y \neq 0$ . Let $z \overset{d}{=} -x \cdot c_i y \cdot -y$ . $z \leq -x \cdot c_i y \leq x \oplus c_i y = c_i x \oplus c_i y \leq c_i(x \oplus y)$ by $x = c_i x$ and by $Bo_\alpha \models (c_i x \oplus c_i y \leq c_i(x \oplus y))$. Therefore $0 \neq z \leq c_i(x \oplus y) \cdot -(x \oplus y)$ shows that $i \in \Delta(x \oplus y)$ .

QED(Lemma 0.10.)

We note that Lemma 0.10 above becomes false if we replace $\mathsf{S}Cr_\alpha$ by "normal $Bo_\alpha$ ". By the above proof, in every $Bo_\alpha$ satisfying $(C_1)$, $(C_3)$ the conclusion of Lemma 0.10 holds. None of these two conditions can be dropped. Actually, in any $Bo_\alpha$ $\mathfrak{A}$ the condition $\mathfrak{A} \models \{c_i 0=0\ ,\ (c_i x=x \rightarrow c_i -x=-x) : i\in\alpha\}$ is equivalent with the conclusion of Lemma 0.10.

## 1. Regular cylindric set algebras

Definition 1.1. Let $V$ be a $Crs_\alpha$-unit. Let $x\subseteq V$ and $H\subseteq\alpha$ . Then $x$ is H-regular in $V$ iff

$(\forall q\in x)(\forall k\in V)[(H\cup\Delta x)1q\subseteq k \Rightarrow k\in x]$.

$x$ is regular in $V$ iff [either $\alpha=0$ or $x$ is 1-regular in $V$].
Let $\mathfrak{A}\in Crs_\alpha$ . Then $x$ is H-regular in $\mathfrak{A}$ iff it is H-regular in $1^{\mathfrak{A}}$ . $\mathfrak{A}$ is said to be H-regular iff every element of $A$ is H-regular in $\mathfrak{A}$ and $\mathfrak{A}$ is regular iff every element of $A$ is regular in $\mathfrak{A}$. Instead of "$x$ is regular in $V$" we shall say "$x$ is regular" when $V$ is understood from context.

The above definition of regularity agrees with [HMTI] 1.1(ix).

A natural question about $Gws_\alpha$-s is: do regular elements generate regular ones? [HMTI] 4.1, 4.2 and section 4 here deal with this question. By [HMTI] 4.1, 4.2 the question arises: which collections of not necessarily finite dimensional elements do generate regular subalgebras. Theorems 1.3, 4.6, 4.7 and 4.9 concern this question. They will be frequently used in constructing regular algebras.

Definition 1.2. Let $\mathfrak{A}\in CA_\alpha$ . Then $x\in A$ is said to be small in

$\mathfrak{A}$ iff for every infinite $K \subseteq \Delta^{\mathfrak{A}}x$ we have

$(\forall\Gamma \subseteq_\omega \alpha)(\exists\Theta \subseteq_\omega K) c^\partial_{(\Theta)} c_{(\Gamma)} x = 0$ .

$Sm^{\mathfrak{A}} \overset{d}{=} \{x \in A : x$ is small in $\mathfrak{A}\}$.

<u>Theorem 1.3.</u> Let $\mathfrak{A} \in Gws_\alpha^{norm}$ be generated by $X \subseteq Sm^{\mathfrak{A}}$. Assume every element of $X$ is regular. Then $\mathfrak{A}$ is regular.

Before proving Theorem 1.3, we shall prove some lemmas which might be of interest in themselves.

<u>Definition 1.3.1.</u> Let $H \subseteq \alpha$ and $\mathfrak{A} \in CA_\alpha$ . Then

$I_H^{\mathfrak{A}} \overset{d}{=} \{x \in A : (\forall\Gamma \subseteq_\omega \alpha)(\exists\Theta \subseteq_\omega \alpha{\sim}H) c^\partial_{(\Theta)} c_{(\Gamma)} x = 0\}$ and

$Dm_H^{\mathfrak{A}} \overset{d}{=} \{x \in A : |\Delta^{\mathfrak{A}}(x){\sim}H| < \omega\}$ . By [HMT] 2.1.4, $Dm_H^{\mathfrak{A}} \in Su\mathfrak{A}$.

$\mathfrak{Dm}_H^{\mathfrak{A}}$ denotes the subalgebra of $\mathfrak{A}$ with universe $Dm_H^{\mathfrak{A}}$ .

<u>Notation:</u> The superscript $\mathfrak{A}$ will be dropped if there is no danger of confusion. Sometimes we shall write $\mathfrak{Dm}_H(\mathfrak{A})$ or $Sm(\mathfrak{A})$ instead of $\mathfrak{Dm}_H^{\mathfrak{A}}$ or $Sm^{\mathfrak{A}}$ .

<u>Lemma 1.3.2.</u> Let $H \subseteq \alpha$ and $\mathfrak{A} \in CA_\alpha$ .

(i) $I_H$ is an ideal of $\mathfrak{A}$ .

(ii) $I_H \cap Dm_H = \{0\}$.

(iii) $Sm{\sim}Dm_H \subseteq I_H$ .

<u>Proof.</u> <u>Proof of (i):</u> Let $x \in I_H$ . Then $(\forall y \le x) y \in I_H$ and $(\forall\Delta \subseteq_\omega \alpha)$ $c_{(\Delta)} x \in I_H$ follow immediately by the definition of $I_H$ . It remains to show that $x+y \in I_H$ whenever $\{x,y\} \subseteq I_H$ . Let $x,y \in I_H$ . Let $\Gamma \subseteq_\omega \alpha$ be arbitrary. We have to show that $(\exists\Theta \subseteq_\omega \alpha{\sim}H) c^\partial_{(\Theta)} c_{(\Gamma)}(x+y) = 0$ . $x \in I_H$ implies that $(\exists\Theta_1 \subseteq_\omega \alpha{\sim}H) c^\partial_{(\Theta_1)} c_{(\Gamma)} x = 0$ . $y \in I_H$ implies that $(\exists\Theta_2 \subseteq_\omega \alpha{\sim}H) c^\partial_{(\Theta_2)} c_{(\Theta_1 \cup \Gamma)} y = 0$ . Let $\Theta \overset{d}{=} \Theta_1 \cup \Theta_2$ . Clearly, $\Theta \subseteq_\omega \alpha{\sim}H$ .

$c^{\partial}_{(\Theta)}c_{(\Gamma)}(x+y) = c^{\partial}_{(\Theta_2)}c^{\partial}_{(\Theta_1)}(c_{(\Gamma)}x + c_{(\Gamma)}y) \le c^{\partial}_{(\Theta_2)}(c^{\partial}_{(\Theta_1)}c_{(\Gamma)}x + c_{(\Theta_1)}c_{(\Gamma)}y) = c^{\partial}_{(\Theta_2)}c_{(\Theta_1\cup\Gamma)}y = 0$, by $CA_\alpha \models c_i^{\partial}(x+y) \le c_i^{\partial}x + c_i y$. We have seen that $x+y\in I_H$. The above proves that $I_H \in Il\mathfrak{A}$, by [HMT] 2.3.7.

Proof of (ii): Let $y \in I_H \cap Dm_H$ be arbitrary. We have to show that $y=0$. $y \in Dm_H$ means that $\Gamma \overset{d}{=} \Delta y\sim H$ is finite. By $y\in I_H$ we have that $(\exists\Theta\subseteq_\omega \alpha\sim H)c^{\partial}_{(\Theta)}c_{(\Gamma)}y=0$. But $c^{\partial}_{(\Theta)}c_{(\Gamma)}y = c_{(\Gamma)}y$ since $\Delta(c_{(\Gamma)}y)\subseteq H$ and $\Theta\subseteq\alpha\sim H$. Therefore $c_{(\Gamma)}y=0$, i.e. $y=0$.

Proof of (iii): Let $x\in Sm\sim Dm_H$ be arbitrary. Then $|\Delta x\sim H|\ge\omega$ by $x\notin Dm_H$. Let $\Gamma\subseteq_\omega \alpha$ be fixed. Then $(\exists\Theta\subseteq_\omega \Delta x\sim H)c^{\partial}_{(\Theta)}c_{(\Gamma)}x=0$ by $x\in Sm$ and by $|\Delta x\sim H|\ge\omega$. Then $\Theta\subseteq_\omega \alpha\sim H$, i.e. we have seen that $x\in I_H$.

QED(Lemma 1.3.2.)

Lemma 1.3.3. Let $\mathfrak{A}\in CA_\alpha$ be generated by $X\subseteq Sm^{\mathfrak{A}}$. Let $H\subseteq\alpha$. Then $Dm_H^{\mathfrak{A}} = Sg^{(\mathfrak{A})}(X\cap Dm_H)$.

Proof. First we state a fact about ideals and generator sets of $CA_\alpha$ -s which is an easy consequence of the definition of $Il\mathfrak{A}$ ([HMT]2.3.5).

Fact(*): Let $\mathfrak{A}\in CA_\alpha$ be generated by $X\subseteq A$. Let $B \in Su\mathfrak{A}$ and $I \in Il\mathfrak{A}$ be such that $B\cap I=\{0\}$. Then $B \subseteq Sg^{(\mathfrak{A})}(X\sim I)$.

(Fact(*) is true because $\mathfrak{A}/I$ is generated by $\{b/I : b\in X\sim I\}$ and the function $\langle b/I : b\in B\rangle$ is one-one.)

Now we return to the proof of Lemma 1.3.3. Let $\mathfrak{A}$ be generated by $X\subseteq Sm$ and let $H\subseteq\alpha$ be fixed. Then $I_H \in Il\mathfrak{A}$, $Dm_H \in Su\mathfrak{A}$, $Dm_H\cap I_H=\{0\}$, and $X\sim I_H\subseteq Dm_H$ by Lemma 1.3.2. Hence by Fact(*) $Dm_H\subseteq Sg(X\cap Dm_H)$ and thus $Dm_H = Sg(X\cap Dm_H)$, by $Dm_H \in Su\mathfrak{A}$.

QED(Lemma 1.3.3.)

<u>Lemma 1.3.4.</u> Let $\alpha>0$ , $\mathfrak{A}\in Gws_{\alpha}$ and $x\in A$ .

(i) Statements a.-e. below are <u>equivalent</u>.

a. $x$ is regular (in $1^{\mathfrak{A}}$).

b. $x$ is $\{i\}$-regular for some $i\in\alpha$ .

c. $x$ is $H$-regular for some $H\subseteq_{\omega}\alpha$ .

d. $x$ is $H$-regular for all nonempty $H\subseteq_{\omega}\alpha$ .

e. $(\forall q\in x)(\forall k\in 1^{\mathfrak{A}})([Rgq\cap Rgk\neq 0 \text{ and } \Delta x 1 q\subseteq k] \Rightarrow k\in x)$ .

(ii) Let $H$ and $G$ be nonempty subsets of $\alpha$ . Suppose that the symmetric difference $H\oplus G$ of $H$ and $G$ is finite. Then $x$ is $H$-regular iff $x$ is $G$-regular.

<u>Proof.</u> <u>Proof of (ii):</u> It is enough to prove the following:

$(\forall F\subseteq_{\omega}\alpha)(\forall H\neq 0)(x \text{ is } H\text{-regular} \iff x \text{ is } H\cup F\text{-regular})$.

The direction "H-regular $\Rightarrow$ $(H\cup F)$-regular" is obvious by observing that $(\forall L\subseteq R\subseteq\alpha)[x \text{ is } L\text{-regular} \Rightarrow x \text{ is } R\text{-regular}]$ . Let $F\subseteq_{\omega}\alpha$ and $0\neq H\subseteq\alpha$ and suppose that $x$ is $H\cup F$-regular. We prove that $x$ is $H$-regular. Let $k\in V$ and $(H\cup\Delta x)1k \subseteq q\in x$ . Since $H\neq 0$ there exists $b\in Rgk\cap Rgq$ . Let this $b$ be fixed. Let $P \stackrel{d}{=} F\sim(H\cup\Delta x)$ .

<u>Notation:</u> $f_b^{(P)} \stackrel{d}{=} (\alpha\sim P)1f\cup P\times\{b\}$. Since $P$ is finite, we have $k_b^{(P)}\in V$ and $q_b^{(P)}\in x$ by $P\subseteq_{\omega}(\alpha\sim\Delta x)$ . By $(\Delta x\cup H\cup F)1k_b^{(P)} \subseteq q_b^{(P)}\in x$ and by $H\cup F$-regularity of $x$ we have $k_b^{(P)}\in x$ . Then $k\in x$ since $P\subseteq_{\omega}(\alpha\sim\Delta x)$ . This proves that $x$ is $H$-regular.

Note that in this direction the assumption $H\neq 0$ was essential (namely: every $0$-regular element is $F$-regular, but while ${}^{\omega}1$ is $1$-regular in ${}^{\omega}1\cup{}^{\omega}\{1\}$ , it is not $0$-regular).

<u>Proof of (i):</u> c. $\Rightarrow$ d. holds by (ii) of the present lemma and by the fact that if $G\subseteq H\subseteq\alpha$ then $G$-regularity of $x$ implies $H$-regularity of $x$ . Then a. - d. are equivalent since a. $\Rightarrow$ b. $\Rightarrow$ c. and d. $\Rightarrow$ a. hold by $\alpha>0$ . e. $\Rightarrow$ d. holds by the definition of $H$-regularity. If $\Delta x\neq 0$ then d. $\Rightarrow$ e. can be seen by choosing $H = \Delta x$ .

Suppose $\Delta x=0$ . Let $q\in x$ and $k\in V$ . Then $Rgq\cap Rgk\neq 0$ implies the existence of an element $b$ such that $q_b^0\in x$ , $k_b^0\in V$ and $k_b^0\in x \iff$ $\iff k\in x$ . The above prove that e. $\iff$ d..

QED(Lemma 1.3.4.)

Lemma 1.3.5. Let $\mathfrak{A}\in Gws_\alpha$ and let $H\subseteq\alpha$ be nonempty.

(i) $H$ -regular elements generate $H$ -regular ones in $\mathfrak{Bl}\,\mathfrak{Dm}_H^{\mathfrak{A}}$ .

(ii) If $\mathfrak{A}$ is normal then $H$ -regular elements generate $H$ -regular ones in $\mathfrak{Dm}_H$ .

Proof. Let $\mathfrak{A}\in Gws_\alpha$ and let $0\neq H\subseteq\alpha$ . By definition, if an element $x\in A$ is $H$ -regular then $-x$ is $H$ -regular too, in any $\mathfrak{A}\in Crs_\alpha$ . Let $x,y\in Dm_H$ both be $H$ -regular. Let $G \stackrel{d}{=} H\cup\Delta x\cup\Delta y$ . Now $G \oplus H$ is finite by $x,y\in Dm_H$ . Then by Lemma 1.3.4 (ii) and by $G\supseteq H\neq 0$ it is enough to show $G$ - regularity instead of $H$ - regularity. Both $x$ and $y$ are $G$ -regular by $G\supseteq H$ . Let $q\in(x\cap y)$ and $k\in V$ be such that $G1q\subseteq k$ . Since both $x$ and $y$ are $G$ -regular, we have $k\in(x\cap y)$. Thus $x\cap y$ is $G$ -regular. By this, (i) is proved. Obviously, if $\mathfrak{A}\in Gws_\alpha$ then $d_{ij}$ is $H$ -regular, for every $H\subseteq\alpha$ , $i,j\in\alpha$ . Suppose $\mathfrak{A}$ is normal and $i\in\alpha$ . Let $k\in V$ and $G1k\subseteq q\in c_ix$ . Then $G1k_b^i\subseteq q_b^i\in x$ for some $b$ . Further, $k_b^i\in V$ because $Rgk\cap Rgq\neq 0$ (by $H\neq 0$), $q_b^i\in V$ , and $V$ is normal. Then $k_b^i\in x$ by $G$ -regularity of $x$ , and hence $k\in c_ix$ . Therefore $c_ix$ is $G$ -regular.

QED(Lemma 1.3.5.)

Lemma 1.3.6. Let $\mathfrak{A}\in Gws_\alpha^{norm}$ be generated by $X\subseteq A$ . Suppose $(\forall H\subseteq\alpha)Dm_H^{\mathfrak{A}} = Sg^{(\mathfrak{A})}(X\cap Dm_H)$ . Then $\mathfrak{A}$ is regular if every element of $X$ is regular.

Proof. Assume the hypotheses. Let $y\in Sg^{(\mathfrak{A})}X$ be arbitrary. Let $H=1\cup\Delta y$ . Clearly $y\in Dm_H$ . Then $y\in Sg^{(\mathfrak{A})}(X\cap Dm_H)$, by our assumption. Every element of $X\cap Dm_H$ is $H$ -regular, by Lemma 1.3.4 and by $H\neq 0$.

Then $y$ is $H$-regular by Lemma 1.3.5. Since $H=1\cup\Delta y$, this means that $y$ is regular.

QED(Lemma 1.3.6.)

Now Theorem 1.3 follows immediately from Lemmas 1.3.3, 1.3.6.

QED (Theorem 1.3.)

Let $H\subseteq\alpha$, $\mathfrak{A}\in CA_\alpha$ and $x\in A$. Recall from [HMT] that $x$ is an $H$-atom of $\mathfrak{A}$ if it is an atom of the Boolean algebra $\mathfrak{Cl}_H\mathfrak{A}$ of $H$-closed elements of $\mathfrak{A}$.

Corollary 1.4.

(i) Every normal $Gws_\alpha$ generated by its atoms is regular. (Therefore every $Gs_\alpha$ generated by its atoms is regular.)

(ii) Let $\mathfrak{A}\in Gws_\alpha^{norm}$ and let $H\subseteq\alpha$. Suppose $A=Sg\ At\ Cl_H\mathfrak{A}$. Then a.-b. below hold.

a. If $H$ is finite then $\mathfrak{A}$ is regular.

b. Let $Y\subseteq At\ Cl_H\mathfrak{A}$ be a set of regular elements. Then $\mathfrak{Sg}^{(\mathfrak{A})}Y$ is regular.

Proof. Since $A = Cl_0\mathfrak{A}$ we have that (i) is a special case of (ii). Hence it is enough to prove (ii). Let $\mathfrak{A}\in Gws_\alpha^{norm}$. By Theorem 1.3 it is enough to prove the following claim.

Claim 1.4.1.

(i) $At\ Cl_H\mathfrak{A} \subseteq Sm^{\mathfrak{A}}$ for every $H\subseteq\alpha$ and $\mathfrak{A}\in CA_\alpha$.

(ii) Let $H\subseteq_\omega\alpha$ and $\mathfrak{A}\in Gws_\alpha^{norm}$. Then every $H$-atom of $\mathfrak{A}$ is regular.

Proof. Proof of(i): Let $y$ be an $H$-atom of $\mathfrak{A}$. Let $K\subseteq\Delta y$ be infinite and let $\Gamma\subseteq_\omega\alpha$. Then $c_{(\Gamma)}y$ is a $H\cup\Gamma$-atom by [HMT] 1.10.3(i). $|\alpha\sim(H\cup\Gamma)|\geq\omega$ and $\Delta y\neq 0$ by $|K|\geq\omega$. Then by [HMT]

1.10.5(i), $y \not\leq c_0^\partial d_{01}$ which implies $c_{(\Gamma)}y \not\leq c_0^\partial d_{01}$ and therefore $\Delta(c_{(\Gamma)}y) = \alpha \sim (H \cup \Gamma)$. Thus $K \cap \Delta(c_{(\Gamma)}y) \neq 0$. Let $i \in K \cap \Delta(c_{(\Gamma)}y)$. Then $c_{(\Gamma)}y > c_i^\partial c_{(\Gamma)}y \in Cl_{(H\cup\Gamma)}\mathfrak{A}$ which implies $c_i^\partial c_{(\Gamma)}y = 0$, by $c_{(\Gamma)}y \in At\ Cl_{(H\cup\Gamma)}\mathfrak{A}$. We have seen that $y \in Sm^{\mathfrak{A}}$. (Note that [HMT] 1.10.3, 1.10.5 hold for $H$-atoms where $H$ is infinite, too. This is true because in their proofs no condition implying $|H|<\omega$ is used.)

<u>Proof of (ii):</u> If $\alpha<\omega$ then we are done by [HMTI] 1.17. Suppose $\alpha \geq \omega$. Let $H \subseteq_\omega \alpha$. Let $y$ be an $H$-atom. Then either $\Delta y = 0$ or else $\Delta y = \alpha \sim H$; and if $\Delta y = 0$ then $y \leq c_0^\partial d_{01}$, by [HMT] 1.10.5 (i) and by $\alpha \geq \omega$. If $\Delta y = \alpha \sim H$ then $y$ is regular by $|H|<\omega$, since every cofinite dimensional element is regular. If $\Delta y = 0$ then by $y \leq c_0^\partial d_{01}$, all the elements of $y$ are contained in one-element subbases which implies regularity, since $\mathfrak{A}$ is <u>normal</u>. (This last implication is <u>not</u> true for $Gws_\alpha$ in general.)

<u>QED(Claim 1.4.1 and Corollary 1.4.)</u>

<u>Remark 1.5.</u> (1) Lemma 1.3.5 becomes false if we omit the condition $H \neq 0$. Namely, let $\mathfrak{L}$ be the full $Gs_\alpha$ with unit $V \stackrel{d}{=} {}^\alpha 2 \cup {}^\alpha\{2\}$. Let $x = \{f \in {}^\alpha 2 : f(0)=0\}$. Clearly, $x$ is $0$-regular in $V$ but $c_0 x = {}^\alpha 2$ is <u>not</u> $0$-regular in $V$, since $\Delta({}^\alpha 2) = 0$.

(2) None of 1.3, 1.3.5 (ii), 1.3.6, 1.4 and 1.4.1 (ii) is true for $Gws_\alpha$ in general. This follows from the following. (See also Prop. 4.2.) Let $\alpha \geq \omega$, $\bar{1} \stackrel{d}{=} \alpha \times \{1\}$ and $V \stackrel{d}{=} {}^\alpha 1 \cup {}^\alpha 2^{(\bar{1})}$. Let $\mathfrak{M} \stackrel{d}{=} \mathfrak{Mn}(\mathfrak{Sb}\,V)$. Then ${}^\alpha 1 \in At\,\mathfrak{M}$ is not regular in $V$.

(3) The condition in Cor. 1.4 (ii)b that the elements of $Y$ be regular cannot be omitted (even if we suppose that $\alpha \sim H$ is infinite): there exists a $Cs_\alpha$ with some non-regular $H$-atoms. Namely, let $\alpha \geq \omega$ and $\mathfrak{L} \stackrel{d}{=} \mathfrak{Sb}\,{}^\alpha\omega$. Let $b \stackrel{d}{=} \{f \in {}^\alpha\omega^{(\bar{0})} : (\alpha \sim H) \upharpoonright f \subseteq \bar{0}\}$ where $\bar{0} = \alpha \times \{0\}$. Clearly $b$ is an $H$-atom of $\mathfrak{L}$. Now $b$ is regular iff $H$ is finite.

Remark 1.6. (Notions of regularity)

Definition 1.6.1. Let $K \subseteq Crs_\alpha$ . Then

$K^{oreg} \overset{d}{=} \{\mathfrak{A} \in K : \mathfrak{A}$ is 0-regular$\}$.

$K^{zdreg} \overset{d}{=} \{\mathfrak{A} \in K : (\forall a \in A \sim Zd\ A)$ a is 0-regular$\}$.

$K^{ireg} \overset{d}{=} \{\mathfrak{A} \in K : (\forall a \in A)(\exists i \in \alpha)$ a is $\{i\}$-regular$\}$.

$K^{creg} \overset{d}{=} \{\mathfrak{A} \in K : (\forall a \in A)(\forall i \in \alpha)$ a is $\{i\}$-regular$\}$.

It seems that among the above notions of regularity the most interesting ones are cregularity and zdregularity.

Proposition 1.6.2. Let $\alpha \geq 2$ .

(i) $Gws_\alpha^{reg} = Gws_\alpha^{zdreg} = Gws_\alpha^{ireg} = Gws_\alpha^{creg} \supset Gws_\alpha^{oreg}$ .

(ii) $Gws_\alpha^{comp\ reg} = Gws_\alpha^{comp\ oreg}$ .

(iii) $\mathsf{I}Crs_\alpha \supset \mathsf{I}Crs_\alpha^{ireg} \supset \mathsf{I}Crs_\alpha^{reg} \supset \mathsf{I}Crs_\alpha^{creg} \supset \mathsf{I}Crs_\alpha^{oreg}$ and

$Crs_\alpha \supset Crs_\alpha^{zdreg} \supset Crs_\alpha^{creg}$ .

(iv) $\mathsf{I}Crs_\alpha^{creg} = \mathsf{I}Crs_\alpha^{zdreg} = \mathbf{SP}Crs_\alpha^{oreg}$.

Proof. Let $\alpha \geq 2$ . Proof of (i): $Gws_\alpha^{reg} = Gws_\alpha^{ireg} = Gws_\alpha^{creg}$ follows from 1.3.4(i). Now we show $Gws_\alpha^{zdreg} \subseteq Gws_\alpha^{ireg}$. Let $\mathfrak{A} \in$ $\in Gws_\alpha \sim Gws_\alpha^{ireg}$. We show $\mathfrak{A} \notin Gws_\alpha^{zdreg}$. By $\mathfrak{A} \notin Gws_\alpha^{ireg}$ there is $x \in A$ such that $(\forall i \in \alpha)$ x is not $\{i\}$-regular. Then $\Delta x = 0$ and x is not 1-regular. Then there are $f, k \in 1^{\mathfrak{A}}$ such that $f0 = k0$, $f \in x$ and $k \in -x$. By $f \neq k$ and $f0 = k0$ we have that there is $W \in Subu(\mathfrak{A})$, $|base(W)| \geq 2$ such that either $f \in W$ or $k \in W$. Since $\Delta -x = 0$ and $-x$ is not 0-regular, we may assume $f \in W$. Let $d \in \{d_{01}^{\mathfrak{A}}, -d_{01}^{\mathfrak{A}}\}$ be such that $f \in d$. Then $\Delta(d \cdot x) = 2$. By $\mathfrak{A} \in Gws_\alpha$ we have $h \overset{d}{=} k_{k1}^{0} \in 1^{\mathfrak{A}}$ and $h \notin x$ by $\Delta x = 0$. Then $\Delta(d \cdot x) 1 h \subseteq f \in d \cdot x$ proves that $d \cdot x$ is not 0-regular. Hence $\mathfrak{A} \notin Gws_\alpha^{zdreg}$ by $\Delta(d \cdot x) \neq 0$. $Crs_\alpha^{creg} \subseteq Crs_\alpha^{zdreg}$ can be seen as follows. Let $\mathfrak{A} \in Crs_\alpha^{creg}$ and let $x \in A \sim Zd\ A$. Let $i \in \Delta x$. Then x is $\{i\}$-regular by $\mathfrak{A} \in Crs_\alpha^{creg}$. Thus x is 0-regular by $i \in \Delta x$. We show $Gs_\alpha^{oreg} \nsubseteq Gs_\alpha^{reg}$. Let $\bar{0} \overset{d}{=} \alpha \times 1, \bar{1} \overset{d}{=} \alpha \times \{1\}$, $V \overset{d}{=} \{\bar{0}\} \cup \{\bar{1}\}$ and $\mathfrak{L} \overset{d}{=} \mathfrak{Sb}V$.

Then $\mathfrak{C}$ is regular by $(\forall f,k\in 1^{\mathfrak{C}})[f0=k0 \Rightarrow f=k]$. $\mathfrak{C} \notin Gws_\alpha^{oreg}$ since $\{\bar{0}\}\in C$ is not 0-regular.

Proof of (ii): Let $\mathfrak{A}\in Gws_\alpha^{comp\ reg}$. We show $\mathfrak{A}\in Gws_\alpha^{oreg}$. Let $x\in A$, $f,k\in 1^{\mathfrak{A}}$, and suppose $\Delta x1k\subseteq f\in x$. Let $h \overset{d}{=} f^0_{k0}$. Then $h\in 1^{\mathfrak{A}}$ by $\mathfrak{A}\in Gws_\alpha^{comp}$ and $h\in x$ by $\Delta x1k\subseteq f$. Now $1\cup\Delta x1k\subseteq h\in x$ and regularity of $x$ imply $k\in x$. We have seen that $x$ is 0-regular.

To save space, we omit the proofs of (iii)and (iv).

QED(Proposition 1.6.2.)

By Prop.1.6.2 we have the following connections with earlier papers. The notion of "i-finiteness" as defined in [AGN1],[AN1],[AGN2] coincides with regularity in $Cs_\alpha\cap Lf_\alpha$. In $Gws_\alpha\cap Lf_\alpha$ "i-finiteness" of [AN1] ,[AGN2] coincides with regularity. In $Crs_\alpha\cap Lf_\alpha$ "i-finiteness" of [AGN1] coincides with 0-regularity.

## 2. Relativization

Definition 2.1. Let $\mathfrak{A}\in Crs_\alpha$. Let $Z\subseteq 1^{\mathfrak{A}}$ and let $\mathfrak{C} = \mathfrak{Sb}\,Z$.

(i) $rl_Z^{\mathfrak{A}} \overset{d}{=} rl_Z^A \overset{d}{=} rl^A(Z) \overset{d}{=} \langle x\cap Z : x\in A\rangle$.

(ii) $Rl_Z\mathfrak{A} \overset{d}{=} Rl_Z\,A \overset{d}{=} Rl(Z)\mathfrak{A} \overset{d}{=} Sg^{(\mathfrak{C})}\ rl_Z^*\,A$ and

$\mathfrak{Rl}_Z\mathfrak{A} \overset{d}{=} \mathfrak{Rl}(Z)\mathfrak{A} \overset{d}{=} \mathfrak{Sg}^{(\mathfrak{C})}\ rl_Z^*\,A$.

We shall omit the superscipts $\mathfrak{A}$ and $A$ if there is no danger of confusion.

The above definition of $rl_Z^{\mathfrak{A}}$ agrees with [HMTI]6.1. We shall frequently use the fact that $rl_Z^A\in Ho(\mathfrak{A},\mathfrak{Rl}_Z\mathfrak{A})$ for any $\mathfrak{A}\in Crs_\alpha$ and $Z\in Zd\ Sb\ 1^{\mathfrak{A}}$. This fact follows by the proof of [HMT]2.3.26, a detailed proof can be found in [N1].

Prop.2.2 below says that regularity can be destroyed by $rl_Z$, unless both $\Delta^{[V]}Z=0$ and $Z\in A$ hold. Both of these conditions are needed by Proposition 2.2(ii) and (iii).

Proposition 2.2.

(i) Let $\mathfrak{A}\in Crs_\alpha^{reg}$ and let $Z \in Zd\mathfrak{A}$. Then $\mathfrak{Rl}_Z\mathfrak{A}$ is regular.

(ii) For every $\alpha\geq\omega$ there are $\mathfrak{A}\in Cs_\alpha^{reg}$ and $Z\in A$ such that $\mathfrak{Rl}_Z\mathfrak{A} \in Cs_\alpha \sim I\,Cs_\alpha^{reg}$.

(iii) For every $\alpha\geq\omega$ and $\varkappa\geq 2$ there are an $\mathfrak{A}\in {}_\varkappa Cs_\alpha^{reg}$ and $Z \in Zd\ Sb\ 1^{\mathfrak{A}}$ such that $\mathfrak{Rl}_Z\mathfrak{A}$ is not regular.

Proof. Proof of (i): Let $\mathfrak{A}\in Crs_\alpha^{reg}$ and let $Z \in Zd\mathfrak{A}$. Let $\mathfrak{R} \overset{d}{=} \mathfrak{Rl}_Z\mathfrak{A}$. Let $y\in R$ be arbitrary. Then $y\in A$ by $Z \in A$. Let $i\in\alpha$. Then $c_i^{\mathfrak{A}}y \leq c_i^{\mathfrak{A}}Z = Z$ by $Z \in Zd\mathfrak{A}$ and therefore $c_i^{\mathfrak{R}}y = (c_i^{\mathfrak{A}}y)\cap Z = c_i^{\mathfrak{A}}y$. Thus $\Delta^{(\mathfrak{R})}y = \Delta^{(\mathfrak{A})}y$. We show that $y$ is regular in $\mathfrak{R}$. Let $p\in y$ and $q \in 1^{\mathfrak{R}} = Z$ be such that $(1\cup\Delta^{(\mathfrak{R})}y)1p\subseteq q$. Then $q \in 1^{\mathfrak{A}}$ by $Z\subseteq 1^{\mathfrak{A}}$ and $(1\cup\Delta^{(\mathfrak{A})}y)1p\subseteq q$ by $\Delta^{\mathfrak{A}}y = \Delta^{\mathfrak{R}}y$. By $y\in A$ and $\mathfrak{A}\in Crs_\alpha^{reg}$ we have that $y$ is regular in $\mathfrak{A}$, and therefore $q\in y$. We have seen that $y$ is regular in $\mathfrak{R}$, too. This proves that $\mathfrak{R}$ is regular, since $y$ was chosen arbitrarily.

Proof of (ii): Let $\alpha\geq\omega$. Let $p \overset{d}{=} \alpha\times 1$, $X \overset{d}{=} {}^\alpha 2^{(p)}$ and $Z \overset{d}{=} {}^\alpha 2$. Let $\mathfrak{A}$ be the $Cs_\alpha$ with base 3 and generated by $\{X,Z\}$. Then $\{X,Z\} \subseteq Sm^{\mathfrak{A}}$ since $(\forall\Gamma\subseteq_\omega\alpha)(\forall i\notin\Gamma)\ c_i^\partial c_{(\Gamma)}Z=0$, and the same holds for $X$ since $X\subseteq Z$. $X$ and $Z$ are regular by $\Delta X=\Delta Z=\alpha$. Therefore $\mathfrak{A}$ is regular by Theorem 1.3. Let $\mathfrak{R} \overset{d}{=} \mathfrak{Rl}_Z\mathfrak{A}$. Then $\mathfrak{R}\in Cs_\alpha$ since $1^{\mathfrak{R}} = Z = {}^\alpha 2$. Clearly $X\in R$ and $\Delta^{\mathfrak{R}}X=0$ (despite the fact that $\Delta^{\mathfrak{A}}X = \alpha$). Since $X\neq 0$ and $X\neq 1^{\mathfrak{R}}$ by $\alpha\geq\omega$ we have that $\mathfrak{R}$ is not regular, by [HMTI]4.3.

(iii) of Proposition 2.2 is a consequence of Prop.4.11.

QED(Proposition 2.2.)

About Proposition 2.3 below see [HMTI]2.10.

Proposition 2.3. $Crs_2 \neq R\ell Cs_2$.

Proof. Let $V \stackrel{d}{=} \{\langle 0,1\rangle, \langle 1,2\rangle, \langle 2,3\rangle\}$ and $\mathfrak{L} \stackrel{d}{=} \mathfrak{Rl}_V \mathfrak{Gb}^2 4$. Let $\mathfrak{A} \stackrel{d}{=} \mathfrak{Sg}^{(\mathfrak{L})}\{\langle 0,1\rangle\}$. Then $\mathfrak{A} \in Crs_2$. We show that $\mathfrak{A} \notin R\ell Cs_2$. We have $A = \{V, 0, \{\langle 0,1\rangle\}, \{\langle 1,2\rangle, \langle 2,3\rangle\}\}$. Let $\mathfrak{N} \in Cs_2$ and suppose $A \subseteq N$. Then $x \stackrel{d}{=} \{\langle 0,1\rangle\} \in N$ and $y \stackrel{d}{=} \{\langle 1,2\rangle, \langle 2,3\rangle\} \in N$. Then $z \stackrel{d}{=}$ $\stackrel{d}{=} y \cdot c_1(d_{01} \cdot c_0 x) \in N$. But $z = \{\langle 1,2\rangle\}$ since $\mathfrak{N} \in Cs_2$. Hence $\{\langle 1,2\rangle\} \in Rl_V \mathfrak{N}$ for arbitrary $\mathfrak{N} \in Cs_2$ if $A \subseteq Rl_V \mathfrak{N}$. Then $z \notin A$ shows $\mathfrak{A} \notin R\ell\ Cs_2$.

QED(Proposition 2.3.)

## 3. Change of base

About Propositions 3.4-3.5 below see [HMTI]3.11(1)-(3). Prop.3.5 (iii) says that [HMTI]3.6 does not generalize to $Gs_\alpha$-s, but Prop.3.4 says that under some additional hypotheses [HMTI]3.6 does generalize to $Gs_\alpha$. Prop.3.5(i) implies that the condition that one of the bases be finite cannot be removed from [HMTI]3.6, and 3.5(iv) implies that Lf cannot be replaced with Dc in [HMTI]3.6.

Definition 3.1.

1. Let f be a function. Then we define
   $\tilde{f} \stackrel{d}{=} \langle \{f \circ q : q \in x \text{ and } Rgq \subseteq Dof\} : x$ is a set of functions $\rangle$.
   Note that $\tilde{f}({}^\alpha U) \subseteq {}^\alpha(Rgf)$ for any sets U and α.
2. Let $\mathfrak{A}, \mathfrak{B} \in Crs_\alpha$. Let $F \in Is(\mathfrak{A}, \mathfrak{B})$ with $DoF = A$.
   (i) F is a base-isomorphism if ($F = A 1 \tilde{f}$ and $base(\mathfrak{A}) \subseteq Dof$) for some one-one f.
   (ii) F is a strong ext-isomorphism if $F = rl^A({}^\alpha U)$ for some $U \subseteq base(\mathfrak{A})$.
   (iii) F is an ext-isomorphism if $F = rl^A(V)$ for some $V \subseteq 1^{\mathfrak{A}}$.

(iv) F is a <u>(strong) ext-base-isomorphism</u> if $F = g\circ h$ for some (strong) ext-isomorphism $g$ and base-isomorphism $h$.

(v) <u>sub</u> is the dual of ext, that is, F is a (strong) sub (-base)--isomorphism if $F^{-1}$ is a (strong) ext(-base)-isomorphism.

(vi) F is a <u>lower base-isomorphism</u> if $F = k^{-1}\circ h\circ t$ for some base--isomorphism $h$ and some strong ext-isomorphisms $k,t$.

The above definition agrees with [HMTI]3.5 and [HMTI]3.15.

<u>Lemma 3.2.</u>

(i) Let $f$ be a one-one function and let $\mathfrak{A}\in Crs_\alpha$. Assume $\tilde{f}\in Is\,\mathfrak{A}$ and $\tilde{f}^*\mathfrak{A}\in Crs_\alpha$. Let $K\in\{Gws_\alpha, Gs_\alpha, Ws_\alpha, Crs_\alpha^{reg}, Gws_\alpha^{norm}, Gws_\alpha^{wd}, Gws_\alpha^{comp}\}$. Then $A1\tilde{f}$ is a strong ext-base-isomorphism, and $\mathfrak{A}\in K$ implies $\tilde{f}^*\mathfrak{A}\in K$.

(ii) Let $\alpha>0$. There is $\mathfrak{A}\in Cs_\alpha^{reg}\cap Lf_\alpha$ and $h\in Is(\mathfrak{A},\mathfrak{A})$ such that $h$ is not a base-isomorphism.

<u>Proof.</u> (i) follows easily from the definitions.

<u>Proof of (ii):</u> Let $\alpha>0$. Let $\varkappa \overset{d}{=} |\omega\cup\alpha|$. $U \overset{d}{=} \varkappa^+$. $\mathfrak{L} \overset{d}{=} \mathfrak{Gb}^\alpha U$. $x \overset{d}{=} \{q\in{}^\alpha U : q_0<\varkappa\}$. $\mathfrak{A} \overset{d}{=} \mathfrak{Sg}^{(\mathfrak{L})}\{x\}$. First we show that there is $h\in Is(\mathfrak{A},\mathfrak{A})$ such that $h(x)= -x$. $\mathfrak{A}$ and $\varkappa$ satisfy the conditions of [HMTI]3.18(i)c) because $|A|\le\varkappa<|U|$, $\mathfrak{A}\in Cs_\alpha^{reg}\cap Lf_\alpha$ and $\varkappa = \sum_{\mu<\omega}\varkappa^\mu$. Then by [HMTI]3.18(i)c) there is $(\varkappa+\varkappa)\subseteq W\subseteq U$ such that $|W| = \varkappa$ and $rl(^\alpha W)\in Is(\mathfrak{A},\mathfrak{B})$ for some $Cs_\alpha$ $\mathfrak{B}$. Let $z \overset{d}{=} x\cap{}^\alpha W$. Clearly $z = \{q\in{}^\alpha W : q_0\in\varkappa\}$ and $B = Sg\{z\}$. Since $\varkappa+\varkappa\subseteq W$ we have $|W\sim\varkappa| = \varkappa$. Let $f\in{}^W W$ be such that $f^*\varkappa = W\sim\varkappa$ and $f\circ f\subseteq Id$. Then $\tilde{f}$ is a base- automorphism of $\mathfrak{Gb}^\alpha W$ by [HMTI]3.1. By $\tilde{f}(z) = W\sim z$ we have $\tilde{f}\in Is(\mathfrak{B},\mathfrak{B})$. Let $h = rl^A(^\alpha W)^{-1}\circ\tilde{f}\circ rl(^\alpha W)$. Then $h\in Is(\mathfrak{A},\mathfrak{A})$ and $h(x)= -x$ and $h\circ h\subseteq Id$. Assume that $h$ is a base--isomorphism. Then there is $k : U \rightarrowtail\!\!\!\!\rightarrow U$ such that $-x = h(x) = \{k\circ q : q\in x\}$. Then $(\forall u\in U\sim\varkappa)(\exists q\in x)(k\circ q)_0=u$. Then

$(\forall u\in U\sim\varkappa)(\exists i<\varkappa)k(i)=u$. This contradicts $\varkappa<|U|$. Thus h is not a base-isomorphism.

QED(Lemma 3.2.)

By Lemma 3.2(ii) it is meaningful to ask which isomorphisms of base-isomorphic algebras are actually base-isomorphisms themselves. From the proof of [HMTI]3.6 it follows that every isomorphism between $Cs_\alpha^{reg}\cap Lf_\alpha$ -s is a base-isomorphism, if one of them has base of power $<\alpha\cap\omega$. Proposition 3.4 below generalizes this form of [HMTI]3.6 to $Gs_\alpha$-s.

Definition 3.3.

(i) Let $\mathfrak{A}\in CA_\alpha$. $\mathfrak{A}$ is said to be of residually nonzero characteristic iff $\mathfrak{A}\cong|\subseteq P_{i\in I}\mathfrak{B}_i$ for some $\mathfrak{B}\in {}^I CA_\alpha$ such that each $\mathfrak{B}_i$ has a nonzero characteristic.

(ii) Let $\mathfrak{A}\in Crs_\alpha$. We say that $\mathfrak{A}$ is base-minimal if $\mathfrak{A}$ is not strongly ext-isomorphic to any $Crs_\alpha$ except itself.

Note that if every subbase of $\mathfrak{B}$ has power $<\alpha\cap\omega$ then $\mathfrak{B}$ is of residually nonzero characteristic.

Proposition 3.4. (generalization of [HMTI]3.6 to $Gs_\alpha$-s) Let $\mathfrak{A},\mathfrak{B}\in$ $\in Gs_\alpha^{reg}\cap Lf_\alpha$ be of residually nonzero characteristic.

(1) Assume that either $\mathfrak{A}$ is finitely generated or $Zd\mathfrak{A}$ is atomic. Then $\mathfrak{A}$ is strongly ext-isomorphic to some base-minimal $Gs_\alpha$.

(2) Assume $\mathfrak{A}$ and $\mathfrak{B}$ are base-minimal. Then every isomorphism between $\mathfrak{A}$ and $\mathfrak{B}$ is a base-isomorphism.

(3) Assume that either $\mathfrak{A}$ is finitely generated or $Zd\mathfrak{A}$ is atomic. Then every isomorphism between $\mathfrak{A}$ and $\mathfrak{B}$ is a lower base--isomorphism.

To prove 3.4 we shall use the following lemmas.

Lemma 3.4.1. Let $\mathfrak{A}\in Gs_\alpha^{reg}\cap Lf_\alpha$ be finitely generated and of residually nonzero characteristic. Then $Zd\mathfrak{A}$ is atomic. Further, $Zd\mathfrak{A}$ is finite if $\mathfrak{A}$ has a characteristic.

To prove 3.4.1, first we establish two other lemmas.

Lemma 3.4.1.1. Let $\mathfrak{A}\in Gs_\alpha^{reg}\cap Lf_\alpha$ be of characteristic $n>0$. Then $\mathfrak{A}\subseteq\mathfrak{B}\in Gs_\alpha^{reg}$ for some monadic-generated $\mathfrak{B}$. Moreover, if $\mathfrak{A}$ is finitely generated then so is $\mathfrak{B}$. In fact, there is a function $G : A \to$ $\to Sb\, Cl_{\alpha\sim 1}\mathfrak{B}$ such that $|G_x|\le n+n^{|\Delta x|}$ and $x\in Sg^{(\mathfrak{B})}G_x$ for every $x\in A$.

Proof. Let $\mathfrak{A}\in Gs_\alpha^{reg}\cap Lf_\alpha$ be of characteristic $n>0$. Let $\mathfrak{L}\overset{d}{=}\mathfrak{Gb}\,1^{\mathfrak{A}}$. Let $U : I \rightarrowtail Subb(\mathfrak{A})$ for some I. Then $(\forall i\in I)|U_i|=n$. Let $k\in$ $\in {}^I({}^n base(\mathfrak{A}))$ be such that $(\forall i\in I)k_i : n \rightarrowtail U_i$. Let $x\in A$ be fixed. Let $m\overset{d}{=}\Delta x$. If $m=0$ then let $G(x)\overset{d}{=}\{x\}$ if $x\notin\{0,1^{\mathfrak{A}}\}$, otherwise let $G(x)\overset{d}{=}0$. Suppose $m>0$. Then $\alpha>0$, hence $0\in\alpha$. Let $x^+\overset{d}{=}\{m1q : q\in x\}$. Let $r\in{}^m n$. Set $N_r\overset{d}{=}\cup\{{}^\alpha U_i : k_i\circ r\in x^+\}$. For every $v<n$ define $y_v\overset{d}{=}\{q\in 1^{\mathfrak{A}} : (\exists i\in I)q_0=k_i v\}$. Let $G(x)\overset{d}{=}\{y_v : v<n\}\cup$ $\cup\{N_r : r\in{}^m n\}$. Then $|G(x)|\le n+n^{|\Delta x|}$ and $(\forall g\in G)\Delta^{(\mathfrak{L})}g\subseteq 1$. We show that $x\in Sg^{(\mathfrak{L})}G(x)$. For every $r\in{}^m n$ we set $z_r\overset{d}{=}\Pi\{s_i^0 y_i : i\in m\}$. Clearly, $z_r = \{q\in 1^{\mathfrak{A}} : m1q\in\{k_i\circ r : i\in I\}\}$. We show that $x =$ $=\sum\{z_r\cdot N_r : r\in{}^m n\}$. Let $q\in x$. Say $q\in{}^\alpha U_i$. Set $r\overset{d}{=}k_i^{-1}\circ(m1q)\in{}^m n$. Thus $k_i\circ r\in x^+$, so $q\in N_r$. Clearly $q\in z_r$ also. Thus $q\in z_r\cdot N_r$. Conversely, suppose $r\in{}^m n$ and $q\in z_r\cdot N_r$. Since $q\in N_r$, there is an $i\in I$ with $q\in{}^\alpha U_i$ and $k_i\circ r\in x^+$. By $q\in z_r$ we get $m1q = k_j\circ r$ for some $j\in I$. Since $m\ne 0$ it follows that $i=j$ and so $m1q\in x^+$. By regularity of $x$ we get $q\in x$. We have seen $x\in Sg^{(\mathfrak{L})}G(x)$. Assume $A = SgX$. Let $\mathfrak{B}\overset{d}{=}\mathfrak{Sg}^{(\mathfrak{L})}\cup\{Gx : x\in X\}$. Clearly, G and $\mathfrak{B}$ satisfy the requirements of Lemma 3.4.1.1.

QED(Lemma 3.4.1.1.)

Remark: If in Lemma 3.4.1.1 we replace "of characteristic $n>0$" with "of residually nonzero characteristic" then the conclusion becomes false. Namely let $\alpha\geq\omega$. Then there is $\mathfrak{A}\in Gs_\alpha^{reg}\cap Lf_\alpha$ with all subbases finite such that no $Crs_\alpha$ $\mathfrak{B}\supseteq\mathfrak{A}$ is monadic generated. (Being monadic-generated is not preserved under $\mathbf{P}$.)

Lemma 3.4.1.2. Let $\mathfrak{A}\in CA_\alpha$ be finitely- and monadic-generated. Suppose $\mathfrak{A}$ is of nonzero characteristic. Then $|Zd\,\mathfrak{A}|<\omega$.

Proof. Let $X\subseteq A$ be such that $A = Sg\, X$, $|X|<\omega$ and $(\forall x\in X)\Delta x\subseteq 1$. By [HMT]2.2.24 we have that $Zd\,\mathfrak{A} = Sg^{(\mathfrak{Bl}\,\mathfrak{A})}C$ where $C =$ $= \{a_\varkappa(Y,Z) : Y\cup Z \subseteq X,\ \varkappa<(\alpha+1)\cap\omega\}$. Let the characteristic of $\mathfrak{A}$ be $n>0$. Then $c_{(\varkappa)}\bar{d}(\varkappa\times\varkappa)=0$ for every $\varkappa>n$. Let $\varkappa<(\alpha+1)\cap\omega$, $\varkappa>n$. Then $a_\varkappa(Y,Z)=0$ for every $Y,Z\subseteq X$ since $a_\varkappa(Y,Z)\subseteq c_{(\varkappa)}\bar{d}(\varkappa\times\varkappa)$. Therefore $|C|<\omega$ by $|X|<\omega, n<\omega$. Then $|Zd\mathfrak{A}| = |Sg^{(\mathfrak{Bl}\,\mathfrak{A})}C|<\omega$.

QED(Lemma 3.4.1.2.)

Now we turn to the proof of Lemma 3.4.1. Let $n<\alpha\cap\omega$. We define the constant term $\sigma_n$ of the discourse language of $CA_\alpha$-s as $\sigma_n \overset{d}{=} c_{(n)}\bar{d}(n\times n)-c_{(n+1)}\bar{d}((n+1)\times(n+1))$. Note that $CA_\alpha \models \sigma_0=1$ and if $n\neq 0$ and $\mathfrak{A}\in Gws_\alpha$ then $\sigma_n^{\mathfrak{A}} = \cup\{V : |base(V)|=n$ and $V\in Subu(\mathfrak{A})\}$. Let $\mathfrak{A}\in CA_\alpha$. Then $\mathfrak{A}$ is of residually nonzero characteristic iff $(\forall a\in A\sim\{0\})(\exists n\in(\alpha\cap\omega)\sim 1)\quad a\cap\sigma_n^{\mathfrak{A}}\neq 0$.

Let $\mathfrak{A}\in Gs_\alpha^{reg}\cap Lf_\alpha$ be finitely generated and be of residually nonzero characteristic. Let $n<\alpha\cap\omega$. Let $\mathfrak{B}_n \overset{d}{=} \mathfrak{Rl}_{\sigma(n)}\mathfrak{A}$. Then $\mathfrak{B}_n \in$ $\in Gs_\alpha^{reg}\cap Lf_\alpha$ since $\mathfrak{B}_n\in Gs_\alpha$ by $\mathfrak{A}\in Gs_\alpha$ which implies that $\sigma_n^{\mathfrak{A}}$ is a $Gs_\alpha$-unit and $\mathfrak{B}_n$ is regular by 2.2(i) since $\mathfrak{A}$ is regular and $\sigma_n^{\mathfrak{A}}\in Zd\,\mathfrak{A}$. Also, $\mathfrak{A}\cong\mathfrak{L}\subseteq P\langle\mathfrak{B}_n : n\in(\alpha\cap\omega)\sim 1\rangle$ for some $\mathfrak{L}$ since $\mathfrak{A}$ is of residually nonzero characteristic. Let $n<\alpha\cap\omega, n\neq 0$. Then $\mathfrak{B}_n\subseteq\mathfrak{I}$ for some finitely and monadic-generated $\mathfrak{I}$, by 3.4.1.1. The characteristic of $\mathfrak{I}$ is nonzero and therefore $|Zd\,\mathfrak{I}|<\omega$ by 3.4.1.2. Therefore $|Zd\,\mathfrak{B}_n|<\omega$ by $\mathfrak{B}_n\subseteq\mathfrak{I}$. Hence $Zd\,\mathfrak{B}_n$ is atomic.

We have $\mathfrak{Zd}\,\mathfrak{L} \subseteq P\langle \mathfrak{Zd}\,\mathfrak{B}_n : n\in(\alpha\cap\omega)\sim 1\rangle$ by [HMT]2.4.3. By $\{\sigma_i^{\mathfrak{A}} : i<\alpha\cap\omega\} \subseteq A$ we have $C \supseteq \{\langle 0 : n\in(\alpha\cap\omega)\sim 1\rangle(i/b) : i\in(\alpha\cap\omega)\sim 1,\ b\in B_i\}$. Therefore $(\forall n\in(\alpha\cap\omega)\sim 1)(\mathrm{Zd}\,\mathfrak{B}_n$ is atomic) implies that $\mathfrak{Zd}\,\mathfrak{L}$ is atomic. Then $\mathrm{Zd}\,\mathfrak{A}$ is atomic by $\mathfrak{A}\cong\mathfrak{L}$.

QED(Lemma 3.4.1.)

Now we turn to the proof of Prop.3.4. Let $\mathfrak{A},\mathfrak{B}\in Gs_\alpha^{reg}\cap Lf_\alpha$ be of residually nonzero characteristic.

Proof of(1): Suppose either $\mathfrak{A}$ is finitely generated or $\mathrm{Zd}\,\mathfrak{A}$ is atomic. Then $\mathrm{Zd}\,\mathfrak{A}$ is atomic in both cases, by 3.4.1. Let $T \stackrel{d}{=}$ At Zd A. Let $Y : T \to \mathrm{Subb}(\mathfrak{A})$ be such that $(\forall a\in T)[{}^\alpha Y(a) \subseteq a$ and $|Y(a)|<\alpha\cap\omega]$. Such a $Y$ exists by the following. Let $a\in T$. Then $a\neq 0$, hence $(\exists n\in(\alpha\cap\omega)\sim 1)\, a\cap\sigma_n^{\mathfrak{A}}\neq 0$ since $\mathfrak{A}$ is of residually nonzero characteristic. Thus $a\cap{}^\alpha Y(a)\neq 0$ for some $Y(a)\in\mathrm{Subb}(\mathfrak{A})$, $|Y(a)|=n$. Then ${}^\alpha Y(a) \subseteq a$ since $a$ is regular and $\Delta a=0$. Let $U \stackrel{d}{=} \cup Rg Y$. Let $W={}^\alpha U$. We show that $rl_W\in Is\,\mathfrak{A}$. Let $\mathfrak{R} \stackrel{d}{=} \mathfrak{Rl}_W\mathfrak{A}$. Then $rl_W\in HO(\mathfrak{A},\mathfrak{R})$ by $W\in \mathrm{Zd}\ \mathrm{Sb}\ 1^{\mathfrak{A}}$. Let $b\in A\sim\{0\}$ be arbitrary. Let $z \stackrel{d}{=} c_{(\Delta b)}b$. Then $z\in \mathrm{Zd}\ A$, by $\mathfrak{A}\in Lf_\alpha$. Then $a\leq z$ for some $a\in$ At Zd A since Zd A is atomic. Then $a=c_{(\Delta b)}(a\cdot b)$, hence $b\cap{}^\alpha Y(a)\neq 0$ by $Y(a)\in\mathrm{Subb}(\mathfrak{A})$ and ${}^\alpha Y(a)\subseteq a$. Thus $b\cap W\neq 0$. We have seen that $rl_W$ is a strong ext-isomorphism. Clearly, $\mathfrak{R}\in Gs_\alpha$. We show that $\mathfrak{R}$ is base-minimal. Suppose that $V\subseteq U$ is such that $rl({}^\alpha V)\in Is\,\mathfrak{R}$. Let $\mathfrak{B} \stackrel{d}{=} \mathfrak{Rl}({}^\alpha V)\mathfrak{R}$. Let $a\in T$ and $a' \stackrel{d}{=} rl({}^\alpha V)rl(W)a$. Then $a' = {}^\alpha V\cap W\cap a = {}^\alpha V\cap{}^\alpha Y(a) = {}^\alpha(V\cap Y(a))$ and $V\cap Y(a)\in\mathrm{Subb}(\mathfrak{B})$ by $Y(a)\in\mathrm{Subb}(\mathfrak{A})$. Let $n \stackrel{d}{=} |Y(a)|$. We have $0\neq a'\leq\sigma_n^{\mathfrak{B}}$ by $0\neq a\leq\sigma_n^{\mathfrak{A}}$. Then $|V\cap Y(a)|=n$ which means $Y(a)\subseteq V$ by $n<\omega$. We have seen that $(\forall a\in T)Y(a)\subseteq V$. Then $V=U$.

Proof of(2): Suppose $\mathfrak{A}\in Gs_\alpha^{reg}\cap Lf_\alpha$ is base-minimal. We show that At Zd A $=\{{}^\alpha Y : Y\in\mathrm{Subb}(\mathfrak{A})\}$. Let $Y\in\mathrm{Subb}(\mathfrak{A})$ and $U \stackrel{d}{=} \mathrm{base}(\mathfrak{A})\sim Y$. Then $rl({}^\alpha U)\notin Is\,\mathfrak{A}$, hence ${}^\alpha U\cap b=0$ for some $b\in A\sim\{0\}$. Then $0\neq b\subseteq{}^\alpha Y$. Then $c_{(\Delta b)}b={}^\alpha Y$ since $\mathfrak{A}\in Gs_\alpha^{reg}\cap Lf_\alpha$. Then At Zd A $= \{{}^\alpha Y : Y\in\mathrm{Subb}(\mathfrak{A})\}$ by $\mathfrak{A}\in Gs_\alpha^{reg}\cap Lf_\alpha$.

Now let $\mathfrak{A},\mathfrak{B}\in Gs_\alpha^{reg}\cap Lf_\alpha$ be of residually nonzero characteristic.

Assume that $\mathfrak{A}$ and $\mathfrak{B}$ are base-minimal. Let $h \in Is(\mathfrak{A},\mathfrak{B})$. Let $Y \in Subb(\mathfrak{A})$ and let $a \stackrel{d}{=} {}^{\alpha}Y$. Then $a \in At\, Zd\, A$, hence $h(a) \in At\, Zd\, B$ and therefore $h(a) = {}^{\alpha}W$ for some $W \in Subb(\mathfrak{B})$, by the above. Then $\mathfrak{Rl}_a\mathfrak{A}, \mathfrak{Rl}_{h(a)}\mathfrak{B} \in Cs_{\alpha}^{reg} \cap Lf_{\alpha}$ by 2.2(i). Clearly, $h \in Is(\mathfrak{Rl}_a\mathfrak{A}, \mathfrak{Rl}_{h(a)}\mathfrak{B})$. We have $|Y| < \alpha \cap \omega$ since $\mathfrak{A}$ is base-minimal and is of residually nonzero characteristic. Then by [HMTI]3.6 there is $F_a : Y \rightarrowtail\!\!\!\!\twoheadrightarrow W$ such that $Rl_a A 1 h \subseteq \tilde{F}_a$. Let $F \stackrel{d}{=} \cup\{F_a : a \in At\, Zd\, A\}$. Then $F : base(\mathfrak{A}) \rightarrowtail\!\!\!\!\twoheadrightarrow base(\mathfrak{B})$ since $h : At\, Zd\, A \rightarrowtail\!\!\!\!\twoheadrightarrow At\, Zd\, B$, and $h = A 1 \tilde{F}$. We have seen that $h$ is a base-isomorphism.

(3) of Prop.3.4 is a consequence of (1) and (2).

QED(Proposition 3.4.)

Proposition 3.4 above and Proposition 3.10 at the end of this section are in a kind of dual relationship to each other. Prop.3.5(i) below exhibits an asymmetry in this duality.

Proposition 3.5. (discussion of the conditions in Proposition 3.4(3))

(i) The condition " $\mathfrak{A}$ be of residually nonzero characteristic" is necessary in Prop.3.4(2),(3). Namely: Let $\alpha \geq \omega$ and $\kappa \geq \omega$. Then there are two finitely generated $\mathfrak{A}, \mathfrak{B} \in {}_{\kappa}Cs_{\alpha}^{reg} \cap Lf_{\alpha}$ satisfying a.-c. below.

a. $\mathfrak{A} \cong \mathfrak{B}$

b. $\mathfrak{A}$ is not lower base-isomorphic to $\mathfrak{B}$.

c. There is no $Gws_{\alpha}$ sub-base-isomorphic to both $\mathfrak{A}$ and $\mathfrak{B}$.

(ii) The condition " $\mathfrak{A}$ is finitely generated or $Zd\mathfrak{A}$ is atomic" is necessary in Prop.3.4(3). Namely: Let $\alpha > 1$ and $\kappa > 0$. There are isomorphic $\mathfrak{A}, \mathfrak{B} \in {}_{\kappa}Gs_{\alpha}^{reg} \cap Lf_{\alpha}$ which are not lower base-isomorphic. Moreover, no $Crs_{\alpha}$ is sub-base-isomorphic to both $\mathfrak{A}$ and $\mathfrak{B}$ (further $\mathfrak{A}$ is hereditarily nondiscrete if $\kappa > 1$).

(iii) "lower base-isomorphism" cannot be replaced with "base-iso-

-morphism" in Prop.3.4.(3). Namely: Let $\varkappa\in\omega\sim 5$ be arbitrary. Then there are $\mathfrak{A},\mathfrak{B}\in {}_\varkappa Gs_\alpha^{reg}\cap Mn_\alpha$ such that $\mathfrak{A}\cong\mathfrak{B}$ but they are not base-isomorphic. If $\varkappa<5\leq\alpha$ then every isomorphism between ${}_\varkappa Gs_\alpha^{reg}\cap Lf_\alpha$ -s is a base-isomorphism.

(iv) "Lf" cannot be replaced by "Dc" in Prop.3.4(2),(3). Namely: For every $\alpha\geq\omega$ there are isomorphic ${}_3Cs_\alpha^{reg}\cap Dc_\alpha$ -s which are not base-isomorphic.

<u>Proof.</u> <u>Proof of 3.5(i):</u> First we prove a lemma.

<u>Lemma 3.5.1.</u> Let $\mathfrak{A}$ be a $Cs_\alpha^{reg}\cap Lf_\alpha$ with base U. Let F be an ultrafilter on I. Let $ud_F^A \overset{d}{=} ud_F \overset{d}{=}$

$\overset{d}{=} ud \overset{d}{=} \langle\{q\in{}^\alpha({}^IU/\bar{F}) : (\exists k\in Pq)\{j\in I : \langle k(i)_j : i<\alpha\rangle\in a\}\in F : a\in A\rangle$.

Then $ud\in Is\mathfrak{A}$, $ud^*\mathfrak{A}\in Cs_\alpha^{reg}$ with base ${}^IU/\bar{F}$ and $ud$ is a strong sub-base-isomorphism. Moreover, let $\mathfrak{A}^+ \overset{d}{=} ud^*\mathfrak{A}$ and $\varepsilon \overset{d}{=}$ $\overset{d}{=} \langle\langle u : i\in I\rangle/\bar{F} : u\in U\rangle$. Let $e = \varepsilon^{-1}$. Then $ud^{-1}\subseteq\tilde{e}\in Is(\mathfrak{A}^+,\mathfrak{A})$ is the strong ext-base-isomorphism induced by $e$ from $\mathfrak{A}^+$ to $\mathfrak{A}$.

<u>Proof.</u> Let everything be as in the hypotheses. Let c be an $(F,\langle U : i\in I\rangle,\alpha)$-choice function such that $c(i,\varepsilon u)=u$ for all $i<\alpha$ and $u\in U$. Let $f,\delta,g$ be as in [HMTI]7.12. We claim that $g=ud$. In fact, for any $a\in A$ we have

$$ga = \{q\in{}^\alpha({}^IU/\bar{F}) : \{j\in I : \langle c(i,q_i)_j : i<\alpha\rangle\in a\}\in F\},$$

so it is clear that $ga\subseteq ud(a)$. Now suppose $k\in Pq$ and $\{j\in I : \langle(ki_j) : i<\alpha\rangle\in a\}\in F$. For all $i<\alpha$ we have $ki/\bar{F} = c(i,qi)/\bar{F}$, so $\{j\in I : (\forall i\in\Delta a)((ki)_j=c(i,qi)_j)\}\in F$. Hence by the regularity of a, $\{j\in I : \langle c(i,qi)_j : i<\alpha\rangle\in a\}\in F$, so $q\in ga$, as desired.

Now by [HMTI]7.4(i) and 7.6 we have $\mathfrak{A}^+\in Cs_\alpha^{reg}$. Also $\tilde{\varepsilon}({}^\alpha U) = {}^\alpha W$. Hence the other conclusions follow from [HMTI]7.12.

<u>QED(Lemma 3.5.1.)</u>

We continue the proof of 3.5(i). Let $\alpha\geq\omega$ and $\varkappa\geq\omega$. Z denotes the set of integers. $K \stackrel{d}{=} Sb_{\omega}Z\times\varkappa$ and $U \stackrel{d}{=} Z\times K$. We define $+ : U\times Z \to U$ as follows: Let $u\in U$ and $z\in Z$. Then $u+z \stackrel{d}{=} \langle u(0)+z, u(1)\rangle$.

$x \stackrel{d}{=} \{q\in{}^{\alpha}U : q_1 = q_0+1\}$,

$y \stackrel{d}{=} \{q\in{}^{\alpha}U : q_0(0) \in q_0(1)0\}$,

$Q \stackrel{d}{=} y\cup\{q\in{}^{\alpha}U : q_0(0)^2 > \max\{z^2 : z\in q_0(1)0\}$ and $q_0(0)$ is odd$\}$.

$\mathfrak{A} \stackrel{d}{=} \mathfrak{Sg}^{(\mathfrak{Sb}\,{}^{\alpha}U)}\{x,y\}$, $\mathfrak{B} \stackrel{d}{=} \mathfrak{Sg}^{(\mathfrak{Sb}\,{}^{\alpha}U)}\{x,Q\}$.

Clearly, $x,y$ and $Q$ are locally finite dimensional regular elements. Therefore $\mathfrak{A},\mathfrak{B}\in Cs_{\alpha}^{reg}\cap Lf_{\alpha}$ by [HMTI]4.2. base$(\mathfrak{A})$ = = base$(\mathfrak{B})$ = U and $|U| = \varkappa$. We shall show that $\mathfrak{A}\cong\mathfrak{B}$ and there is no $Gws_{\alpha}$ sub-base-isomorphic to both $\mathfrak{A}$ and $\mathfrak{B}$. Since obviously c. $\Rightarrow$ b. in 3.5.(i), this will prove Proposition 3.5.(i).

Throughout the proof of 3.5.(i) we shall use the following notations. F is a fixed nonprincipal ultrafilter on $\omega$. $U^+ \stackrel{d}{=} {}^{\omega}U/\bar{F}$. Let ud be as in the formulation of Lemma 3.5.1. Let $\mathfrak{A}^+ = ud^*\mathfrak{A}$ and $\mathfrak{B}^+ = ud^*\mathfrak{B}$. Then $\mathfrak{A}^+, \mathfrak{B}^+ \in Cs_{\alpha}^{reg}\cap Lf_{\alpha}$, base$(\mathfrak{A}^+)$ = base$(\mathfrak{B}^+) = U^+$ and $ud \in Is(\mathfrak{A},\mathfrak{A}^+)$, $ud \in Is(\mathfrak{B},\mathfrak{B}^+)$. $A^+$ and $B^+$ are the universes of $\mathfrak{A}^+$ and $\mathfrak{B}^+$. For every $b \in A\cup B$ we shall use the notation $b^+ = ud(b)$. Thus $A^+ = Sg\{x^+,y^+\}$ and $B^+ = Sg\{x^+,Q^+\}$. Let $g\in{}^{\omega}U$. Then $\bar{g} \stackrel{d}{=} g/\bar{F}$. We define $+ : U^+\times Z \to U^+$ as follows: Let $g\in{}^{\omega}U$ and $z\in Z$. Then $\bar{g}+z \stackrel{d}{=} \langle g(n)+z : n\in\omega\rangle/\bar{F}$.

Claim 3.5.2. There is no $Gws_{\alpha}$ sub-base-isomorphic to both $\mathfrak{A}$ and $\mathfrak{B}$.

To prove this, we first establish two other claims.

Claim 3.5.2.1. Let $h \in Is(\mathfrak{B},\mathfrak{A})$. Then $(\forall q\in h(x))q_0(1) = q_1(1)$.

Proof. We have $\mathfrak{B} \models x\cdot s_2^1x \le d_{12}$ and $\Delta^{(\mathfrak{B})}x = 2$. Therefore $\mathfrak{A}\models$ $\models h(x)\cdot s_2^1h(x) \le d_{12}$ and $\Delta^{(\mathfrak{A})}h(x) = 2$. Let $b\in A$ be arbitrary such that $\Delta^{(\mathfrak{B})}b = 2$. Suppose $q\in b$ and $q_01 \ne q_11$. We show that $\mathfrak{A} \models b\cdot s_2^1b \not\le d_{12}$. Let $\langle H,\gamma\rangle \stackrel{d}{=} M \stackrel{d}{=} q_11 \ne q_01$. For every $z\in Z$ let $z^+ \stackrel{d}{=} \langle\langle z,\langle H\cup\{i\},\gamma\rangle\rangle : i\in\omega\rangle/\bar{F}$. Let $k : U^+ \rightarrowtail U^+$ be a permutation of $U^+$ such that $k = \{\langle\varepsilon(z,M),z^+\rangle,\langle z^+,\varepsilon(z,M)\rangle$ :

$: z\in Z\} \cup R1Id$, for some $R$. Such a $k$ exists.

Now $\tilde{k}y^+=y^+$, since for every $p\in {}^\alpha U^+$ we have $p(0/\varepsilon(z,M)) \in y^+$ iff $p(0,z^+)\in y^+$ (and hence $p\in y^+$ iff $k\circ p\in y^+$ since $\Delta y^+=1$ and $y^+$ is regular, by 3.5.1). To see this equivalence, from the definition of $ud$ it is easy to check that $p(0/\varepsilon(z,M))\in y^+$ iff $\{j\in\omega : z\in H\}\in F$, and $p(0/z^+)\in y^+$ iff $\{j\in\omega : z\in H\cup\{j\}\}\in F$, so the equivalence follows.

Also $\tilde{k}x^+=x^+$ since if $p\in x^+$ then $(p(0)=\varepsilon(z,M)$ iff $p(1)=\varepsilon(z+1,M))$ and $(p(0)=z^+$ iff $p(1)=(z+1)^+)$.

Therefore $A^+1\tilde{k} \subseteq Id$ by $A^+ = Sg\{x^+,y^+\}$. Let $p \stackrel{d}{=} \varepsilon\circ q$. Then $p\in b^+$ by $q\in b$ and $k(p_0)=p_0=\varepsilon(q_0)$ by $q_0 1\neq M$ and $k(p_1)\neq p_1=\varepsilon(q_1)$ by $q_1 1=M$. Also, $k\circ p\in b^+$. Thus since $\mathfrak{A}^+$ is regular and $\Delta b^+=2$, we have $p^2_{kp_1} \in b^+\cdot s^1_2 b^+-d_{12}$, and hence $\mathfrak{A}^+ \models b^+\cdot s^1_2 b^+ \nleq d_{12}$. Therefore $\mathfrak{A} \models b\cdot s^1_2 b \nleq d_{12}$ by $ud \in Is(\mathfrak{A},\mathfrak{A}^+)$.

QED(Claim 3.5.2.1.)

Claim 3.5.2.2. Let $V \subseteq {}^\alpha U$ be a $Gws_\alpha$-unit and assume that $rl_V \in Is(\mathfrak{B}, \mathfrak{Rl}_V\mathfrak{B})$. Then $(\forall W\in Subb(V))(\exists M\in K)\ Z\times\{M\} \subseteq W$.

Proof. Let $V \subseteq {}^\alpha U$ be a $Gws_\alpha$-unit and let $\mathfrak{R} \stackrel{d}{=} \mathfrak{Rl}_V\mathfrak{B}$. Assume that $rl_V \in Is(\mathfrak{B},\mathfrak{R})$. Let ${}^\alpha W^{(p)} \in Subu(V)$. Let $\langle m,M\rangle\in W$. It is enough to show that this implies $\{\langle m-1,M\rangle,\langle m+1,M\rangle\} \subseteq W$. Let $q \stackrel{d}{=} p(0/\langle m,M\rangle)(1/\langle m,M\rangle)$. Then $q\in V$. We have $c_0^{\mathfrak{B}}x=c_1^{\mathfrak{B}}x=1^{\mathfrak{B}}$. Therefore $c_0^{\mathfrak{R}}(rl_V x)=c_1^{\mathfrak{R}}(rl_V x)=1^{\mathfrak{R}}$, so $c_0^{[V]}(x\cap V) = c_1^{[V]}(x\cap V) = V$. Then $q \in c_0^{[V]}(x\cap V)$ by $q\in V$ and this means that $q^0_u \in x\cap V$ for some $u$. Then $u = \langle m-1,M\rangle$ by $q^0_u\in x$ and $u\in W$ by $q^0_u\in V$. Similarly, $q \in c_1^{[V]}(x\cap V)$ implies $\langle m+1,M\rangle\in W$.

QED(Claim 3.5.2.2.)

Now we prove 3.5.2=3.5(1)(c).

Assume that there is a $Gws_\alpha$ sub-base-isomorphic to both $\mathfrak{A}$ and $\mathfrak{B}$. Then there is an $\mathfrak{R}\in Gws_\alpha$ with unit $V$ and with base $Y \subseteq U$ such that $rl_V \in Is(\mathfrak{B},\mathfrak{R})$ and there is $f : Y \rightarrowtail U$ such that $rl(\tilde{f}V) \in Is(\mathfrak{A},\tilde{f}^*\mathfrak{R})$. Let $h \stackrel{d}{=} rl(\tilde{f}V)^{-1}\circ f\circ rl_V$. Then $h \in Is(\mathfrak{B},\mathfrak{A})$.

By Claim 3.5.2.2 we have that $(\exists M\in K) Z\times\{M\} \subset | \in Subb(V)$. We show that $(\exists L\in K)\ f^*(Z\times\{M\}) \subseteq Z\times\{L\}$. It is enough to show that $(\forall z\in Z) f(z,M)(1) = f(z+1,M)(1)$. Let ${}^{\alpha}W^{(p)} \in Subu(V)$ be such that $Z\times\{M\} \subseteq W$. Let $z\in Z$. Let $q \stackrel{d}{=} p(0/\langle z,M\rangle)(1/\langle z+1,M\rangle)$. Then $q \in V\cap x$ and therefore $f\circ q \in h(x)$. Then $f(z,M)(1)=f(z+1,M)(1)$ by Claim 3.5.2.1. Let $L\in K$ be such that $f^*(Z\times\{M\}) \subseteq Z\times\{L\}$. Let $Y' \stackrel{d}{=} Z\times\{M\}$. Consider the generator element $Q\in B$. Let

$T \stackrel{d}{=} \{u\in Y' : (\exists q\in V) q^0_u \in V\cap Q\}$ and

$N \stackrel{d}{=} \{u\in Y' : (\exists q\in V) q^0_u \in V\sim Q\}$ .

Then $N = Y'\sim T$ since $\mathcal{B}$ is regular and $\Delta Q = 1$. $|T|=|N|=\omega$ by the definition of $Q$ and by $Y' = Z\times\{M\}$. Since $h(Q) \in A$ is regular and $\Delta^{(\mathfrak{A})}h(Q) = 1$, by the definition of $h$ we have $(\forall q\in {}^{\alpha}U)[(\forall u\in f^*T) q^0_u \in h(Q)$ and $(\forall u\in f^*N) q^0_u \in -h(Q)]$. Let $t : \omega \rightarrowtail f^*T$ and $n : \omega \rightarrowtail f^*N$ be two one-one mappings such that $(\forall z \in Z) |\{i\in\omega : t_i+z=n_i\}| < \omega$. Such $t$ and $n$ exist. In fact, well-order $f^*T$ and $f^*N$ in type $\omega$. Define $t_i$ and $n_i$ inductively. For $i$ even, let $t_i$ = "least element of $f^*T\sim\{t_j : j<i\}$", $n_i$ = "least $w\in f^*N$ such that $(\forall z\in Z)[|z|\le i \Rightarrow t_i+z\ne w]$". For $i$ odd, interchange the roles of $T$ and $N$. Let $\bar{t} \stackrel{d}{=} t/\bar{F}$ and $\bar{n} \stackrel{d}{=} n/\bar{F}$. Then $\bar{t},\bar{n} \in U^+ = {}^{\omega}U/\bar{F}$. $\mathfrak{A}^+ \in Cs^{reg}_{\alpha}$ by Lemma 3.5.1 and therefore by the properties of $t$ and $n$ we have $(\forall q\in {}^{\alpha}U^+)[q(0/\bar{t})\in h(Q)^+$ and $q(0/\bar{n})\notin h(Q)^+]$. Let $\bar{T} \stackrel{d}{=} \{\bar{t}+z : z\in Z\}$ and $\bar{N} \stackrel{d}{=} \{\bar{n}+z : z\in Z\}$. Then $\bar{T}\cap\bar{N} = 0$ by $(\forall z\in Z)|\{i\in\omega : t_i+z=n_i\}|<\omega$. Let $d : U^+ \rightarrowtail\!\!\!\twoheadrightarrow U^+$ be a permutation of $U^+$ such that $d = \{\langle\bar{t}+z,\bar{n}+z\rangle,\langle\bar{n}+z,\bar{t}+z\rangle : z\in Z\} \cup (U^+\sim(\bar{T}\cup\bar{N}))1Id$. We show that $A^+1\breve{d} \subseteq Id$.

Since $t$ and $n$ are one-one and $Rg\,t\cup Rg\,n \subseteq Z\times\{L\}$ for some $L\in K$ we have $(\forall q\in {}^{\alpha}U^+)(\forall z\in Z)\{q(0/\bar{t}+z),q(0/\bar{n}+z)\} \cap y^+ = 0$ since $\{u\in Z\times\{L\} : (\exists q\in {}^{\alpha}U) q^0_u \in y\}|<\omega$ by the definition of $y$. Therefore $\breve{d}\, y^+ = y^+$.

By the definition of $x$ we have $(\forall q\in {}^{\alpha}U^+)(\forall u,v\in U^+)(q^{01}_{uv}\in x^+$ iff

$v=u+1$). Let $u,v\in U^+$. By the definitions of $\bar{T},\bar{N}$ and $d$ we have $v=u+1$ iff $d(v)=d(u)+1$. Therefore $\tilde{d}\, x^+ = x^+$.

By $A^+ = Sg\{x^+,y^+\}$ then we have $A^+1\tilde{d} \subseteq Id$. Therefore $\tilde{d}\, h(Q)^+ = h(Q)^+$. Now $(\forall q\in{}^{\alpha}U^+)[q(0/\bar{t})\in h(Q)^+$ and $q(0/\bar{n})\notin h(Q)^+]$ and $d(\bar{t})=\bar{n}$ contradict $\tilde{d}\, h(Q)^+ = h(Q)^+$. This contradiction establishes 3.5.2.
QED(Claim 3.5.2.)

Claim 3.5.3. $\mathfrak{A} \cong \mathfrak{B}$.

Proof. Let $H \overset{d}{=} \{h\in{}^{Z}U^+ : h$ is one-one and $(\forall z\in Z)h(z+1)=h(z)+1\}$. Let $M \subseteq Z$. Define

$G_M^A \overset{d}{=} \{h\in H : (\forall z\in Z)(\forall q\in{}^{\alpha}U^+)[q(0/h(z))\in y^+$ iff $z\in M]\}$.

$G_M^B \overset{d}{=} \{h\in H : (\forall z\in Z)(\forall q\in{}^{\alpha}U^+)[q(0/h(z))\in Q^+$ iff $z\in M]\}$.

Claim 3.5.3.1. $(\forall M\subseteq Z)\,|G_M^A|=|G_M^B|=\varkappa^{\omega}$.

Proof. For every $n\in\omega$ define $L_n \overset{d}{=} \{z\in Z : z^2\le n^2\}$. Then $L\in{}^{\omega}Sb_{\omega}Z$ is such that $\cup RgL=Z$. Let $M\subseteq Z$ be fixed. Let $k\in{}^{\omega}(\omega\times\varkappa)$ be arbitrary. Let $t \overset{d}{=} pj_0\circ k$. Let $T \overset{d}{=} \langle\{t(n)+z : z\in M\cap L_n\}\cup\{(t(n)+n)^2\} : n\in\omega\rangle$. Then $T\in{}^{\omega}Sb_{\omega}Z$. Let $h(k,M) \overset{d}{=} \langle\langle(t(n)+z, \langle Tn,k(n)1\rangle) : n\in\omega\rangle/\bar{F} : z\in Z\rangle$. Then $h(k,M)\in H$ since $h(k,M)\in{}^{Z}U^+$ is a one-one mapping such that $(\forall z\in Z)\ h(k,M)(z)+1 = h(k,M)(z+1)$.
We show that $h(k,M)\in G_M^A\cap G_M^B$. Recall that $y^+$ and $Q^+$ are regular elements and $\Delta y^+ = \Delta Q^+ = 1$. Let $q\in{}^{\alpha}U^+$. Let $z\in M$. Then $t(n)+z\notin T_n$ implies $z\notin L_n$, i.e. $z^2>n^2$. Now $|\{n\in\omega : n^2<z^2\}|<\omega$ shows $q(0/h(k,M)z)\in y^+$. Suppose $z\notin M$. Then $q(0/h(k,M)z)\notin y^+$ since $|\{n\in\omega : t(n)+z\in T_n\}| = |\{n\in\omega : t(n)+z=(t(n)+n)^2\}|\le 1$.

Therefore $h(k,M)\in G_M^A$. Next we show $h(k,M)\in G_M^B$. By $y^+\subseteq Q^+$ it is enough to show $(\forall z\in Z)q(0/h(k,M)z)\notin Q^+\sim y^+$. Let $z\in Z$. Then $|\{n\in\omega : (t(n)+z)^2 > (t(n)+n)^2\}|<\omega$ shows $q(0/h(k,M)z)\notin Q^+\sim y^+$. Therefore $h(k,M)\in G_M^B$.

Let $k,d\in{}^{\omega}(\omega\times\varkappa)$ be such that $\{i\in\omega : k_i\ne d_i\}\in F$, i.e. $k/\bar{F}\ne d/\bar{F}$. We show that $h(k,M)\ne h(d,M)$. Let $D \overset{d}{=} \{i\in\omega : k_i\ne d_i\}$. Then

$(\forall n\in D)\langle k(n)0,(T^k_n,k(n)1)\rangle \neq \langle d(n)0,(T^d_n,d(n)1)\rangle$ shows $h(k,M)0 \neq$ $\neq h(d,M)0$. By Prop.4.3.7 of [CK] we have $|{}^{\omega}\varkappa/\bar{F}|=\varkappa^{\omega}$ since every nonprincipal ultrafilter on $\omega$ is $\omega$-regular. Therefore $|G^A_M| =$ $= |G^B_M| = \varkappa^{\omega}$.

QED(Claim 3.5.3.1.)

Now we define an equivalence $\equiv$ on Sb Z. Let $L,M\in$ Sb Z. Then we define $L \equiv M$ iff $(\exists z\in Z)L=\{r+z : r\in M\}$.

Claim 3.5.3.2. Let $L,M \in$ Sb Z be such that $L \not\equiv M$. Then $(\forall h\in G^A_L)(\forall k\in G^A_M)$ $Rgh\cap Rgk = 0$. Similarly for $G^B$.

Proof. Assume $Rgh\cap Rgk \neq 0$. Then $Rgh = Rgk$ by $h,k\in H$. Then there is $z\in Z$ such that $h(0)=k(z)=k(0)+z$. Let this $z\in Z$ be fixed. Then by the definition of $G^A_L$ we obtain for every $r\in Z$ and $q\in{}^{\alpha}U^+$ $r\in L \iff q(0/h(r))\in y^+ \iff q(0/k(r)+z)\in y^+ \iff q(0/k(r+z))\in y^+ \iff r+z\in M$. Then $M = \{r+z : r\in L\}$ showing $M \equiv L$. The proof for $G^B$ is entirely analogous.

QED(Claim 3.5.3.2.)

Let $W \subseteq$ Sb Z be a set of representatives for the partition Sb Z / $\equiv$. For every $L\in W$ let $G^{A+}_L \subseteq G^A_L$ be such that $(\forall h\in G^A_L)(\exists k\in G^{A+}_L)Rgh=Rgk$ and $(\forall h,k\in G^{A+}_L)[Rgh=Rgk \Rightarrow h=k]$. We define $G^{B+}_L \subseteq G^B_L$ similarly. Then still $|G^{B+}_L| = |G^{A+}_L| = \varkappa^{\omega}$ (since for a given $h\in G^A_L$ there are only countably many $k\in G^A_L$ with $Rgh=Rgk$ see the proof of 3.5.3.2).

Claim 3.5.3.3. $(\forall u\in U^+)(\exists! h\in\cup\{G^{A+}_L : L\in W\})u\in Rgh$. The same for $G^B$.

Proof. Let $u\in U^+$. Let $q\in{}^{\alpha}U^+$ and define $L \stackrel{d}{=} \{z\in Z : q(0/u+z)\in y^+\}$. Then $L\subseteq Z$ and hence $(\exists! M\in W)M\equiv L$. Then $M = \{z+r : z\in L\}$ for some $r\in Z$. Let $k \stackrel{d}{=} \langle u+(z-r) : z\in Z\rangle$. Then $k \in G^A_M$. Then there is $h \in$ $\in G^{A+}_M$ such that $Rgh=Rgk$. This proves existence. Uniqueness follows from the definitions of $W$ and $G^{A+}_L$. The proof for $G^B$ is entirely analogous.

QED(3.5.3.3.)

For every $L\in W$ let $g_L : G_L^{A+} \rightarrowtail\!\!\!\!\twoheadrightarrow G_L^{B+}$ be a one-one and onto function. Let $p \overset{d}{=} \cup\{g_L : L\in W\}$. Then $p : \cup\{G_L^{A+} : L\in W\} \rightarrowtail\!\!\!\!\twoheadrightarrow \cup\{G_L^{B+} : L\in W\}$ is a one-one and onto funtion by Claim 3.5.3.2. Let $R \overset{d}{=} \{\langle h(z), p(h)z\rangle : h\in\cup\{G_L^{A+} : L\in W\}$ and $z\in Z\}$.

Claim 3.5.3.4. $R : U^+ \rightarrowtail\!\!\!\!\twoheadrightarrow U^+$ is a permutation of $U^+$ and $\tilde{R} \in \mathrm{Is}(\mathfrak{A}^+, \mathfrak{B}^+)$.

Proof. Let $u\in U^+$ be arbitrary. By Claim 3.5.3.3 there is a unique $h\in\cup\{G_L^{A+} : L\in W\}$ such that $u\in Rgh$. Let $u=h(z)$. Let $v\in U^+$. Then $\langle u,v\rangle\in R$ iff $v = p(h)z$. Therefore $R$ is a function with domain $U^+$. A similar argument, using $G^B$ instead of $G^A$, shows that $R^{-1}$ is a function with domain $U^+$. These statements prove that $R : U^+ \rightarrowtail\!\!\!\!\twoheadrightarrow U^+$ is a permutation of $U^+$. Therefore $\tilde{R} \in \mathrm{Is}(\mathfrak{Gb}\,{}^{\alpha}U^+, \mathfrak{Gb}\,{}^{\alpha}U^+)$. We show that $\tilde{R}\, x^+ = x^+$ and $\tilde{R}\, y^+ = \Omega^+$. By Lemma 3.5.1 we have $x^+ = \{q\in{}^{\alpha}U^+ : q_1=q_0+1\}$. Therefore $\tilde{R}\, x^+ = x^+$ since $u=v+1$ iff $Ru=Rv+1$ holds by the definition of $R$.

Let $q\in{}^{\alpha}U^+$. Let $u=q_0$. Then $(\exists L\subseteq Z)(\exists h\in G_L^A)(\exists k\in G_L^B)(\exists z\in Z)[u=h(z)$ and $Ru=k(z)]$. By the definition of $G_L^A$ we have $q\in y^+ \iff z\in L \iff q(0/k(z))\in\Omega^+ \iff q(0/Ru)\in\Omega^+ \iff R\circ q\in\Omega^+$ since $y^+$ and $\Omega^+$ are regular elements. Therefore $\tilde{R}\, y^+ = \Omega^+$.

Thus $R \in \mathrm{Is}(\mathfrak{A}^+, \mathfrak{B}^+)$ by $A^+ = Sg\{x^+, y^+\}$, $B^+ = Sg\{x^+, \Omega^+\}$.

QED(Claim 3.5.3.4.)

We have that $\tilde{R} \in \mathrm{Is}(\mathfrak{A}^+, \mathfrak{B}^+)$ is a base-isomorphism between $ud^*\mathfrak{A} = \mathfrak{A}^+$ and $ud^*\mathfrak{B} = \mathfrak{B}^+$. Then $\mathfrak{A}^+$ is ext-base-isomorphic to both $\mathfrak{A}$ and $\mathfrak{B}$. Thus $\mathfrak{A} \cong \mathfrak{B}$. Actually, $ud_F^{B-1} \circ \tilde{R} \circ ud_F^A \in \mathrm{Is}(\mathfrak{A}, \mathfrak{B})$.

QED(Claim 3.5.3.)

By these claims, Proposition 3.5.(i) is proved.

Proof of 3.5(ii): We shall need lemmas 3.5.4, 3.5.5 below.

Lemma 3.5.4. Let $\gamma=|\gamma|\geq\omega$. There are $\mathfrak{A},\mathfrak{B}\in BA$ with $\mathfrak{A}\cong\mathfrak{B}$ such that $|1^{\mathfrak{A}}|=\gamma$ and $(\forall V\subseteq 1^{\mathfrak{B}})[|V|\leq\gamma \Rightarrow rl_V \notin Is\mathfrak{B}]$. That is, $\mathfrak{B}$ is not ext-

-isomorphic to any $\mathfrak{C}$ with $|1^{\mathfrak{C}}| \le |1^{\mathfrak{A}}|$.

Proof. Notation: For any set H we define $\mathfrak{P}(H)$ to be the unique Boolean set algebra with universe Sb H. We base the proof on the well known result (*) below, see Hausdorff[H].

(*) Let $\kappa=|\kappa|$ and $\beta=2^{\kappa}$. Then $\mathfrak{Fr}_{\beta}\,BA \cong \mathfrak{A} \subseteq \mathfrak{P}(\kappa)$ for some $\mathfrak{A}$ such that $(\forall x\in A\sim\{0\})|x|=\kappa$.

Let $\gamma\ge\omega$ be a cardinal. By (*) there exists an $\mathfrak{A} \subseteq \mathfrak{P}(\gamma)$ with $\mathfrak{A} \cong \mathfrak{Fr}_{(\gamma^+)}\,BA$. Let $\mathfrak{C} \subseteq \mathfrak{P}(\gamma^+)$ be BA-freely generated by $\{x_\alpha : \alpha\in\gamma^+\}$ with $|C|=\gamma^+$ and $(\forall z\in C\sim\{0\})|z|=\gamma^+$. For each $\alpha\in\gamma^+$ let $y_\alpha \overset{d}{=} x_\alpha\sim\alpha$. Clearly $\{y_\alpha : \alpha\in\gamma^+\}$ BA-freely generates some $\mathfrak{B} \subseteq \mathfrak{P}(\gamma^+)$, since $(\forall z\in C\sim\{0\})|z|>\gamma$. Let $V\subseteq\gamma^+$ with $|V|\le\gamma$. Then there is an $\alpha\in\gamma^+$ such that $V\subseteq\alpha$ and hence $V\cap y_\alpha=0$. Thus $rl_V\notin Is\,\mathfrak{B}$. Observing $\mathfrak{A} \cong \mathfrak{Fr}_{(\gamma^+)}BA \cong \mathfrak{B}$ completes the proof of 3.5.4.

QED(Lemma 3.5.4.)

Lemma 3.5.5. Let $n\in\omega$ and $\mathfrak{A},\mathfrak{C}$ be two $CA_\alpha$-s both of characteristic n. Let $\mathfrak{B} \subseteq \mathfrak{Zd}\,\mathfrak{A}$ and $f \in Ism(\mathfrak{B},\mathfrak{Zd}\,\mathfrak{C})$. Then there exists $F\in Ism(\mathfrak{Sg}^{(\mathfrak{A})}B,\mathfrak{C})$ with $f \subseteq F$. Hence $\mathfrak{Sg}^{(\mathfrak{A})}Dof \cong \mathfrak{Sg}^{(\mathfrak{C})}Rgf$.

Proof. Assume the hypotheses. Let $G\subseteq_\omega B$ be arbitrary. Let $L \overset{d}{=} At\,\mathfrak{Sg}^{(\mathfrak{B})}G$. Then $|L|<\omega$ and $L\subseteq Zd\,\mathfrak{A}$. Let $\mathfrak{Q} \overset{d}{=} \mathfrak{Sg}^{(\mathfrak{A})}L$ and $\mathfrak{R} \overset{d}{=} \mathfrak{Sg}^{(\mathfrak{C})}f^*L$. Then $G\subseteq Q$. Since $\sum^{\mathfrak{A}}L=1^{\mathfrak{A}}$ by [HMT]2.4.7 we have $\langle\langle x\cdot a : a\in L\rangle : x\in Q\rangle \in Is(\mathfrak{Q},P_{a\in L}\,\mathfrak{Rl}_a\mathfrak{Q})$ and $\langle\langle x\cdot f(a) : a\in L\rangle : x\in R\rangle \in Is(\mathfrak{R},P_{a\in L}\,\mathfrak{Rl}_{f(a)}\mathfrak{R})$. Let $a\in L$. Then $\mathfrak{Rl}_a\mathfrak{Q},\ \mathfrak{Rl}_{fa}\mathfrak{R} \in Mn_\alpha$ are both of characteristic n, hence by [HMT]2.5.25 there exists $h_a \in Is(\mathfrak{Rl}_a\mathfrak{Q},\mathfrak{Rl}_{fa}\mathfrak{R})$. By [HMT]0.3.6(iii), $\langle\langle h_a(p_a) : a\in L\rangle : p\in P_{a\in L}(Rl_a\mathfrak{Q})\rangle \in Is(P_{a\in L}\,\mathfrak{Rl}_a\mathfrak{Q},P_{a\in L}\,\mathfrak{Rl}_{fa}\mathfrak{R})$ and hence $F \overset{d}{=} \langle\sum\{h_a(x\cdot a) : a\in L\} : x\in Q\rangle \in Is(\mathfrak{Q},\mathfrak{R})$. Let $a\in L$. Then $F(a)=h_a(a)=f(a)$ proves $L1f \subseteq F$ and hence $G1f \subseteq F$. We have proved statement (*) below.

(*) $(\forall G\subseteq_\omega B)(\exists F \in Ism(\mathfrak{Sg}^{(\mathfrak{A})}G,\mathfrak{C}))G1f \subseteq F$.

By [HMT]0.2.14, statement (*) implies the existence of $F \in$
$\in \mathrm{Ism}(\mathfrak{Sg}^{(\mathfrak{A})}B, \mathfrak{L})$ with $f \subseteq F$.

QED(Lemma 3.5.5.)

Now we turn to the proof of 3.5(ii). Let $\alpha>1$ and $\varkappa>0$. Let $\gamma=|\gamma|\geq\varkappa+\omega$. By Lemma 3.5.4 there are Boolean set algebras $\mathfrak{A}\cong\mathfrak{B}$ with $|1^{\mathfrak{A}}|=\gamma$ and such that (**) below holds.

(**) $(\forall U\subseteq 1^{\mathfrak{B}})[|U|\leq\gamma \Rightarrow rl_U \notin Is\mathfrak{B}]$.

Define $f \overset{d}{=} \langle \cup\{{}^{\alpha}(\{y\}\times\varkappa) : y\in x\} : x\in A \rangle$ and $k \overset{d}{=} \langle \cup\{{}^{\alpha}(\{y\}\times\varkappa) : y\in x\} :$ $: x\in B \rangle$. Let $\mathfrak{L} \overset{d}{=} \mathfrak{Sb}\, f(1^{\mathfrak{A}})$ and $\mathfrak{G} \overset{d}{=} \mathfrak{Sb}\, k(1^{\mathfrak{B}})$. Then $f \in$ $\in \mathrm{Ism}(\mathfrak{A}, \mathfrak{P}\mathfrak{L})$, $k \in \mathrm{Ism}(\mathfrak{B}, \mathfrak{P}\mathfrak{G})$ and $\mathfrak{L}, \mathfrak{G} \in {}_{\varkappa}Gs_{\alpha}$. There is $n$ such that both $\mathfrak{L}$ and $\mathfrak{G}$ are of characteristic $n$. Hence $\mathfrak{N} \overset{d}{=}$ $\overset{d}{=} \mathfrak{Sg}^{(\mathfrak{L})}f^{*}A \cong \mathfrak{Sg}^{(\mathfrak{G})}k^{*}B \overset{d}{=} \mathfrak{P}$, by Lemma 3.5.5, since $\mathfrak{P}\mathfrak{L} \supseteq f^{*}\mathfrak{A} \cong$ $\cong k^{*}\mathfrak{B} \subseteq \mathfrak{P}\mathfrak{P}$.

Assume that some $Crs_{\alpha}$ $\mathfrak{Q}$ is sub-base-isomorphic to both $\mathfrak{N}$ and $\mathfrak{P}$. Then $|base(\mathfrak{Q})|\leq|base(\mathfrak{N})|=|base(f1^{\mathfrak{A}})|=|1^{\mathfrak{A}}|\cdot\varkappa=\gamma\cdot\varkappa=\gamma$. Hence there is a $V\subseteq 1^{\mathfrak{P}}$ with $|base(V)|\leq\gamma$ and $rl_V\in Is\mathfrak{P}$. Then $rl_V\circ k \in$ $\in Is\mathfrak{B}$. Let $U \overset{d}{=} (pj_0)^{*}base(V)$. If $rl_V k(x)\neq 0$ then $x\cap U\neq 0$ since then $(\exists y\in x)V\cap{}^{\alpha}(\{y\}\times\varkappa)\neq 0$. Therefore $rl_U\in Is\mathfrak{B}$. By $|U|\leq|base(V)|\leq\gamma$, this contradicts property (**) of $\mathfrak{B}$ formulated above. We have derived a contradiction from the assumption that $\mathfrak{Q}$ is sub-base--isomorphic to both $\mathfrak{N}$ and $\mathfrak{P}$. Hence no $Crs_{\alpha}$ is sub-base--isomorphic to both $\mathfrak{N}$ and $\mathfrak{P}$.

Let $f$ be a one-one function and $\mathfrak{A},\mathfrak{B}\in Crs_{\alpha}$. The <u>base-relation</u> $f^{AB}$ induced by $f$ on $A\times B$ is defined to be $f^{AB} \overset{d}{=} \{\langle x,y\rangle\in A\times B :$ $: \tilde{f}(x) \subseteq y$ and $(\widetilde{f^{-1}})y \subseteq x\}$. Now:

(***) $\mathfrak{A}$ and $\mathfrak{B}$ are lower base-isomorphic iff there is a one--one function $f$ with $f^{AB} \in Is(\mathfrak{A},\mathfrak{B})$.

By (***) and by the above, $\mathfrak{N}$ and $\mathfrak{P}$ are not lower base-isomorphic.

<u>Proof of 3.5(iii):</u> Let $5\leq\varkappa<\omega$. Define $V \overset{d}{=} {}^{\alpha}2\cup{}^{\alpha}\{2,3\}\cup{}^{\alpha}\{4\}\cup{}^{\alpha}(\varkappa\sim 5)$

and $W \overset{d}{=} {}^{\alpha}2 \cup {}^{\alpha}\{2\} \cup {}^{\alpha}\{3\} \cup {}^{\alpha}\{4\} \cup {}^{\alpha}(\varkappa \sim 5)$. Let $\mathfrak{A} \overset{d}{=} \mathfrak{Mn}(\mathfrak{Sb}\,V)$ and $\mathfrak{B} \overset{d}{=} \mathfrak{Mn}(\mathfrak{Sb}\,W)$. Then $\mathfrak{A},\mathfrak{B} \in {}_{\varkappa}Gs_{\alpha}^{reg}$ by Prop.4.2, $\mathfrak{A} \cong \mathfrak{B}$ by [HMT]2.5.25 and it is obvious that $\mathfrak{A}$ and $\mathfrak{B}$ are not base-isomorphic.

Let $\varkappa < 5 \le \alpha$ and $\mathfrak{A},\mathfrak{B} \in {}_{\varkappa}Gs_{\alpha}^{reg} \cap Lf_{\alpha}$, $\mathfrak{A} \cong \mathfrak{B}$. Suppose $\varkappa = 4$. Then there are 5 cases: $\nabla\mathfrak{A} \in \{\{4\},\{3,1\},\{2\},\{2,1\},\{1\}\}$. By $\mathfrak{A} \cong \mathfrak{B}$ we have $\nabla\mathfrak{A} = \nabla\mathfrak{B}$. If $\nabla\mathfrak{A} = \{4\}$ then $\mathfrak{A},\mathfrak{B} \in Cs_{\alpha}$ and we are done by [HMTI]3.6. If $\nabla\mathfrak{A} = \nabla\mathfrak{B} = \{3,1\}$ then $\mathfrak{A}, \mathfrak{B}$ are base-minimal and we are done by 3.4(2). Let $\nabla\mathfrak{A} = \{2\}$. If $|At\ Zd\ A| = 2$ then $\mathfrak{A}$ and $\mathfrak{B}$ are base-minimal, and we are done. Suppose $|At\ Zd\ A| = 1$. Then $(\forall Y \in Subb(\mathfrak{A}))\, rl({}^{\alpha}Y) \in$ $\in Is\mathfrak{A}$ and similarly for $\mathfrak{B}$. Hence $(\forall Y \in Subb(\mathfrak{A}))(\forall W \in$ $\in Subb(\mathfrak{B}))\mathfrak{Rl}({}^{\alpha}Y)\mathfrak{A} \cong \mathfrak{Rl}({}^{\alpha}W)\mathfrak{B}$. $\mathfrak{Rl}({}^{\alpha}Y)\mathfrak{A} \in Cs_{\alpha}^{reg}$ by $\mathfrak{A} \in Gs_{\alpha}^{reg} \cap Lf_{\alpha}$ and similarly for $\mathfrak{Rl}({}^{\alpha}W)\mathfrak{B}$. Hence we are done by [HMTI]3.6. The cases $\nabla\mathfrak{A} \in \{\{2,1\},\{1\}\}$ and $\varkappa < 4$ are similar to the above ones.

<u>Proof of 3.5(iv)</u>: Let $\alpha \ge \omega$. Let $H \subseteq \alpha$ be such that $|H| \cap |\alpha \sim H| \ge \omega$. Let $p \overset{d}{=} H \times 1$ and $\mathfrak{L} \overset{d}{=} \mathfrak{Sb}\,{}^{\alpha}3$. Define $X \overset{d}{=} \{f \in {}^{\alpha}3 : H1f \in {}^{H}2^{(p)}\}$ and $Y \overset{d}{=} \{f \in {}^{\alpha}3 : H1f \in {}^{H}2\}$. We let $\mathfrak{A} \overset{d}{=} \mathfrak{Sg}^{(\mathfrak{L})}\{X\}$ and $\mathfrak{B} \overset{d}{=} \mathfrak{Sg}^{(\mathfrak{L})}\{Y\}$. Then $\mathfrak{A},\mathfrak{B} \in {}_{3}Cs_{\alpha} \cap Dc_{\alpha}$ by $|\alpha \sim H| \ge \omega$ and by [HMT]2.1.7.

Now we show that $\mathfrak{A}$ and $\mathfrak{B}$ are regular. To this end, we use Theorem 1.3. Clearly, $\mathfrak{L} \in Gws_{\alpha}^{norm}$. Let $x \in \{X,Y\}$. Now $x$ was defined so that $x$ is regular. Now we check that $x$ is small. Let $K \subseteq \Delta x$ be infinite and let $\Gamma \subseteq_{\omega} \alpha$. Let $i \in K \sim \Gamma$. Then $i \in H \sim \Gamma$ since $\Delta x = H$. Then $g_2^i \notin c_{(\Gamma)}x$ for any $g \in {}^{\alpha}3$. Hence $c_i^{\partial} c_{(\Gamma)} x = 0$. We have seen that $x$ is small. Then Theorem 1.3 implies that $\mathfrak{A}$ and $\mathfrak{B}$ are regular.

Next we show $\mathfrak{A} \cong \mathfrak{B}$. Let $x \in \{X,Y\}$, $q \in x$ and $Q \overset{d}{=} {}^{\alpha}3^{(q)}$. We shall use Prop.4.7 to show $rl_Q \in Is(\mathfrak{Sg}^{(\mathfrak{L})}\{x\})$. Clearly $\Delta^{(\mathfrak{L})}Q = 0$ and $x$ is $Q$-wsmall by Remark 4.10 since we have seen that $x$ is small. Let $\Gamma \subseteq_{\omega} \alpha$, $f \in x$ and $s \in {}^{\alpha}3$. Then $f[\Gamma/s] \in x$ iff $s^{*}(H \cap \Gamma) \subseteq 2$ iff $q[\Gamma/s] \in x$. Thus condition (ii) of 4.7 is satisfied. Cond.(i) is satisfied since $|\{x\}| = 1$. $\mathfrak{Mn}(\mathfrak{L})$ is simple by $\mathfrak{L} \in Cs_{\alpha}$ (see [HMTI]5.3) and $|\Delta x| = |H| \ge \omega$. Then $rl_Q \in Is(\mathfrak{Sg}^{(\mathfrak{L})}\{x\})$, by Prop.4.7(II).

Let $q \overset{d}{=} \alpha\times 1$. Then $rl_\Omega \in Is\mathfrak{A}$ and $rl_\Omega \in Is\mathfrak{B}$, by the above. $rl_\Omega{}^*\mathfrak{A}=rl_\Omega{}^*\mathfrak{B}$ by $\Omega\cap X=\Omega\cap Y$. Therefore $\mathfrak{A}\cong\mathfrak{B}$.

It remains to show that $\mathfrak{A}$ and $\mathfrak{B}$ are not base-isomorphic. Suppose $f : 3 \rightarrowtail\!\!\!\rightarrow 3$ is such that $\tilde{f} \in Is(\mathfrak{A},\mathfrak{B})$. Let $Z \overset{d}{=} \tilde{f}X$. Then $Z\in B$ and, clearly, $Z\neq Y$. For every $n<3$ let $Wn \overset{d}{=} {}^{\alpha}3(\langle n : i<\alpha\rangle)$. Above, we have seen that $rl(Wn) \in Is\mathfrak{B}$ for $n\in 2$. Hence $f2=2$ since $W(f2)\cap Z=0$ (by $W2\cap X=0$). Then $f0\in 2$ and $f^*2\subseteq 2$. Thus $rl(Wf0)\in$ $\in Is\mathfrak{B}$ and $W(f0)\cap Z=W(f0)\cap Y$. A contradiction, since $Z\neq Y$.

QED Proposition(3.5.)

The algebras $\mathfrak{N}, \mathfrak{P}$ in the proof of 3.5(ii) were generated by uncountably many elements.

Problem 3.6. Let $\alpha\geq\omega$. Do there exist two countably generated isomorphic $\mathfrak{A},\mathfrak{B}\in Gs_\alpha^{reg}\cap Lf_\alpha$ both with all subbases finite such that they are not lower base-isomorphic?

[HMTI]3.19 refers to Propositions 3.7, 3.8 below. Prop.3.7 below implies that the condition $\sum_{\mu<\lambda} \varkappa^\mu=\varkappa$ cannot be replaced with any weaker condition in [HMTI]3.18.

With respect to the condition $|\alpha|\leq\varkappa$ in 3.7, note that it follows from $|A|\leq\varkappa$ for non-discrete $\mathfrak{A}$. The condition $|A|\leq\varkappa$ was noted (in [HMTI]3.19) to be necessary in [HMTI]3.18.

Proposition 3.7. Let $U$ be a cardinal. Let the cardinals $\varkappa$ and $\lambda$ be such that $\omega\leq\lambda\leq|\alpha|^+$ and $|\alpha|\leq\varkappa< U$. Assume $\sum_{\mu<\lambda} \varkappa^\mu\neq\varkappa$. Then there is an $\mathfrak{A}\in{}_U Cs_\alpha^{reg}\cap Dc_\alpha$ such that (i) - (ii) below hold.

(i) $|A|\leq\varkappa$ and $\lambda=\cup\{|\Delta x|^+ : x\in A\}$.

(ii) For any $W\subseteq U$ if $|W|=\varkappa$ then $rl({}^\alpha W) \notin Hom(\mathfrak{A},\mathfrak{B})$ for any $Crs_\alpha$ $\mathfrak{B}$ and hence $\mathfrak{A}$ is not ext-isomorphic to any ${}_\varkappa Cs_\alpha$.

Proof. Let $U, \varkappa$ and $\lambda$ be as in the hypotheses. Then there is $|\nu|=\nu<\lambda$ such that $\varkappa^{\nu}\neq\varkappa$. Let $H\subseteq\nu$ be such that $0\notin H$ and $|H|=|\nu\sim H|=\nu$. Let $\mathfrak{L} \overset{d}{=} \mathfrak{Gb}^{\alpha}U$. Let $K\subseteq U$ be such that $|K|=\varkappa$. Such a $K$ exists by $\varkappa<|U|$. Let $k\in U\sim K$ be fixed and let $U' \overset{d}{=} U\sim\{k\}$. Then $K\subseteq U'$. Let $n : {}^{H}U' \to U'$ be such that $|n^{*}({}^{H}K)|>\varkappa$. Such a function $n$ exists because $|{}^{H}K| = \varkappa^{\nu}>\varkappa$ and $|U'|>\varkappa$. We define

$x \overset{d}{=} \{q\in{}^{\alpha}U : q_0 = n(H1q) \text{ and } H1q\in{}^{H}U'\}$.

$T \overset{d}{=} \{\{q\in{}^{\alpha}U : q_0=u\} : u\in K\}$.

For every cardinal $\mu$ such that $\nu<\mu<\lambda$ let $L_{\mu}\subseteq\mu$ be such that $\nu\cap L_{\mu}\subseteq H$ and $|L_{\mu}|=\mu$. Let

$y_{\mu} \overset{d}{=} \{q\in{}^{\alpha}U : (\forall i\in L_{\mu})q_i=k\}$.

$Y \overset{d}{=} \{y_{\mu} : \nu<\mu<\lambda \text{ and } \mu=|\mu|\}$.

$G \overset{d}{=} \{x\}\cup T\cup Y$.

Let $\mathfrak{A} \overset{d}{=} \mathfrak{Sg}^{(\mathfrak{L})}G$. Clearly $\mathfrak{A}\in Dc_{\alpha}$ since $(\forall z\in G)\,\nu\sim\Delta z\supseteq\nu\sim(H\cup 1)$ and $|\nu\sim(H\cup 1)|\geq\omega$.

First we show that $\mathfrak{A}$ is regular. We shall use Thm 1.3. Clearly, every element of $G$ is regular. Clearly, $T\cup Y\subseteq Sm^{\mathfrak{A}}$. We show that $x\in Sm^{\mathfrak{A}}$, too. Clearly $\Delta x = H\cup 1$. Let $\Gamma\subseteq_{\omega}\alpha$ and $i\in(H\cup 1)\sim\Gamma$. Let $q\in c_{(\Gamma)}x$. Then $q^{i}_{k}\notin c_{(\Gamma)}x$ since $x\subseteq\{f\in{}^{\alpha}U : f^{*}(H\cup 1)\subseteq U\sim\{k\}\}$. Thus $x\in Sm^{\mathfrak{A}}$. Then by Thm 1.3 we have that $\mathfrak{A}$ is regular.

Next we prove $\lambda = \cup\{|\Delta a|^{+} : a\in A\}$, and $|A|\leq\varkappa$. For every cardinal $\nu<\mu<\lambda$ we have $|\Delta y_{\mu}|=|L_{\mu}|=\mu$. Therefore by $Y\subseteq G$ and by $|\Delta x|=\nu$ we have $\cup\{|\Delta a|^{+} : a\in A\}\geq\lambda$. Next we show that $(\forall a\in A)|\Delta a|<\lambda$. Let $a\in A$. Then $a\in Sg\,G_0$ for some $G_0\subseteq_{\omega}G$. For every finite subset $G_0$ of $G$ there is a cardinal $\mu<\lambda$ such that $(\forall g\in G_0)\Delta g\subseteq\mu$, i.e. $G_0\subseteq Dm_{\mu}$. (For the notation $Dm_{\mu}$ see section 0.) Then $a\in Dm_{\mu}$ by $Dm_{\mu}\in Su\mathfrak{A}$, i.e. $|\Delta a\sim\mu|<\omega$. Then $|\Delta a|\leq\mu<\lambda$. It remains to show that $|A|\leq\varkappa$. $|G|=\varkappa$ since $|T| = |K| = \varkappa$ and $|Y| \leq |\alpha| \leq \varkappa$ by $\lambda \leq |\alpha|^{+}$. Therefore $|A| \leq \varkappa\cup\omega\cup|\alpha| = \varkappa$, by [HMT]0.1.19, 0.1.20.

It remains to show that for any $W\subseteq U$, if $|W|=\varkappa$ then $rl^{A}({}^{\alpha}W)\notin Hom(\mathfrak{A}, \mathfrak{Rl}({}^{\alpha}W)\mathfrak{L})$. Let $W\subseteq U$ be such that $|W| = \varkappa$. Let $V \overset{d}{=} {}^{\alpha}W$.

Let $\mathfrak{R} \stackrel{d}{=} \mathfrak{Rl}_V\mathfrak{A}$. First we show that if $K \not\subseteq W$ then $rl_V \notin Hom(\mathfrak{A},\mathfrak{R})$. Suppose $u \in K{\sim}W$. Let $z \stackrel{d}{=} \{q\in{}^\alpha U : q_0=u\}$. Then $z\in T\subseteq A$ by $u\in K$, and $c_0z=1$. However, $rl_V(z)=0$ by $u\notin W$ and hence $c_0^{[V]}rl_V\, z \neq \neq rl_V(c_0z)$. Thus $rl_V\notin Hom(\mathfrak{A},\mathfrak{R})$ if $K\not\subseteq W$. Assume $K\subseteq W$. We show that $c_0^{[V]}(rl_V(x)) \neq rl_V(c_0x)$. Clearly $c_0x= \{q\in{}^\alpha U : H1q\in{}^HU'\}$. Therefore ${}^HK \subseteq \{H1q : q\in{}^\alpha W\cap c_0x\}$ by $K\subseteq W\cap U'$. Thus there is $q\in \in V\cap c_0x$ such that $n(H1q)\notin W$, by $|n^*({}^HK)|>\varkappa=|W|$. Therefore $q \notin \notin c_0^{[V]}(rl_Vx)$ since $(\forall f\in{}^\alpha W\cap x)H1f\notin q$. Therefore $q\in rl_V(c_0x){\sim}c_0^{[V]}(rl_Vx)$ showing that $rl_V\notin Hom(\mathfrak{A},\mathfrak{R})$.

<u>QED(Proposition 3.7.)</u>

Proposition 3.8 below shows that the hypothesis $\sum_{\mu<\lambda} \varkappa^\mu=\varkappa$ cannot be omitted from [HMTI]3.18 even if we replace ext-isomoprhism with <u>ordinary</u> isomorphism or homomorphism.

<u>Proposition 3.8.</u> For each $\alpha\geq\omega$ and cardinal $\varkappa>\alpha$ there is an $\mathfrak{A}\in {}_\varkappa Cs_\alpha^{reg}\cap Dc_\alpha$ of power $|\alpha|$ such that $(\forall\mathfrak{B}\in Gs_\alpha)$ $[Hom(\mathfrak{A},\mathfrak{B})\neq 0 \Rightarrow \Rightarrow (\forall W\in subb(\mathfrak{B}))|W|>\alpha]$.

<u>Proof.</u> Let $\alpha\geq\omega$ and let $U$ be a cardinal such that $U>\alpha$. Let $H\subseteq\alpha{\sim}1$ be such that $|H|=\alpha$ and $|\alpha{\sim}H|\geq\omega$. Let $X \stackrel{d}{=} \{q\in{}^\alpha U : q0\notin q^*H\}$. Let $\mathfrak{L} \stackrel{d}{=} \mathfrak{Gb}\,{}^\alpha U$ and $\mathfrak{A} \stackrel{d}{=} \mathfrak{Sg}^{(\mathfrak{L})}\{X\}$. Then $|A|=|\alpha|$ and $\mathfrak{A}\in Cs_\alpha\cap Dc_\alpha$ since $\Delta X=1\cup H$.

Next we show that for every $\mathfrak{B}\in Gs_\alpha$ the hypothesis $Hom(\mathfrak{A},\mathfrak{B})\neq 0$ implies that every subbase of $\mathfrak{B}$ has power $>\alpha$. Suppose $h : \mathfrak{A} \to \to \mathfrak{B} \in Gs_\alpha$. Let $y \stackrel{d}{=} h(X)$. Then $(\forall i\in H{\sim}1)y \leq -d_{0i}^{(\mathfrak{B})}$ since $(\forall i\in H{\sim}1)X\leq \leq -d_{0i}^{(\mathfrak{A})}$ and $h \in Hom(\mathfrak{A},\mathfrak{B})$. Suppose there is a subbase $W$ of $\mathfrak{B}$ such that $|W|\leq\alpha$. (Then $W\neq 0$ since no subbase is empty if $\alpha\neq 0$.) Then there exists a $q\in{}^\alpha W\subseteq 1^{\mathfrak{B}}$ such that $q^*(H{\sim}1)=W$, since $\mathfrak{B}\in Gs_\alpha$. Now, $q\notin c_0y$ since $(\forall w\in W)[(\exists i\in H{\sim}1)q_i=w$ and thus $q_w^0\notin y\subseteq -d_{0i}^{(\mathfrak{B})}]$. Therefore $c_0^{(\mathfrak{B})}y\neq 1^{\mathfrak{B}}$. But we have $c_0^{(\mathfrak{A})}X=1^{\mathfrak{A}}$ (by $|U|>\alpha$), contradicting

$h \in Hom(\mathfrak{A},\mathfrak{B})$. We have seen that $Hom(\mathfrak{A},\mathfrak{B}) \neq 0$ and $\mathfrak{B} \in Gs_\alpha$ imply that every subbase of $\mathfrak{B}$ is of power $>\alpha$.

It remains to show that $\mathfrak{A}$ is regular. $\mathfrak{A} \in Cs_\alpha^{reg}$ will be proved in section Reducts, in Prop.8.24, because it uses methods of that section.

QED(Proposition 3.8.)

We conjecture that [HMTI]3.18 remains true if the condition "$|A| \leq \varkappa$" is replaced by the weaker condition " $\mathfrak{A}$ can be generated by $\leq \varkappa$ elements and $(\forall x \in A) \varkappa \geq |\Delta x|$". In particular:

Conjecture 3.9. Let $K \in \{Cs_\alpha^{reg}, Ws_\alpha\}$. Let $\mathfrak{A} \in K$, $S \subseteq base(\mathfrak{A})$, $A = Sg^{(\mathfrak{A})}G$ and $|\omega \cup G \cup S| \leq \varkappa \leq |base(\mathfrak{A})|$. Let $\lambda \stackrel{d}{=} \cup\{|\Delta x|^+ : x \in A\}$. Assume $\varkappa = \sum_{\mu<\lambda} \varkappa^\mu$. Then we conjecture that $\mathfrak{A}$ is strongly ext-isomorphic to some $\mathfrak{B} \in {}_\varkappa K$ with $S \subseteq base(\mathfrak{B})$.

We note that the above conjecture is interesting only in the case when $\varkappa < |\alpha|$.

Proposition 3.10(i) below was quoted in [HMTI]7.30(g). Proposition 3.10 is an algebraic version of the various model theoretic theorems to the effect that elementarily equivalent structures have isomorphic elementary extensions. Note that (i) of Prop.3.10 below is stronger than the quoted model theoretic result since the elementary extensions in 3.10(i) are identical (and not only isomorphic). In this connection see also Problems 3.11, 3.13.

Proposition 3.10. Let $\mathfrak{A},\mathfrak{B} \in Crs_\alpha$ be such that $\mathfrak{A} \cong \mathfrak{B}$. Then statements (i)-(vi) below hold.

(i) Assume $\mathfrak{A},\mathfrak{B} \in {}_\infty Cs_\alpha$ and $base(\mathfrak{A}) \cap base(\mathfrak{B}) = 0$. Let $\alpha \geq \omega$. Then there is a $\mathfrak{C} \in {}_\infty Cs_\alpha$ ext-isomorphic to both $\mathfrak{A}$ and $\mathfrak{B}$.

(ii) Assume $\mathfrak{A},\mathfrak{B}\in Cs_\alpha^{reg}\cap Lf_\alpha$ and $\alpha\geq\omega$. Then some $\mathfrak{C}\in Cs_\alpha^{reg}\cap Lf_\alpha$ is ext-base-isomorphic to both $\mathfrak{A}$ and $\mathfrak{B}$.

(iii) Assume $base(\mathfrak{A})\cap base(\mathfrak{B})=0$ and $\alpha\neq 1$. Then there is some $\mathfrak{C}\in Crs_\alpha$ strongly ext-isomorphic to both $\mathfrak{A}$ and $\mathfrak{B}$. Let $K\in\{Gs_\alpha, Gws_\alpha, Gws_\alpha^{norm}, Gws_\alpha^{wd}, Crs_\alpha^{reg}\}$. If $\mathfrak{A},\mathfrak{B}\in K$ then $\mathfrak{C}\in K$.

(iv) Assume $\mathfrak{A},\mathfrak{B}\in Gws_\alpha^{comp}$ have disjoint units and $base(\mathfrak{A})=base(\mathfrak{B})$. Then there is $\mathfrak{C}\in Gws_\alpha^{comp}$, with the same base, ext-isomorphic to both $\mathfrak{A}$ and $\mathfrak{B}$.

(v) Let $\alpha\geq\omega$. There are isomorphic $\mathfrak{A},\mathfrak{B}\in Ws_\alpha\cap Lf_\alpha$ such that no $Ws_\alpha$ is ext-base-isomorphic to both $\mathfrak{A}$ and $\mathfrak{B}$.

(vi) Let $h\in Is(\mathfrak{A},\mathfrak{B})$ and let $\mathfrak{A},\mathfrak{B}$ be as in (i). Then $h=k\circ t^{-1}$ for some ext-isomorphisms $k\in Is(\mathfrak{C},\mathfrak{B})$, $t\in Is(\mathfrak{C},\mathfrak{A})$ and $\mathfrak{C}\in Cs_\alpha$.

Proof. Proof of (ii): First we prove a technical lemma.

Lemma 3.10.1. (Existence of partial base-isomorphisms) Let $\alpha\geq\omega$. Let $\mathfrak{A},\mathfrak{B}\in Cs_\alpha^{reg}\cap Lf_\alpha$ and let $f\in Is(\mathfrak{A},\mathfrak{B})$. Let $x_0,\ldots,x_n\in A$ and $q_0\in x_0,\ldots,q_n\in x_n$. Let $R \overset{d}{=} \cup\{q_i{}^*\Delta x_i : i\leq n\}$.

Then there is a one-one function $t : R \rightarrowtail base(\mathfrak{B})$ such that $(\forall p\in 1^{\mathfrak{B}})(\forall i\leq n)[\Delta x_i 1 p \subseteq t\circ q_i \Rightarrow p\in f(x_i)]$.

Proof. Assume the hypotheses. We shall prove the lemma by induction on n.

1. Assume n=0. Let $q\in x\in A$. $d \overset{d}{=} \Pi\{d_{ij} : q(i)=q(j) \text{ and } i,j\in\Delta x\}\cdot \cdot\Pi\{-d_{ij} : q(i)\neq q(j) \text{ and } i,j\in\Delta x\}$. Now $d$ is a term in the language of $CA_\alpha$-s, since $|\Delta x|<\omega$ by $\mathfrak{A}\in Lf_\alpha$, and therefore $d^{(\mathfrak{A})}\in A$ and $d^{(\mathfrak{B})}\in B$. Clearly, $q\in d^{(\mathfrak{A})}\cap x$. Thus $f(d^{(\mathfrak{A})}\cap x) = d^{(\mathfrak{B})}\cap f(x)\neq 0$ by $f\in Is(\mathfrak{A},\mathfrak{B})$. Let $k\in d^{(\mathfrak{B})}\cap f(x)$ be arbitrary. Define $t \overset{d}{=} \overset{d}{=} \{\langle q(i),k(i)\rangle : i\in\Delta x\}$. Then $t : Rg(\Delta x 1 q) \rightarrowtail base(\mathfrak{B})$ is a one-one function. Let $p\in 1^{\mathfrak{B}}$ be such that $\Delta x 1 p\subseteq t\circ q$. Then $\Delta x 1 p\subseteq k$, by $\Delta x 1 t\circ q\subseteq k$. Thus $p\in f(x)$ by [HMTI]1.13 since $\Delta x=\Delta f(x)$, $k\in f(x)$, and $f(x)\in B$ is regular by $\mathfrak{B}\in Cs_\alpha^{reg}$.

<u>2. Let $n\in\omega$ and assume</u> that 3.10.1 holds for $n$. Let $x_0,\dots,x_{n+1}\in A$, $q_0\in x_0,\dots,q_{n+1}\in x_{n+1}$. Let $\Delta \overset{d}{=} \cup\{\Delta x_i : i\le n\}$. Let $\tau$ be a finite permutation of $\alpha$ such that $\tau^*(\Delta x_{n+1})\cap\Delta=0$. Such a $\tau$ exists since $\Delta$ and $\Delta x_{n+1}$ are finite by $\mathfrak{A}\in Lf_\alpha$. Let $D \overset{d}{=} \tau^*\Delta x_{n+1}$. By $\mathfrak{A},\mathfrak{B}\in Dc_\alpha$ and $\alpha\ge\omega$ it follows that $s_\tau$ is a derived operation both in $\mathfrak{A}$ and $\mathfrak{B}$; see [HMT]1.11.9. By [HMT]1.11.10, and by $\mathfrak{A}\in Cs_\alpha\cap Dc_\alpha$ we have $p\in x_{n+1}$ iff $p\circ\tau^{-1}\in s_\tau x_{n+1}$. By [HMT]1.11.12(x) we have $\Delta(s_\tau x_{n+1})\subseteq\tau^*\Delta x_{n+1} = D$. Let $i\le n$. Recall the notation $f[H/k]$ from section 0. We define

$$y_i \overset{d}{=} x_i\cdot s_\tau x_{n+1} \quad\text{and}\quad h_i \overset{d}{=} q_i[D/q_{n+1}\circ\tau^{-1}].$$

Then $h_i\in y_i$ since $\mathfrak{A}$ is regular. Let $R_1 \overset{d}{=} \cup\{Rg(\Delta y_i\upharpoonright h_i) : i\le n\}$. Then, by our induction hypothesis, we have a one-one function $t_1 :$ $: R_1 \rightarrowtail base(\mathfrak{B})$ with the property that $(\forall p\in 1^{\mathfrak{B}})(\forall i\le n)[\Delta y_i\upharpoonright p\subseteq t_1\circ h_i \Rightarrow$ $\Rightarrow p\in f(y_i)]$. Let $R = \cup\{Rg(\Delta x_i\upharpoonright q_i) : i\le n+1\}$ and let $t : R \rightarrowtail$ $\rightarrowtail base(\mathfrak{B})$ be any one-one function such that $(R\cap R_1)\upharpoonright t_1\subseteq t$. Let $i\le n+1$ and let $p\in 1^{\mathfrak{B}}$ be such that $\Delta x_i\upharpoonright p\subseteq t\circ q_i$. We show that $p \in$ $\in f(x_i)$.

Suppose first $i\le n$. Let $p' \overset{d}{=} p[\Delta y_i/t_1\circ h_i]$. Now $Do\,p'=\alpha$ since $\Delta y_i\subseteq Do(t_1\circ h_i)$ by $Rg(\Delta y_i\upharpoonright h_i)\subseteq Do\,t_1$. Therefore $p'\in 1^{\mathfrak{B}}$ by $\mathfrak{B}\in Cs_\alpha$, $p\in 1^{\mathfrak{B}}$ and $Rg\,t_1\subseteq base(\mathfrak{B})$. Then $p'\in f(y_i)$ since $\Delta y_i\upharpoonright p'\subseteq t_1\circ h_i$. Thus $p'\in f(x_i)$ since $f(y_i) = f(x_i)\cdot s_\tau f(x_{n+1})$ by $f\in Is(\mathfrak{A},\mathfrak{B})$. We show $\Delta x_i\upharpoonright p\subseteq p'$. Let $H \overset{d}{=} \Delta y_i\cap\Delta x_i$. It is enough to show $H\upharpoonright p\subseteq p'$. Now, $H\upharpoonright h_i\subseteq q_i$ by $H\cap D\subseteq\Delta x_i\cap D=0$ and $h_i = q_i[D/q_{n+1}\circ\tau^{-1}]$. Therefore $H\upharpoonright t_1\circ h_i\subseteq t\circ h_i$ by $Rg(H\upharpoonright h_i)\subseteq Rg(\Delta x_i\upharpoonright q_i)\cap Rg(\Delta y_i\upharpoonright h_i)\subseteq R\cap R_1$. Then $H\upharpoonright p' =$ $= H\upharpoonright t_1\circ h_i = H\upharpoonright t\circ h_i = H\upharpoonright t\circ q_i = H\upharpoonright p$ (as desired) by the above and by $H = \Delta y_i\cap\Delta x_i$, $p' = p[\Delta y_i/t_1\circ h_i]$, $\Delta x_i\upharpoonright p\subseteq t\circ q_i$.

Suppose next $i=n+1$. We have $D\upharpoonright(t\circ q_{n+1}\circ\tau^{-1}) \subseteq p\circ\tau^{-1}$ by $\Delta x_{n+1}\upharpoonright p\subseteq t\circ q_{n+1}$ and by $D = \tau^*\Delta x_{n+1}$. Let $p' \overset{d}{=} (p\circ\tau^{-1})[\Delta y_0/t_1\circ h_0]$. Now $p'\in f(y_0)= f(x_0)\cdot s_\tau f(x_{n+1})$ just as in the case $i\le n$. We show $D\upharpoonright p'\subseteq p\circ\tau^{-1}$. Let $H \overset{d}{=} \Delta y_0\cap D$. Now $H\upharpoonright h_0\subseteq q_{n+1}\circ\tau^{-1}$ and therefore

$H1p' = H1t_1 \circ h_0 = H1t \circ h_0 = H1t \circ q_{n+1} \circ \tau^{-1} = H1p \circ \tau^{-1}$ by $Rg(H1h_0) \subseteq$ $\subseteq R \cap R_1$. Then $p \circ \tau^{-1} \in s_\tau f(x_{n+1})$ by $\Delta(s_\tau f(x_{n+1})) \subseteq D$, $p' \in s_\tau f(x_{n+1}) \in B$ and $\mathfrak{B} \in Cs_\alpha^{reg}$. Then $p \in f(x_{n+1})$ by [HMT]1.11.10.

QED(Lemma 3.10.1.)

Now we turn to the proof of 3.10(ii). Let $\alpha \geq \omega$. Let $\mathfrak{A}, \mathfrak{B} \in$ $\in Cs_\alpha^{reg} \cap Lf_\alpha$ be of bases $U, W$ respectively. Let $f \in Is(\mathfrak{B}, \mathfrak{A})$. Let $I$ be any set such that $|I| \geq |B \cup W \cup \omega|$. Let $F$ be a regular ultrafilter on $I$. Let $U^+ = {}^I U/\bar{F}$. Let $\varepsilon : U \rightarrowtail U^+$, $\mathfrak{A}^+$ and $ud =$ $= ud_F^A \in Is(\mathfrak{A}, \mathfrak{A}^+)$ be defined as in Lemma 3.5.1. Let $e \overset{d}{=} \varepsilon^{-1}$. Then $ud^{-1} \subseteq \tilde{e}$ is a strong ext-base-isomorphism and $\mathfrak{A}^+ \in Cs_\alpha^{reg}$ by Lemma 3.5.1. Let $K \overset{d}{=} \{\langle x, \Delta x 1 q\rangle : q \in x \in B\}$. Then $|K| \leq |B \cup W \cup \omega| \leq |I|$ by $\mathfrak{B} \in Lf_\alpha$. Let $E \subseteq F$ be such that $|E| = |I|$ and $(\forall i \in I)(\{Z \in E : i \in Z\}$ is finite). Such an $E$ exists since $F$ is $|I|$-regular. Let $m : K \rightarrowtail E$ be one-one. Define $G_i \overset{d}{=} \{k \in K : i \in m(k)\}$ for all $i \in I$. Fix $i \in I$. Then $G_i$ is finite. Let $R_i \overset{d}{=} \cup\{Rg(k_1) : k \in G_i\}$. Then $R_i \subseteq_\omega W$ since $\mathfrak{B} \in Lf_\alpha$. By Lemma 3.10.1 there is $t_i : R_i \rightarrowtail U$ such that $(\forall p \in {}^\alpha U)(\forall \langle x, q\rangle \in G_i)[t_i \circ q \subseteq p \Rightarrow p \in f(x)]$. Let $\bar{b} \in {}^W({}^I U)$ be such that $(\forall i \in I)(\forall w \in R_i)\bar{b}(w)_i = t_i(w)$. Let $b \overset{d}{=} \langle \bar{b}(w)/\bar{F} : w \in W\rangle$. Then $b : W \to U^+$. Next we show that $(\forall x \in B)\ \tilde{b}(x) \subseteq ud(fx)$. Let $x \in B$ and $q \in x$. Then $b \circ q \in {}^\alpha(U^+)$ and $\bar{b} \circ q \in P(b \circ q)$. To prove $b \circ q \in ud(fx)$ it is enough to prove $\{i \in I : \langle \bar{b}q(j)_i : j < \alpha\rangle \in fx\} \in F$ by the definition of $ud_F^B$ in Lemma 3.5.1. Let $Z \overset{d}{=} m(\langle x, \Delta x 1 q\rangle)$. Then $Z \in E \subseteq F$ and $(\forall i \in Z)\langle x, \Delta x 1 q\rangle \in G_i$. Thus $(\forall i \in Z) q^*(\Delta x) \subseteq R_i$. Let $i \in Z$ and $j \in \Delta x$. Then $q_j \in R_i$ and hence $\bar{b}q(j)_i = t_i(q(j))$. Thus $t_i \circ (\Delta x 1 q) \subseteq \langle \bar{b}q(j)_i : j < \alpha\rangle$ and therefore $\langle \bar{b}q(j)_i : j < \alpha\rangle \in fx$ by the definition of $t_i$. Then $b \circ q \in ud(fx)$ by $Z \in F$. We have proved Statement (*):

(*) $(\forall x \in B)\ \tilde{b}(x) \subseteq ud\ f(x)$.

Since $ud \circ f$ is a homomorphism, (*) implies that $b$ is one-one as the following computation shows. Let $v, w \in W$ be different. Then there is $p\in{}^{\alpha}W$ such that $p_0=v$ and $p_1=w$. Thus $p\in -d_{01}^{\mathfrak{B}}$. By (*) then $b\circ p\in ud\ f(-d_{01}) = -d_{01}^{(\mathfrak{A}^+)}$ showing that $b(v)\neq b(w)$. Then by [HMTI]3.1 we have $\tilde{b}^*\mathfrak{B}\in Cs_\alpha$ and $\tilde{b}\in Is(\mathfrak{B},\tilde{b}^*\mathfrak{B})$ is a base-isomorphism.

Let $V \stackrel{d}{=} \tilde{b}(1^{\mathfrak{B}}) = \tilde{b}({}^{\alpha}W)$. Let $x\in B$ be fixed. Then $\tilde{b}x \subseteq V\cap ud(f(x))$. By (*) we have $\tilde{b}(-x)\subseteq ud\ f(-x) = -ud\ f(x)$. By $\tilde{b}(-x)=V\sim\tilde{b}(x)$ we have $V\sim\tilde{b}(x) \subseteq -ud\ f(x)$. Thus $\tilde{b}x = V\cap ud\ f(x)$. This proves $B1\tilde{b} = = rl_V\circ ud\circ f\in Is(\mathfrak{B},\tilde{b}^*\mathfrak{B})$. Thus $rl_V^{A+} = \tilde{b}\circ(ud\circ f)^{-1}\in Is(\mathfrak{A}^+,\tilde{b}^*\mathfrak{B})$. Let $d \stackrel{d}{=} \stackrel{d}{=} b^{-1}$. Then $\tilde{d} = rl_V\circ(B1\tilde{b})^{-1}\in Is(\mathfrak{A}^+,\mathfrak{B})$ is a strong ext-base-isomorphism.

We have proved (ii) since $\mathfrak{A}^+\in Cs_\alpha^{reg}$ is strongly ext-base-isomorphic to both $\mathfrak{A}$ and $\mathfrak{B}$. We have also proved Statement 3.10.2 below, which shows the "(ii)-part" of 3.10.(vi).

<u>Statement 3.10.2.</u> Let $\mathfrak{A},\mathfrak{B}\in Cs_\alpha^{reg}\cap Lf_\alpha$ and $f\in Is(\mathfrak{B},\mathfrak{A})$. Then there is a $Cs_\alpha^{reg}$ $\mathfrak{A}^+$ and two strong ext-base-isomorphisms $\tilde{e}\in Is(\mathfrak{A}^+,\mathfrak{A})$ and $\tilde{d}\in Is(\mathfrak{A}^+,\mathfrak{B})$ such that $f = (A^+1\tilde{e})\circ\tilde{d}^{-1}$.

<u>Proof of (iii) and (iv)</u>: Let $\mathfrak{A},\mathfrak{B}\in Crs_\alpha$ have units $V$ and $W$ and have bases $U$ and $Y$ respectively. Assume $\mathfrak{A}\cong\mathfrak{B}$ and $V\cap W=0$. If $\alpha=0$ then $\mathfrak{A}\cong\mathfrak{B}$ implies $\mathfrak{A}=\mathfrak{B}$ and we are done.

Let $f\in Is(\mathfrak{A},\mathfrak{B})$. Let $\mathfrak{D}$ be the full $Crs_\alpha$ with unit $V\cup W$. Let $X \stackrel{d}{=} \{x\cup f(x) : : x\in A\}$. Then $X\in Su\,\mathfrak{D}$ by $f\in Hom(\mathfrak{A},\mathfrak{B})$ and $V\cap W=0$. Let $\mathfrak{C}$ be the $Crs_\alpha$ with unit $V\cup W$ and with universe $X$.

Suppose that either ($\alpha\geq 2$ and $U\cap Y=0$) or ($\mathfrak{A},\mathfrak{B}\in Gws_\alpha^{comp}$ and $U=Y$). Then $\Delta^{(\mathfrak{D})}V = \Delta^{(\mathfrak{D})}W = 0$. Therefore $rl_V^{\mathfrak{C}}\in Is(\mathfrak{C},\mathfrak{A})$ and $rl_W^{\mathfrak{C}}\in Is(\mathfrak{C},\mathfrak{B})$ since by $f\in Is(\mathfrak{A},\mathfrak{B})$ we have $x=0$ iff $x\cup f(x)=0$ iff $f(x)=0$. I.e., $\mathfrak{C}$ is ext-isomorphic to both $\mathfrak{A}$ and $\mathfrak{B}$. If $\mathfrak{A},\mathfrak{B}\in$ $\in Gws_\alpha^{comp}$ and $U=Y$ then $\mathfrak{C}\in Gws_\alpha^{comp}$ with base $U$. So far, (iv) has been proved.

Suppose $\alpha\geq 2$ and $U\cap Y=0$. Then $rl_V^{\mathfrak{C}} = rl^{\mathfrak{C}}({}^{\alpha}U)$ and $rl_W^{\mathfrak{C}} = = rl^{\mathfrak{C}}({}^{\alpha}Y)$ showing that $\mathfrak{C}$ is <u>strongly</u> ext-isomorphic to both $\mathfrak{A}$ and $\mathfrak{B}$. Let $K\in\{Gs_\alpha, Gws_\alpha, Gws_\alpha^{norm}, Gws_\alpha^{wd}\}$. If $\mathfrak{A},\mathfrak{B}\in K$ then clearly $\mathfrak{L}\in K$. Let $\mathfrak{A},\mathfrak{B}\in Crs_\alpha^{reg}$. We show that $\mathfrak{L}$ is regular. Let $y\in C$ and $p\in y$, $q\in V\cup W$ be such that $1\cup\Delta y1p\subseteq q$. There is $x\in A$ such that $y=x\cup f(x)$. By $rl_V^{\mathfrak{L}}\in Is(\mathfrak{L},\mathfrak{A})$ and $rl_W^{\mathfrak{L}}\in Is(\mathfrak{L},\mathfrak{B})$, $\Delta y=\Delta x=\Delta f(x)$. Suppose $p\in x$. Then $q\in V$ by $p(0)=q(0)$ and $base(V)\cap base(W)=0$. Therefore $q\in x$ since $x$ is regular and $1\cup\Delta x1p\subseteq q$. Similarly $p\in f(x)$ implies $q\in f(x)$.

Proof of (v): Let $\alpha\geq\omega$, $p\overset{d}{=}\alpha 1 Id$, $q\overset{d}{=}\langle 0 : i\in\alpha\rangle$, $V\overset{d}{=}{}^{\alpha}{}_{\alpha}{}^{(p)}$, $W\overset{d}{=}{}^{\alpha}{}_{\alpha}{}^{(q)}$. Let $\mathfrak{A}\overset{d}{=}\mathfrak{M}\mathfrak{n}(\mathfrak{Sb}V)$, $\mathfrak{B}\overset{d}{=}\mathfrak{M}\mathfrak{n}(\mathfrak{Sb}W)$. Then $\mathfrak{A}\cong\mathfrak{B}$ by [HMTI]3.22. Suppose $\mathfrak{L}\in Ws_\alpha$ is ext-base-isomorphic to both $\mathfrak{A}$ and $\mathfrak{B}$. We shall derive a contradiction. Let $1^{\mathfrak{L}}={}^{\alpha}U^{(r)}$. Let the corresponding sub-base-isomorphisms be induced by $f:\alpha\rightarrowtail U$ and $g:\alpha\rightarrowtail U$, i.e. let $rl^{\mathfrak{L}}(\tilde{f}V)\in Is(\mathfrak{L},\tilde{f}^*\mathfrak{A})$ and $rl^{\mathfrak{L}}(\tilde{g}W)\in Is(\mathfrak{L},\tilde{g}^*\mathfrak{B})$. Note that $base(\mathfrak{A})\subseteq Dof$ and $base(\mathfrak{B})\subseteq Dog$. Then ${}^{\alpha}U^{(r)}\cap\tilde{f}V\neq 0$ and ${}^{\alpha}U^{(r)}\cap\tilde{g}W\neq 0$. Let $p'\in{}^{\alpha}U^{(r)}\cap\tilde{f}V$ and $q'\in{}^{\alpha}U^{(r)}\cap\tilde{g}W$. Then $p'=f\circ p''$ for some $p''\in V={}^{\alpha}{}_{\alpha}{}^{(p)}$ and therefore $|Rgp'|<\omega$ by $|Rgp|<\omega$. Also, $q'=g\circ q''$ for some $q''\in W={}^{\alpha}{}_{\alpha}{}^{(q)}$ and therefore $|Rgq'|\geq\omega$ by $|Rgq|=|\alpha|\geq\omega$ and since $g$ is one-one. By $|Rgp'|<\omega$ and $|Rgq'|\geq\omega$ we have $|\{i\in\alpha : p'(i)\neq q'(i)\}|\geq\omega$, and this contradicts $p', q'\in{}^{\alpha}U^{(r)}$.

Proof of (i): Let $\mathfrak{A},\mathfrak{B}\in{}_{\infty}Cs_\alpha$ be of bases $U,Y$ respectively. Assume $\mathfrak{A}\cong\mathfrak{B}$, $U\cap Y=0$ and $\alpha\geq\omega$. Let $\kappa$ be a cardinal such that $\kappa=\kappa^{|\alpha|}$ and $\kappa>|A\cup B\cup U\cup Y|$. Then by [HMTI]7.25(ii), $\mathfrak{A}$ and $\mathfrak{B}$ are sub-isomorphic to $\mathfrak{A}^+,\mathfrak{B}^+\in Cs_\alpha$ respectively such that $|base(\mathfrak{A}^+)|=|base(\mathfrak{B}^+)|=\kappa$. Then $rl({}^{\alpha}U)\in Is(\mathfrak{A}^+,\mathfrak{A})$ and $rl({}^{\alpha}Y)\in Is(\mathfrak{B}^+,\mathfrak{B})$. Let $H\overset{d}{=}U\cup Y\cup\kappa$. Then $|H|=\kappa$ and therefore there are $\mathfrak{D}',\mathfrak{G}'\in Cs_\alpha$ ext-isomorphic to $\mathfrak{A}$, $\mathfrak{B}$ respectively and such that $base(\mathfrak{D}')=base(\mathfrak{G}')=H$. Let $V\overset{d}{=}\cup\{{}^{\alpha}H^{(p)} : p\in{}^{\alpha}U\}$ and $W\overset{d}{=}\cup\{{}^{\alpha}H^{(p)} : p\in{}^{\alpha}Y\}$. Since $V\in Zd\,Sb^{\alpha}H$ we have $rl_V^{D'}\in Ho(\mathfrak{D}',\mathfrak{D})$ for some $Crs_\alpha$ $\mathfrak{D}$ with unit $V$. Then $rl^{D}({}^{\alpha}U)\in Is(\mathfrak{D},\mathfrak{A})$ by [HMT]0.2.10(ii) since by ${}^{\alpha}U\subseteq V$ we have $rl^{D}({}^{\alpha}U)\circ rl_V^{D'}=rl^{D'}({}^{\alpha}U)\in Is(\mathfrak{D}',\mathfrak{A})$. That is, $\mathfrak{A}$ is sub-isomorphic to $\mathfrak{D}\in Gws_\alpha^{comp}$. Similarly, $\mathfrak{B}$ is sub-isomorphic to some $\mathfrak{G}\in Gws_\alpha^{comp}$ with unit $W$. By $U\cap Y=0$ and $\alpha\geq\omega$ we have

$V\cap W=0$. By $\mathfrak{A}\cong\mathfrak{B}$ we have $\mathfrak{D}\cong\mathfrak{G}$, and base($\mathfrak{D}$)=base($\mathfrak{G}$)=H. Therefore there is $\mathfrak{N}\in Gws_\alpha^{comp}$ with base H ext-isomorphic to both $\mathfrak{D}$ and $\mathfrak{G}$; this was proved as 3.10(iv).

By (1) in the proof of [HMTI]7.17, every $Gws_\alpha^{comp}$ is sub-isomorphic to some $Cs_\alpha$. Therefore, by $|H|\geq\omega$, there exists $\mathfrak{L}\in Cs_\alpha$ ext-isomorphic to $\mathfrak{N}\in{}_\infty Gws_\alpha^{comp}$. Then $\mathfrak{L}$ is ext-isomorphic to both $\mathfrak{A}$ and $\mathfrak{B}$, since $\mathfrak{N}$ is ext-isomorphic to both $\mathfrak{A}$ and $\mathfrak{B}$ and the composition of ext-isomorphisms is again an ext-isomorphism.

QED(Proposition 3.10.)

Problem 3.11. Let $\alpha\geq\omega$. Let $\mathfrak{A},\mathfrak{B}\in{}_\infty Cs_\alpha^{reg}$ and let $\mathfrak{A}\cong\mathfrak{B}$. Does there exist $\mathfrak{L}\in Cs_\alpha^{reg}$ ext-base-isomorphic to both $\mathfrak{A}$ and $\mathfrak{B}$ ?

We shall need the following definition in formulating Problem 3.13 as well as in subsequent parts of this paper.

Definition 3.12. Let $\mathfrak{A}\in Crs_\alpha$ with base U. Let F be an ultrafilter on I. Let $c : \alpha\times{}^I U/F \to {}^I U$. Then $ud_{cF}^A \overset{d}{=} ud_c \overset{d}{=} \langle\{q\in{}^\alpha({}^I U/F) : \{i\in I : \langle c(j,q_j)_i : j<\alpha\rangle\in a\}\in F\} : a\in A\rangle$.

Problem 3.13. Let $\alpha\geq\omega$, $\mathfrak{A},\mathfrak{B}\in Cs_\alpha$. Assume $\mathfrak{A}\cong\mathfrak{B}$. Are there ultrafilters F,D, an (F,$\langle$base($\mathfrak{A}$) : $i\in\cup F\rangle,\alpha$)-choice function c, and an (D,$\langle$base($\mathfrak{B}$) : $i\in\cup D\rangle,\alpha$)-choice function d such that $ud_{cF}^{A}{}^*\mathfrak{A}$ and $ud_{dD}^{B}{}^*\mathfrak{B}$ are base-isomorphic and $ud_{cF}\in Is\,\mathfrak{A}$, $ud_{dD}\in Is\,\mathfrak{B}$?

We know that the answer is yes if $\mathfrak{A},\mathfrak{B}\in Cs_\alpha^{reg}\cap Lf_\alpha$. A positive answer to this problem would be an algebraic counterpart of the Keisler-Shelah Isomorphic Ultrapowers Theorem. See also 3.10 and 3.11.

Let $h\in Is(\mathfrak{A},\mathfrak{B})$. Are there F,D,c,d as above such that $h = (ud_{dD}^{B})^{-1}\circ f\circ ud_{cF}^{A}$ for some base-isomorphism f between $ud_{cF}{}^*\mathfrak{A}$ and $ud_{dD}{}^*\mathfrak{B}$?

By (1) in the proof of [HMTI]7.17 and by the proof of [HMTI]7.13

we have that every $Gws_\alpha^{comp}$ is sub-isomorphic to some $Cs_\alpha$ and every $Ws_\alpha$ is sub-isomorphic to some $Cs_\alpha^{reg}$. Below we give a rather direct and simple construction of sub-isomorphisms from $Gws_\alpha^{comp}$ -s into $Cs_\alpha$- s and from $Ws_\alpha$-s into $Cs_\alpha^{reg}$-s. This construction is useful e.g. when to a given concrete $\mathfrak{A}\in {}_\varkappa Gws_\alpha^{comp}$, $\varkappa<\omega$ we want to see clearly the concrete structure of a $Cs_\alpha$ $\mathfrak{B}$ ext-isomorphic to $\mathfrak{A}$ in such a way that the structure of $\mathfrak{B}$ would not be much more complicated than that of $\mathfrak{A}$. If $\mathfrak{A}\in Ws_\alpha$ then $\mathfrak{B}\in Cs_\alpha^{reg}$. The construction below works only for $\varkappa<\omega$, it is an open problem to find a construction meeting the above (somewhat vague) requirements for $\varkappa\geq\omega$ or to improve the construction for finite $\varkappa$ from the above point of view. In this line we note that ${}_\omega Ws_\alpha \not\subseteq \mathbf{I}\,{}_\omega Cs_\alpha$ if $\alpha\geq\omega$, by [HMTI]7.30a).

<u>Theorem 3.14.</u> Let $\alpha,\varkappa$ be ordinals. Let $W$ be a $Gws_\alpha^{comp}$ -unit with base $\varkappa$ and let $p\in W$. Let $F$ be an ultrafilter on $I \stackrel{d}{=} Sb_\omega\alpha$. Define

$$h^W_{F,p} \stackrel{d}{=} h \stackrel{d}{=} \langle x\cup\{q\in{}^\alpha\varkappa\sim W : \{\Gamma\in I : p[\Gamma/q]\in x\}\in F\} \ : \ x\subseteq W\rangle .$$

Assume $F\supseteq\{\{\Delta\in I : \Gamma\subseteq\Delta\} \ : \ \Gamma\in I\}$. Then (i)-(iii) below hold.

(i) Let $\varkappa<\omega$. Then $h\in Ism(\mathfrak{Sb}\,W, \mathfrak{Sb}\,{}^\alpha\varkappa)$ is a sub-isomorphism.

(ii) Let $\varkappa<\omega$. Suppose $W={}^\alpha\varkappa^{(p)}$. Then $h^*\mathfrak{Sb}W\in{}_\varkappa Cs_\alpha^{reg}$ and $h = \langle\{q\in{}^\alpha\varkappa : \{\Gamma\in I : p[\Gamma/q]\in x\}\in F\} \ : \ x\subseteq W\rangle$.

(iii) Let $\mathfrak{A}\subseteq\mathfrak{Sb}\,{}^\alpha\varkappa^{(p)}$ and let $V\subseteq{}^\alpha\varkappa$ be such that $\Delta^{(\varkappa)}V=0$, $V\neq 0$, and $\{p[\Delta x/s] : x\in A,\ s\in V\}\subseteq{}^\alpha\varkappa^{(p)}$. Let $f\stackrel{d}{=} rl_V\circ h$. Then $f\in Ism(\mathfrak{A},\mathfrak{Sb}V)$, $f$ is a sub-isomorphism if $p\in V$, $f^*\mathfrak{A}$ is regular and $f=\langle\{q\in V : p[\Delta x/q]\in x\} \ : \ x\in A\rangle$.

<u>Proof.</u> Assume the hypotheses. Let $Q\stackrel{d}{=}{}^\alpha\varkappa^{(p)}$ and $g\stackrel{d}{=}\langle\{q\in{}^\alpha\varkappa : \{\Gamma\in I : p[\Gamma/q]\in x\}\in F\} \ : \ x\subseteq Q\rangle$.

<u>Claim 3.14.1.</u> Let $x\subseteq Q$ and $i,j\in\alpha$.

(i) $x\subseteq gx$.

(ii) $g\in Hom(\langle Sb\,Q,\cap,{}_Q\sim,D^{[Q]}_{ij}\rangle,\langle Sb\,{}^\alpha\varkappa,\cap,{}_{{}^\alpha\varkappa}\sim,D^{(\varkappa)}_{ij}\rangle)$.

(iii) If $\varkappa<\omega$ then $g\in Ism(\mathfrak{Gb}\,Q, \mathfrak{Gb}\,{}^{\alpha}\varkappa)$ is a sub-isomorphism and $g^{*}\mathfrak{Gb}\,Q\in{}_{\varkappa}Cs_{\alpha}^{reg}$.

Proof. (i). Let $x\subseteq Q$ and $q\in x$. Let $\Gamma \overset{d}{=} Do(p\sim q)$. Then $|\Gamma|<\omega$ by $q\in Q = {}^{\alpha}\varkappa^{(p)}$. Let $Z \overset{d}{=} \{\Delta\in I : \Gamma\subseteq\Delta\}$. Then $Z\in F$ and $p[\Delta/q]=q\in x$ for all $\Delta\in Z$. Thus $q\in gx$. (ii). Let $x,y\in SbQ$ and $i,j\in\alpha$. Let $q\in{}^{\alpha}\varkappa$. Then $q\in gx\cap gy$ iff $q\in g(x\cap y)$ since $F$ is a filter. $q\in gx$ iff $\{\Gamma\in I : p[\Gamma/q]\in x\}\in F$ iff $\{\Gamma\in I : p[\Gamma/q]\in Q\sim x\}\notin F$ iff $q\in{}^{\alpha}\varkappa\sim g(Q\sim x)$. $g(D_{ij}^{[Q]})=D_{ij}^{(\varkappa)}$ since $\{\Gamma\in I : \{i,j\}\subseteq\Gamma\}\in F$. (iii). Suppose $\varkappa<\omega$. Then $q\in c_i gx$ iff $(\exists b\in\varkappa)\{\Gamma\in I : p[\Gamma/q_b^i]\}\in F$ iff $\{\Gamma\in I : (\exists b\in\varkappa)p[\Gamma/q_b^i]\in x\}\in F$ iff $\{\Gamma\in I : i\in\Gamma$ and $(\exists b\in\varkappa)p[\Gamma/q]_b^i\in x\}\in F$ iff $\{\Gamma\in I : i\in\Gamma$ and $p[\Gamma/q]\in c_i x\}\in F$ iff $q\in g(c_i x)$, by $\varkappa<\omega$ and $\{\Gamma\in I : i\in\Gamma\}\in F$. By (i)-(ii) then $g\in Ism(\mathfrak{Gb}\,Q, \mathfrak{Gb}\,{}^{\alpha}\varkappa)$. Let $\mathfrak{B} \overset{d}{=} g^{*}\mathfrak{Gb}\,Q$. Then $rl_{Q}^{B} = g^{-1}$ since $x \subseteq Q\cap gx$, $-x \subseteq Q\cap g{-x} = Q\sim gx$ imply $x=Q\cap gx$. Next we prove $\mathfrak{B}\in Cs_{\alpha}^{reg}$. Let $x \subseteq Q$, $q\in gx$, $f\in{}^{\alpha}\varkappa$ be such that $\Delta^{(\mathfrak{B})}gx1f\subseteq q$. By $\Delta gx = \Delta x$ and by $Ws_{\alpha} = Ws_{\alpha}^{reg}$ we have $(\forall\Gamma\in I)[p[\Gamma/q]\in x \Rightarrow p[\Gamma/f]\in x]$. Then by the definition of $g$ we have $f\in gx$.

QED(Claim 3.14.1.)

Proof of (i): Let $\varkappa<\omega$. Let $\mathfrak{L} \overset{d}{=} \mathfrak{Gb}\,W$. By 3.14.1(iii) we have $g\in Ism(\mathfrak{Rl}_{Q}\mathfrak{L}, \mathfrak{Gb}\,{}^{\alpha}\varkappa)$. Then $g\circ rl_{Q}\in Hom(\mathfrak{Gb}\,W, \mathfrak{Gb}\,{}^{\alpha}\varkappa)$ since $\Delta^{[W]}Q=Q$. Let $Z \overset{d}{=} {}^{\alpha}\varkappa\sim W$. Then $\Delta^{(\varkappa)}Z=0$, hence $f \overset{d}{=} rl_{Z}\circ g\circ rl_{Q}\in Hom(\mathfrak{Gb}\,W, \mathfrak{Gb}\,Z)$. By [HMTI] 6.2, $k \overset{d}{=} \langle x\cup fx : x\subseteq W\rangle\in Hom(\mathfrak{L}, \mathfrak{Gb}\,{}^{\alpha}\varkappa)$. Observing $k = h_{F,p}^{W}$ completes the proof of (i).

Proof of (ii): Let $\varkappa<\omega$. Suppose $W = Q$. Let $x\subseteq Q$. Then $x = g(x)\cap Q$ by 3.14.1(iii), hence $gx = x\cup\{q\in{}^{\alpha}\varkappa\sim Q : \{\Gamma\in I : p[\Gamma/q]\in x\}\in F\} = hx$. Now 3.14.1(iii) completes the proof of (ii).

Proof of (iii): Let $\mathfrak{A} \subseteq \mathfrak{Gb}\,Q$, $V \subseteq {}^{\alpha}\varkappa$ and $f$ be as in the hypotheses of 3.14(iii). First we show that $fx = V\cap hx = \{q\in V : p[\Delta x/q]\in x\}$. Let $q\in V$. Let $s \overset{d}{=} p[\Delta x/q]$. Then $s\in Q$ by the hypotheses and therefore $\Gamma \overset{d}{=} Do(p\sim s)$ is finite. Let $Z \overset{d}{=} \{\Delta\in I : \Gamma\subseteq\Delta\}$. Then $Z\in F$ and $(\forall\Delta\in Z)p[\Delta/s]=s$. By $\Delta x1s\subseteq q$ and regularity of $\mathfrak{A}$ we have $p[\Delta/s]\in x$

iff $p[\Delta/s]\in x$ for every $\Delta\in I$. Now, $q\in hx$ iff $\{\Delta\in I : p[\Delta/q]\in x\}\in F$ iff $\{\Delta\in I : \Gamma\subseteq\Delta$ and $p[\Delta/s]\in x\}\in F$ iff $s=p[\Delta x/q]\in x$. We have seen $fx=\{q\in V : p[\Delta x/q]\in x\}$. Next we show $f\in Hom(\mathfrak{A}, \mathfrak{Sb}V)$. By 3.14.1(ii) and by $\Delta^{(\varkappa)}V=0$ it is enough to show that $f$ is a homomorphism w.r.t. the cylindrifications. Let $i\in\alpha$, $x\in A$ and $q\in V$. Then $q\in f(c_i x)$ iff $p[\Delta c_i x/q]\in c_i x$ iff $p[\{i\}\cup\Delta x/q]\in c_i x$ iff $(\exists b\in\varkappa)p[\{i\}\cup\Delta x/q]^i_b\in x$ iff $(\exists b\in\varkappa)p[\Delta x/q^i_b]\in x$ iff $(\exists b\in\varkappa)q^i_b\in fx$ iff $q\in c_i fx$, by $C_i^{(\varkappa)}V=V$. Let $q\in V$ and $s\in x\in A$. Let $\Gamma \overset{d}{=} \{i\in\Delta x : q_i\neq s_i\}$. Then $|\Gamma|<\omega$ by $p[\Delta x/q]\in Q$, $s\in Q$. Thus $q' \overset{d}{=} q[\Gamma/s]\in V$ by $\Delta^{(\varkappa)}V=0$. Now $p[\Delta x/q']\in x$ by $\Delta x 1 s \subseteq$ $\subseteq p[\Delta x/q']$. I.e. $q'\in fx$. We have seen that $f\in Ism(\mathfrak{A}, \mathfrak{Sb}V)$. If $p\in V$ then $Q\subseteq V$ and thus $fx\supseteq x$, which implies that $f$ is a sub-isomorphism. It remains to show that $f^*\mathfrak{A}$ is regular. Let $x\in A$, $s\in fx$ and $q\in V$ be such that $\Delta fx 1 s \subseteq q$. By $\Delta fx=\Delta x$ we have $p[\Delta x/s] =$ $= p[\Delta x/q]$ and therefore $q\in fx$ by $s\in fx = \{g\in V : p[\Delta x/g]\in x\}$.

QED(Theorem 3.14)

Corollary 3.15b) below is a generalization of [HMTI]3.22 and 7.27. For the necessity of the conditions in a) and b) see [HMTI]7.30a)b).

Corollary 3.15.

a) Every $Ws_\alpha\cap Lf_\alpha$ is sub-isomorphic to some $Cs_\alpha^{reg}$ with the same base; and every $Cs_\alpha^{reg}\cap Lf_\alpha$ with nonempty base is ext-isomorphic to some $Ws_\alpha\cap Lf_\alpha$. Thus $\mathbf{H}(Ws_\alpha\cap Lf_\alpha) = \mathbf{I}\, Cs_\alpha^{reg}\cap Lf_\alpha$.

b) Let $\mathfrak{A}\in Ws_\alpha$ have unit ${}^\alpha U^{(p)}$. Let $H\subseteq\alpha$ be such that $(\forall x\in A)$ $|\Delta x\sim H|<\omega$ and let $q\in{}^\alpha U$ be such that $H1p\subseteq q$. Then $\mathfrak{A}$ is isomorphic to some $\mathfrak{B}\in Ws_\alpha$ with unit ${}^\alpha U^{(q)}$. Moreover $\langle\{s\in{}^\alpha U^{(q)} :$ $: p[\Delta x/s]\in x\} : x\in A\rangle\in Is(\mathfrak{A}, \mathfrak{B})$.

Remark 3.16. (i) Thm. 3.14(ii) becomes false if the condition $\varkappa<\omega$ is dropped, namely for any $\varkappa\geq\omega\leq\alpha$ and $p\in{}^\alpha\varkappa$, $V \overset{d}{=} {}^\alpha\varkappa^{(p)}$ we have $h_F\notin Hom(\mathfrak{Sb}V, \mathfrak{Sb}h_F(V))$.

(ii) By Prop.5.6(i) we have that for all $1<\varkappa<\omega\le\alpha$ there is $V\in$ $\in Zd\ Sb^{\alpha}\varkappa$ such that $h_F$ does not preserve regularity. Hence Thm.3.14 (ii)-(iii) do not extend to $(Gws_\alpha^{comp})^{reg}$ from $Ws_\alpha$.

Remark 3.17.

(1) Let $\omega\le\varkappa=|\varkappa|\le\alpha$. Let $L\subseteq\alpha$. Then there are $\mathfrak{A}\in{}_\varkappa Ws_\alpha\cap Dc_\alpha$ and $\mathfrak{B}\in{}_\varkappa Ws_\alpha$ such that (i)-(iv) below hold.

(i) ${}_\mu Cs_\alpha\cap \mathbf{H}\mathfrak{A} = {}_0Cs_\alpha$ for all $\mu\le\alpha$.

(ii) $(\forall\mathfrak{N}\in Gws_\alpha^{comp}\cap\mathbf{H}\mathfrak{B})(\forall q\in 1^{\mathfrak{N}})|base(\mathfrak{N})\sim q^*L|\ge\varkappa$.

(iii) If $|\alpha\sim L|\ge\omega$ then $\mathfrak{B}\in Dc_\alpha$.

(iv) $|A|+|B|\le|\alpha|$.

The proof is an easy modification of [HMTI]7.30a). The basic change is to replace "$\max\{wq_\mu : 0<\mu<\alpha\}$" with "$\max\{wq_\mu : 0<\mu\in L\}$" everywhere in the proof.

(2) If we replace "$Ws_\alpha$" by "K" in [HMTI]7.30a)(4) then we obtain statement (*) below.

(*) There is $\mathfrak{A}\in{}_\varkappa K$ with $|A|\le\alpha$ such that for all $\mathfrak{B}\in K\cap\mathbf{H}\mathfrak{A}$ and $q\in 1^{\mathfrak{B}}$, $|base(\mathfrak{B})\sim Rgq|\ge\varkappa$.

Let $\alpha\ge\varkappa\ge\omega$. Then (*) is true for $K\in\{Ws_\alpha, Gws_\alpha^{comp}\}$ by the proof of [HMTI]7.30a). Let $K = Ws_\alpha\cap Dc_\alpha$. Then (*) is true iff $\varkappa\ge\omega^+$. In particular: let $|\alpha|=\omega$, and $\mathfrak{A}\in Ws_\alpha\cap Dc_\alpha$ with $|A|\le\alpha$. Then there are $\mathfrak{B}\in Ws_\alpha\cap\mathbf{H}\mathfrak{A}$ and $q\in 1^{\mathfrak{B}}$, with $Rgq=base(\mathfrak{B})$. The proof goes by iterating the proof of [HMTI]3.18 and using $Ws_\alpha\subseteq\mathbf{I}\,Cs_\alpha^{reg}$. We omit it. The case of $\varkappa\ge\omega^+$ is immediate by 3.17(1)(ii)-(iii) above, choosing $|\alpha\sim L|=\omega$.

## 4. Subalgebras

About Propositions 4.1-4.3 below see [HMTI]4.2 statements (1)-(4).

Proposition 4.1. Let $\mathfrak{L}$ be a full $Gws_\alpha$. Then (i) and (ii) below are equivalent.

(i) $\mathfrak{L}$ is normal.

(ii) Every subalgebra of $\mathfrak{L}$ generated by a set of locally finite dimensional regular elements is regular.

Proof. (i) ⇒ (ii) follows from Thm 1.3. Next we prove (ii) ⇒ (i). Suppose $\mathfrak{L}$ is a full $Gws_\alpha$ and $\mathfrak{L}$ is not normal. We shall exhibit a regular $x \in C$ such that $\Delta x=1$ and $c_0 x$ is not regular. Since $\mathfrak{L}$ is not normal we have $\alpha \geq \omega$, by [HMTI]1.6; furthermore, there are two subunits ${}^\alpha Y^{(p)}$ and ${}^\alpha W^{(q)}$ of $\mathfrak{L}$ such that $Y \cap W \neq 0$ and $Y \sim W \neq 0$. Let $b \in Y \cap W$ and $a \in Y \sim W$. Let the unit of $\mathfrak{L}$ be V. Define $x \stackrel{d}{=} \{k \in V : k(0)=a\}$. Then $x \in C$ since $\mathfrak{L}$ is full, and $\Delta x=1$ since $p^0_a \in x$ and $p^0_b \in V \sim x$. Clearly, $x$ is regular.

Now we show that $c_0 x$ is not regular. $\Delta(c_0 x)=0$, by [HMT]1.6.8. $q^0_b \in V$ by $b \in W$ and $q^0_a \notin V$ since $a \notin W$ and $\Delta^{[V]}({}^\alpha W^{(q)})=0$ by the definition of subunits (see Def.0.1). Thus $q^0_b \notin c_0 x$. At the same time $p^0_b \in c_0 x$ since $p^0_a \in x$. Now the two sequences $q^0_b$ and $p^0_b$ are both in V, they coincide on $1 \cup \Delta(c_0 x)$ but one of them is in $c_0 x$ while the other is not. Thus $c_0 x$ is not regular.

QED(Proposition 4.1.)

Proposition 4.2 below implies that there is a $Gws_\alpha$ the minimal subalgebra of which is not regular. It also implies that the condition "full" is necessary in Prop.4.1.

Proposition 4.2. Let $\mathfrak{A} \in Gws_\alpha$. Then (i)-(iii) below are equivalent.

(i) $\mathfrak{Mn}(\mathfrak{A})$ is regular.

(ii) $(\forall Y,W \in Subb(\mathfrak{A}))[Y \cap W \neq 0 \Rightarrow |Y| \cap \omega = |W| \cap \omega]$.

(iii) $c_{(\kappa)} \bar{d}(\kappa \times \kappa)$ is regular (in $\mathfrak{A}$) for every $\kappa < \alpha \cap \omega$.

Proof. We may suppose $\alpha\geq\omega$ by [HMTI]1.17, 1.6. Let $\mathfrak{A}\in Gws_\alpha$ with unit V. Let $Subu(V) = \{{}^\alpha Y_i^{(pi)} : i\in I\}$. Let $\varkappa<\omega$. Let $a_\varkappa \overset{d}{=} c_{(\varkappa)}\bar{d}(\varkappa\times\varkappa)$. Then $a_\varkappa\in Mn(\mathfrak{A})$, $\Delta(a_\varkappa)=0$ and $a_\varkappa = \cup\{{}^\alpha Y_i^{(pi)} : i\in I, |Y_i|\geq\varkappa\}$.

First we prove (ii) $\Leftrightarrow$ (iii). Suppose that (ii) holds. Let $\varkappa<\omega$. Let $k\in a_\varkappa$, $q\in V$ and $k(0)=q(0)$. Suppose $k\in{}^\alpha Y_i^{(pi)}$, $q\in{}^\alpha Y_n^{(pn)}$. Then $|Y_i|\geq\varkappa$ by $k\in a_\varkappa$. Then $|Y_n|\geq\varkappa$ by (ii) since $Y_i\cap Y_n\neq 0$ by $k(0)=q(0)$. Therefore $q\in a_\varkappa$, i.e. $a_\varkappa$ is regular. Suppose that (ii) fails. Then there are two subunits ${}^\alpha Y^{(r)}$ and ${}^\alpha W^{(s)}$ of V such that $Y\cap W\neq 0$ and $|W|>|Y|<\omega$. Let $b\in Y\cap W$ and $\varkappa \overset{d}{=} |Y|+1$. Then $s_b^0\in a_\varkappa$ by $|W|\geq\varkappa$, $b\in W$ and $r_b^0\in V\sim a_\varkappa$ by $b\in Y$, $|Y|<\varkappa$. Therefore $a_\varkappa$ is not regular.

Next we prove (iii) $\Longleftrightarrow$ (i). (i) $\Rightarrow$ (iii) holds trivially by $(\forall\varkappa<\omega)$ $a_\varkappa\in Mn(\mathfrak{A})$. Suppose (iii), i.e. suppose that $a_\varkappa$ is regular for every $\varkappa<\omega$. By $\mathfrak{A}\in Gws_\alpha$ we have that $d_{ij}$ is regular, for every $i,j\in\alpha$. Let $X \overset{d}{=} \{a_\varkappa, d_{ij} : \varkappa<\omega, i,j\in\alpha\}$. Let $H \overset{d}{=} 1$. Then $X\subseteq Dm_H$ and every element of X is H-regular, by 1.3.4(i). Therefore $\mathfrak{Sg}^{(\mathfrak{Rd}\,\mathfrak{A})}X$ is regular, by 1.3.5(i), 1.3.4(i). By [HMT]2.2.24 we have $Mn(\mathfrak{A}) = Sg^{(\mathfrak{Rd}\,\mathfrak{A})}X$.

QED(Proposition 4.2.)

Proposition 4.3. Let $\alpha\geq\omega$. Then (i)-(ii) below hold.

(i) The greatest regular Lf subuniverse of a $Gws_\alpha$ need not exist in general. Namely: There are an $\mathfrak{A}\in Gws_\alpha$ and elements x,y of $\mathfrak{A}$ such that $\Delta x=\Delta y=1$ and both {x} and {y} generate regular subalgebras in $\mathfrak{A}$, but {x,y} does not.

(ii) The greatest regular Lf subalgebra of a $Gws_\alpha$ may exist even if regular elements do not generate regular ones. Namely: There are an $\mathfrak{A}\in Gws_\alpha$ and $x\in A$ such that $\Delta x=1$, x is regular, $c_0x$ is not regular and $\mathfrak{Mn}(\mathfrak{A})$ is the greatest regular subalgebra of $\mathfrak{A}$.

Proof. Let $\alpha \geq \omega$, $p \stackrel{d}{=} \langle 0 : \varkappa < \alpha \rangle$, $r \stackrel{d}{=} \langle \alpha : \varkappa < \alpha \rangle$ and let $V \stackrel{d}{=}$ $\stackrel{d}{=} {}^{\alpha}\alpha^{(p)} \cup {}^{\alpha}(\alpha+\alpha)^{(r)}$, $\mathfrak{L} \stackrel{d}{=} \mathfrak{Gb}\, V$.

Proof of (i): Let $x \stackrel{d}{=} \{q \in V : q_0$ is even$\}$ and $y \stackrel{d}{=} \{q \in V :$ $: (q_0 \in \alpha \Rightarrow q_0$ is odd$)$ and $(q_0 \geq \alpha \Rightarrow q_0$ is even$)\}$. Now $Sg\{x\}$ and $Sg\{y\}$ are regular. This can be seen by using 1.3.5(i) and [HMT]2.2.24 since $\Delta x = \Delta y = 1$. However, $\{x,y\}$ is not contained in any regular subalgebra, since $x \cap y = \{q \in V : q_0 \geq \alpha$ and $q_0$ is even$\}$ and thus $c_0(x \cdot y) =$ $= {}^{\alpha}(\alpha+\alpha)^{(r)}$ which is not regular. (i) is proved.

Proof of (ii): Let $x \stackrel{d}{=} \{q \in V : q_0 \geq \alpha\}$ and $\mathfrak{A} \stackrel{d}{=} \mathfrak{Sg}^{(\mathfrak{L})}\{x\}$. Now $\Delta x = 1$, $x$ is regular and $c_0 x = {}^{\alpha}(\alpha+\alpha)^{(r)}$ is not regular. We show that $\mathfrak{Mn}(\mathfrak{A})$ is the only regular subalgebra of $\mathfrak{A}$. $\mathfrak{Mn}(\mathfrak{A})$ is regular by Prop.4.2, since $|\alpha+\alpha| = |\alpha|$. Let $y \in A$, $y \notin Mn(\mathfrak{A})$. We show that $y$ generates an irregular element. By [HMT]2.2.24, $A =$ $= Sg^{(\mathfrak{Bl}\,\mathfrak{A})}\{x_i, d_{ij} : i,j \in \alpha\}$, where $x_i \stackrel{d}{=} \{q \in V : q_i \geq \alpha\}$. Then $y \in A$ implies that

$$y = \sum_{j<n} \Big( \prod_{i<m_j} x_{\mu(i,j)} \cdot \prod_{i<n_j} -x_{\nu(i,j)} \cdot d^j \Big)$$

for some $n, m_j, n_j, \mu(i,j), \nu(i,j) \in \omega$, and $d^j = \prod_{i<\rho} \delta_i$ where $\{\delta_i : i < \rho\} \subseteq \{d_{ij}, -d_{ij} : i,j \in \alpha\}$. We may assume $(\forall j<n) d^j = d^0 \stackrel{d}{=} d$, since $y$ generates such a $y'$. Let $H$ denote the set of indices occurring in $y$, i.e. let $H \stackrel{d}{=} \{\mu(i,j), \nu(i,j) : j<n,\ i<(m_j \cup n_j)\} \cup \Delta(d)$. Then $|H| < \omega$.

Case 1 $(\forall j<n) m_j \neq 0$. Then $c_{(H)} y = {}^{\alpha}(\alpha+\alpha)^{(r)}$, since $y \subseteq {}^{\alpha}(\alpha+\alpha)^{(r)}$ and $y \neq 0$.

Case 2 $(\exists j<n) m_j = 0$. Then we may suppose $n_j \neq 0$. Then $c_{(H)}(-y \cdot d) =$ $= {}^{\alpha}(\alpha+\alpha)^{(r)}$, since $0 \neq (-y \cdot d) \subseteq {}^{\alpha}(\alpha+\alpha)^{(r)}$.

Thus, in both cases, $y$ generates ${}^{\alpha}(\alpha+\alpha)^{(r)}$ which is a non-regular element. Thus (ii) is proved.

QED(Proposition 4.3.)

Remark 4.4. By Prop.4.1 we have that locally finite dimensional regular elements generate regular ones in every normal $Gws_\alpha$.

If we do not suppose normality of $Gws_\alpha$-s then cylindrifications can destroy regularity of locally finite dimensional elements, but only cylindrifications.: See the counterexamples in 4.1-4.3 and see 1.3.5 which implies that locally finite dimensional regular elements always generate regular ones in the "cylindrifications-free reduct" $\langle A,+,\cdot,-,0,1,d_{ij}\rangle_{i,j\in\alpha}$ of $\mathfrak{A}\in Gws_\alpha$. (Note that in $Crs_\alpha$-s only negation (apart from 0,1) preserves regularity of locally finite dimensional elements.)

If we do not require locally finite dimensionality then again the Boolean operations can destroy regularity, already in $Cs_\alpha$.:

Proposition 4.4.1. Let $\alpha\geq\omega$ and $\varkappa>1$. There is $\mathfrak{A}\in Cs_\alpha\cap Dc_\alpha$ of base $\varkappa$ such that regular elements generate nonregular ones in $\mathfrak{Bl}\,\mathfrak{A}$, in fact there are disjoint regular $x,y\in A$ such that $x\cup y = x\oplus y$ is not regular.

Proof. Let $\alpha\geq\omega$ and $\varkappa>1$. Let $H\subseteq\alpha$ be such that $|H|\cap|\alpha\sim H|\geq\omega$. Set $R \overset{d}{=} \{q\in{}^\alpha\varkappa : \{i\in H : q_i\neq 0\}$ is finite$\}$ and $x \overset{d}{=} \{q\in R : (\forall i\in H)q_i=0\}$. $y \overset{d}{=} R\sim x$. Let $\mathfrak{A}$ be the $Cs_\alpha$ with base $\varkappa$ generated by $\{x,y\}$. $\Delta x=H$, $\Delta y=H$ and hence $x$ and $y$ are regular. However $x\oplus y = x\cup y = R$ is not regular since $\Delta R=0$ and $R\notin\{1^{\mathfrak{A}},0\}$.

QED(Proposition 4.4.1.)

To construct regular algebras we shall frequently need Propositions 4.6, 4.7 and 4.9 below. They are closely related to Thm.1.3 and they address the question "which (not necessarily finite dimensional) regular elements generate regular ones".

Definition 4.5. Let $V$ be a $Gws_\alpha$-unit and let $x\subseteq V$, $Q\subseteq V$. Then $x$ is defined to be Q-weakly small (Q-wsmall) in $V$ iff for every infinite $K\subseteq\Delta^{[V]}x$ we have

$(\forall\Gamma \subseteq_\omega \alpha)(\forall q\in Q)(\exists\theta \subseteq_\omega K)\ c_{(\theta)}\{q\} \not\leq c_{(\Gamma)}x$.

$x$ is said to be <u>weakly small</u> (wsmall) in $V$ if $x$ is $V$-wsmall.

Note that weakly smallness is a weaker property than smallness, since $x$ is small in $V$ iff for every infinite $K\subseteq\Delta^{[V]}x$ we have $(\forall\Gamma \subseteq_\omega \alpha)(\exists\theta \subseteq_\omega K)(\forall q\in V)\ c_{(\theta)}\{q\} \not\leq c_{(\Gamma)}x$.

Theorem 1.3 says that small regular elements generate regular ones in normal $Gws_\alpha$-s. The next Proposition 4.6 says that weakly small regular elements generate regular ones in normal $Gws_\alpha$-s, if they (the generator elements) are "very disjoint".

<u>Proposition 4.6.</u> Let $\mathfrak{A}\in Gws_\alpha^{norm}$ be generated by a set $G$ of weakly small regular elements. Assume that $(\forall x,y\in G)(x\neq y \Rightarrow$ $\Rightarrow (\forall\Gamma \subseteq_\omega \alpha)c_{(\Gamma)}x\cap c_{(\Gamma)}y=0)$. Then $\mathfrak{A}$ is regular.

Proposition 4.6 is a special case of the next Prop.4.7. Recall the notation $Dm_H$ from def.1.3.1.

<u>Proposition 4.7.</u> Let $\mathfrak{A}\in Gws_\alpha^{norm}$ with unit $V$ be generated by $G\subseteq A$. Let $Q\subseteq V$, $\Delta^{[V]}Q=0$ and suppose that every element of $G$ is $Q$-wsmall. Assume conditions (i)-(ii) below, for every $y\in G$ and $\Gamma \subseteq_\omega \alpha$.

(i) $(\forall x\in G\sim\{y\})\ c_{(\Gamma)}x\cap c_{(\Gamma)}y=0$.

(ii) $(\forall f\in y)(\exists q\in Q)(\forall p)[f[\Gamma/p]\in y$ iff $q[\Gamma/p]\in y]$.

Then statements (I)-(III) below hold.

(I) $\mathfrak{A}$ is regular if every element of $G$ is regular.

(II) $rl_Q^{\mathfrak{A}}$ is an isomorphism, if $\mathfrak{Mn}(\mathfrak{A})$ is simple, and if $(\forall y\in$ $\in G)|\Delta y|\geq\omega$.

(III) $(\forall H\subseteq\alpha)Dm_H\cap Ig^{(\mathfrak{A})}(G\sim Dm_H)=\{0\}$.

To prove this proposition, we need two lemmas.

Lemma 4.7.1. Let $\alpha \geq \omega$ and let $\mathfrak{A} \in Gws_\alpha^{norm}$ be generated by $G \subseteq A$. Let $Q \subseteq y \subseteq 1^{\mathfrak{A}}$ satisfy conditions (i) and (ii) of Prop.4.7. Then for every $z \in A$, statements (I) and (II) below hold.

(I) $(\exists \Gamma \subseteq_\omega \alpha) z \cap c_{(\Gamma)} y \neq 0 \Rightarrow (\exists \Gamma \subseteq_\omega \alpha) z \cap c_{(\Gamma)} Q \neq 0$.

(II) There is $\Theta \subseteq_\omega \alpha$ such that for every $\Theta \subseteq \Gamma \subseteq_\omega \alpha$ we have $(\forall f \in y)(\exists q \in Q)(\forall p)[f[\Gamma/p] \in z \text{ iff } q[\Gamma/p] \in z]$.

To prove Lemma 4.7.1, we need in turn a definition and two lemmas. 4.7.1.1, 4.7.1.2 below will be used in subsequent parts of this paper too.

Definition 4.7.1.1. Let $\rho : \beta \rightarrowtail \alpha$ be one-one. Define

$rb^\rho \stackrel{d}{=} \langle \langle\langle f_{\rho i}, (\alpha \sim Rg\rho) 1 f \rangle : i \in \beta \rangle : f$ is a function and $Rg\rho \subseteq Do f \rangle$.

$rd^\rho \stackrel{d}{=} rb^{\rho *}$.

Lemma 4.7.1.2. Let $\rho : \beta \rightarrowtail \alpha$ be one-one. Let $V$ be a $Crs_\alpha$-unit. Then (i) and (ii) below hold and $rd^\rho V$ is a $Crs_\beta$-unit.

(i) $rd^\rho \in Is(\mathfrak{Rd}^\rho \mathfrak{Sb} V, \mathfrak{Sb}\, rd^\rho V)$ and $rb^\rho : V \rightarrowtail\!\!\!\!\rightarrow rd^\rho V$, if $\beta \neq 0$.

(ii) If $V$ is a $Gws_\alpha$-unit with subunits $\{{}^\alpha Y_i^{(pi)} : i \in I\}$ then $rd^\rho V$ is a $Gws_\beta$-unit with subunits $\{{}^\beta(Y_i \times \{(\alpha \sim Rg\rho) 1 g\})^{(rb^\rho pi)} : i \in I,\ g \in {}^\alpha Y_i^{(pi)}\}$.

Proof. Let $\rho : \beta \rightarrowtail \alpha$ be one-one. Let $H \stackrel{d}{=} Rg\rho$. Let $V$ be a $Crs_\alpha$-unit. Notation: For every $g \in V$ we denote $g' \stackrel{d}{=} (\alpha \sim H) 1 g$.

Proof of (i): Let $q \in V$ be fixed. Define $V(q) \stackrel{d}{=} \{g \in V : g' \subseteq q\}$ and $Y(q) \stackrel{d}{=} \{g \circ \rho : g \in V(q)\}$. Let $f(q) \stackrel{d}{=} \langle \{g \circ \rho : g \in x\} : x \in SbV(q) \rangle$. Then $f(q) \in Ho(\mathfrak{Rd}^\rho \mathfrak{Sb} V(q), \mathfrak{Sb} Y(q))$ by [HMTI]8.1., since $(\forall g \in V(q))(g \circ \rho)^+ = g$ and $Rg f(q) = SbY(q)$. Let $b(q) \stackrel{d}{=} \langle (u, q') : u \in base(Y(q)) \rangle$. Then $b(q)$ is a one-one function on $base(Y(q))$ and therefore $b(q)$ defines a base-isomorphism $\widetilde{b(q)}$, see [HMTI]3.1.

Let $W(q) \overset{d}{=} \widetilde{b(q)}(Y(q))$ and $h(q) \overset{d}{=} \widetilde{b(q)} \circ f(q)$. Now $h(q) \in Ho(\mathfrak{Rd}^{\rho}\mathfrak{Gb}V(q), \mathfrak{Gb}W(q))$ and $h(q) = SbV(q)1rd^{\rho}$ since $h(q)x = \widetilde{b(q)}(f(q)x) = \widetilde{b(q)}\{g \circ \rho : g \in x\} = \{b(q) \circ g \circ \rho : g \in x\} = \{rb^{\rho}(g) : g \in x\} = rd^{\rho}x$ for every $x \subseteq V(q)$. Let $W \overset{d}{=} \cup\{W(q) : q \in V\}$, $\mathfrak{R} \overset{d}{=} \mathfrak{Rd}^{\rho}\mathfrak{Gb}V$, $\mathfrak{N} \overset{d}{=} \mathfrak{Gb}W$. Let $g,q \in V$. Then clearly $q' = g'$ iff $V(q) \cap V(g) \neq 0$ iff $V(q) = V(g)$ iff $W(q) = W(g)$ iff $base(W(q)) \cap base(W(g)) \neq 0$. Therefore $\Delta^{(\mathfrak{N})}W(q) = 0$ because if $i \in \beta$, $g \in W(q)$, $g^i_a \in W$ then $g^i_a \in W(q)$ since $Rgg \cap Rgg^i_a \neq 0$ by $\alpha > 1$. Also, $\Delta^{(\mathfrak{R})}V(q) = 0$ since if $i \in H$, $g \in V(q)$, $g^i_a \in V$ then $g^i_a \in V(q)$ by $g' = (g^i_a)'$. Therefore we may apply [HMTI]6.2 to $\mathfrak{R}$ and $\mathfrak{N}$. By [HMTI]6.2, [HMT]0.3.6(iii) and by $(\forall q \in V)\ SbV(q)1rd^{\rho} \in Ho(\mathfrak{Rd}^{\rho}\mathfrak{Gb}V(q), \mathfrak{Gb}W(q))$ we obtain that $rd^{\rho} \in Ho(\mathfrak{R}, \mathfrak{N})$. $rd^{\rho}$ is one-one because $rd^{\rho} = rb^{\rho *}$ and $rb^{\rho}$ is one-one on $V$.

<u>Proof of (ii):</u> Let $\{{}^{\alpha}Y_i^{(pi)} : i \in I\} = Subu(V)$. Let $q \in V$. Let $J \overset{d}{=} \{i \in I : q[H/p_i] \in {}^{\alpha}Y_i^{(pi)}\}$. Then $V(q) = \{q[H/g] : (\exists i \in J) g \in {}^{H}Y_i^{(H1pi)}\}$ and therefore $Y(q) = \cup\{{}^{\beta}Y_i^{(pi \circ \rho)} : i \in J\}$. Let $W_i \overset{d}{=} {}^{\beta}Y_i^{(pi \circ \rho)}$ for $i \in J$. Let $g \in W_i \cap W_k$ for some $i,k \in J$. Then $q[H/g \circ \rho^{-1}] \in {}^{\alpha}Y_i^{(pi)} \cap {}^{\alpha}Y_k^{(pk)}$ and therefore ${}^{\alpha}Y_i^{(pi)} = {}^{\alpha}Y_k^{(pk)}$ which implies $W_i = W_k$. This shows that $Y(q)$ is a $Gws_{\alpha}$-unit with subbases $\{{}^{\beta}Y_i^{(pi \circ \rho)} : i \in J\}$. This immediately yields Lemma 4.7.1.2(ii).

<u>QED(Lemma 4.7.1.2.)</u>

Let $V \subseteq {}^{\alpha}U$, $f,q \in V$ and $H \subseteq \alpha$. Define the function $t(f,q,H) : V \to {}^{\alpha}U$ as follows. Let $s \in V$. Then

$$t(f,q,H)(s) \overset{d}{=} \begin{cases} f[H/s] & \text{if } (\alpha \sim H)1q \subseteq s \\ q[H/s] & \text{if } (\alpha \sim H)1f \subseteq s. \\ s & \text{otherwise} \end{cases}$$

<u>Lemma 4.7.1.3.</u> Let $V$ be a $Gws_{\alpha}$-unit, $\alpha \geq \omega$. Let $f,q \in V$ be such that the bases of the (unique) subunits of $V$ containing $f$ and $q$ coincide. Let $H \subseteq_{\omega} \alpha$ and $\rho : |H| \rightarrowtail\!\!\!\rightarrow H$. Then $t(f,q,H)^{*} \in Is(\mathfrak{Rd}^{\rho}\mathfrak{Gb}V, \mathfrak{Rd}^{\rho}\mathfrak{Gb}V)$.

Proof. Let $V, f, q, H$ and $\rho$ be as in the hypotheses. Let $Subu(V) = \{{}^{\alpha}Y_i^{(pi)} : i \in I\}$. Notation: For every $g \in V$ we denote $g' = (\alpha \sim H) 1 g$.

Let $W \stackrel{d}{=} rd^{\rho}V$. For every $w \in base(W)$ define $b(w) \stackrel{d}{=} \begin{cases} \langle u, q' \rangle & \text{if } w = \langle u, f \rangle \\ \langle u, f' \rangle & \text{if } w = \langle u, q \rangle \\ w & \text{otherwise} \end{cases}$.

Then $b : base(W) \rightarrowtail base(W)$ and $\tilde{b}(W) = W$ since $f[H/p] \in V$ iff $q[H/p] \in V$ and $W = \cup\{{}^{|H|}(Y_i \times \{g'\}) : i \in I, \ g \in {}^{\alpha}Y_i^{(pi)}\}$ by Lemma 4.7.1.2 (ii) and by $|H| < \omega$. Therefore $\tilde{\tilde{b}} \in Is(\mathfrak{Cs}\, W, \mathfrak{Cs}\, W)$ by [HMTI]3.1. By Lemma 4.7.1.2(i) we have that $rd^{\rho} \in Is(\mathfrak{Rd}^{\rho}\, \mathfrak{Cs}\, V, \mathfrak{Cs}\, W)$. Therefore it is enough to show $t(f,q,H)^* = rd^{\rho^{-1}} \circ \tilde{b} \circ rd^{\rho}$. By $rd^{\rho} = rb^{\rho *}$ it is enough to show $t(f,q,H)g = rb^{\rho^{-1}}(b \circ rb^{\rho}g)$ for every $g \subseteq V$. Let $g \in V$ be such that $g' \notin \{f', q'\}$. Then $rb^{\rho^{-1}}(b \circ rb^{\rho}g) = rb^{\rho^{-1}}(b \circ \langle\langle g_{\rho i}, g'\rangle : i \in |H|\rangle) = rb^{\rho^{-1}}(\langle\langle g_{\rho i}, g'\rangle : i \in |H|\rangle) = g = t(f,q,H)g$. Suppose $g' = f'$. Then $g = f[H/g]$ and $rb^{\rho^{-1}}(b \circ rb^{\rho}g) = rb^{\rho^{-1}}(b \circ \langle\langle g_{\rho i}, f'\rangle : i \in |H|\rangle) = rb^{\rho^{-1}}(\langle\langle g_{\rho i}, q'\rangle : i \in |H|\rangle) = q[H/g] = t(f,q,H)g$. The case of $g' = q'$ is entirely analogous.

QED(Lemma 4.7.1.3.)

Now we turn to the proof of Lemma 4.7.1. Let $\alpha \geq \omega$, $\mathfrak{A} \in Gws_{\alpha}^{norm}$ with unit $V$, $A = Sg\, G$ and assume that $Q \subseteq y \subseteq V$ satisfy conditions (i),(ii) of 4.7.1. Let $z \in A$. It is enough to prove (II) for $z$, since (I) follows from (II). Let $\Theta$ be a finite nonempty subset of $\alpha$ such that $z \in Sg^{(\mathfrak{Rd}_{\Theta}\mathfrak{A})}G$. Let $\Theta \subseteq \Gamma \subseteq_{\omega} \alpha$ be arbitrary. Let $f \in y$ and let $q \in Q$ be such that $(\forall p)[f[\Gamma/p] \in y$ iff $q[\Gamma/p] \in y]$. Then $f[\Gamma/q] \in y \subseteq V$ since $q[\Gamma/q] = q \in Q \subseteq y$. Then the bases of the subunits of $V$ containing $f$ and $q$ coincide since $V$ is normal and $\Gamma \neq 0$, $\alpha \sim \Gamma \neq 0$. Let $\rho : |\Gamma| \rightarrowtail \Gamma$. Then $t(f,q,\Gamma)^* \in Is(\mathfrak{Rd}^{\rho}\mathfrak{Cs}V, \mathfrak{Rd}^{\rho}\mathfrak{Cs}V)$ by Lemma 4.7.1.3. By definition of $t(f,q,\Gamma)$ we have that $t(f,q,\Gamma)^*x = x$ iff $(\forall p)[f[\Gamma/p] \in x$ iff $q[\Gamma/p] \in x]$. Therefore $G 1 t(f,q,\Gamma)^* \subseteq Id$ since $t(f,q,\Gamma)^*y = y$ by the choice of $q$, and if $x \in G \sim \{y\}$ then $(\forall p)\{f[\Gamma/p], q[\Gamma/p]\} \cap x = 0$ by $c_{(\Gamma)}y \cap c_{(\Gamma)}x = 0$. Let $\mathfrak{R}$ be the subalgebra of $\mathfrak{Rl}^{\rho}\,\mathfrak{Sb}\,V$ generated by G. Then $Rlt(f,q,\Gamma)^{*}\subseteq Id$ and $z\in R$ by $z\in Sg^{(\mathfrak{Rl}^{\rho}\mathfrak{A})}G$. Then $t(f,q,\Gamma)^{*}z=z$ i.e. $(\forall p)[f[\Gamma/p]\in z$ iff $q[\Gamma/p]\in z]$.

QED(Lemma 4.7.1.)

Lemma 4.7.2. Let $\mathfrak{A}\in Gws_{\alpha}^{norm}$ with unit V be generated by $G\subseteq A$. Let $Q\subseteq V$, $\Delta^{[V]}Q=0$. Assume conditions a. and b. below.

a. $(\forall z\in Ig^{(\mathfrak{A})}G\sim\{0\})\, z\cap Q\neq 0$.

b. For every $H\subseteq\alpha$ and $z\in Ig^{(\mathfrak{A})}(G\sim Dm_H)$ we have that $(\forall q\in z\cap Q)(\forall\Omega\subseteq_{\omega}\alpha)(\exists\Theta\subseteq_{\omega}\alpha\sim(H\cup\Omega))c_{(\Theta)}\{q\}\not\subseteq z$.

Then statements (I)-(III) below hold.

(I) $\mathfrak{A}$ is regular if every element of G is regular.

(II) $rl_Q^{\mathfrak{A}}\in Is\,\mathfrak{A}$, if $\mathfrak{Mn}(\mathfrak{A})$ is simple and $(\forall y\in G)|\Delta y|\geq\omega$.

(III) $(\forall H\subseteq\alpha)Ig^{(\mathfrak{A})}(G\sim Dm_H)\cap Dm_H=\{0\}$.

Proof. Assume the hypotheses.

Proof of (II): Suppose $\mathfrak{Mn}(\mathfrak{A})$ is simple and $(\forall y\in G)|\Delta y|\geq\omega$. Let $x\in A\sim\{0\}$. Then $x=d\oplus z$ for some $d\in Mn(\mathfrak{A})$ and $z\in Ig^{(\mathfrak{A})}G$, by [HMTI]5.1.

Case 1 $z\in Mn(\mathfrak{A})$. Then $x\in Mn(\mathfrak{A})$. By $\Delta^{[V]}Q=0$ we have $rl_Q\in$ $\in Ho(\mathfrak{Mn}(\mathfrak{A}))$. We have that $\mathfrak{Mn}(\mathfrak{A})$ is simple. Therefore $x\cap Q=$ $=rl_Q(x)\neq 0$ by $x\in Mn(\mathfrak{A})\sim\{0\}$.

Case 2 $z\notin Mn(\mathfrak{A})$. If $d=0$ then $x=z\in Ig^{(\mathfrak{A})}G\sim\{0\}$. By condition a. then $x\cap Q\neq 0$ and we are done. Suppose $d\neq 0$. Then $d\cap Q\neq 0$ by $d\in$ $\in Mn(\mathfrak{A})\sim\{0\}$ and by Case 1. Let $q\in d\cap Q$. If $q\notin z$ then $q\in d-z\subseteq$ $\subseteq d\oplus z=x$ and we are done. Suppose $q\in z$. Then $q\in z\cap Q$ and by condition b. we have that $(\exists\Theta\subseteq_{\omega}\alpha\sim\Delta d)c_{(\Theta)}\{q\}\not\subseteq z$. Let $q[\Theta/p]\notin z$. Now $q[\Theta/p]\in$ $\in d$ since $q\in d$ and $\Theta\cap\Delta d=0$. Therefore $q[\Theta/p]\in d-z\subseteq d\oplus z=x$. By $q\in Q$ and $c_{(\Theta)}Q=Q$ we have $q[\Theta/p]\in Q$. Therefore $q[\Theta/p]\in x\cap Q$ showing that $x\cap Q\neq 0$. (II) is proved.

Proof of (III): Let $H\subseteq\alpha$. Let $x\in Ig^{(\mathfrak{A})}(G\sim Dm_H)$, $x\neq 0$. Then $x\cap Q\neq 0$ by condition a. Let $q\in x\cap Q$. Let $\Omega\subseteq_{\omega}\alpha$. Then

$(\exists\Theta\subseteq_\omega \alpha\sim(H\cup\Omega))c_{(\Theta)}\{q\}\not\leq x$, by condition b. Therefore $\Theta\cap\Delta x\neq 0$ by $q\in x$, and thus $\Delta x\not\subseteq H\cup\Omega$. Since $\Omega\subseteq_\omega\alpha$ was arbitrary, this shows $|\Delta x\sim H|\geq\omega$, i.e. $x\notin Dm_H$. This proves (III).

To prove (I) we need a lemma:

Lemma 4.7.2.1. Let $\mathfrak{A}\in CA_\alpha$ be generated by $G\subseteq A$ and let $H\subseteq\alpha$. Consider conditions (i)-(iii) below.

(i) $Dm_H^{\mathfrak{A}}\cap Ig^{(\mathfrak{A})}(G\sim Dm_H^{\mathfrak{A}})=\{0\}$.

(ii) $Dm_H^{\mathfrak{A}} = Sg^{(\mathfrak{A})}(G\cap Dm_H^{\mathfrak{A}})$.

(iii) $Sg(G\cap H\text{-dim})\subseteq H\text{-dim}$, where $H\text{-dim}\overset{d}{=}Mn(\mathfrak{A})\cup\{x\in A : |\Delta x\oplus H|<\omega\}$.

Then (i) $\Rightarrow$ (ii) and $[(\forall H\in Sb\alpha)$(i) holds$]\Rightarrow(\forall H\in Sb\alpha)$(iii) holds.

Proof. Let $\mathfrak{A}\in CA_\alpha$ and let $H\subseteq\alpha$.

Proof of (i) $\Rightarrow$ (ii): Let $I\overset{d}{=}Ig^{(\mathfrak{A})}(G\sim Dm_H)$. Assume $Dm_H\cap I=\{0\}$. By Fact(*) in the proof of 1.3.3 we then have $Dm_H\subseteq Sg^{(\mathfrak{A})}(G\sim I)$. By $G\sim Dm_H\subseteq I$ we have $G\sim I\subseteq Dm_H$ and then by $Dm_H\in Su\,\mathfrak{A}$ we have $Dm_H = Sg(G\cap Dm_H)$.

Proof of (i) $\Rightarrow$ (iii): Suppose $\mathfrak{A}$ and $G$ satisfy (i), i.e. $Dm_H^{\mathfrak{A}}\cap Ig^{(\mathfrak{A})}(G\sim Dm_H^{\mathfrak{A}})=\{0\}$. Let $G'\subseteq G$ be arbitrary and let $\mathfrak{A}'\overset{d}{=}\mathfrak{Sg}^{(\mathfrak{A})}G'$. First we show that $G'$ and $\mathfrak{A}'$ satisfy (i), too. Let $Dm_H'\overset{d}{=}Dm_H(\mathfrak{A}')$. $G'\sim Dm_H' = \{y\in G' : |\Delta y\sim H|\geq\omega\}\subseteq G\sim Dm_H$, and by $\mathfrak{A}'\subseteq\mathfrak{A}$ then $Ig^{(\mathfrak{A}')}(G'\sim Dm_H')\subseteq Ig^{(\mathfrak{A})}(G\sim Dm_H)$. By $Dm_H'\subseteq Dm_H$ and by $Dm_H\cap Ig^{(\mathfrak{A})}(G\sim Dm_H) = \{0\}$ then we have $Dm_H'\cap Ig^{(\mathfrak{A}')}(G'\sim Dm_H') = \{0\}$, as desired.

Now we turn to the proof of (iii). Let $G'\overset{d}{=}G\cap H\text{-dim}$. Let $x\in Sg\,G'$ be arbitrary and let $K\overset{d}{=}\Delta x$. We show that $|H\oplus K|\geq\omega$ implies $x\in Mn(\mathfrak{A})$. Suppose $|H\oplus K|\geq\omega$. Now

(1) $G'\cap Dm_K\subseteq Mn(\mathfrak{A})$.

For, note that $H\text{-dim}\subseteq Dm_H$, and hence $Sg(H\text{-dim})\subseteq Dm_H$. So $x\in Dm_H$ and so $K\sim H$ is finite and hence $H\sim K$ is infinite. If $y\in G'\cap Dm_K$ then $|\Delta y\sim K|<\omega$, so $|H\sim\Delta y|$ is infinite. Hence $y\in Mn(\mathfrak{A})$ by

$y \in G' \subseteq H$-dim. Thus (1) holds. Since $G'$ and $\mathfrak{A}' = \mathfrak{S}\mathfrak{g}G'$ satisfy (i), we have that $Dm_K(\mathfrak{A}') = Sg^{(\mathfrak{A}')}(G' \cap Dm_K(\mathfrak{A}')) \subseteq Mn(\mathfrak{A}')$, by (i) $\Rightarrow$ $\Rightarrow$ (ii). By $x \in Dm_K(\mathfrak{A}')$ then $x \in Mn(\mathfrak{A}')$. We have seen $Sg(G \cap H\text{-dim}) \subseteq$ $\subseteq$ H-dim.

QED(Lemma 4.7.2.1.)

Now we can prove (I) of 4.7.2. By (III) and by Lemma 4.7.2.1 we have that $(\forall H \subseteq \alpha) Dm_H = Sg(G \cap Dm_H)$. Then by $\mathfrak{A} \in Gws_\alpha^{norm}$ we can apply Lemma 1.3.6 which yields that $\mathfrak{A}$ is regular if every element of $G$ is regular.

QED(Lemma 4.7.2.)

Now we turn to the proof of Prop.4.7. Let $\mathfrak{A} \in Gws_\alpha^{norm}$ be generated by $G \subseteq A$. Let $Q \in Zd\, \mathfrak{S}\mathfrak{b}\, 1^{\mathfrak{A}}$. First we show that conditions (i)-(ii) of 4.7 imply that 4.7.2.a holds, i.e. $(\forall z \in Ig^{(\mathfrak{A})} G \sim \{0\}) z \cap Q \neq 0$. Let $z \in Ig^{(\mathfrak{A})} G$, $z \neq 0$. Then $z \cap c_{(\Gamma)} y \neq 0$ for some $\Gamma \subseteq_\omega \alpha$ and $y \in G$. The conditions of Lemma 4.7.1 are satisfied by $Q \cap y$ and $y$. Therefore 4.7.1 (I) says that $(\exists \Gamma \subseteq_\omega \alpha) z \cap c_{(\Gamma)} Q \neq 0$. By $\Delta^{[V]} Q = 0$ we have $Q =$ $= c_{(\Gamma)} Q$, therefore $z \cap Q \neq 0$.

Assume now condition (i) of 4.7 and assume that every element of $G$ is $Q$-wsmall. We show that then condition b. of 4.7.2 is satisfied. Let $H \subseteq \alpha$ and $z \in Ig^{(\mathfrak{A})}(G \sim Dm_H)$. Let $q \in z \cap Q$ and $\Omega \subseteq_\omega \alpha$ be arbitrary. By $z \in Ig^{(\mathfrak{A})}(G \sim Dm_H)$ we have that $z \leq c_{(\Gamma)} \sum Y$ for some $\Gamma \subseteq_\omega \alpha$ and $Y \subseteq_\omega G \sim Dm_H$. By $q \in z \cap Q$ then $q \in c_{(\Gamma)} y \cap Q$ for some $y \in Y$. Let $K \stackrel{d}{=} \Delta y \sim (H \cup \Omega)$. By $y \notin Dm_H$ we have $|K| \geq \omega$. Then $(\exists \Theta \subseteq_\omega$ $\subseteq_\omega K) c_{(\Theta)}\{q\} \nsubseteq c_{(\Gamma)} y$, since $y$ is $Q$-wsmall. By condition (i) we then have $c_{(\Theta)}\{q\} \nsubseteq z$, since $z \subseteq \sum\{c_{(\Gamma)} y : y \in Y\}$ and $c_{(\Theta)}\{q\} \subseteq$ $\subseteq c_{(\Theta \cup \Gamma)} y$ by $q \in c_{(\Gamma)} y$. We have seen that condition b. of 4.7.2 is satisfied.

Hence 4.7.2 yields the conclusion of 4.7.

QED(Proposition 4.7.)

Definition 4.8. Let V be a $Gws_\alpha$-unit and let $x \subseteq V$.

(i) Let $q \in V$, $\Gamma \subseteq_\omega \alpha$ and $K \subseteq \alpha$. Then x is defined to be $(q,\Gamma,K)$-small iff

$$(\exists \Theta \subseteq_\omega \alpha\sim\Gamma)(\exists h \in c^{[V]}_{(K\cap\Theta)}\{q\})(\forall \Omega \subseteq_\omega \alpha\sim\Theta)\ h \notin c_{(\Omega)}x.$$

(ii) x is defined to be irreversibly-small (i-small) in V iff

$$(\forall q \in x)(\forall \Gamma \subseteq_\omega \alpha)(\forall K \subseteq \Delta^{[V]}x)[|K| \geq \omega \Rightarrow x \text{ is } (q,\Gamma,K)\text{-small}].$$

Note that i-smallness is a stronger property than wsmallness. Proposition 4.9 below shows that the condition of disjointness can be eliminated from Prop.4.6 if we change from wsmall to i-small.

Proposition 4.9. Every normal $Gws_\alpha$ generated by i-small regular elements is regular.

Proof. We may assume $\alpha \geq \omega$ by [HMTI]1.17. Before giving the proof of 4.9 we need a definition and a claim.

Definition 4.9.1. Let $\mathfrak{A} \in Gws_\alpha$ and $x \in A$. Tnen x is said to be (*)-small iff

$$(\forall q \in 1^{\mathfrak{A}})(\forall \Gamma \subseteq_\omega \alpha)(\forall K \subseteq \Delta^{(\mathfrak{A})}x)[|K| \geq \omega \Rightarrow x \text{ is } (q,\Gamma,K)\text{-small}].$$

Note that (*)-smallness appears to be a stronger property than i-smallness.

Claim 4.9.2. Let $\mathfrak{A} \in Gws_\alpha$ and let $x \in A$. Then x is i-small iff x is (*)-small.

Proof. It is enough to prove that i-smallness implies (*)-smallness. Suppose x is i-small. Let $q \in 1^{\mathfrak{A}}$, $\Gamma \subseteq_\omega \alpha$ and $K \subseteq \Delta x$, $|K| \geq \omega$. We have to prove that x is $(q,\Gamma,K)$-small. If $(\forall \Delta \subseteq_\omega \alpha) q \notin c_{(\Delta)}x$ then we are done. Assume $q \in c_{(\Delta)}\{p\}$ for some $\Delta \subseteq_\omega \alpha$ and $p \in x$.

Then $x$ is $(p,\Gamma\cup\Delta,K)$-small and therefore $(\exists\Theta \subseteq_\omega \alpha\sim(\Gamma\cup\Delta))(\exists k\in c_{(K\cap\Theta)}\{p\})(\forall\Omega \subseteq_\omega \alpha\sim\Theta)k \notin c_{(\Omega)}x$. There is $h \in$ $\in c_{(K\cap\Theta)}\{q\}$ such that $k \in c_{(\Delta)}\{h\}$ since $k \in c_{(K\cap\Theta)}\{p\} \subseteq$ $\subseteq c_{(K\cap\Theta)}c_{(\Delta)}\{q\} = c_{(\Delta)}c_{(K\cap\Theta)}\{q\}$. It is enough to show that $(\forall\Omega \subseteq_\omega$ $\subseteq_\omega \alpha\sim\Theta)h \notin c_{(\Omega)}x$. Let $\Omega \subseteq_\omega \alpha\sim\Theta$. Then $k \notin c_{(\Omega\cup\Delta)}x$ since $\Omega\cup\Delta \subseteq_\omega$ $\subseteq_\omega \alpha\sim\Theta$ by $\Theta \subseteq \alpha\sim\Delta$. Then $h \notin c_{(\Omega)}x$ by $k \in c_{(\Delta)}\{h\}$. We have seen that $x$ is $(q,\Gamma,K)$-small.

<u>QED(Claim 4.9.2.)</u>

Now we turn to the proof of 4.9. Let $\mathfrak{A}\in Gws_\alpha^{norm}$ be generated by $G\subseteq A$ and assume that every element of $G$ is i-small. Let $H\subseteq\alpha$. Let $L\subseteq\alpha$ be such that $|L\sim H|<\omega$. Define

$$S(L) \stackrel{d}{=} S \stackrel{d}{=} \{x\in A : (\forall q\in 1^{\mathfrak{A}})(\forall\Gamma \subseteq_\omega \alpha)\ x \text{ is } (q,\Gamma,\alpha\sim L)\text{-small}\}.$$

We shall prove the following statements about $S$.:

(1) $S$ is closed under $+$.

(2) $G\sim Dm_H \subseteq S$.

(3) $(\forall z\in Ig\ S)(\forall q\in 1^{\mathfrak{A}})(\exists\Theta \subseteq_\omega \alpha\sim L)\ c_{(\Theta)}\{q\} \nsubseteq z$.

<u>Proof of (1):</u> Let $x,y\in S$. Let $q\in 1^{\mathfrak{A}}$ and $\Gamma \subseteq_\omega \alpha$ be arbitrary. Then $(\exists\Theta \subseteq_\omega \alpha\sim\Gamma)(\exists h\in c_{(\Theta\sim L)}\{q\})(\forall\Omega \subseteq_\omega \alpha\sim\Theta)\ h \notin c_{(\Omega)}x$, by $x\in S$, and $(\exists\Delta \subseteq_\omega \alpha\sim(\Gamma\cup\Theta))(\exists k\in c_{(\Delta\sim L)}\{h\})(\forall\Omega \subseteq_\omega \alpha\sim\Delta)\ k \notin c_{(\Omega)}y$, by $y\in S$. Then $\Theta\cup\Delta \subseteq_\omega \alpha\sim\Gamma$, and $k \in c_{((\Theta\cup\Delta)\sim L)}\{q\}$. Let $\Omega \subseteq_\omega \alpha\sim(\Theta\cup\Delta)$. Then $k \notin$ $\notin c_{(\Omega)}y$ by $\Omega \subseteq_\omega \alpha\sim\Delta$. Also $k \notin c_{(\Omega)}x$ since $h \in c_{(\Delta)}\{k\}$ and $h \notin$ $\notin c_{(\Omega\cup\Delta)}x$ by $\Omega\cup\Delta \subseteq_\omega \alpha\sim\Theta$. This shows that $k \notin c_{(\Omega)}(x+y)$, thus $x+y\in S$.

<u>Proof of (2):</u> Let $x\in G\sim Dm_H$. Then $|\Delta x\sim H|\geq\omega$ by $x\notin Dm_H$ and therefore $|\Delta x\sim L|\geq\omega$ by $|L\sim H|<\omega$. Therefore $x$ is $(q,\Gamma,\Delta x\sim L)$-small for every $q\in 1^{\mathfrak{A}}$ and $\Gamma \subseteq_\omega \alpha$, since $x$ is $(*)$-small by $x\in G$. Then $x\in S$ since if $x$ is $(q,\Gamma,K)$-small then $x$ is $(q,\Gamma,M)$-small for every $M\supseteq K$.

<u>Proof of (3):</u> Let $z\in Ig\ S$. Then $z\leq c_{(\Gamma)}y$ for some $\Gamma \subseteq_\omega \alpha$ and $y\in S$, by (1). Let $q\in 1^{\mathfrak{A}}$. Then $(\exists\Theta \subseteq_\omega \alpha\sim\Gamma)(\exists h\in$ $\in c_{(\Theta\sim L)}\{q\})(\forall\Omega \subseteq_\omega \alpha\sim\Theta)\ h \notin c_{(\Omega)}y$, by $y\in S$. Then $h \notin c_{(\Gamma)}y$ since

$\Gamma \subseteq_\omega \alpha\sim\Theta$. Then $h\notin z$, $h\in c_{(\Theta\sim L)}\{q\}$ complete the proof of (3).

Now we show that the conditions of Lemma 4.7.2 are satisfied with $Q=1^{\mathfrak{A}}$. Only condition b. is not immediate. Let $H\subseteq\alpha$, $z\in Ig^{(\mathfrak{A})}(G\sim Dm_H)$, $q\in 1^{\mathfrak{A}}$ and $\Omega \subseteq_\omega \alpha$. Then $z\in Ig^{(\mathfrak{A})}S(H\cup\Omega)$ by (1) and (2) and then $(\exists\Theta \subseteq_\omega \alpha\sim(H\cup\Omega))\ c_{(\Theta)}\{q\} \nleq z$ by (3). Thus b. holds, so 4.7.2 yields the conclusion of 4.9.

QED(Proposition 4.9.)

Remark 4.10. Let $V$ be a $Crs_\alpha$-unit and let $x\subseteq V$. Consider conditions (i)-(iii) below.

(i) $x$ is small in $V$.

(ii) $x$ is weakly small in $V$.

(iii) $x$ is irreversibly small in $V$.

Then (i) $\Rightarrow$ (ii), (iii) $\Rightarrow$ (ii), but (i) $\not\Rightarrow$ (iii), (iii) $\not\Rightarrow$ (i). In fact, let $\alpha\geq\omega$, let $x = \{q\in{}^\alpha\omega : (\exists n\in\omega)[q_n\neq 0,\ (\forall k<n)q_k\leq n,\ (\forall k>n)q_k=0]\}$, and let $y = \{q\in{}^\alpha\omega : (\exists n\in\omega)(\forall k>n)q_k=n\}$. Then $x$ is small but not i-small, and $y$ is i-small but not small.

4.10.1. $G$ is said to be a set of hereditarily disjoint elements of $\mathfrak{A}$ iff $(\forall x,y\in G)[x\neq y \Rightarrow (\forall\Gamma \subseteq_\omega \alpha)c_{(\Gamma)}x\cap c_{(\Gamma)}y=0]$ and $G\subseteq A$. Let $G$ be a set of hereditarily disjoint small elements of $Sb\ V$ with $(\forall x\in G)\,|\Delta(\cup G)\sim\Delta x|<\omega$. Then $\cup G$ is wsmall in $V$. Indeed, let $G$ be a set of hereditarily disjoint wsmall elements of $Sb\ V$. Assume $\cup G \in B \overset{d}{=} \cap\{Dm_{\Delta x} : x\in G\}$. To prove that $\cup G$ is wsmall, let $q\in V$, $\Gamma \subseteq_\omega \alpha$, $K \subseteq \Delta(\cup G)$ with $|K|\geq\omega$. If $q \notin c_{(\Gamma)}\cup G$ then we are done. Assume $q \in c_{(\Gamma)}\cup G$. Then $q \in c_{(\Gamma)}x$ for some $x\in G$. Let $L \overset{d}{=} K\cap\Delta x$. Since $\cup G\in B$, $|L|\geq\omega$. By wsmallness of $x$ there are $\Theta \subseteq_\omega L$ and $p\in c_{(\Theta)}\{q\}$ such that $p \notin c_{(\Gamma)}x$. Let $y \in G\sim\{x\}$. By hereditary disjointness $p \notin c_{(\Gamma\cup\Theta)}y$. Thus $p \notin c_{(\Gamma)}\cup G$ proving that $\cup G$ is wsmall. We have proved a strenghtened version of 4.10.1.

However, not every wsmall element can be obtained from small ones by using 4.10.1. An example of this is the following wsmall element: $x = \{q\in{}^{\omega}2 : (\exists n\in\omega)[q_n=q_{n+1}=1,\ (\forall k>n+1)q_k=0]\}$. Indeed, if $0\notin\{y,z\}\subseteq Sb\,x$ then $y$ and $z$ are not hereditarily disjoint.

Propositions 4.11, 4.13 below are applications of Propositions 4.6, 4.7, and 4.9. For more applications see sections 5 and 6.

About Propositions 4.11-4.13 below see [HMTI]5.6(15). In [HMTI]5.6 (15) it is announced that the construction given in [HMTI]5.6(4) can be modified to show that $(\forall\alpha\geq\omega)(\forall\kappa\geq 2)\ (\exists\ \mathfrak{A}\in{}_{\kappa}Cs_{\alpha}^{reg})(\exists\,\mathfrak{B}\in\mathbf{H}\mathfrak{A})|Zd\,\mathfrak{B}|>2$. This modified construction is given in Prop.4.11 below. (Actually, $\mathfrak{A}\in Cs_{\alpha}^{reg}\cap Dc_{\alpha}$ there.) Prop.4.11 also says that regularity can be destroyed by relativization with a zero-dimensional element; see Prop.2.2(iii).

Proposition 4.11. Let $\alpha\geq\omega$ and $\kappa\geq 2$ be arbitrary. There are an $\mathfrak{A}\in{}_{\kappa}Cs_{\alpha}^{reg}\cap Dc_{\alpha}$ and a $W\in Zd\ \mathfrak{Sb}\ 1^{\mathfrak{A}}$ such that $\mathfrak{Rl}_W\mathfrak{A}\in(Gws_{\alpha}^{comp}\cap\mathbf{H}\mathfrak{A})\sim Gws_{\alpha}^{reg}$, and $|Zd\,\mathfrak{Rl}_W\mathfrak{A}|>2$.

Proof. Let $\alpha\geq\omega$ and $\kappa\geq 2$. Let $H\subseteq\alpha$ be such that $|H|=\omega$ and $|\alpha\sim H|\geq\omega$. Let $h:\omega\rightarrowtail H$ be one-one and onto. Let $p\in{}^{\omega+1}({}^{H}\kappa)$ be such that $(\forall i<k\leq\omega)pi\notin{}^{H}\kappa^{(pk)}$. Such $H,h$ and $p$ exist by $|H|=\omega$ and $\kappa\geq 2$. Let $i\leq\omega$. Define

$R_i\stackrel{d}{=}\{q\in{}^{\alpha}\kappa : H1q\in{}^{H}\kappa^{(pi)}$ and $(H\sim h^{*}i)1q\subseteq pi\}$. $x\stackrel{d}{=}\bigcup\{R_i : i\leq\omega\}$.

Let $\mathfrak{C}\stackrel{d}{=}\mathfrak{Sb}\,{}^{\alpha}\kappa$ and $\mathfrak{A}\stackrel{d}{=}\mathfrak{Sg}^{(\mathfrak{C})}\{x\}$. Note that $(\forall i<\omega)R_i\in Sm^{\mathfrak{C}}$, $\bigcup\{R_i : i<\omega\}\notin Sm^{\mathfrak{C}}$ and $R_{\omega}=\{q\in{}^{\alpha}\kappa : H1q\in{}^{H}\kappa^{(p\omega)}\}$.

Claim 4.11.1. $\mathfrak{A}\in Cs_{\alpha}^{reg}\cap Dc_{\alpha}$.

Proof. $\Delta x=H$, and therefore $\mathfrak{A}\in Dc_{\alpha}$ by $|\alpha\sim H|\geq\omega$. We show that $\mathfrak{A}$ is regular by using Prop.4.7. Let $Q\stackrel{d}{=}\{q\in{}^{\alpha}\kappa : (\exists i<\omega)H1q\in{}^{H}\kappa^{(pi)}\}$. Then $Q\in Zd\,\mathfrak{C}$. Let $G\stackrel{d}{=}\{x\}$. Then $\mathfrak{A}\in Gws_{\alpha}^{norm}$ is generated by G. Condition (i) of 4.7 is satisfied, since $|G|=1$. We show that condition (ii) is satisfied by $x$ and $Q$.: Let $f\in x$ and $\Gamma\subseteq_\omega\alpha$. If $f\in Q$ then we are done. Assume $f\notin Q$. Then $f\in R_\omega$ and for every $g$ we have $f[\Gamma/g]\in x$ iff $\Gamma 1 g\in {}^\Gamma\varkappa$. Let $n\in\omega$ be such that $H\cap\Gamma\subseteq h^*n$. Then $\Gamma\cap\Delta R_n=0$ and therefore for every $g$ we have $p_n[\Gamma/g]\in x$ iff $\Gamma 1 g\in{}^\Gamma\varkappa$. By $p_n\in Q$ the above shows that condition (ii) of 4.7 is satisfied. We show that $x$ is Q-wsmall.: Let $K\subseteq H$, $|K|\geq\omega$ and let $\Gamma\subseteq_\omega\alpha$, and $q\in Q$. If $q\notin c_{(\Gamma)}x$ then we are done. Suppose $q\in Q\cap c_{(\Gamma)}x$. Then $(\exists n\in\omega)q\in c_{(\Gamma)}R_n$. Let $L\overset{d}{=}K\sim(\Gamma\cup h^*n)$. $L\neq 0$ by $|K|\geq\omega$ and $|\Gamma\cup h^*n|<\omega$. Let $i\in L$. Let $k\in\varkappa\sim\{p_n(i)\}$. Such a $k$ exists by $\varkappa\geq 2$. Now $q^i_k\notin c_{(\Gamma)}x$. We have seen that $x$ is Q-wsmall. $x$ is defined so that $x$ is regular (since $\Delta x=H$). Then $\mathfrak{A}$ is regular, by Prop.4.7.

QED(Claim 4.11.1.)

Returning to the proof of 4.11, let $W\overset{d}{=}{}^\alpha\varkappa\sim Q$. Then $W$ is a compressed $Gws_\alpha$-unit with base $\varkappa$, and therefore $W\in Zd\,\mathfrak{L}$. Thus $rl_W\in Ho\,\mathfrak{A}$ and therefore $\mathfrak{Rl}_W\mathfrak{A}=rl_W{}^*\mathfrak{A}\in Gws_\alpha^{comp}\cap \mathbf{H}\,\mathfrak{A}$. $rl_W(x)=$ $=x\cap W=x\sim Q=R_\omega=\{q\in{}^\alpha\varkappa : H1q\in{}^H\varkappa^{(p\omega)}\}\neq W$. Therefore $rl_W(x)\in$ $\in Zd(\mathfrak{Rl}_W\mathfrak{A})\sim\{0,W\}$ since $\Delta R_\omega=0$, showing that $|Zd(\mathfrak{Rl}_W\mathfrak{A})|>2$. Since $rl_W(x)=R_\omega$ is not regular we have $\mathfrak{Rl}_W\mathfrak{A}\notin Gws_\alpha^{reg}$.

QED(Proposition 4.11.)

Remark 4.12. In [HMTI]6.16(2) it is shown that for any $\alpha\geq\omega$ and $\varkappa\geq 2$ there is a ${}_\varkappa Ws_\alpha$ having a homomorphic image with more than two zero dimensional elements. The construction given there can be modified to work for $Cs_\alpha^{reg}\cap Dc_\alpha$. This construction is as follows. Let $\alpha\geq\omega$ and $\varkappa\geq 2$. Let $p\in{}^\alpha\varkappa$, and let $H\subseteq\alpha$ be such that $|H|=\omega$ and $|\alpha\sim H|\geq\omega$. Let $h:\omega\rightarrowtail\!\!\!\twoheadrightarrow H$ be one-one and onto. Let $x\overset{d}{=}$ $\overset{d}{=}\{q\in{}^\alpha\varkappa : (\exists n\in\omega)[q_{hn}\neq p_{hn}, q_{h(n+1)}\neq p_{h(n+1)}, (H\sim h^*(n+1))1q\subseteq p]\}$, and let $\mathfrak{A}$ be the $Cs_\alpha$ with base $\varkappa$ and generated by $\{x\}$. Then $\mathfrak{A}\in Cs_\alpha\cap Dc_\alpha$ since $\Delta x=H$ and $|\alpha\sim H|\geq\omega$. $\mathfrak{A}$ is regular by Prop.4.6

since $x$ is wsmall and regular. The proof of $(\exists \mathfrak{B} \in H\mathfrak{A})|Zd\mathfrak{B}|>2$ is nearly the same as in [HMTI] 6.16(2) therefore we omit it.

Proposition 4.13 below is an application of Prop.4.9.

Proposition 4.13. Let $\alpha \geq \omega$ and $\varkappa \geq 2$. Then there is an $\mathfrak{A} \in {}_{\varkappa}Cs_{\alpha}^{reg} \cap Dc_{\alpha}$ such that $(\exists \mathfrak{B} \in H\mathfrak{A})|Zd\mathfrak{B}| = |{}^{({}^{\alpha}\varkappa)}2|$, i.e. the largest possible (obtainable from a ${}_{\varkappa}Cs_{\alpha}$). Moreover, $B = Sg\, Zd\,\mathfrak{B}$, hence $\mathfrak{B} \in Lf_{\alpha}$.

Proof. Let $\alpha \geq \omega$ and $\varkappa \geq 2$. Let $H \subseteq \alpha$ be such that $|H| = |\alpha|$ and $|\alpha \sim H| \geq \omega$. Let $\gamma \overset{d}{=} |{}^{\alpha}\varkappa| = |{}^{H}\varkappa|$. Let $P \subseteq {}^{H}\varkappa$ be such that $|P| = \gamma$ and $(\forall f,g \in P)[f \neq g \Rightarrow f \notin {}^{H}\varkappa^{(g)}]$. Such a $P$ exists. Let $S \subseteq SbP$ be a partition of $P$ such that $|S| = \gamma$ and $(\forall s \in S)|s| \geq |H|$. Let $K \subseteq SbP$ be such that $|K| = 2^{\gamma}$, $0 \notin K$ and $(\forall x \in K)(\forall s \in S)[s \subseteq x$ or $s \cap x = 0]$. Such a $K$ exists. Let $\nu : P \to Sb_{\omega}H$ be such that $(\forall s \in S)\nu^{*}(s) = Sb_{\omega}H$. Let $\mu : H \rightarrowtail\!\!\!\twoheadrightarrow (Sb_{\omega}H) \sim \{0\}$ be one-one and onto. Define $N : P \to Sb_{\omega}H$ as $N \overset{d}{=} \langle \{j \in H : \mu(j) \subseteq \nu(f)\} : f \in P \rangle$. For every $x \in K$ let $y_x \overset{d}{=}$ $\overset{d}{=} \{q \in {}^{\alpha}\varkappa : (\exists f \in x)(H \sim N(f)) 1 f \subseteq q\}$. $G \overset{d}{=} \{y_x : x \in K\}$, and $\mathfrak{A} \overset{d}{=} \mathfrak{S}\mathfrak{g}^{(\mathfrak{L})}G$ where $\mathfrak{L} = \mathfrak{Sb}\,{}^{\alpha}\varkappa$.

Claim 4.13.1. $\mathfrak{A} \in Cs_{\alpha}^{reg} \cap Dc_{\alpha}$.

Proof. For every $z \in G$ we have $\Delta z = H$ since $(\forall x \in K)(\forall i \in H)(\exists f \in x) i \notin$ $\notin N(f)$. Therefore $\mathfrak{A} \in Dc_{\alpha}$ by $|\alpha \sim H| \geq \omega$. Clearly, every element of $G$ is regular. Now we check the conditions of Prop.4.9. We have to show that $y_x$ is i-small, for every $x \in K$. Let $x \in K$. Then $\Delta(y_x) = H$. Let $q \in y_x$, $\Gamma \subseteq_{\omega} \alpha$, and $L \subseteq H$ be such that $|L| \geq \omega$. By $q \in y_x$ we have $(H \sim N(f)) 1 f \subseteq q$ for some $f \in x$. Let $i \in L \sim (N(f) \cup \Gamma)$ and let $r \in \varkappa \sim \{f(i)\}$. Then $q_r^i \notin c_{(\Omega)} y_x$ for every $\Omega \subseteq_{\omega} \alpha \sim \{i\}$. Thus $y_x$ is i-small and hence the generator $G$ of $\mathfrak{A}$ satisfies the conditions of Prop.4.9. Therefore $\mathfrak{A}$ is regular, by Prop.4.9.

QED(Claim 4.13.1.)

Let $I \overset{d}{=} Ig\{(c_i z) - z : i \in \alpha,\ z \in G\}$. Let $\mathfrak{B} \overset{d}{=} \mathfrak{A}/I$.

Claim 4.13.2. $|\mathrm{Zd}\,\mathfrak{B}|=2^{\gamma}$.

Proof. Let $\Gamma \subseteq_{\omega} H$. Let $H_{\Gamma} \overset{d}{=} \{j\in H : \mu(j) \supseteq \cup\mu^{*}\Gamma\}\sim\Gamma$. Thus $H_{\Gamma}\neq 0$. Let $R_{\Gamma} \overset{d}{=} \{q\in{}^{\alpha}\varkappa : (\exists f\in P) H_{\Gamma}1f\subseteq q\}$. First we show that $(\forall z\in G) c_{(\Gamma)} z\sim z\subseteq R_{\Gamma}$.

Let $x\in K$ and let $q\in c_{(\Gamma)}y_x\sim y_x$. By $q\in c_{(\Gamma)}y_x$ we have $(H\sim(N(f)\cup \cup\Gamma))1f\subseteq q$ for some $f\in x$. By $q\notin y_x$ we have $(H\sim N(f))1f\not\subseteq q$. Therefore $\Gamma \not\subseteq N(f)$. Let $i\in\Gamma\sim N(f)$. Then $\mu(i) \not\subseteq \nu(f)$ and therefore $(\forall j\in H)[\mu(j) \supseteq \cup\mu^{*}\Gamma \Rightarrow \mu(j) \not\subseteq \nu(f)]$. I.e. $H_{\Gamma} \subseteq H\sim(N(f)\cup\Gamma)$ and therefore $H_{\Gamma}1f\subseteq q$. I.e. $q\in R_{\Gamma}$, by $f\in P$. We have seen $(\forall z\in G)(\forall\Gamma \subseteq_{\omega} H) c_{(\Gamma)} z\sim z \subseteq R_{\Gamma}$.

This implies $(\forall z\in I)(\exists\Gamma \subseteq_{\omega} H)\ z \subseteq R_{\Gamma}$, as follows. Let $z\in I$. Then $z\leq c_{(\Omega)}\sum \{c_{ij}z_j\sim z_j : j\in J\}$, for some $\Omega \subseteq_{\omega} \alpha$, $|J|<\omega$, $\{ij : j\in J\} \subseteq \alpha$, $\{z_j : j\in J\} \subseteq G$. Let $\Gamma \overset{d}{=} H\cap(\{ij : j\in J\}\cup\Omega)$. Then $z\leq c_{(\Gamma)}\sum \{c_{(\Gamma)}z_j\sim z_j : j\in J\}$, by $(\forall z\in G)\Delta z=H$. Now $(\forall j\in J) c_{(\Gamma)}z_j\sim z_j \subseteq R_{\Gamma}$ as we have seen above, therefore $z\subseteq c_{(\Gamma)}R_{\Gamma} = R_{\Gamma}$.

Let $x,w\in K$ and suppose $x\sim w\neq 0$. We show that $y_x\sim y_w \notin I$. To this end, by the above it is enough to show that $(\forall\Gamma \subseteq_{\omega} H) y_x\sim y_w \not\subseteq R_{\Gamma}$. Let $\Gamma \subseteq_{\omega} H$. Let $j\in H_{\Gamma}$ be arbitrary. By $x\sim w\neq 0$ we have $s\subseteq x\sim w$ for some $s\in S$. Then $\nu(f) \supseteq \mu(j)$ for some $f\in s$ by $\nu^{*}(s) = Sb_{\omega}H$. Then $j\in N(f)\cap H_{\Gamma}$. Let $q\in{}^{\alpha}\varkappa$ be such that $\{i\in\alpha : q(i)=f(i)\}=\alpha\sim N(f)$. Such a $q$ exists by $\varkappa\geq 2$. Now $q\in y_x\sim y_w$ by $f\in x\sim w$, and $q\notin R_{\Gamma}$ by $j\in H_{\Gamma}$, $|H_{\Gamma}|\geq\omega$ and $q(j)\neq f(j)$. We have seen $y_x\sim y_w \notin I$.

Then $(\forall x,w\in K)[x\neq w \Rightarrow (y_x\oplus y_w) \notin I]$, and therefore $(\forall x,w\in K)[x\neq w \Rightarrow y_x/I \neq y_w/I]$. Let $G' \overset{d}{=} \{z/I : z\in G\}$. Then $G'\subseteq B$ and $|G'| = |G| = 2^{\gamma}$. By the definition of $I$ we have $G'\subseteq \mathrm{Zd}\,\mathfrak{B}$. By $|B| \leq |A| \leq 2^{\gamma}$ then $|\mathrm{Zd}\,\mathfrak{B}| = 2^{\gamma}$.

QED(Claim 4.13.2 and Proposition 4.13.)

About Theorems 4.14-4.15 below see [HMTI]4.8.

Notation: Let K be a class of similar algebras. Then **Ud** K denotes the class of unions of directed (under $\subseteq$) nonempty subsets of K.

In [HMTI]7.11 it is proved that $\mathbf{Ud}(\mathbf{I}Cs_\alpha^{reg} \cap Lf_\alpha) = \mathbf{I}Cs_\alpha^{reg} \cap Lf_\alpha$. Theorem 4.14 below shows that "reg" cannot be dropped and "Lf" cannot be replaced by "Dc" in this theorem.

<u>Theorem 4.14.</u> Let $\alpha \geq \omega$. Then (i)-(ii) below hold.

(i) $\mathbf{Ud}(\mathbf{I}Cs_\alpha^{reg} \cap Dc_\alpha) \nsubseteq \mathbf{I}Cs_\alpha$.

(ii) $\mathbf{Ud}(\mathbf{I}Cs_\alpha \cap Lf_\alpha) \nsubseteq \mathbf{I}Cs_\alpha$.

<u>Theorem 4.15.</u> Let $\alpha \geq \omega$. Then (i)-(iii) below hold.

(i) $\mathbf{Ud}(\mathbf{I}Cs_\alpha \cap Lf_\alpha) = \{\mathfrak{A} \in Lf_\alpha : \mathfrak{Nr}(\mathfrak{A})$ is nondiscrete and simple or $|A| \leq 2\}$.

(ii) $\mathbf{Ud}(\mathbf{I}Cs_\alpha \cap Dc_\alpha) = \{\mathfrak{A} \in Dc_\alpha : \mathfrak{Nr}(\mathfrak{A})$ is nondiscrete and simple or $|A| \leq 2\}$ iff $\alpha \leq 2^\omega$.

(iii) $\mathbf{Ud}(\mathbf{I}Cs_\alpha) \subset \{\mathfrak{A} \in \mathbf{I}Gs_\alpha : \mathfrak{Nr}(\mathfrak{A})$ is nondiscrete and simple or $|A| \leq 2\}$.

To prove Theorems 4.14-4.15 we shall need the following propositions which are closely related to [HMTI]7.28-29 and which deal with the question under which conditions a $Gws_\alpha$ is isomorphic to a $Cs_\alpha$. They show that under additional hypotheses on the $Gws_\alpha$ the criteria given in [HMTI]7.28-29 can be improved.

<u>Proposition 4.16.</u> Let $\alpha \geq \omega$. Let $\mathfrak{A} \in Gws_\alpha$ be non-discrete and suppose that $\mathfrak{A}$ has a characteristic. Then (i)-(iv) below hold.

(i) Let $\mathfrak{A} \in Lf_\alpha$. Then $\mathfrak{A} \in \mathbf{I}Cs_\alpha$ if $\mathfrak{A}$ has $\leq 2^{|\alpha|}$ subunits.

(ii) Let $\mathfrak{A} \in Dc_\alpha$ be finitely generated. Then $\mathfrak{A} \in \mathbf{I}Cs_\alpha$ if $\mathfrak{A}$ has $\leq 2^\omega$ subunits.

(iii) Let $H \subseteq \alpha$ be such that $|\alpha \sim H| \geq \omega$ and $(\forall x \in A)|\Delta x \sim H| < \omega$. Then $\mathfrak{A} \in \mathbf{I}Cs_\alpha$ if $\mathfrak{A}$ has $\leq 2^{|\alpha \sim H|}$ subunits.

(iv) For every $\alpha$ and $H \subseteq \alpha$ such that $\alpha > 2^{|\alpha \sim H|}$ there is an $\mathfrak{A} \in Gs_\alpha \sim \mathbf{I}Cs_\alpha$ such that $\mathfrak{Nr}(\mathfrak{A})$ is simple, $\mathfrak{A}$ is generated by a single element, $(\forall x \in A)|\Delta x \sim H| < \omega$ and $\mathfrak{A}$ has $(2^{|\alpha \sim H|})^+$ subbases

Corollary 4.17.

(i) Every nondiscrete $\mathfrak{A}\in Lf_\alpha$ with a characteristic and with $|A|\le 2^{|\alpha|}$ is isomorphic to a $Cs_\alpha$, if $\alpha\ge\omega$.

(ii) Every nondiscrete finitely generated $\mathfrak{A}\in Dc_\alpha$ with a characteristic and with $|A|\le 2^\omega$ is isomorphic to a $Cs_\alpha$.

(iii) Let $\mathfrak{A}\in CA_\alpha$ have a characteristic and be nondiscrete. Let $H\subseteq\alpha$ be such that $|\alpha\sim H|\ge\omega$, $(\forall x\in A)|\Delta x\sim H|<\omega$ and $|A|\le 2^{|\alpha\sim H|}$. Then $\mathfrak{A}$ is isomorphic to a $Cs_\alpha$.

Problem 4.18. Let $\alpha\ge\omega$. Is every nondiscrete $Gs_\alpha\cap Dc_\alpha$ with a characteristic and with finitely many subbases isomorphic to a $Cs_\alpha$ ?

Proof of 4.16.: Let $\alpha\ge\omega$. Statements 4.16(i) and (ii) follow from 4.16(iii). ((i) is a special case of (iii) where $H=0$ and if $\mathfrak{A}\in Dc_\alpha$ is finitely generated then $\mathfrak{A}$ satisfies the conditions of (iii) with some $H\subseteq\alpha$, $|\alpha\sim H|\ge\omega$ by [HMT]1.11.4.

Proof of 4.16(iii): Let $\mathfrak{A}\in Gws_\alpha$ have a characteristic. Assume that $\mathfrak{A}$ is nondiscrete. By [HMTI]7.21 we may assume that $\mathfrak{A}\notin{}_\infty Gws_\alpha$. Then the set $\{{}^\alpha Y_i^{(pi)} : i\in I\}$ of subunits of $\mathfrak{A}$ is such that $I\ne 0$ and $(\forall i,j\in I)\ |Y_i|=|Y_j|$. Let $n \overset{d}{=} |Y_i|$ for $i\in I$. Let $H\subseteq\alpha$ be such that $|\alpha\sim H|\ge\omega$ and $(\forall x\in A)|\Delta x\sim H|<\omega$. Suppose $|I|\le 2^{|\alpha\sim H|}$. We have to show $\mathfrak{A}\in I\,Cs_\alpha$. For every $i\in I$ let $\mathfrak{Rl}({}^\alpha Y_i^{(pi)})\mathfrak{A}\cong\mathfrak{L}_i$ where $1^{\mathfrak{L}i}={}^\alpha n^{(qi)}$. Such $\mathfrak{L}_i$ exists by $|Y_i|=n$. Then $\mathfrak{A}\cong I\subseteq P_{i\in I}\mathfrak{L}_i$, by [HMTI]6.2. Let $J$ and $\{r_i : i\in J\}\subseteq{}^\alpha n$ be such that $I\subseteq J$, $(\forall i\in I)H1r_i\subseteq q_i$, $(\forall i,j\in J)[i\ne j \Rightarrow r_j\notin{}^\alpha n^{(ri)}]$ and $\cup\{{}^\alpha n^{(ri)} : i\in J\}={}^\alpha n$. To see that such $\{r_i : i\in J\}$ exists observe that $\mathfrak{A}$ is nondiscrete and hence $n>1$ and $|I|\le 2^{|\alpha\sim H|}$, $|\alpha\sim H|\ge\omega$. For every $i\in I$ let $\mathfrak{D}_i\in Ws_\alpha$ with unit ${}^\alpha n^{(ri)}$ be isomorphic to $\mathfrak{L}_i$. Such $\mathfrak{D}_i$ exists by Cor. 3.15.b) since $H1r_i\subseteq q_i$, $1^{\mathfrak{L}i}={}^\alpha n^{(qi)}$ and $(\forall x\in C_i)|\Delta x\sim H|<\omega$. For every $j\in J\sim I$ let $\mathfrak{D}_j\in Ws_\alpha$ with unit ${}^\alpha n^{(rj)}$ be such that $\mathfrak{D}_j$ is a homomorphic image of some $\mathfrak{L}_i$, $i\in I$. Such $\mathfrak{D}_j$ exists by [HMTI]

7.27, $n<\omega$, $I\neq 0$. By the choice of $\langle \mathfrak{D}_j : j\in J\rangle$ we have $P_{i\in I}\,\mathfrak{L}_i \cong$ $\cong \mathsf{I}\subseteq P_{j\in J}\,\mathfrak{D}_j$. By [HMTI]6.2 and by the choice of $\{r_j : j\in J\}$ we have $P_{j\in J}\,\mathfrak{D}_j \cong \mathsf{I}\subseteq \mathfrak{Sb}\,{}^{\alpha}n\in Cs_{\alpha}$. Therefore $\mathfrak{A}\in\mathsf{I}\,Cs_{\alpha}$ by $\mathfrak{A}\cong\mathsf{I}\subseteq$ $\subseteq P_{i\in I}\,\mathfrak{L}_i \cong \mathsf{I}\subseteq P_{j\in J}\,\mathfrak{D}_j$.

<u>Proof of 4.16(iv):</u> Let $\alpha\geq\omega$ and $H\subseteq\alpha$ be such that $|\alpha|>2^{|\alpha\sim H|}$. Then $|H|=|\alpha|$ and therefore there exists a $G\subset H$, $G\neq H$ such that $|G|=(2^{|\alpha\sim H|})^{+}$. For every $i\in G$ let $Y_i \overset{d}{=} \{\langle 0,i\rangle,\langle 1,i\rangle\}$ and $p_i \overset{d}{=}$ $\overset{d}{=} \langle\langle 0,i\rangle : \kappa<\alpha\rangle\,(i/\langle 1,i\rangle)$. Let $x \overset{d}{=} \{q : (\exists i\in G)[q\in{}^{\alpha}Y_i$ and $H1q\subseteq p_i]\}$. Let $\mathfrak{A}$ be the $Gs_{\alpha}$ with unit $\cup\{{}^{\alpha}Y_i : i\in G\}$ and generated by $x$. Then $\mathfrak{A}\in\mathsf{I}\,Gs_{\alpha}$ with $(2^{|\alpha\sim H|})^{+}$ subbases, $\mathfrak{Mn}(\mathfrak{A})$ is simple, $\mathfrak{A}$ is generated by a single element and $(\forall y\in A)\,|\Delta y\sim H|<\omega$ by $\Delta x=H$. We have to show $\mathfrak{A}\notin\mathsf{I}\,Cs_{\alpha}$. Let $k\in H\sim G$. For every $i\in G$ let $y_i \overset{d}{=}$ $\overset{d}{=} c_i(x-d_{ik})\cdot d_{ik}$. Then $y_i\in A$ and $y_i = \{q\in{}^{\alpha}Y_i : (\forall j\in H)q_j = \langle 0,i\rangle\}$. Then $\{y_i : i\in G\}\subseteq A$ is a set of disjoint nonzero elements, such that $(\forall i\in G)y_i \leq \Pi\{d_{kj} : k,j\in H\}$. In every $Cs_{\alpha}$ with finite base there are only $2^{|\alpha\sim H|}$ such elements. Thus $|G|>2^{|\alpha\sim H|}$ implies $\mathfrak{A}\notin\mathsf{I}\,Cs_{\alpha}$.

<u>QED(Proposition 4.16.)</u>

<u>Proof of 4.17.:</u> It is enough to prove (iii). Let $\mathfrak{A}\in CA_{\alpha}$ and $H\subseteq\alpha$ be such that $|\alpha\sim H|\geq\omega$, $(\forall x\in A)\,|\Delta x\sim H|<\omega$ and $|A|\leq 2^{|\alpha\sim H|}$. Assume that $\mathfrak{A}$ is nondiscrete. Then $\mathfrak{A}\in Dc_{\alpha}$ by $|\alpha\sim H|\geq\omega$, and therefore $\mathfrak{A}\cong\mathfrak{B}\in$ $\in Gws_{\alpha}$ since $Dc_{\alpha}\subseteq\mathsf{I}\,Gws_{\alpha}$, by Cor.3.14(a) of [AGN2] together with [HMTI]7.16, or by [AN1]. Let $1^{\mathfrak{B}} \overset{d}{=} V \overset{d}{=} \cup\{{}^{\alpha}Y_i^{(pi)} : i\in I\}$ where $\{{}^{\alpha}Y_i^{(pi)} : i\in I\} = Subu(\mathfrak{B})$. Let $i : B\to I$ be such that $(\forall a\in B\sim\{0\})a\cap$ $\cap{}^{\alpha}Y_{i(a)}^{(pi(a))}\neq 0$. Let $W \overset{d}{=} \cup\{{}^{\alpha}Y_{i(a)}^{(pi(a))} : a\in B\}$. Then $rl_W\in Is\,\mathfrak{B}$ since $\Delta^{[V]}W=0$ and $(\forall a\in B\sim\{0\})a\cap W\neq 0$. Let $\mathfrak{L} \overset{d}{=} \mathfrak{Rl}_W\mathfrak{B}$. Now $\mathfrak{L}\in Gws_{\alpha}$ has $\leq|A|\leq 2^{|\alpha\sim H|}$ subunits, and $|\alpha\sim H|\geq\omega$, $(\forall x\in C)\,|\Delta x\sim H|<\omega$. Then $\mathfrak{L}\in$ $\mathsf{I}\,Cs_{\alpha}$ by 4.16(iii). By $\mathfrak{A}\cong\mathfrak{L}$ then $\mathfrak{A}\in\mathsf{I}\,Cs_{\alpha}$ too.

<u>QED(Corollary 4.17.)</u>

<u>Proof of 4.15.:</u> Let $\alpha\geq\omega$. The inclusions $\subseteq$ in (i)-(iii) hold

because $\mathfrak{Mn}(\mathfrak{A})$ is simple for every $\mathfrak{A}\in Cs_\alpha$ by [HMTI]5.3, and the minimal subalgebra remains the same in directed union; and further, $\mathbf{Ud}\,K=K$ if $K\in\{Lf_\alpha, Dc_\alpha, \mathbf{I}\,Gs_\alpha\}$ by [HMTI]7.10. Every discrete $Cs_\alpha$ is of cardinality $\le 2$.

<u>Proof of 4.15(i)-(ii):</u> Let $\mathfrak{A}\in Dc_\alpha$, $\mathfrak{Mn}(\mathfrak{A})$ simple. Then $\mathfrak{A}$ is the directed union of its finitely generated subalgebras, and if $\mathfrak{B}\in CA_\alpha$ is finitely generated then $|B|\le|\alpha|\cup\omega=|\alpha|$. Now 4.17(i)-(ii) imply that $\mathbf{Ud}(\mathbf{I}\,Cs_\alpha\cap Lf_\alpha) = \{\mathfrak{A}\in Lf_\alpha : \mathfrak{Mn}(\mathfrak{A})$ is nondiscrete and simple or $|A|\le 2\}$ and if $\alpha\le 2^\omega$ then $\mathbf{Ud}(\mathbf{I}\,Cs_\alpha\cap Dc_\alpha) = \{\mathfrak{A}\in Dc_\alpha : \mathfrak{Mn}(\mathfrak{A})$ is nondiscrete and simple or $|A|\le 2\}$. Suppose $\alpha>2^\omega$. Let $H\subseteq\alpha$ be such that $|\alpha\sim H|=\omega$. Then there is an $\mathfrak{A}\in Dc_\alpha\sim\mathbf{I}\,Cs_\alpha$ such that $\mathfrak{Mn}(\mathfrak{A})$ is simple, and $\mathfrak{A}$ is generated by a single element, by 4.16(iv). Then $\mathfrak{A}\notin\mathbf{Ud}\,\mathbf{I}\,Cs_\alpha$ since $\mathfrak{A}\notin\mathbf{I}\,Cs_\alpha$ and $\mathfrak{A}$ is generated by a single element. This shows $\mathbf{Ud}(\mathbf{I}\,Cs_\alpha\cap Dc_\alpha) \ne \{\mathfrak{A}\in Dc_\alpha : \mathfrak{Mn}(\mathfrak{A})$ is nondiscrete and simple or $|A|\le 2\}$, if $\alpha>2^\omega$.

<u>Proof of 4.15(iii)</u>: Let $\mathfrak{C}$ be the full $Gs_\alpha$ with unit ${}^\alpha Y_0\cup{}^\alpha Y_1$ where $Y_0=\{0,1\}$ and $Y_1=\{2,3\}$. Let $a_j \overset{d}{=} \{\langle j : i<\alpha\rangle\}$. Let $\mathfrak{A}\subseteq\mathfrak{C}$, $\{a_j : j<4\}\subseteq A$. Then $\mathfrak{Mn}(\mathfrak{A})$ is simple and $\mathfrak{A}$ has characteristic 2, by [HMTI]5.4. Therefore $\mathfrak{A}\notin\mathbf{Ud}\,\mathbf{I}\,Cs_\alpha$ since no subalgebra of $\mathfrak{A}$ containing $\{a_j : j<4\}$ is in $\mathbf{I}\,Cs_\alpha$.
<u>QED(Theorem 4.15.)</u>

<u>Proof of 4.14.:</u> Let $\alpha\ge\omega$.

<u>Proof of 4.14(ii):</u> Let $I\ge 2^{2^{|\alpha|}}$ and let $\mathfrak{A}$ be the greatest locally finite dimensional subalgebra of ${}^I(\mathfrak{Sb}\,{}^\alpha 2)$. Then $\mathfrak{A}\in Lf_\alpha$ and $\mathfrak{A}$ has characteristic 2 by [HMTI]5.4 and [HMT]2.4.64 , therefore $\mathfrak{Mn}(\mathfrak{A})$ is simple. Then $\mathfrak{A}\in\mathbf{Ud}(\mathbf{I}\,Cs_\alpha\cap Lf_\alpha)$ by 4.15(i). But $\mathfrak{A}\notin\mathbf{I}\,Cs_\alpha$ since the characteristic of $\mathfrak{A}$ is 2 and $|A|\ge 2^{|I|}>|I|\ge 2^{2^{|\alpha|}}$.

<u>Proof of 4.14(i):</u> Let $\alpha\ge\omega$. Let $I$ be a set such that $|I|>2^{|\alpha|}$. Let $H\subseteq\alpha$ be such that $|H|=|\alpha\sim H|=|\alpha|$. Let $x \overset{d}{=} \{s\in{}^\alpha 3 : H1s\in{}^H 2\}$. For every $i\in I$ let $Y_i \overset{d}{=} \{\langle 0,i\rangle,\langle 1,i\rangle,\langle 2,i\rangle\}$ and let $b_i \overset{d}{=} \langle\langle j,i\rangle : j<3\rangle$. Then $b_i : 3 \rightarrowtail\!\!\!\twoheadrightarrow Y_i$ is one-one and onto. Let $V \overset{d}{=} \cup\{{}^\alpha Y_i : i\in I\}$ and

let $\mathfrak{A}(J) \overset{d}{=} \mathfrak{Sg}^{(\mathfrak{Sb}V)}\{\tilde{b}_i x : i\in J\}$. Then $\mathfrak{A}(J)\in Dc_\alpha$ by $(\forall i\in J)\Delta(\tilde{b}_i x)=H$ and $|\alpha\sim H|\geq\omega$.

Claim 4.14.1. Let $J\subseteq I$. Then (i)-(iii) below are equivalent.

(i) $\mathfrak{A}(J) \in \mathbf{I}Cs_\alpha^{reg}$ .

(ii) $\mathfrak{A}(J) \in \mathbf{I}Cs_\alpha$ .

(iii) $|J| \leq 2^{|\alpha|}$ .

Proof of (iii) ⇒ (i): Let $J\subseteq I$, $|J|\leq 2^{|\alpha|}$. We may assume $|J|=2^{|\alpha|}$, since if $J\subseteq G\subseteq I$ then $\mathfrak{A}(J)\subseteq\mathfrak{A}(G)$. Let $p\in {}^J({}^H2)$ be such that ${}^H2 = \cup\{{}^H2(pi) : i\in J\}$ is a disjoint union. Such a $p$ exists by $|J| = 2^{|H|} = 2^{|\alpha|}$. Let $i\in J$. Define

$Z_i \overset{d}{=} \{s\in {}^\alpha 3 : H{\upharpoonright}s\in {}^H3(pi)\}$,

$z_i \overset{d}{=} \{s\in {}^\alpha 3 : H{\upharpoonright}s\in {}^H2(pi)\} = x\cap Z_i$,

$y_i \overset{d}{=} \tilde{b}_i x$,

$W \overset{d}{=} \cup\{Z_i : i\in J\}$, $Q \overset{d}{=} \cup\{{}^\alpha Y_i : i\in J\}$,

$\mathfrak{M} \overset{d}{=} \mathfrak{Sg}^{(\mathfrak{Sb}\,{}^\alpha 3)}\{x\}$ and $\mathfrak{L} \overset{d}{=} \mathfrak{Sg}^{(\mathfrak{Sb}\,{}^\alpha 3)}\{z_i : i\in J\}$.

Now $\mathfrak{L}\in Cs_\alpha^{reg}$ by Thm 1.3 since $z_i$ is a small regular element for every $i\in J$. Let $i\in J$. We show, by 4.7(II), that $rl_{Z_i}\in Is\,\mathfrak{M}$, $rl_W\in Is\,\mathfrak{L}$ and $rl_Q\in Is\,\mathfrak{A}(J)$. We check the conditions of 4.7(II) for each of the three cases. $\Delta^{(3)}Z_i = \Delta^{(3)}W = \Delta^{[V]}Q = 0$. $x\in Sm^{\mathfrak{M}}$, $\{z_i : i\in J\}\subseteq Sm^{\mathfrak{L}}$ $\{y_i : i\in J\}\subseteq Sm^{\mathfrak{A}(J)}$ and therefore these generators are weakly small, too. Let $\Gamma\subseteq_\omega \alpha$ and $i,j\in J$, $i\neq j$. Then $c_{(\Gamma)}z_i\cap c_{(\Gamma)}z_j = c_{(\Gamma)}y_i\cap c_{(\Gamma)}y_j = 0$ shows that condition (i) of 4.7 is satisfied. $(\forall f\in x)(\forall\Gamma\subseteq_\omega\alpha)(\forall s)[f[\Gamma/s]\in x$ iff $pi[\Gamma/s]\in x]$ and $pi\in z_i$; $\cup\{z_i : i\in J\}\subseteq W$ and $\cup\{y_i : i\in J\}\subseteq Q$ show that condition (ii) of 4.7 is satisfied. $\mathfrak{M}$, $\mathfrak{L}$, $\mathfrak{A}(J)$ have characteristics and $|\Delta x| = |\Delta z_i| = |\Delta y_i| = |H| \geq \omega$.

We have seen that the conditions of 4.7(II) are satisfied. Therefore $rl_{Z_i}\in Is\,\mathfrak{M}$, $rl_W\in Is\,\mathfrak{L}$ and $rl_Q\in Is\,\mathfrak{A}(J)$ by 4.7(II). Let $i\in J$ and $f_i \overset{d}{=} rl(Zi)\circ\tilde{b}_i^{-1}$. Then $f_i\in Is(\mathfrak{Sg}^{(\mathfrak{Sb}\,{}^\alpha Y_i)}\{y_i\}, \mathfrak{Sg}^{(\mathfrak{Sb}\,Z_i)}\{z_i\})$ such that $f_i(y_i)=z_i$ since $\tilde{b}_i\in Is(\mathfrak{M}, \mathfrak{Sg}^{(\mathfrak{Sb}\,{}^\alpha Y_i)}\{y_i\})$ and $y_i = \tilde{b}_i x$. Therefore by [HMTI]

6.2, $\Delta^{[Q]\alpha}Y_i = \Delta^{[W]}Z_i = 0$, $Q = \cup\{{}^{\alpha}Y_i : i\in J\}$, $W = \cup\{Z_i : i\in J\}$ and by $\mathfrak{Rl}_Q\, \mathfrak{A}(J) = \mathfrak{Sg}^{(\mathfrak{Sb}Q)}\{Y_i : i\in J\}$, $\mathfrak{Rl}_W\mathfrak{L} = \mathfrak{Sg}^{(\mathfrak{Sb}W)}\{Z_i : i\in J\}$ we have that $\mathfrak{Rl}_Q\,\mathfrak{A}(J) \cong \mathfrak{Rl}_W\mathfrak{L}$. Therefore $\mathfrak{A}(J) \in \mathbf{I}Cs_{\alpha}^{reg}$ by $\mathfrak{A}(J) \cong \mathfrak{Rl}_Q\,\mathfrak{A}(J) \cong \mathfrak{Rl}_W\mathfrak{L} \cong \mathfrak{L} \in Cs_{\alpha}^{reg}$.

Proof of (ii) $\Rightarrow$ (iii): Let $J\subseteq I$ be such that $|J|>2^{|\alpha|}$. We have to show $\mathfrak{A}(J) \notin \mathbf{I}Cs_{\alpha}$. The characteristic of $\mathfrak{A}(J)$ is 3 and $\{Y_i : i\in J\} \subseteq A(J)$ is a set of disjoint nonzero elements. Therefore $|J|>2^{|\alpha|}$ implies $\mathfrak{A}(J) \notin \mathbf{I}Cs_{\alpha}$ since in no $Cs_{\alpha}$ with finite characteristic are there more than $2^{|\alpha|}$ disjoint nonzero elements (by $\alpha\geq\omega$).

Since (i) $\Rightarrow$ (ii) is obvious, we have seen (i) $\Leftrightarrow$ (ii) $\Leftrightarrow$ (iii).

QED(Claim 4.14.1.)

We return to the proof of 4.14. $\mathfrak{A}(I)$ is the union of the directed set $\{\mathfrak{A}(J) : J \subseteq_{\omega} I\}$ of algebras. For every $J \subseteq_{\omega} I$ we have $\mathfrak{A}(J) \in \mathbf{I}Cs_{\alpha}^{reg}\cap Dc_{\alpha}$, by Claim 4.14.1 and by $|\alpha\sim H|\geq\omega$. But $\mathfrak{A}(I) \notin \mathbf{I}Cs_{\alpha}$ by $|I|>2^{|\alpha|}$ and by Claim 4.14.1. Therefore $\mathbf{Ud}(\mathbf{I}Cs_{\alpha}^{reg}\cap Dc_{\alpha}) \not\subseteq \mathbf{I}Cs_{\alpha}$.

QED(Theorem 4.14.)

## 5. Homomorphisms

By [HMTI]5.2, the members of $(Cs_{\alpha}^{reg}\cup Ws_{\alpha})\cap Lf_{\alpha}$ are simple. Below we show $\mathfrak{A}\in Cs_{\alpha}^{reg}\cup Ws_{\alpha}$ and $\mathfrak{A}$ simple do not imply that $\mathfrak{A}\in Lf_{\alpha}$. However, "$\cap Lf_{\alpha}$" cannot be replaced by "$\cap Dc_{\alpha}$" ; see Corollary 5.4 (iii) and [HMTI]5.5.

Proposition 5.1. Let $\alpha\geq\omega$ and $\varkappa\geq 2$.

(i) There are a $Cs_{\alpha}^{reg}\sim Dc_{\alpha}$ and a $Ws_{\alpha}\sim Dc_{\alpha}$ such that both are simple.

(ii) ${}_{\varkappa}Ws_{\alpha}\cap Ss_{\alpha} \not\subseteq Dc_{\alpha}$, ${}_{\varkappa}Ws_{\alpha} \not\subseteq Ss_{\alpha}$.

Proof. We have a direct construction for an $\mathfrak{A}\in Cs_\alpha^{reg}\sim Dc_\alpha$ such that $\mathfrak{A}$ is simple, but to save space instead of this construction we give here the following proof. Q denotes the set of rational numbers. $\bar{0}\stackrel{d}{=}\langle 0 : i<\alpha\rangle$ and $V\stackrel{d}{=}{}^{\alpha}Q^{(\bar{0})}$. $x\stackrel{d}{=}\{q\in V : 0=\sum\{q_i : i<\alpha\}\}$. $\mathfrak{B}\stackrel{d}{=}$ $\stackrel{d}{=}\mathfrak{Sg}^{(\mathfrak{Sb}V)}\{x\}$. Clearly, $\mathfrak{B}\in Ws_\alpha\sim Dc_\alpha$.

Claim 1: $\mathfrak{B}$ is simple. The proof of Claim 1 goes by eliminating cylindrifications. Claim 1 here is an immediate consequence of Claim 1 of [AN7]. Therefore we omit the proof here.

By [HMTI]7.13, $\mathfrak{B}\in\mathsf{I}\,Cs_\alpha^{reg}$ proving the rest of (i). Let $\varkappa\geq 2$ and let $\mathfrak{B}$ be a full $Ws_\alpha$ with base $\varkappa$. Then by [HMTI]6.11, $\mathfrak{B}$ is subdirectly indecomposable and clearly $\mathfrak{B}$ is not simple. Hence $\mathfrak{B}\notin Ss_\alpha$.

QED(Proposition 5.1.)

Remark 5.2. (About representing homomorphisms by relativizations.) Let $\alpha\geq\omega$. There are $\mathfrak{A}\in Cs_\alpha^{reg}$ and $\mathfrak{B}\in Ws_\alpha$, $I\in Il\,\mathfrak{A}$ and $J\in Il\,\mathfrak{B}$ such that $|Zd\,\mathfrak{A}/I|>2$ and $|Zd\,\mathfrak{B}/J|>2$, cf. 4.13 and [HMTI]6.16(2). But a difference between $Cs_\alpha^{reg}$ and $Ws_\alpha$ is the following. By 4.11 we have $|Zd(rl_V{}^*\mathfrak{A})|>2$ for some $\mathfrak{A}\in Cs_\alpha^{reg}\cap Dc_\alpha$ and $V\in Zd\,\mathfrak{Sb}\,1^{\mathfrak{A}}$. However, $|Zd(rl_V{}^*\mathfrak{B})|\leq 2$ for all $\mathfrak{B}\in Ws_\alpha$ and for all $V$ such that $rl_V\in Ho\,\mathfrak{B}$. The latter statement can be seen as follows. Let $\mathfrak{B}\in Ws_\alpha$, $rl_V\in Ho\,\mathfrak{B}$ and let $\mathfrak{R}\stackrel{d}{=}rl_V{}^*\mathfrak{B}$. Suppose $\Delta^{(\mathfrak{R})}x=0$ and $x\notin\{0,1^{\mathfrak{R}}\}$. Then there are $q\in x$ and $f\in 1^{\mathfrak{R}}\sim x$. By $\mathfrak{B}\in Ws_\alpha$ and $1^{\mathfrak{R}}\subseteq 1^{\mathfrak{B}}$ we have $(\exists\Gamma\subseteq_\omega\alpha)f\in c_{(\Gamma)}\{q\}$. Let $x=V\cap y$, $y\in B$. Then $f\in V\cap c^{\mathfrak{B}}_{(\Gamma)}y=c^{\mathfrak{R}}_{(\Gamma)}x=x$, since $rl_V\in Ho\,\mathfrak{B}$. A contradiction. Analogously to ext-isomorphisms, one could say that some $Cs_\alpha^{reg}$-s are "ext-homomorphic" to directly decomposable CA-s, while $Ws_\alpha$-s are not.

Notation: $Dind_\alpha\stackrel{d}{=}\{\mathfrak{A}\in CA_\alpha : |Zd\,\mathfrak{A}|\leq 2\}$. "Dind" is an abbreviation of "directly indecomposable or one-element".

Corollary 5.4 of 4.13 and Theorem 5.3 below together with a part of Theorem 5.3 itself are quoted in (6),(7),(12),(15), (16) of [HMTI] 5.6 and in [HMTI]5.5.

Theorem 5.3. Let $\kappa\geq 2$ and $\alpha\geq\omega$.

(1) There is $\mathfrak{A}\in {}_\kappa Cs_\alpha^{reg}\cap Dc_\alpha$ such that (i)-(iii) below hold.

(2) There is $\mathfrak{A}\in {}_\kappa Cs_\alpha^{reg}$ such that (i)-(v) below hold.

(3) There is $\mathfrak{A}\in {}_\kappa Ws_\alpha$ such that (i)-(v) below hold.

(i) $\mathbf{H}\mathfrak{A} \nsubseteq \mathbf{I}\{\mathfrak{B}\in Crs_\alpha : |base(\mathfrak{B})|\leq\kappa\}$.

(ii) $\mathbf{H}\mathfrak{A} \subseteq Dind_\alpha$.

(iii) $\mathbf{H}\mathfrak{A} \subseteq \mathbf{I}Cs_\alpha$ iff $\kappa\geq\omega$.

(iv) $\mathbf{H}\mathfrak{A} \subseteq \mathbf{I}Cs_\alpha^{reg}$ iff $\kappa\geq\omega$.

(v) $Cs_\alpha\cap\mathbf{H}\mathfrak{A} \subseteq Cs_\alpha^{reg}$ and $Gws_\alpha^{norm}\cap\mathbf{H}\mathfrak{A} \subseteq Gws_\alpha^{reg}$.

Proof. Let $\kappa\geq 2$ and $\alpha\geq\omega$. Let $\mathfrak{A}\in {}_\kappa Gws_\alpha$ and suppose that $\mathfrak{A}$ satisfies (i). Then $\mathfrak{A}$ satisfies (iii) too, since if $\kappa<\omega$ then $\mathbf{I}Cs_\alpha\cap\mathbf{H}\mathfrak{A} \subseteq \mathbf{I}\,{}_\kappa Cs_\alpha$ and if $\kappa\geq\omega$ then $\mathbf{H}\mathfrak{A} \subseteq \mathbf{I}Cs_\alpha$ by [HMTI]7.21-22. If $\mathfrak{A}$ satisfies (iii) and (v) then $\mathfrak{A}$ satisfies (iv) too. Therefore we have to concentrate on statements (i), (ii) and (v) only. Recall the notations $Sm^{\mathfrak{A}}$ and $Dm_H^{\mathfrak{A}}$ from 1.2, 1.3.1.

Lemma 5.3.1. Let $\mathfrak{A},\mathfrak{B}\in CA_\alpha$.

(i) $(\forall h\in Hom(\mathfrak{A},\mathfrak{B}))(\forall x\in Sm^{\mathfrak{A}})[h(x)\in Sm^{\mathfrak{B}}$ and $(h(x)\neq 0 \Rightarrow |\Delta x\sim\Delta h(x)|<\omega)]$.

(ii) Suppose $\mathfrak{A} = \mathfrak{Sg}^{(\mathfrak{A})}\{x\in Sm^{\mathfrak{A}} : |\alpha\sim\Delta x|<\omega\}$. Then $Gws_\alpha^{norm}\cap\mathbf{H}\mathfrak{A} \subseteq Gws_\alpha^{reg}$.

(iii) Suppose $\mathfrak{A} = \mathfrak{Sg}^{(\mathfrak{A})}(Sm^{\mathfrak{A}}\sim Dm_0^{\mathfrak{A}})$ and $\mathfrak{A}$ has a characteristic. Then $\mathbf{H}\mathfrak{A} \subseteq Dind_\alpha$.

(iv) Suppose $\mathfrak{A} = \mathfrak{Sg}^{(\mathfrak{A})}Sm^{\mathfrak{A}}$ and $\mathfrak{A}\in\mathbf{I}(Gws_\alpha^{comp})^{reg}$. Then $\mathbf{H}\mathfrak{A}\subseteq Dind_\alpha$.

Proof. Let $\mathfrak{A}\in CA_\alpha$. Proof of (i): Let $h\in Hom(\mathfrak{A},\mathfrak{B})$. Then $h^*(Sm^{\mathfrak{A}}) \subseteq Sm^{\mathfrak{B}}$ follows immediately from the definition of Sm. Let $x\in Sm^{\mathfrak{A}}$ and $K \overset{d}{=} \Delta x\sim\Delta h(x)$. Suppose $|K|\geq\omega$. Then $(\exists\Theta\subseteq_\omega K)c^\partial_{(\Theta)}x=0$ and hence $c^\partial_{(\Theta)}h(x) = h(x) = 0$.

Proof of (ii): Suppose $\mathfrak{A} = \mathfrak{Sg}^{(\mathfrak{A})}\{x\in Sm^{\mathfrak{A}} : |\alpha\sim\Delta x|<\omega\}$ and $\mathfrak{B}\in$

$\in Gws_\alpha^{norm} \cap \mathbf{H}\mathfrak{A}$. Then $\mathfrak{B} = \mathfrak{Sg}^{(\mathfrak{B})}\{x \in Sm^{\mathfrak{B}} : |\alpha \sim \Delta x| < \omega\}$ by (i). Then $\mathfrak{B}$ is regular by Theorem 1.3 since every cofinite dimensional element is regular in $\mathfrak{B}$.

Proof of (iii): Suppose $\mathfrak{A} = \mathfrak{Sg}^{(\mathfrak{A})}(Sm^{\mathfrak{A}} \sim Dm_0^{\mathfrak{A}})$ and $\mathfrak{A}$ has a characteristic. Let $\mathfrak{B} \in \mathbf{H}\mathfrak{A}$, $|B|>1$. Then $\mathfrak{Mn}(\mathfrak{B})$ is simple and $\mathfrak{B} = \mathfrak{Sg}^{(\mathfrak{B})}(Sm^{\mathfrak{B}} \sim Dm_0^{\mathfrak{B}})$ by (i). Thus $Dm_0^{\mathfrak{B}} = Mn(\mathfrak{B})$ by Lemma 1.3.3. Then $Zd\mathfrak{B} \subseteq Mn(\mathfrak{B})$ and thus $|Zd\mathfrak{B}| \le 2$ since $\mathfrak{Mn}(\mathfrak{B})$ is simple.

Proof of (iv): Let $\mathfrak{A} = \mathfrak{Sg}^{(\mathfrak{A})} Sm^{\mathfrak{A}} \in \mathbf{I}(Gws_\alpha^{comp})^{reg}$. Let $\mathfrak{B} \in \mathbf{H}\mathfrak{A}$. Then $B = Sg\ Sm^{\mathfrak{B}}$ by (i). Thus by Lemma 1.3.3, $\mathfrak{Dm}_0^{\mathfrak{B}} = \mathfrak{Sg}^{(\mathfrak{B})}(Sm^{\mathfrak{B}} \cap \cap Dm_0^{\mathfrak{B}}) \preceq \mathfrak{Sg}^{(\mathfrak{A})}(Sm^{\mathfrak{A}} \cap Dm_0^{\mathfrak{A}}) \in \mathbf{I}(Gws_\alpha^{comp})^{reg} \cap Lf_\alpha$. Therefore $|Zd\mathfrak{B}| \le 2$ since every $(Gws_\alpha^{comp})^{reg} \cap Lf_\alpha$ is simple.

QED(Lemma 5.3.1.)

We return to the proof of Theorem 5.3.

Proof of (1) and (2): Let $H \in {}^\alpha Sb\alpha$ be such that $(\forall n \in \alpha)|H_n| \ge 2$ and $(\forall n,m \in \alpha)[n \ne m \Rightarrow H_n \cap H_m = 0]$. Let $M \overset{d}{=} \cup RgH$. Let $T \subseteq M$ be such that $(\forall n \in \alpha)|T \cap H_n| = 1$. Then $|T| = |\alpha|$. Let $\gamma \overset{d}{=} |{}^\alpha \varkappa|$ and $p \in {}^\gamma({}^T\varkappa)$ be such that $(\forall i<j<\gamma) pi \notin {}^T\varkappa(pj)$. Define $R \subseteq {}^\alpha\varkappa$ and $x : Sb\gamma \sim \{0\} \rightarrow$ $\rightarrow Sb\ R$ as follows:

$R \overset{d}{=} \{q \in {}^\alpha\varkappa : (\forall n \in \alpha)|q^* H_n| = 1\}$,

$x_Y \overset{d}{=} \{q \in R : T1q \in \cup\{{}^T\varkappa(pi) : i \in Y\}\}$ for all $Y \subseteq \gamma$, $Y \ne 0$.

$\mathfrak{L} \overset{d}{=} \mathfrak{Sb}\,{}^\alpha\varkappa$.

Claim 1. $Rgx \subseteq Sm^{\mathfrak{L}} \cap Dm_M^{\mathfrak{L}}$ and every element of $Rgx$ is regular in $\mathfrak{L}$.

Proof. Let $Y \subseteq \gamma$, $Y \ne 0$. Then it is easy to see that $\Delta(x_Y) = M$ and $x_Y$ is regular. Let $K \subseteq M$ be infinite and let $\Gamma \subseteq_\omega \alpha$. Let $i \in$ $\in (H_n \cap K) \sim \Gamma$ and $j \in H_n \sim (\Gamma \cup \{i\})$ for some $n \in \alpha$. Let $q \in x_Y$ and let $u \in \varkappa$, $u \ne q(j)$. Then $q(i/u) \notin c_{(\Gamma)} x_Y$. Thus $c_i^\partial c_{(\Gamma)} x_Y = 0$.

QED(Claim 1)

Let $L \subseteq Sb\ \gamma$ be such that $|L| > \gamma$ and $(\forall Z, Y \in L)[Z \ne Y \Rightarrow |Z \cap Y| < \gamma = |Z|]$. The existence of such an $L$ is a theorem of set theory. Let $\mathfrak{A} \overset{d}{=}$ $\overset{d}{=} \mathfrak{Sg}^{(\mathfrak{Sb}\,{}^\alpha\varkappa)}\{x_Y : Y \in L\}$. Now $\mathfrak{A}$ is regular and $\mathbf{H}\mathfrak{A} \subseteq Dind_\alpha$ by Thm 1.3,

Claim 1, $|M|\geq\omega$ and Lemma 5.3.1. If $|\alpha\sim M|\geq\omega$ then $\mathfrak{A}\in Dc_\alpha$ and if $\alpha=M$ then $\mathfrak{A}$ satisfies (v) by Claim 1 and Lemma 5.3.1. Next we show that $(\exists I\in Il\mathfrak{A})(\forall\mathfrak{B}\in Crs_\alpha\cap I\{\mathfrak{A}/I\})|base(\mathfrak{B})|>\varkappa$. Let $W \overset{d}{=}$ $\overset{d}{=} \{{}^T\varkappa^{(p)} : p\in{}^T\varkappa\}$. Define $J \overset{d}{=} \{y\in C : |\{w\in W : (\exists q\in y)T1q\in w\}|<\gamma\}$ and $I \overset{d}{=} A\cap J$. Clearly, $J\in Il\mathfrak{C}$ and therefore $I\in Il\mathfrak{A}$. Let $Y\subseteq\gamma$. Then $x_Y\in J$ iff $|Y|<\gamma$ since $|\{w\in W : (\exists q\in x_Y)T1q\in w\}| = |Y|$. Let $Y,Z\in L$. Then $x_Y\notin I$ by $|Y|=\gamma$ and $x_Y\cdot x_Z=x_{Y\cap Z}\in I$ by $|Y\cap Z|<\gamma$. These facts show that $\{x_Y/I : Y\in L\}$ is an antichain of cardinality $|L|>\gamma$ in $\mathfrak{A}/I$. Let $\mathfrak{A}/I\cong\mathfrak{B}\in Crs_\alpha$. Since $\mathfrak{B}$ contains an antichain of cardinality $>\gamma$ and the members of an antichain are mutually disjoint in any $Crs_\alpha$ we have $|1^{\mathfrak{B}}|>\gamma=|{}^\alpha\varkappa|$. Since $1^{\mathfrak{B}}\subseteq{}^\alpha base(\mathfrak{B})$ this means $|base(\mathfrak{B})|>\varkappa$. We have seen that $\mathfrak{A}$ satisfies (i).

<u>Proof of (3):</u> Let $\bar{0}\overset{d}{=}\alpha\times 1$, $V\overset{d}{=}{}^\alpha\varkappa^{(\bar{0})}$ and $\gamma\overset{d}{=}|V|$. Then $\gamma=$ $=\varkappa\cdot|\alpha|$. Let $p\in{}^\gamma({}^\alpha\varkappa)$ be such that $(\forall i\in\gamma)(\exists H\subseteq\alpha)[|H|\geq\omega$ and $(\forall j\in$ $\in\gamma\sim\{i\})(\forall m\in H)p_j(m)\neq p_i(m)\neq 0]$ and $(\forall i<j<\gamma)Rg(p_i\cap p_j)\subseteq 1$. Such a $p$ exists. Define $R\subseteq V$ and $x : Sb\gamma\to SbR$ as follows:

$R\overset{d}{=}\{q\in V : |\{i\in\alpha : q_i\neq 0\}|\leq 1\}$ and

$x_Y\overset{d}{=}\{q\in R : q\subseteq(\bar{0}\cup\cup\{p_j : j\in Y\})\}$ for all $Y\subseteq\gamma$.

Let $\mathfrak{C}\overset{d}{=}\mathfrak{Sb}\,V$. Let $J\overset{d}{=}Ig^{(\mathfrak{C})}\{x_Y : Y\subseteq\gamma$ and $|Y|<\gamma\}$.

<u>Claim 2:</u> Let $Y\subseteq\gamma$. Then $(x_Y\in J$ iff $|Y|<\gamma)$ and $(\forall i<j<\alpha)x_Y\sim d_{ij}\in J$.

<u>Proof.</u> Let $Y\subseteq\gamma$ and let $i<j<\alpha$. Then $x_Y\sim d_{ij}\subseteq\{\bar{0}^j_u,\bar{0}^i_u : u\in\varkappa\sim 1\}\subseteq$ $\subseteq c_ic_j\{\bar{0}\}\in J$ since $(\forall Z\subseteq\gamma)\bar{0}\in x_Z$. Suppose $|Y|=\gamma$. We show $x_Y\notin J$. By $x_Z\cup x_W=x_{Z\cup W}$ we have only to show $(\forall Z\subseteq\gamma)[|Z|<\gamma\Rightarrow(\forall\Gamma\subseteq_\omega\alpha)x_Y\nsubseteq$ $\nsubseteq c_{(\Gamma)}x_Z]$. Let $Z\subseteq\gamma$, $|Z|<\gamma$ and let $\Gamma\subseteq_\omega\alpha$. Let $i\in Y\sim Z$. By the construction of $p$, there is an infinite $H\subseteq\alpha$ such that $(\forall j\in Z)$ $(\forall m\in H)p_j(m)\neq p_i(m)\neq 0$. Then for every $m\in H\sim\Gamma$ we have $\bar{0}(m/p_i(m))\in$ $\in x_Y\sim c_{(\Gamma)}x_Z$.

<u>QED(Claim 2)</u>

Let $L\subseteq Sb\gamma$ be such that $|L|>\gamma$ and $(\forall Z,Y\in L)[Z\neq Y\Rightarrow|Z\cap Y|<\gamma=|Z|]$. Let $\mathfrak{A}\overset{d}{=}\mathfrak{Sg}^{(\mathfrak{C})}\{x_Y : Y\in L\}$. Let $I\overset{d}{=}J\cap A$. By the second condition on p we have $x_Z \cap x_Y \subseteq x_{Z\cap Y}$. Hence by Claim 2 we have that $\{x_Y/I : Y\in L\}$ is an antichain of cardinality $|L|$ in $\mathfrak{A}/I$ such that $(\forall i,j<\alpha)(\forall Y\in L)\ x_Y/I \le d_{ij}/I$. Let $\mathfrak{A}/I \cong \mathfrak{B}\in Crs_\alpha$ for some $\mathfrak{B}$. Then $|base(\mathfrak{B})| \ge |L|>\gamma$ by the above. Therefore $\mathfrak{A}\in {}_\varkappa Ws_\alpha$ satisfies (i). To prove that $\mathfrak{A}$ satisfies (ii) and (v), observe that $Rgx \subseteq Sm^{\mathfrak{A}} \cap Dm_\alpha^{\mathfrak{A}}$ and $\mathfrak{A}$ has characteristic $\varkappa$, and then apply Lemma 5.3.1.

QED(Theorem 5.3)

By relativizing the proof of 5.3(3) to a $T\subseteq\alpha$ with $|\alpha\sim T|=\omega$ as it was done in the proof of (1)-(2) of 5.3, we obtain that if $\varkappa^\omega \le |\alpha|$ then there is $\mathfrak{B}\in {}_\varkappa Ws_\alpha \cap Dc_\alpha$ satisfying (i)-(iii) of Theorem 5.3. In case $\varkappa^\omega > |\alpha|$ we do not know whether there is $\mathfrak{B}\in {}_\varkappa Ws_\alpha \cap Dc_\alpha$ satisfying (i) of Theorem 5.3.

In connection with Corollary 5.4 below see also figures [HMTI]5.7 and [HMTI]6.10 keeping in mind $Cs_\alpha^{reg} \subseteq Dind_\alpha$. In connection with (v) see Problem 4 in [HMTI]9.

Corollary 5.4. Let $\varkappa\ge 2$ and $\alpha\ge\omega$. Then (i)-(iii) below hold.

(i) $(\exists\, \mathfrak{A}\in {}_\varkappa Cs_\alpha^{reg} \cap Dc_\alpha)\, |A| = |{}^{({}^\alpha\varkappa)}2|$.

(ii) $\mathbf{H}({}_\varkappa Cs_\alpha^{reg} \cap Dc_\alpha) \nsubseteq Dind_\alpha \supseteq \mathbf{I} Cs_\alpha^{reg}$ and

$\mathbf{H}({}_\varkappa Ws_\alpha \cap Dc_\alpha) \nsubseteq Dind_\alpha \supseteq \mathbf{I} Cs_\alpha^{reg} \cup \mathbf{I} Ws_\alpha$.

(iii) $(\exists\, \mathfrak{A}\in {}_\varkappa Cs_\alpha^{reg} \cap Dc_\alpha)[\ \mathfrak{A}$ is not simple and $\mathbf{H}\mathfrak{A} \subseteq Dind_\alpha$ and $|A|>1]$.

Suppose further $\varkappa<\omega$. Then (iv)-(v) below hold.

(iv) $Dind_\alpha \cap \mathbf{H}({}_\varkappa Cs_\alpha^{reg} \cap Dc_\alpha) \nsubseteq \mathbf{I} Cs_\alpha$ and

$Dind_\alpha \cap \mathbf{H}\, {}_\varkappa Ws_\alpha \nsubseteq \mathbf{I} Cs_\alpha$.

(v) ${}_\varkappa Cs_\alpha^{reg} \cap Dc_\alpha \nsubseteq \mathbf{H} Ws_\alpha$.

Proof. **(i)** and the first part of (ii) are immediate by 4.13. The second part of (ii) follows from inspecting the proof of [HMTI]6.16(2) because there $\mathfrak{A}\in {}_\varkappa Ws_\alpha$ and $I\in Il\,\mathfrak{A}$ were constructed with $\mathfrak{A}/I \notin Dind_\alpha$

and it is easy to check that if $\alpha\geq\omega+\omega$ then $\mathfrak{A}\in Dc_{\alpha}$. But then [HMTI] 8.4 implies the result for any $\alpha\geq\omega$. (We note that the construction in the proof of [HMTI]6.16(2) can be amended to obtain an $\mathfrak{A}\in {}_{\varkappa}Ws_{\alpha}\cap Dc_{\alpha}$ and an $I\in Il\,\mathfrak{A}$ such that $|Zd\ \mathfrak{A}/I| = 2^{\max(\alpha,\varkappa)}$. The amendment we know of is not completely obvious.) (iii) is a consequence of Thm.5.3(1) since if $\mathfrak{A}\in {}_{\varkappa}Cs_{\alpha}$ then condition (i) of Thm.5.3 implies that $\mathfrak{A}$ is not simple. (v) follows from (i) since if $2\leq\varkappa<\omega$ then $\mathbf{I}\,{}_{\varkappa}Cs_{\alpha}\cap \mathbf{H}Ws_{\alpha}\subseteq \mathbf{H}\,{}_{\varkappa}Ws_{\alpha}$ and if $\mathfrak{B}\in \mathbf{H}\,{}_{\varkappa}Ws_{\alpha}$ then $|B|\leq 2^{|\alpha|}$. (iv) is an immediate consequence of Thm.5.3.

QED(Corollary 5.4.)

Later we shall frequently use Proposition 5.6 below which is in constrast with [HMTI]7.13, 7.17 which state that $Ws_{\alpha}\subseteq \mathbf{I}\,Cs_{\alpha}^{reg}$ and $\mathbf{I}\,Gws_{\alpha}^{comp} = \mathbf{I}\,Cs_{\alpha}$.

$\mathbf{I}\,Gws_{\alpha}^{wd} = \mathbf{I}\,Gws_{\alpha}^{norm} = \mathbf{I}\,Gs_{\alpha} = \mathbf{HSP}\ Gws_{\alpha}$.

|

$\mathbf{I}\ Cs_{\alpha} = \mathbf{I}\,Gws_{\alpha}^{comp}$

|

$\mathbf{I}\ Cs_{\alpha}\cap Dind_{\alpha}$

|

$\mathbf{I}\ Gws_{\alpha}^{comp\ reg}$

|

$\mathbf{I}\ Cs_{\alpha}^{reg}$

|

$\mathbf{I}\ Ws_{\alpha}$

Figure 5.5 ($\alpha\geq\omega$)

Proposition 5.6 below implies that of the seven classes $\{Gws^{wd}, Gws^{norm}, Gws^{comp}, Gws, Gs, Cs, Ws\}$ only $K\in\{Cs, Gws^{comp}\}$ are such that $\mathbf{I}\,K_{\alpha}\neq \mathbf{I}\,K_{\alpha}^{reg}$.

<u>Proposition 5.6.</u> Let $\varkappa\geq 2$ and $\alpha\geq\omega$.

(i) ${}_{\varkappa}Gws_{\alpha}^{comp\ reg}\cap Dc_{\alpha} \nsubseteq \mathbf{I}\,Cs_{\alpha}^{reg}$ but

$$\mathbf{I}\,Gws_{\alpha}^{comp\ reg}\cap Lf_{\alpha} = \mathbf{I}\,Cs_{\alpha}^{reg}\cap Lf_{\alpha} = \mathbf{H}Ws_{\alpha}\cap Lf_{\alpha}.$$

(ii) The classes on Figure 5.5 are all different and the indicated inclusions hold.

(iii) Let $K\in\{Gws^{wd}, Gws^{norm}, Gws, Gs\}$. Then $\mathbf{I}\,K_{\alpha} = \mathbf{I}\,K_{\alpha}^{reg} = \mathbf{I}\,Gs_{\alpha}$.

<u>Proof.</u> Let $\alpha\geq\omega$ and $\varkappa\geq 2$. <u>Proof of (i)</u>: Let $H,L\subseteq\alpha$ be such that $|H|\cap|L|\cap|\alpha\sim(H\cup L)|\geq\omega$, $H\cap L=0$. Let $\bar{0}\overset{d}{=}\alpha\times 1$ and $\bar{1}\overset{d}{=}\alpha\times\{1\}$, $V\overset{d}{=}{}^{\alpha}{}_{\varkappa}{}^{(\bar{0})}\cup{}^{\alpha}{}_{\varkappa}{}^{(\bar{1})}$. Let $x\overset{d}{=}\{q\in V : H1q\subseteq\bar{0}\}$, $y\overset{d}{=}\{q\in V : L1q\subseteq\bar{1}\}$. $\mathfrak{A}\overset{d}{=}\mathfrak{Sg}^{(\mathfrak{Sb}\,V)}\{x,y\}$. Then $\mathfrak{A}\in Gws_{\alpha}^{comp\ reg}\cap Dc_{\alpha}$ by Thm 1.3. We show that $\mathfrak{A}\notin\mathbf{I}\,Cs_{\alpha}^{reg}$. Let $\mathfrak{B}\in Cs_{\alpha}^{reg}$ and $a,b\in B$ be such that $\Delta a\cap\Delta b=0$ and $a\neq 0$, $b\neq 0$. Then $a\cdot b\neq 0$ since $(\forall p\in a)(\forall q\in b)p[\Delta b/q]\in a\cdot b$. Now $\mathfrak{A}\notin\mathbf{I}\,Cs_{\alpha}^{reg}$ since $\Delta x\cap\Delta y=0$, $x\neq 0$, $y\neq 0$ and $x\cdot y=0$.

Let $\mathfrak{A}\in Gws_{\alpha}^{comp\ reg}\cap Lf_{\alpha}$. Then $|Zd\,\mathfrak{A}|\leq 2$, thus $\mathfrak{A}$ is simple by [HMT]2.3.14. Let $p\in 1^{\mathfrak{A}}$ and $V\overset{d}{=}{}^{\alpha}base(\mathfrak{A})^{(p)}$. Then $\mathfrak{A}\cong\mathfrak{Rl}_V\mathfrak{A}\in Ws_{\alpha}\cap Lf_{\alpha}\subseteq\mathbf{I}\,Cs_{\alpha}^{reg}\cap Lf_{\alpha}$ by 3.15(a).

<u>Proof of (ii):</u> $\mathbf{I}\,Cs_{\alpha}^{reg}\neq\mathbf{I}\,Gws_{\alpha}^{comp\ reg}$ follows from (i). $Cs_{\alpha}\cap Dind_{\alpha}\nsubseteq\mathbf{I}\,Gws_{\alpha}^{comp\ reg}$ will be proved in 5.7(iv). $\mathbf{I}\,Ws_{\alpha}\neq\mathbf{I}\,Cs_{\alpha}^{reg}$ e.g. by 5.4(v). $Ws_{\alpha}\cup Cs_{\alpha}^{reg}\subseteq Gws_{\alpha}^{comp\ reg}\subseteq Dind_{\alpha}$ by the definitions. $\mathbf{I}\,Gws_{\alpha}^{comp}=\mathbf{I}\,Cs_{\alpha}$ by (1) in the proof of [HMTI]7.17 and $Ws_{\alpha}\subseteq\mathbf{I}\,Cs_{\alpha}^{reg}$ by [HMTI]7.13. Let $\mathfrak{A}\in Gws_{\alpha}$. Then $\mathfrak{A}\cong|\subseteq P\mathfrak{B}$ for some $\mathfrak{B}\in{}^{\rho}Ws_{\alpha}$. By [HMTI]3.1 there is $\mathcal{L}\in{}^{\rho}Ws_{\alpha}$ such that $(\forall i<j<\rho)base(\mathcal{L}_i)\cap base(\mathcal{L}_j)=0$ and $P\mathfrak{B}\cong P\mathcal{L}$. Let $V\overset{d}{=}\bigcup_{i<\rho}1^{\mathcal{L}_i}$. Then $V$ is a $Gws_{\alpha}^{wd}$-unit and by [HMTI]6.2 $P\mathcal{L}\cong\mathfrak{N}$ with $1^{\mathfrak{N}}=V$. Thus $\mathfrak{A}\cong|\subseteq P\mathcal{L}\cong\mathfrak{N}\in Gws_{\alpha}^{wd}$ implies $\mathfrak{A}\in\mathbf{I}\,Gws_{\alpha}^{wd}$. Thus $Gws_{\alpha}\subseteq\mathbf{I}\,Gws_{\alpha}^{wd}$ is proved. By $Gws_{\alpha}^{wd}\subseteq Gws_{\alpha}^{norm}\subseteq Gws_{\alpha}$ then $\mathbf{I}\,Gws_{\alpha}^{wd}=\mathbf{I}\,Gws_{\alpha}^{norm}=\mathbf{I}\,Gws_{\alpha}$. By [HMTI]7.14, 7.16 $\mathbf{HSP}\,Gws_{\alpha}=\mathbf{I}\,Gws_{\alpha}=\mathbf{I}\,Gs_{\alpha}$.

<u>Proof of (iii):</u> $Gws_{\alpha}^{wd}=Gws_{\alpha}^{wd\ reg}$ is easy to see by the definitions. Now [HMTI]7.14 together with (ii) of the present theorem complete the proof.

QED(Proposition 5.6.)

Theorem 5.7(i) below was quoted in [HMTI]5.6(11). The contrast between (i) and (ii) of Thm 5.7 implies that the only possible way of making a homomorphic image of a $Cs_\alpha \cap Lf_\alpha$ into a non-$\mathbf{I}\,Cs_\alpha$ is to create new zerodimensional elements. In Corollary 5.4 for two classes $K \subseteq L$ we often stated $Dind_\alpha \cap \mathbf{H}K \not\subseteq L$ instead of the weaker $\mathbf{H}K \not\subseteq L$. Some motivation for this is Thm 5.7(ii) below.

Theorem 5.7. Let $\varkappa \geq 2$ and $\alpha \geq \omega$.

(i) $\mathbf{H}({}_\varkappa Cs_\alpha \cap Lf_\alpha) \not\subseteq \mathbf{I}\,Cs_\alpha$ if $\varkappa < \omega$ and

$\mathbf{H}({}_\varkappa Cs_\alpha \cap Lf_\alpha) \not\subseteq \mathbf{I}\{\mathfrak{B} \in Crs_\alpha : |base(\mathfrak{B})| \leq \varkappa\}$ for all $\varkappa$.

(ii) $Dind_\alpha \cap \mathbf{H}(\,Cs_\alpha \cap Lf_\alpha) \not\subseteq \mathbf{I}\,Cs_\alpha^{reg}$.

(iii) ${}_\varkappa Cs_\alpha \cap Dind_\alpha \cap \mathbf{H}({}_\varkappa Cs_\alpha^{reg} \cap Dc_\alpha) \not\subseteq \mathbf{I}\,Gws_\alpha^{comp\ reg}$.

(iv) ${}_\varkappa Cs_\alpha \cap Dind_\alpha \cap \mathbf{H}({}_\varkappa Ws_\alpha \cap Dc_\alpha) \not\subseteq \mathbf{I}\,Gws_\alpha^{comp\ reg}$.

Proof. Let $\varkappa \geq 2$ and $\alpha \geq \omega$. The following lemma is well known from the theory of BA- s. We quote it without proof.

Lemma 5.7.1. Let $\mathfrak{B} \in BA$ be complete and atomic. Let $\rho = |B| \geq \omega$. There are $I \in Il\mathfrak{B}$ and $x \in {}^\rho B$ such that $(\forall i<j<\rho)\,[x_i/I \cdot x_j/I = 0/I \neq x_i/I]$ in $\mathfrak{B}/I$.

Proof of (i): Let $\mathfrak{L} \overset{d}{=} \mathfrak{Cb}\,{}^\alpha\varkappa$ and $\mathfrak{B} \overset{d}{=} \mathfrak{Zd}\mathfrak{L}$. Let $\rho \overset{d}{=} |{}^{({}^\alpha\varkappa)}2|$. Then $|B| = \rho$ and $\mathfrak{B}$ is a complete and atomic infinite BA. Then by Lemma 5.7.1 there are $I \in Il\mathfrak{B}$ and $x \in {}^\rho B$ such that $(\forall i<j<\rho)$ $[x_i/I \cdot x_j/I = 0/I \neq x_i/I]$ in $\mathfrak{B}/I$. Let $\mathfrak{A} \subseteq \mathfrak{L}$ be arbitrary such that $B \subseteq A$. Let $J \overset{d}{=} Ig^{(\mathfrak{A})}I$. By [HMT]2.3.7 then $I = B \cap J$. Thus $(\forall i<j<\rho)$ $[x_i/J \cdot x_j/J = 0/J \neq x_i/J]$ in $\mathfrak{A}/J$. Suppose $h \in Is(\mathfrak{A}/J, \mathfrak{N})$ for some $\mathfrak{N} \in Crs_\alpha$ such that $|base(\mathfrak{N})| \leq \varkappa$. Then $|1^{\mathfrak{N}}| \leq |{}^\alpha\varkappa| < \rho$ contradicting $(\forall i<j<\rho) h(x_i/J) \cap h(x_j/J) = 0 \neq h(x_i/J)$. This proves

(*) $\mathfrak{A}/J \notin \mathbf{I}\{\mathfrak{N} \in Crs_\alpha : |base(\mathfrak{N})| \leq \varkappa\}$.

Assume next $\kappa<\omega$ and $\mathfrak{N}\in Cs_\alpha$ and $h\in Is(\mathfrak{A}/J,\mathfrak{N})$. By [HMTI]5.3 then $\mathfrak{N}\in {}_\kappa Cs_\alpha$ which is impossible by (*) above. Thus $\mathfrak{A}/J\notin \mathbf{I}Cs_\alpha$ is proved for $\kappa<\omega$.

Proof of (ii): Let $\mathfrak{A}\in H(Cs_\alpha \cap Lf_\alpha)$, $\alpha\geq\omega$. By [HMTI]7.13-16, $\mathfrak{A}\in \mathbf{SP}Cs_\alpha^{reg}$. Assume $\mathfrak{A}\in Dind_\alpha$, that is $|Zd\,\mathfrak{A}|\leq 2$. Then by [HMT]2.4.43 $\mathfrak{A}$ is subdirectly indecomposable or $|A|=1$. Thus $\mathfrak{A}\in \mathbf{I}Cs_\alpha^{reg}$.

To prove (iii) and (iv) we shall need the following lemma.

Lemma 5.7.2. Let $\mathfrak{B}\in CA_\alpha$ be generated by $\{x,y,z\}$. Suppose that $1=\Delta y\subseteq \Delta x=\Delta z$, $|\alpha\sim\Delta x|=|\alpha|$, $y\cdot s_1^0 y\leq d_{01}$, $x\cdot z=0$, and $(\forall i\in\Delta x)x+z\leq y\cdot d_{0i}$. Then (i) and (ii) below hold.

(i) $\mathfrak{B}\notin \mathbf{I}(Gws_\alpha^{comp})^{reg}$.

(ii) Suppose $\mathfrak{B}\in \mathbf{I}Gws_\alpha$ and $\mathfrak{B}$ has characteristic $\kappa$. Then $\mathfrak{B}\in \mathbf{I}\,{}_\kappa Cs_\alpha$ if $\kappa\neq 0$ and $\mathfrak{B}\in \mathbf{I}\,{}_\rho Cs_\alpha$ for every $\rho\geq\omega$ if $\kappa=0$.

Proof. Let $\mathfrak{B}\in CA_\alpha$ and suppose that $\{x,y,z\}\subseteq B$ satisfies the hypotheses of Lemma 5.7.2. Let $H\overset{d}{=}\Delta x$.

Claim 5.7.2.1. Suppose $h:\mathfrak{B}\to\mathfrak{N}\in Gws_\alpha^{comp\ reg}$. Then there is $u\in base(\mathfrak{N})$ such that $h(y)\in\{0,\{q\in {}^{\mathfrak{N}}1 : q(0)=u\}\}$ and $\{h(x),h(z)\}\subseteq$ $\subseteq\{0,\{q\in {}^{\mathfrak{N}}1 : (\forall i\in H)q_i=u\}\}$.

Proof. Let $h:\mathfrak{B}\to\mathfrak{N}\in Gws_\alpha^{comp\ reg}$ and $Y\overset{d}{=}h(y)$. Then $\Delta Y\subseteq 1$ and $Y\cdot s_1^0 Y\leq d_{01}$. Suppose $p,q\in Y$ and $p0\neq q0$. Now $p(0/q0)\in 1^{\mathfrak{N}}$ since $\mathfrak{N}$ is compressed, thus $p(0/q0)\in Y$ since $\mathfrak{N}$ is regular. Now $p,p(0/q0)\in Y$, $p0\neq q0$, $\Delta Y\subseteq 1$ and $\mathfrak{N}\in Gws_\alpha$ imply $Y\cdot s_1^0 Y\nleq d_{01}$. We have seen $(\exists u\in base(\mathfrak{N}))\ h(y)\in\{0,\{q\in 1^{\mathfrak{N}} : q0=u\}\}$. Then $(\forall i\in H)x\leq$ $\leq y\cdot d_{0i}$, $\Delta x=H$ and $\mathfrak{N}\in Gws_\alpha^{reg}$ imply $h(x)\in\{0,\{q\in 1^{\mathfrak{N}} : (\forall i\in H)q_i=u\}\}$. The same argument works for $z$.

QED(Claim 5.7.2.1.)

Now by Claim 5.7.2.1, $x\cdot z=0$, $x\neq 0$, $z\neq 0$ immediately yield $\mathfrak{B}\notin$ $\notin \mathbf{I}Gws_\alpha^{comp\ reg}$. To prove 5.7.2(ii), we shall need the following lemma.

First to every ordinal $\beta\geq 2$ we define a $Cs_\alpha^{reg}$ $\mathfrak{R}_\beta$. Let $\beta\geq 2$. Then $Y_\beta \overset{d}{=} \{q\in{}^\alpha\beta : q0=0\}$ and $X_\beta \overset{d}{=} \{q\in{}^\alpha\beta : (\forall i\in H)q_i=0\}$. Let $\mathfrak{L} \overset{d}{=} \mathfrak{Sb}\,{}^\alpha\beta$ and $\mathfrak{R}_\beta \overset{d}{=} \mathfrak{Sg}^{(\mathfrak{L})}\{Y_\beta,X_\beta\}$. By this the system $\langle \mathfrak{R}_\beta : \beta\in Ord\sim 2\rangle$, where Ord is the class of all ordinals, has been defined. By Thm 1.3, $\mathfrak{R}_\beta \in$ $\in {}_\beta Cs_\alpha^{reg}$ for every $\beta\geq 2$.

Lemma 5.7.2.2.

(i) $\mathfrak{R}_\beta \cong \mathfrak{R}_\omega$ for every $\beta\geq\omega$.

(ii) ${}^\alpha\mathfrak{R}_\beta \in \mathsf{I}\,{}_\beta Cs_\alpha$ for every $\beta\in Ord\sim 2$.

Proof. Proof of (i): Let $\beta\geq\omega$. Let $F$ be an ultrafilter on some ordinal $\rho$ such that $|{}^\rho\beta/\bar{F}| = |{}^\rho\omega/\bar{F}|$. Such an $F$ exists. Let $c : \alpha\times{}^\rho\beta/\bar{F} \to {}^\rho\beta$ be an $(F,\langle \beta : i<\rho\rangle,\alpha)$ -choice function such that $(\forall u\in\beta)(\forall i\in\alpha)\ c(i,\bar{u}/\bar{F})=\bar{u}$ where $\bar{u} = \langle u : j<\rho\rangle$. The homomorphism $ud_{cF}\in$ $\in Ho(\mathfrak{R}_\beta)$ was defined in 3.12. By $\mathfrak{R}_\beta\in Cs_\alpha^{reg}$ and [HMTI]7.6, 7.12, $ud_{cF}\in Is(\mathfrak{R}_\beta,\mathfrak{R}_\beta^+)$ for some $\mathfrak{R}_\beta^+\in Cs_\alpha^{reg}$ such that $base(\mathfrak{R}_\beta^+) = {}^\rho\beta/F$. Let $U = {}^\rho\beta/F$.

$ud_{cF}(Y_\beta) = \{q\in{}^\alpha U : q0=\bar{0}/\bar{F}\}$ and

$ud_{cF}(X_\beta) = \{q\in{}^\alpha U : (\forall i\in H)q_i=\bar{0}/\bar{F}\}$, because $(\forall q\in{}^\alpha U)(\forall i\in H)[q_i=\bar{0}/\bar{F}$ iff $c(i,q_i)=\bar{0}]$. Let $\gamma \overset{d}{=} |U|$. Since $R_\beta^+ = Sg\{ud_{cF}(Y_\beta),\ ud_{cF}(X_\beta)\}$ we have $\mathfrak{R}_\beta^+ \cong \mathfrak{R}_\gamma$. Thus $\mathfrak{R}_\beta \cong \mathfrak{R}_\gamma$. Repeating the above argument for $\mathfrak{R}_\omega$ we obtain $\mathfrak{R}_\omega \cong \mathfrak{R}_\gamma$. Hence $\mathfrak{R}_\beta \cong \mathfrak{R}_\omega$.

Proof of (ii): Let $T \overset{d}{=} \alpha\sim H$. Let $p \in {}^\alpha({}^T\varkappa)$ be such that $(\forall i<j<\alpha)\ p_i \notin {}^T\varkappa^{(pj)}$. Such a $p$ exists since $|T| = |\alpha|$. For all $i \in \alpha\sim 1$ let $V_i \overset{d}{=} \{q\in{}^\alpha\varkappa : T1q\in{}^T\varkappa^{(pi)}\}$. Let $V_0 \overset{d}{=} {}^\alpha\varkappa\sim\bigcup\{V_i : i\in\alpha\sim 1\}$. Let $i\in\alpha$. Then $rl_{Vi}\in Ho(\mathfrak{R}_\varkappa, \mathfrak{M}_i)$ for some $\mathfrak{M}_i\in Gws_\alpha^{comp}$. Let $v\in$ $\in R_\varkappa\sim\{0\}$. Then $T\cap\Delta v$ is finite, hence by regularity of $v$ we have $v\cap V_i\neq 0$. Thus $rl(V_i)\in Is(\mathfrak{R}_\varkappa, \mathfrak{M}_i)$. Then $\mathfrak{B} \cong \mathsf{I} \subseteq P_{i\in\alpha}\ \mathfrak{M}_i$. Since $1^{(\mathfrak{M}_i)} = V_i$ and $\bigcup_{i<\alpha} V_i = {}^\alpha\varkappa$ and $\{V_i : i\in\alpha\} \subseteq Zd\,\mathfrak{Sb}\,{}^\alpha\varkappa$ are disjoint, by [HMTI]6.2 we have $P_{i\in\alpha}\ \mathfrak{M}_i \cong \mathsf{I} \subseteq \mathfrak{Sb}\,{}^\alpha\varkappa$. Hence $\mathfrak{B}\cong \mathsf{I} \subseteq \mathfrak{Sb}\,{}^\alpha\varkappa$ which is equivalent to saying that $\mathfrak{B} \in \mathsf{I}\,{}_\varkappa Cs_\alpha$.

QED(Lemma 5.7.2.2.)

Now we turn to the proof of 5.7.2(ii). Suppose that $\mathfrak{B} \in \mathsf{I}\,Gws_\alpha$, $\mathfrak{B}$ has characteristic $\varkappa$ and $\{x,y,z\} \subseteq B$ satisfies the hypotheses of Lemma 5.7.2. By [HMTI]7.14-16 we have $\mathfrak{B} \in \mathbf{SP}Cs_\alpha^{reg}$ and hence $\mathfrak{B} \cong$ $\cong \mathsf{I} \subseteq_d P\mathfrak{N}$ for some $\mathfrak{N} \in {}^J Cs_\alpha^{reg}$. By $|B| \le \alpha$ we may assume $J=\alpha$. Let $j \in \alpha$. Then there exists $h \in Ho(\mathfrak{B}, \mathfrak{N}_j)$. Let $U \overset{d}{=} base(\mathfrak{N}_j)$. Then by Claim 5.7.2.1 we have that $\mathfrak{N}_j \subseteq \mathfrak{Sg}\{\{q \in {}^\alpha U : q(0)=u\}, \{q \in {}^\alpha U :$ $: (\forall i \in H) q_i = u\}\}$ for some $u \in U$. Hence $\mathfrak{N}_j \cong \mathsf{I} \subseteq \mathfrak{R}_\beta$ for $\beta =$ $= |base(\mathfrak{N}_j)|$. We have proved the existence of $\beta \in {}^\alpha Ord$ such that $\mathfrak{B} \cong \mathsf{I} \subseteq P_{j\in\alpha}\ \mathfrak{R}_{(\beta j)}$.

Case 1 Assume $\varkappa \ne 0$. Then by $Hom(\mathfrak{B}, \mathfrak{R}_{(\beta j)}) \ne 0$ and by [HMTI]5.3 we have $\beta j = \varkappa$ for all $j \in \alpha$. Hence $\mathfrak{B} \cong \mathsf{I} \subseteq {}^\alpha \mathfrak{R}_\varkappa \in \mathsf{I}\,{}_\varkappa Cs_\alpha$, by Lemma 5.7.2.2(ii).

Case 2 Assume $\varkappa = 0$. Let $\rho \ge \omega$. Then by the above argument we have $\beta j \ge \omega$ for all $j \in \alpha$. By Lemma 5.7.2.2(i) then $\mathfrak{R}_\rho \cong \mathfrak{R}_{(\beta j)}$ for all $j \in \alpha$. Thus $\mathfrak{B} \cong \mathsf{I} \subseteq {}^\alpha \mathfrak{R}_\rho \in \mathsf{I}\,{}_\rho Cs_\alpha$ by Lemma 5.7.2.2(ii).

QED(Lemma 5.7.2)

Proof of (iii) and (iv): Let $H \subseteq \alpha$ be such that $0 \in H$, $|H| \ge \omega$ and $|\alpha \sim H| = |\alpha|$. Let $L \subseteq H$ be such that $|L| \cap |H \sim L| \ge \omega$. We let

$y \overset{d}{=} \{q \in {}^\alpha\varkappa : q(0)=0\}$, $\bar{0} \overset{d}{=} \alpha \times 1$,

$x \overset{d}{=} \{q \in {}^\alpha\varkappa : L1q \subseteq \bar{0}$ and $|\{i \in H : q_i \ne 0| = 1\}$ and

$z \overset{d}{=} \{q \in {}^\alpha\varkappa : (H\sim L)1q \subseteq \bar{0}$ and $|\{i \in H : q_i \ne 0\}| = 1\}$.

$\mathfrak{A} \overset{d}{=} \mathfrak{Sg}^{(\mathfrak{C})}\{x,y,z\}$ where $\mathfrak{C} \overset{d}{=} \mathfrak{Cb}\,{}^\alpha\varkappa$.

Now $\mathfrak{A} \in Cs_\alpha^{reg} \cap Dc_\alpha$ and $\mathbf{H}\mathfrak{A} \subseteq Dind_\alpha$ by Thm 1.3 and Lemma 5.3.1(iv) since $\{x,y,z\} \subseteq Sm^{\mathfrak{A}}$. Let $I \overset{d}{=} A \cap Ig^{(\mathfrak{C})}\{v\}$ where $v = \{q \in {}^\alpha\varkappa :$ $: H1q \subseteq \bar{0}\}$. Clearly $\{x-d_{ij}, z-d_{ij}, x-y, z-y : i,j \in H\} \subseteq I$ and $\{x,z\} \cap$ $\cap I = 0$ since $(\forall w \in I)(\exists \Gamma \subseteq_\omega \alpha) w \subseteq \{q \in {}^\alpha\varkappa : (H\sim\Gamma)1q \subseteq \bar{0}\}$. Thus $\mathfrak{A}/I$ and $\{x/I, y/I, z/I\}$ satisfy the hypotheses of Lemma 5.7.2 which then yields $\mathfrak{A}/I \in \mathsf{I}\,{}_\varkappa Cs_\alpha \sim \mathsf{I}\,(Gws_\alpha^{comp})^{reg}$. So far, (iii) has been proved.

Let $V \overset{d}{=} {}^\alpha\varkappa^{(\bar{0})}$ and $\mathfrak{B} \overset{d}{=} \mathfrak{Rl}_V \mathfrak{A}$. Then $\mathfrak{B} \in Ws_\alpha \cap Dc_\alpha$ and $\mathbf{H}\mathfrak{B} \subseteq Dind_\alpha$ by $\mathfrak{B} \in \mathbf{H}\mathfrak{A}$. Let $J = B \cap Ig^{(\mathfrak{Cb}\,V)}\{V \cap v\}$. Again, $\mathfrak{B}/J$ and $\{(x \cap V)/J,$ $(y \cap V)/J, (z \cap V)/J\}$ satisfy the hypotheses of 5.7.2. QED(Theorem 5.7.)

Remark 5.8. Let $\varkappa \geq 2$ and $\alpha \geq \omega$. Statements (i)-(v) below hold by [HMTI]5 and the present section. Corollary 5.4 and Theorem 5.7 here imply (1)-(2) below.

(1) Statements (i)-(iii) below become false if either Lf is replaced by Dc or $Cs^{reg}$ is replaced by Cs.

(2) (iv) becomes false if Lf is replaced by Dc but it remains true if $Cs^{reg}$ is replaced by Cs.

(i) $\mathbf{H}(\ Cs_\alpha^{reg} \cap Lf_\alpha) \subseteq \mathbf{I}\ Cs_\alpha$.

(ii) $\mathbf{H}({}_\varkappa Cs_\alpha^{reg} \cap Lf_\alpha) \subseteq \mathbf{I}\ {}_\varkappa Cs_\alpha \cup {}_0Cs_\alpha$.

(iii) $Cs_\alpha \cap \mathbf{H}({}_\varkappa Cs_\alpha^{reg} \cap Lf_\alpha) \subseteq \mathbf{I}\ {}_\varkappa Cs_\alpha \cup {}_0Cs_\alpha$ if $\varkappa \geq \omega$.

(iv) $Dind_\alpha \cap \mathbf{H}(\ Cs_\alpha^{reg} \cap Lf_\alpha) \subseteq \mathbf{I}\ Cs_\alpha$.

(v) $Cs_\alpha \cap \mathbf{H}\ {}_\varkappa Cs_\alpha \subseteq \mathbf{I}\ {}_\varkappa Cs_\alpha \cup {}_0Cs_\alpha$ iff $\varkappa < \omega$.

Problem 5.9. Let $\alpha \geq \omega$. We know that the inclusions indicated on Figure 5.10 are not equalities, except those indicated by question marks, where we do not have counterexamples. How do Figures [HMTI]5.7, 6.9 and 6.10 look like if $\mathbf{I}\,Gws_\alpha^{comp\ reg}$ is included?

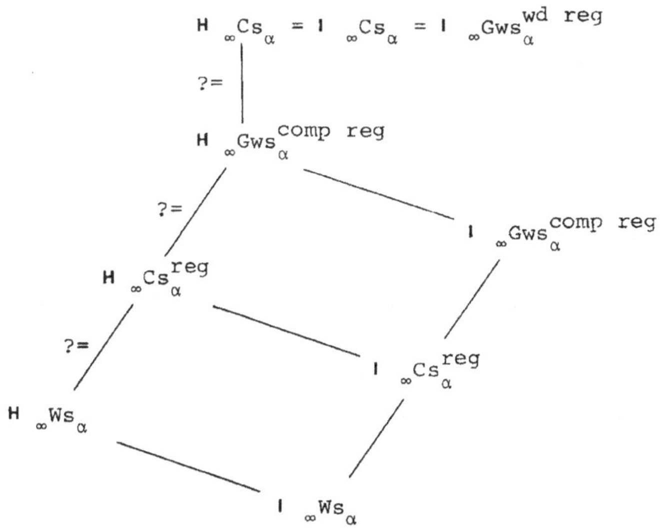


Figure 5.10. ($\alpha \geq \omega$)

Problem 5.11. Let $2\le\alpha$. Let $K\in\{CA_\alpha, Gs_\alpha\}$. Is every epimorphism in K surjective? That is, are there $\mathfrak{B}\subseteq\mathfrak{A}\in K$ such that $(\forall\mathfrak{L}\in K)(\forall f,h\in Hom(\mathfrak{A},\mathfrak{L}))[B1f\subseteq h \Rightarrow f=h]$ but $B\neq A$?

We note that this problem is equivalent to a problem in definability theory.

## 6. Products

For some of our purposes the original definitions of $CA_\alpha$ and related notions are too restrictive. The restriction lies in the assumption that $\alpha$ is an ordinal instead of an arbitrary set.

Definition 6.0.

(i) Let H be a set and let $h : H \to \alpha$ and $\mathfrak{A}$ be an algebra similar to $CA_\alpha$-s. We define $\mathfrak{Rd}^{(h)}\mathfrak{A} \stackrel{d}{=} \langle A,+^{\mathfrak{A}},\cdot^{\mathfrak{A}},-^{\mathfrak{A}},0^{\mathfrak{A}},1^{\mathfrak{A}},c^{\mathfrak{A}}_{h(i)},d^{\mathfrak{A}}_{h(i)h(j)}\rangle_{i,j\in H}$. Let $n : H \rightarrowtail\!\!\!\!\to |H|$. Then $CA_H \stackrel{d}{=} \{\mathfrak{Rd}^{(n)}\mathfrak{A} : \mathfrak{A}\in CA_{|H|}\}$.

(ii) Let H and $L\subseteq T$ be three sets, $h : H \to T$ and let $\mathfrak{A}$ be an algebra similar to $CA_T$-s. We define $\mathfrak{Rd}^{(h)}\mathfrak{A}$ to be the same as in (i) above with the obvious changes. $\mathfrak{Rd}_L\mathfrak{A} \stackrel{d}{=}$ $\stackrel{d}{=} \mathfrak{Rd}^{(L1Id)}\mathfrak{A}$.

(iii) Let H be a set and $n : H \rightarrowtail\!\!\!\!\to |H|$. Recall the function $rd^{(n)}$ from 4.7.1.1. We note that 4.7.1.1 and 4.7.1.2 apply to the present generality since there we did not assume that $\beta$ is an ordinal. Let $K\in\{Ws,Cs,Gs,Gws,Gws^{norm},Gws^{comp},Gws^{wd},Crs^{creg},Crs\}$. We define

$$K_H \stackrel{d}{=} \{(rd^{(n)})^*\mathfrak{Rd}^{(n)}\mathfrak{A} : \mathfrak{A}\in K_{|H|}\}.$$

$K^{reg}_H \stackrel{d}{=} K_H\cap Crs^{creg}_H$ if $K_H\subseteq Gws_H$. See 1.6.1-1.6.2.

(iv) Related notions like $Bo_H$, $Nr_H$ etc. are defined analogously.

Remark Correctness of the above definition follows from [HMT]2.6.2 and section Reducts. Namely: Let $K$ be as in (iii). The definition of $K_H$ and $CA_H$ is independent of the choice of the enumeration $n : H \rightarrowtail |H|$ since for all ordinals $\alpha,\beta$ and $\rho : \alpha \rightarrowtail \beta$ we have $K_\alpha = \{rd^{(\rho)*}\mathfrak{Rl}^{(\rho)}\mathfrak{A} : \mathfrak{A}\in K_\beta\}$ by section Reducts.
We note that for any $Crs_H\ \mathfrak{A}$, $1^{\mathfrak{A}} \subseteq {}^H base(\mathfrak{A})$.

[HMTI]6.2 gives a natural subdirect decomposition of $Gs_\alpha$-s into $Cs_\alpha$-s. Let $\mathfrak{A}\in Gs_\alpha$, then $\langle rl(^\alpha U) : U\in Subb(\mathfrak{A})\rangle$ is a subdirect decomposition of $\mathfrak{A}$ by [HMTI]6.2 and clearly $rl(^\alpha U)^*\mathfrak{A} \in Cs_\alpha$ for all $U\in Subb(\mathfrak{A})$. In view of [HMTI]1.15 one might be tempted to think that if $\mathfrak{A}\in Gs_\alpha^{reg}$ then the natural subdirect decomposition of $\mathfrak{A}$ yields $Cs_\alpha^{reg}$-s or at least some of the natural subdirect factors $rl(^\alpha U)^*\mathfrak{A}$, $U\in Subb(\mathfrak{A})$ will be <u>regular</u>. Thm 6.1 below states that this is very far from being true. Thus by the Fact below, regularity is a property different from the other properties of $Gws_\alpha$-s introduced in [HMTI]1.

<u>Fact:</u> Let $\mathfrak{A}\in K\in\{Gs_\alpha, Cs_\alpha, Ws_\alpha, Gws_\alpha, Gws_\alpha^{wd}, Gws_\alpha^{norm}, Gws_\alpha^{comp}\}$. Then $rl(^\alpha U)^*\mathfrak{A}\in K$ for all $U\in Subb(\mathfrak{A})$.

<u>Proof:</u> Obvious by the definitions. <u>QED</u>

<u>Theorem 6.1.</u> Let $\alpha\geq\omega$ and $\varkappa\geq 2$. There is an $\mathfrak{A}\in {}_\varkappa Gs_\alpha^{reg}$ for which (i)-(iii) below hold.

(i) $rl(^\alpha U)^*\mathfrak{A} \notin I\,Cs_\alpha^{reg}$ for all $U\in Subb(\mathfrak{A})$.

(ii) $rl(^\alpha U)^*\mathfrak{A} \notin Dind_\alpha$ for all $U\in Subb(\mathfrak{A})$.

(iii) $rl(W)^*\mathfrak{A} \notin Cs_\alpha^{reg}$, moreover $rl(W)^*\mathfrak{A} \notin Cs_\alpha\cap Dind_\alpha$ for any non-empty $W\subseteq 1^{\mathfrak{A}}$ such that $rl_W\in Ho\,\mathfrak{A}$.

<u>roof.</u> Let $\alpha\geq\omega$ and $\varkappa\geq 2$. For any set $s$ let $\bar{s} \overset{d}{=} \langle s : i<\alpha\rangle$.
: $\overset{d}{=} Sb_\omega\alpha$. For any $H\in I$ we define

$$_H \overset{d}{=} \{q\in {}^\alpha(\varkappa\times\{H\})^{(\overline{\langle n,H\rangle})} : n\in\varkappa \text{ and } H\times\{\langle n,H\rangle\}\subseteq q\}.$$

$y \overset{d}{=} \cup\{x_H : H\in I\}$, $V \overset{d}{=} \cup\{{}^{\alpha}(\varkappa\times\{H\}) : H\in I\}$ and $\mathfrak{A} \overset{d}{=} \mathfrak{Sg}^{(\mathfrak{Sb}V)}\{y\}$.

First we show that $\mathfrak{A}$ is regular. Let $J \overset{d}{=} Ig^{(\mathfrak{A})}\{y\}$.

<u>Claim 6.1.1.</u> $|\alpha\sim\Delta z|<\omega$ for every $z\in J$, $z\neq 0$.

<u>Proof.</u> Let $z\in J\sim\{0\}$. Then $z\subseteq c_{(\Gamma)}y$ for some $\Gamma\subseteq_{\omega}\alpha$. Let this $\Gamma$ be fixed. Since $z\neq 0$ there is $q\in z$. Then $q\in c_{(\Gamma)}x_H$ for some $H\in I$ by infinite additivity of $c_{(\Gamma)}$. Let this $q$ and $H$ be fixed. There exists a finite $L\subseteq\alpha$ such that $\Gamma\cup H\subseteq L$ and $z\in Sg^{(\mathfrak{Rd}_L\mathfrak{A})}\{y\}$. We show that $\Delta z\supseteq\alpha\sim L$. Let $i\in\alpha\sim L$ and $K \overset{d}{=} H\cup\{i\}$.

<u>Claim 6.1.1.1.</u> $z\cap c_{(\Gamma)}x_K\neq 0$.

<u>Proof.</u> Let $\rho \overset{d}{=} L1Id$ and let the function $rb^{\rho} : V \to Rg\,rb^{\rho}$ be defined as in 4.7.1.1. <u>Notation:</u> $rb^{\rho} \overset{d}{=} rb_L$ and $rd^{\rho} \overset{d}{=} rd_L$. I.e. $rb_L = \langle\, \langle (f_i,(\alpha\sim L)1f) : i\in L\rangle : f\in V\rangle$ and $rd_L = rb_L{}^{*}$. By 4.7.1.2 then $rd_L \in Ism(\mathfrak{Rd}_L\mathfrak{A}, \mathfrak{Sb}\,rd_L V)$. Let $\mathfrak{B} \overset{d}{=} \mathfrak{Sg}^{(\mathfrak{Rd}_L\mathfrak{A})}\{y\}$ and $\mathfrak{N} \overset{d}{=}$ $\overset{d}{=} rd_L{}^{*}\mathfrak{B}$. Then $z\in B$ and $rd_L\in Is(\mathfrak{B},\mathfrak{N})$. Recall that $q\in z\cap c_{(\Gamma)}x_H$. Let $n\in\varkappa$ be such that $q\in{}^{\alpha}(\varkappa\times\{H\})^{(\overline{\langle n,H\rangle})}$. Let $g \overset{d}{=} \langle\langle q_j(0),K\rangle : j<\alpha\rangle$ and $p \overset{d}{=} g^{i}_{\langle n,K\rangle}$. Then $g\in c_i c_{(\Gamma)}x_K$ and $p\in c_{(\Gamma)}x_K$. We show that $p\in z$. Let $Y = (\varkappa\times\{H\})\times\{(\alpha\sim L)1q\}$, $Z = (\varkappa\times\{K\})\times\{(\alpha\sim L)1p\}$. Then $Y,Z\in Subb(\mathfrak{N})$ by 4.7.1.2(ii). Let $y^{+} = rd_L y$. Then by $H=H\cap L=K\cap L$ we have $y^{+}\cap{}^{L}Y =$ $= \{h\in{}^{L}Y : (\forall j\in H)h_j=\langle\langle n,H\rangle,(\alpha\sim L)1q\rangle\}$ and $y^{+}\cap{}^{L}Z = \{h\in{}^{L}Z : (\forall j\in H)h_j=$ $=\langle\langle n,K\rangle,(\alpha\sim L)1p\rangle\}$. Let $W=base(\mathfrak{N})$. Let $k : W \rightarrowtail\!\!\!\twoheadrightarrow W$ be such that $k\circ k=W1Id$ and $(\forall m\in\varkappa)k(\langle m,H\rangle,(\alpha\sim L)1q)=(\langle m,K\rangle,(\alpha\sim L)1p)$ and $(W\sim(Y\cup Z))1k\subseteq$ $\subseteq Id$. Let $\mathfrak{L}$ be the full $Gs_L$ with unit $1^{\mathfrak{N}}$. Then $\tilde{k}\in Is(\mathfrak{L},\mathfrak{L})$ is a base-automorphism of $\mathfrak{L}$. We have $\tilde{k}(y^{+})=y^{+}$ because $\tilde{k}(y^{+}\cap{}^{L}Y)=y^{+}\cap{}^{L}Z$ and $C1(\tilde{k}\circ\tilde{k})\subseteq Id$, $(W\sim(Y\cup Z))1k\subseteq Id$. Then $N1\tilde{k}\subseteq Id$ since $N=Sg^{(\mathfrak{L})}\{y^{+}\}$. Thus $\tilde{k}(rd_L(z))=rd_L(z)$. Since $q\in z$ we have $rb_L(q)\in rd_L(z)$. Then by $i\notin L$, $rb_L(p) = k\circ rb_L(q) \in \tilde{k}(rd_L(z)) = rd_L(z)$. Since $rb_L$ is one-to-one on $V$ by 4.7.1.2(i), $rb_L(p)\in rd_L(z)$ implies $p\in z$. Now $p\in z\cap c_{(\Gamma)}x_K$ by $p\in c_{(\Gamma)}x_K$.

<u>QED(Claim 6.1.1.1.)</u>

Let $p\in z\cap c_{(\Gamma)}x_K$. Then $p\in{}^{\alpha}(\varkappa\times\{K\})^{(\overline{\langle n,K\rangle})}$ for some $n\in\varkappa$. Let

$m\in\varkappa\sim\{n\}$. By $i\in K\sim\Gamma$ we have $p^{i}_{\langle m,K\rangle}\notin c_{(\Gamma)}y$. Then $p\in z$ but $p^{i}_{\langle m,K\rangle}\notin z$ by $z\subseteq c_{(\Gamma)}y$. Hence $i\in\Delta z$. Since $i\in\alpha\sim L$ was chosen arbitrarily we have proved $\Delta z\supseteq\alpha\sim L$ and hence $|\alpha\sim\Delta z|\le|L|<\omega$.

QED(Claim 6.1.1.)

By Claim 6.1.1, 4.7.2.1 and 1.3.6 we have that $\mathfrak{A}$ is regular since $A=Sg\{y\}$ and $y$ is regular.

Next we show that $\mathfrak{A}$ satisfies (i)-(iii). Let $U\in Subb(\mathfrak{A})$. Then $U=\varkappa\times\{H\}$ for some $H\subseteq_{\omega}\alpha$. Let $\mathfrak{R}\overset{d}{=}\mathfrak{Rl}({}^{\alpha}U)\mathfrak{A}$. Then $x_H=y\cap{}^{\alpha}U\in R$ and $c^{\mathfrak{R}}_{(H)}x_H=\cup\{{}^{\alpha}(\varkappa\times\{H\})^{(\overline{\langle n,H\rangle})}: n\in\varkappa\}\in Zd\mathfrak{R}\sim\{0,{}^{\alpha}U\}$. We have seen that $\mathfrak{A}$ satisfies (ii). (i) is a consequence of (ii). Let $W\subseteq 1^{\mathfrak{A}}$, $W\neq 0$. Suppose that $rl_W\in Ho\,\mathfrak{A}$ and $\mathfrak{R}\overset{d}{=}rl_W^{*}\mathfrak{A}\in Cs_{\alpha}$. Then $W={}^{\alpha}Y$ for some $Y\subseteq | \in Subb(\mathfrak{A})$. Then $|Y|\ge 2$ since $\mathfrak{R}\models c_0\text{-}d_{01}=1$ by $\mathfrak{A}\models c_0\text{-}d_{01}=1$ and $rl_W\in Ho\,\mathfrak{A}$. Let $Y\subseteq\varkappa\times\{H\}$ and $z\overset{d}{=}c^{\mathfrak{R}}_{(H)}(W\cap y)$. Then $z=\cup\{{}^{\alpha}Y^{(\bar{t})}: t\in Y\}$, and since $|Y|\ge 2$ we have $z\notin\{0,W\}$, $\Delta^{(\mathfrak{R})}z=0$. Thus $rl_W^{*}\mathfrak{A}\notin Dind_{\alpha}\cap Cs_{\alpha}\supset Cs^{reg}_{\alpha}$.

QED(Theorem 6.1.)

Proposition 6.2(i)-(ii) below was quoted in [HMTI]6.8(3),(11) and in [HMTI]6.10.

Proposition 6.2. Let $\varkappa\ge 2$ and $\alpha\ge\omega$.

(i) ${}_{\varkappa}Cs_{\alpha}\cap Lf_{\alpha}\ \not\subseteq\ \mathbf{P}\,Dind_{\alpha}$.

(ii) ${}_{\varkappa}Cs^{reg}_{\alpha}\cap Dc_{\alpha}\ \not\subseteq\ \mathbf{P}\,Ws_{\alpha}$.

(iii) $\mathbf{H}({}_{\varkappa}Cs^{reg}_{\alpha}\cap Dc_{\alpha})\ \not\subseteq\ \mathbf{P}\,Gws^{comp\ reg}_{\alpha}$.

(iv) $\mathbf{H}({}_{\varkappa}Cs^{reg}_{\alpha}\cap Dc_{\alpha})\ \not\subseteq\ \mathbf{P}\,Cs_{\alpha}$ if $\varkappa<\omega$.

(v) $\mathbf{H}({}_{\varkappa}Ws_{\alpha}\cap Dc_{\alpha})\ \not\subseteq\ \mathbf{P}\,Gws^{comp\ reg}_{\alpha}$.

(vi) $\mathbf{H}\,{}_{\infty}Ws_{\alpha}\ \not\subseteq\ \mathbf{P}\,{}_{\infty}Cs^{reg}_{\alpha}$.

(vii) ${}_{\infty}Cs^{reg}_{\alpha}\ \not\subseteq\ \mathbf{P}\,{}_{\infty}Ws_{\alpha}$.

Proof. Let $\varkappa\ge 2$ and $\alpha\ge\omega$. Let $\mathfrak{L}\overset{d}{=}\mathfrak{Gb}^{\alpha}\varkappa$. Proof of (i): Let $\mathfrak{B}\subseteq\mathfrak{Zd}\,\mathfrak{L}$ be an atomless BA. (Such a $\mathfrak{B}$ exists.) Let $\mathfrak{A}\overset{d}{=}\mathfrak{Sg}^{(\mathfrak{L})}B$.

Then $\mathfrak{A} \in Cs_\alpha \cap Lf_\alpha$ and $Zd\mathfrak{A} = B$ by e.g. [HMT]2.2.24(iii). Now $\mathfrak{A} \notin$ $\notin \mathbf{P}Dind_\alpha$ since $\mathfrak{Z}\mathfrak{N}$ is atomic for every $\mathfrak{N} \in \mathbf{P}Dind_\alpha$.

In the rest of the proof we shall use the following fact.

<u>Fact 6.2.1.</u> Let $K,L \subseteq CA_\alpha$ be such that $Dind_\alpha \cap K \not\subseteq L \cup I\,{}_0Cs_\alpha$. Then $K \not\subseteq$ $\not\subseteq \mathbf{P}L$.

<u>Proof.</u> Let $\mathfrak{A} \in Dind_\alpha \sim I\,{}_0Cs_\alpha$. Then $\mathfrak{A}$ is directly indecomposable and hence $\mathfrak{A} \notin L$ implies $\mathfrak{A} \notin \mathbf{P}L$. <u>QED(Fact 6.2.1.)</u>

<u>Proof of (ii):</u> Let $H \subseteq \alpha$ be such that $|H| \cap |\alpha \sim H| \geq \omega$. Let $x_n \stackrel{d}{=}$ $\stackrel{d}{=} \{q \in {}^\alpha\varkappa : (\forall i \in H) q_i = n\}$, for all $n \in \varkappa$. Let $\mathfrak{A} \stackrel{d}{=} \mathfrak{Sg}^{(\mathfrak{L})}\{x_0, x_1\}$. Then $\mathfrak{A} \in {}_\varkappa Cs_\alpha^{reg} \cap Dc_\alpha$ by 1.3 and $|\alpha \sim H| \geq \omega$. $\mathfrak{A} \notin I\,Ws_\alpha$ since $x_0 \cap x_1 = 0$, $(\forall i,j \in H) x_0 \cup x_1 \subseteq d_{ij}$ and $|H| \geq \omega$. Now $\mathfrak{A} \notin \mathbf{P}Ws_\alpha$ by Fact 6.2.1 since $|A| > 1$ and $Cs_\alpha^{reg} \subseteq Dind_\alpha$.

By Fact 6.2.1, (iii),(v) are corollaries of 5.7(iii),(iv); and (iv) is a corollary of 5.4(iv). (vi) and (vii) follow from (v) and (ii) respectively, by choosing $\varkappa \geq \omega$.

<u>QED(Proposition 6.2.)</u>

The following theorem was quoted in [HMTI]6.16(7).

<u>Theorem 6.3.</u> Let $\varkappa \geq 2$ and $\alpha \geq \omega$. Then some <u>weakly</u> subdirectly indecomposable ${}_\varkappa Cs_\alpha^{reg} \cap Dc_\alpha$ is <u>not</u> subdirectly indecomposable.

<u>Proof.</u> Let $\varkappa \geq 2$ and $\alpha \geq \omega$. Let $\langle H_n : n \in \omega \rangle \in {}^\omega(Sb\alpha)$ be a system of mutually disjoint infinite subsets of $\alpha$ such that $|\alpha \sim \bigcup\{H_n : n \in \omega\}| \geq \omega$. For every $n \in \omega$ we let

$$x_n \stackrel{d}{=} \{f \in {}^\alpha\varkappa : H_n 1 f \subseteq \bar{0}\}, \quad \text{where } \bar{0} \stackrel{d}{=} \alpha \times 1.$$

Let $G \stackrel{d}{=} \{x_n : n \in \omega\}$ and $\mathfrak{A} \stackrel{d}{=} \mathfrak{Sg}^{(\mathfrak{Sb}\,{}^\alpha\varkappa)}G$. For every $n \in \omega$ we have $\Delta(x_n) = H_n$. Thus $\mathfrak{A} \in Dc_\alpha$ by $|\alpha \sim \bigcup\{H_n : n \in \omega\}| \geq \omega$ and by [HMT]2.1.7. Let $Q \stackrel{d}{=} {}^\alpha\varkappa^{(\bar{0})}$. We show that $\mathfrak{A}$, $G$ and $Q$ satisfy the conditions of 4.7.2. Let $z \in Ig^{(\mathfrak{A})}G$, $z \neq 0$. Then there is $n \in \omega \sim 1$ such that

$z \in Ig^{(\mathfrak{A})}\{x_i : i<n\}$ and $z \in Sg^{(\mathfrak{A})}\{x_i : i<n\}$. Let $y \overset{d}{=} \cup\{x_i : i<n\}$. Let $i<n$ and $\Theta \subseteq_\omega \alpha$ be such that $(\forall j<n)[H_j \cap \Theta = 0$ iff $j=i]$. Then $x_i = c^\partial_{(\Theta)} y$. Thus $Sg^{(\mathfrak{A})}\{x_i : i<n\} = Sg^{(\mathfrak{A})}\{y\}$. Let $\mathfrak{B} \overset{d}{=} \mathfrak{S}\mathfrak{g}\{y\}$. Then $z \in Ig^{(\mathfrak{B})}\{y\}$. We show that $\mathfrak{B}, \{y\}$ and $Q \cap y$ satisfy the conditions of 4.7.1. (i) is satisfied since $|\{y\}|=1$. Let $f \in y$ and $\Gamma \subseteq_\omega \alpha$. Let $\Theta \subseteq_\omega \cup\{H_i : i<n\}$ be such that $\Theta \cap \Gamma = 0$ and $(\forall i<n)[(H_i \sim \Gamma)1f \subseteq \bar{0}$ iff iff $(H_i \cap \Theta)1f \subseteq \bar{0}]$. Such a $\Theta$ exists. Let $q \overset{d}{=} \bar{0}[\Theta/f]$. Clearly $q \in Q \cap y$, and for any $p \in {}^\Gamma\varkappa$, $f[\Gamma/p] \in y$ iff $q[\Gamma/p] \in y$. We have seen that (ii) of 4.7.1 is satisfied. By $z \in Ig^{(\mathfrak{B})}\{y\}$ we have $z \cap c_{(\Gamma)} y \neq 0$ for some $\Gamma \subseteq_\omega \alpha$. Thus by 4.7.1(I) we have $z \cap c_{(\Gamma)}(Q \cap y) \neq 0$, i.e. $z \cap Q \neq 0$. We have seen that condition a.) of 4.7.2 is satisfied. Let $M \subseteq \alpha$, $z \in Ig^{(\mathfrak{A})}(G \sim Dm_M)$ and $\Omega \subseteq_\omega \alpha$. Then $z \subseteq c_{(\Gamma)} \Sigma\{x_i : i \in S\}$ for some $\Gamma \subseteq_\omega \alpha$ and $S \subseteq_\omega \omega$ such that $(\forall i \in S)|H_i \sim M| \geq \omega$. Let $\Theta \subseteq_\omega \alpha \sim (M \cup \Omega \cup \Gamma)$ be such that $(\forall i \in S)\Theta \cap H_i \neq 0$. Then $(\forall q) q[\Theta/\bar{1}] \notin z$ where $\bar{1} = \alpha \times \{1\}$. Thus condition b.) of 4.7.2 is satisfied. By this we have seen that all the conditions of 4.7.2 are satisfied. Therefore $\mathfrak{A}$ is regular by 4.7.2(I) since every element of $G$ is regular. $\mathfrak{A}$ is weakly subdirectly indecomposable since $rl_Q \in Is\,\mathfrak{A}$ by 4.7.2(II), thus $\mathfrak{A} \cong rl_Q{}^*\mathfrak{A} \in Ws_\alpha$ and every $Ws_\alpha$ is weakly subdirectly indecomposable by [HMTI]6.13. Next we prove that $\mathfrak{A}$ is not subdirectly indecomposable. Let $y \in A$, $y \neq 0$. Then $y \in Sg^{(\mathfrak{A})}\{x_i : i<n\}$ for some $n<\omega$. Let $N \overset{d}{=} \cup\{H_i : i<n\}$. Then $y \in Dm_N$ by [HMT] 2.1.4. By 4.7.2(III) $y \notin Ig^{(\mathfrak{A})}(G \sim Dm_N)$. Let $m>n$. Then $x_m \in G \sim Dm_N$ and thus $(\forall \Gamma \subseteq_\omega \alpha) y \not\leq c_{(\Gamma)} x_m$. Hence $\mathfrak{A}$ is subdirectly decomposable by [HMT]2.4.44.

QED(Theorem 6.3.)

Remark 6.4. Let $\varkappa \geq 2$ and $\alpha \geq \omega$. By [HMTI]6.15, every nontrivial $Cs_\alpha^{reg}$ is directly indecomposable, in short $Cs_\alpha^{reg} \subseteq Dind_\alpha$. By 4.13 we know that $\mathbf{H}\,{}_\varkappa Cs_\alpha^{reg} \not\subseteq Dind_\alpha$. But we know more: There is $\mathfrak{A} \in Lf_\alpha \cap \mathbf{H}({}_\varkappa Cs_\alpha^{reg} \cap Dc_\alpha)$ having $|{}^{({}^\alpha\varkappa)}2|$ -many different direct factor congruences by 4.13. It remains open whether there is $\mathfrak{A} \in \mathbf{H}\,{}_\varkappa Cs_\alpha^{reg}$ such that

$\mathfrak{A} \cong P\mathcal{L}$ with $\mathcal{L} \in {}^I(CA_\alpha \sim I\,{}_0Cs_\alpha)$ and $|I| \geq |{}^\alpha\varkappa|$, because of the following. Let $\mathfrak{B} \in Lf_\alpha$, $\mathfrak{B} \models c_0 - d_{01} = 1$ and suppose $\mathfrak{B} \cong P\mathcal{L}$ with $\mathcal{L} \in {}^I(CA_\alpha \sim I\,{}_0Cs_\alpha)$. Then $|I| < \omega$ since no infinite direct product of nondiscrete $CA_\alpha$-s is an $Lf_\alpha$ and no member of $(\mathbf{H}\mathfrak{B}) \sim I\,{}_0Cs_\alpha$ is discrete. Note that the algebra $\mathfrak{A} \in Lf_\alpha \cap \mathbf{H}({}_\varkappa Cs_\alpha^{reg} \cap Dc_\alpha)$ constructed in 4.13 is such that $\mathfrak{A} \models c_0 - d_{01} = 1$.

Remark 6.5. (i) below is a generalization of 4.16(i). It could be interesting to replace the condition "simple" in (i) below with a more general but "abstract" condition. However, by (ii), the most obvious candidate for this does not work.

Let $0 < \varkappa < \omega \leq \alpha$. Then (i)-(iii) below hold.

(i) Let $\mathfrak{B} \in {}^\rho\{\mathfrak{A} \in {}_\varkappa Cs_\alpha : \mathfrak{A}$ is simple$\}$. Then $P\mathfrak{B} \in \mathbf{I}Cs_\alpha$ iff $|\rho| \leq 2^{|\alpha|}$.

(ii) There is a subdirectly indecomposable $\mathfrak{A} \in {}_\varkappa Cs_\alpha^{reg}$ such that ${}^{\varkappa+1}\mathfrak{A} \notin \mathbf{I}Cs_\alpha$.

(iii) $\rho \leq \varkappa$ iff for every $\mathfrak{B} \in {}^\rho\{\mathfrak{A} \in {}_\varkappa Cs_\alpha : \mathfrak{A}$ is subdirectly indecomposable$\}$ we have $P\mathfrak{B} \in \mathbf{I}Cs_\alpha$.

Proof. Let $\mathfrak{B} \in {}^\rho({}_\varkappa Cs_\alpha)$ be such that $|\rho| \leq 2^{|\alpha|}$ and $\mathfrak{B}_i$ is simple for every $i < \rho$. We may suppose that $(\forall i < \rho)\ \varkappa = base(\mathfrak{B}_i)$. Let $V \in {}^\rho(Subu({}^\alpha\varkappa))$ be one-one. For every $i < \rho$ we let $\mathfrak{N}_i \overset{d}{=} \mathfrak{Rl}(V_i)\mathfrak{B}_i$. Then $\mathfrak{N}_i \cong \mathfrak{B}_i$ since $\mathfrak{B}_i$ is simple. Then $P\mathfrak{B} \cong P\mathfrak{N} \in \mathbf{I}Gws_\alpha^{comp}$ by [HMTI]6.2. By (1) in the proof of [HMTI]7.17 we have $\mathbf{I}Gws_\alpha^{comp} = \mathbf{I}Cs_\alpha$. Thus $P\mathfrak{B} \in \mathbf{I}Cs_\alpha$. Assume $|\rho| = \rho > |{}^\alpha 2|$ and let $\mathfrak{B} \in {}^\rho({}_\varkappa Cs_\alpha)$. Then $|PB| \geq 2^\rho > |{}^{({}^\alpha\varkappa)}2|$ proving that $P\mathfrak{B} \notin \mathbf{I}\,{}_\varkappa Cs_\alpha$. Thus $P\mathfrak{B} \notin \mathbf{I}Cs_\alpha$ by $\varkappa < \omega \leq \alpha$. Thus (i) is shown. If $\varkappa = 1$ then (ii) is obvious. Let $\varkappa \geq 2$ and let $\mathfrak{A}$ be the subalgebra of $\mathfrak{Sb}\,{}^\alpha\varkappa$ generated by the element $\{\langle 0 : i < \alpha\rangle\}$. By 1.3, $\mathfrak{A} \in {}_\varkappa Cs_\alpha^{reg}$. By the proof of [HMTI]6.16(6), $\mathfrak{A}$ is subdirectly indecomposable. Let $\rho > \varkappa$. There are $\varkappa+1$ disjoint atoms $x_0, \ldots, x_\varkappa$ in ${}^\rho\mathfrak{A}$ such that $(\forall n \leq \varkappa)(\forall i \in \alpha)\, x_n < d_{0i}$ in ${}^\rho\mathfrak{A}$. Thus ${}^\rho\mathfrak{A} \notin \mathbf{I}\,{}_\varkappa Cs_\alpha$ and hence ${}^\rho\mathfrak{A} \notin \mathbf{I}Cs_\alpha$ by $\varkappa < \omega \cap \alpha$. (ii) is proved.

Now (ii) together with [HMTI]7.29 and [HMTI]6.12 proves (iii).
QED(Remark 6.5.)

Problems 6.6. Let $2\le\varkappa<\omega\le\alpha$.

(i) Is $\mathbf{I}\,{}_\varkappa Gs_\alpha = \mathbf{HP}\,{}_\varkappa Cs_\alpha$? (See the proof of [HMTI]6.8(5).)

(ii) How Figure [HMTI]6.9 will look if we replace $K$ with ${}_\varkappa K$ in it for all $K\subseteq Crs_\alpha$ that occur there? Along these lines we note that there is $\mathfrak{A}\in{}_\varkappa Cs_\alpha^{reg}$ such that ${}^2\mathfrak{A}\notin \mathbf{I}\,Cs_\alpha$. Indeed, let $\mathfrak{A}\overset{d}{=}\mathfrak{Sg}^{(\mathfrak{Sb}\,{}^\alpha\varkappa)}At\,\mathfrak{Sb}\,{}^\alpha\varkappa$ and use 1.3.

(iii) Is ${}_\infty Cs_\alpha \subseteq \mathbf{HP}\,{}_\infty Cs_\alpha^{reg}$?

(iv) Is $\mathbf{HP}\,{}_\infty Cs_\alpha^{reg} = \mathbf{H}\,{}_\infty Cs_\alpha^{reg}$?

(v) Find an abstract characterization of $\mathbf{I}\,Cs_\alpha^{reg}$ as a subclass of $\mathbf{I}\,Gs_\alpha$ (or $\mathbf{I}\,Cs_\alpha$).

(vi) Is every weakly subdirectly indecomposable $Gws_\alpha$ (or $Cs_\alpha$) an $\mathbf{I}\,Gws_\alpha^{comp\ reg}$?

(vii) Is $\mathbf{HP}\,Cs_\alpha^{reg} = \mathbf{HP}\,Gws_\alpha^{comp\ reg}$? (Note that $\mathbf{P}\,Gws_\alpha^{comp\ reg} \ne \mathbf{P}\,Cs_\alpha^{reg}$ by 5.6.)

## 7. Ultraproducts

Throughout this section, $\varkappa$ denotes a cardinal.

Definition 7.0. (p.115 of [HMT]) Let $K$ be a class of similar algebras. Then we define

$\mathbf{Uf}\ K \overset{d}{=} \{\mathfrak{A} : (K\cap \mathbf{Up}\{\mathfrak{A}\}) \ne 0\}$ and

$\mathbf{Up}'K \overset{d}{=} \cup\{\mathbf{Up}\{\mathfrak{A}\} : \mathfrak{A}\in K\}$.

That is, $\mathbf{Uf}\ K$ is the class of all "ultraroots" or "ultra-factors" of members of $K$, and $\mathbf{Up}'K$ is the class of all ultrapowers of members of $K$.

<u>Theorem 7.1.</u> Let $\alpha\geq\omega$. For every cardinal $\varkappa$ let $\mathfrak{Rf}(\varkappa)$ denote the greatest regular locally finite-dimensional subalgebra of $\mathfrak{Cs}\,{}^{\alpha}\varkappa$. (It exists by [HMTI]4.1.)

(i) $\mathbf{I}\,{}_{\infty}Cs_{\alpha} = \mathbf{SUp}\{\mathfrak{Rf}(\omega)\} = \mathbf{SUp}({}_{\varkappa}Cs_{\alpha}^{reg}\cap Lf_{\alpha})$ for every $\varkappa\geq\omega$ and $\mathbf{I}\,{}_{\varkappa}Gs_{\alpha} = \mathbf{SUp}\{\mathfrak{Rf}(\varkappa)\}$ if $\varkappa\in\omega\sim 2$.

(ii) $\mathbf{SUp}\,Cs_{\alpha} = \mathbf{SUp}'(Cs_{\alpha}^{reg}\cap Lf_{\alpha}) = \mathbf{HSUp}\,Cs_{\alpha} = \mathbf{HSUp}\,Ws_{\alpha} =$
$= \mathbf{I}\{\mathfrak{A}\in Gs_{\alpha} : \mathfrak{A}$ has characteristic $\varkappa\neq 1$ or $|A|\leq 2\}$.

(iii) $\mathbf{SUp}\,Ws_{\alpha} = \mathbf{SUp}'(Ws_{\alpha}\cap Lf_{\alpha}) = \mathbf{SUp}\,Cs_{\alpha} \sim \mathbf{I}\,{}_{0}Cs_{\alpha}$.

(iv) $\mathbf{SUp}'\,{}_{\varkappa}Cs_{\alpha} = \mathbf{SUp}'\,{}_{\lambda}Cs_{\alpha} = \mathbf{SUp}'\,{}_{\infty}Cs_{\alpha}$ for all infinite $\varkappa,\lambda$.

Let $\alpha\geq\omega$. To prove Thm 7.1 we shall need the following lemmas.

<u>Lemma 7.1.1.</u> $\mathbf{P}\{\mathfrak{Rf}(\varkappa)\} \subseteq \mathbf{SUp}\{\mathfrak{Rf}(\varkappa)\}$ for every $\varkappa\geq 2$.

<u>Proof.</u> Let $\varkappa\geq 2$ and $\alpha\geq\omega$. Let $\mathfrak{R} \overset{d}{=} \mathfrak{Rf}(\varkappa)$. By [HMT]0.3.72(i) and 0.3.9(vi) we have $\mathbf{P}\{\mathfrak{R}\} \subseteq \mathbf{SUp}\{{}^{\alpha}\mathfrak{R}\}$. Therefore it is enough to show ${}^{\alpha}\mathfrak{R} \in \mathbf{SUp}\{\mathfrak{R}\}$ since $\mathbf{SUp}$ is a closure operator by [HMT]0.3.70(i). Let $I \overset{d}{=} Sb_{\omega}\alpha$. Let $F$ be an ultrafilter on $I$ such that $(\forall\Gamma\in I)$ $\{\Delta\in I : \Gamma\subseteq\Delta\}\in F$. Let $g : I \to I$, $f : I \to \alpha$ and $s : I \to Rgs$ be such that $\Gamma\cap g(\Gamma)=0$, $f(\Gamma)\in\Gamma$ and $s(\Gamma) : \Gamma \rightarrowtail g(\Gamma)$ is one-to-one for every $\Gamma\in I\sim\{0\}$. Let $\Gamma\in I$. For every $i\in\Gamma\sim\{f(\Gamma)\}$ we define
$z_{i\Gamma} \overset{d}{=} \{q\in{}^{\alpha}\varkappa : (\forall j\in g(\Gamma))[q_j=0 \text{ iff } j=s(\Gamma)i]\}$.
$z_{f(\Gamma)\Gamma} \overset{d}{=} {}^{\alpha}\varkappa\sim\bigcup\{z_{i\Gamma} : i\in\Gamma\sim\{f(\Gamma)\}\}$ and
$z_{i\Gamma} \overset{d}{=} 0$ for every $i\in\alpha\sim\Gamma$. Clearly, $(\forall i\in\alpha)z_{i\Gamma}\in R$. Let $i\in\alpha$. Thus we have $z_i \overset{d}{=} \langle z_{i\Gamma} : \Gamma\in I\rangle$. By this we have defined $z\in{}^{\alpha}({}^{I}R)$.

<u>Claim 1</u>

(i) $\sum\{z_i : i<\alpha\} = 1^{({}^{I}\mathfrak{R})}$.

(ii) $z_i\cdot z_j = 0$ for every $i<j<\alpha$.

(iii) $\sum\{z_i\cdot x_i : i<\alpha\}$ exists in ${}^{I}\mathfrak{R}$ for every $x\in{}^{\alpha}({}^{I}R)$ and $\sum\{z_i\cdot x_i : i<\alpha\} = \langle\sum\{z_{i\Gamma}\cdot x_{i\Gamma} : i\in\Gamma\} : \Gamma\in I\rangle$.

(iv) $z_i/F \neq 0$ and $\Delta(z_i/F) = 0$ for every $i\in\alpha$.

(v) $\Delta(z_{i\Gamma})\cap\Gamma = 0$ for every $i\in\alpha$ and $\Gamma\in I$.

Proof. Let $\Gamma\in I$. By the construction of $z$ we have that $z_{i\Gamma}\neq 0$ iff $i\in\Gamma$, and $\sum\{z_{i\Gamma} : i\in\Gamma\} = {}^{\alpha}\kappa$. This implies that $z_i/F \neq 0$ and also that (i) and (iii) hold. Also $z_{i\Gamma}\cdot z_{j\Gamma}=0$ for $i\neq j$ since $s(\Gamma)$ is one--one, and $\Delta(z_{i\Gamma})\cap\Gamma = 0$ by $\Delta(z_{i\Gamma}) \subseteq g(\Gamma)$ and $\Gamma\cap g(\Gamma) = 0$. These facts prove (ii), (v). $\Delta(z_i/F)=0$ follows from (v) since $(\forall i<\alpha)$ $\{\Gamma\in I : i\in\Gamma\}\in F$.

QED(Claim 1)

Notation: $\bar{H} \overset{d}{=} \langle H : \Gamma\in I\rangle$ for every set $H$.

Define the following mappings: $d \overset{d}{=} \langle\langle \overline{y_i} : i<\alpha\rangle : y\in{}^{\alpha}R\rangle$, $h \overset{d}{=} \langle \sum\{z_i\cdot x_i : i<\alpha\} : x\in{}^{\alpha}({}^{I}R)\rangle$, $e \overset{d}{=} h\circ d$ and $E \overset{d}{=} \bar{F}^{\star}\circ e = \langle e(y)/\bar{F} : y\in{}^{\alpha}R\rangle$. Then $E : {}^{\alpha}R \to {}^{I}R/\bar{F}$.

Claim 2 $E \in \mathrm{Ism}({}^{\alpha}\mathfrak{R}, {}^{I}\mathfrak{R}/F)$.

Proof. 1. Case of cylindrifications: Let $i<\alpha$ and $y\in{}^{\alpha}R$. We have to show $c_iE(y) = E(c_iy)$. By using Claim 1, $c_ie(y) = c_i\langle \sum\{z_{j\Gamma}\cdot y_j : j\in\Gamma\} : \Gamma\in I\rangle = \langle c_i\sum\{z_{j\Gamma} : j\in\Gamma\} : \Gamma\in I \text{ and } i\notin\Gamma\rangle\cup\langle\sum\{z_{j\Gamma}\cdot c_iy_j : j\in\Gamma\} : \Gamma\in I \text{ and } i\in\Gamma\rangle \in \langle\sum\{z_{j\Gamma}\cdot c_iy_j : j\in\Gamma\} : \Gamma\in I\rangle/\bar{F} = e(c_iy)/\bar{F} = E(c_iy)$. Thus $c_iE(y) = c_ie(y)/\bar{F} = E(c_iy)$.

2. Case of diagonal elements: Let $k,n\in\alpha$. $E(d_{kn}^{({}^{\alpha}\mathfrak{R})}) = \sum\{z_i\cdot d_{kn}^{({}^{I}\mathfrak{R})} : i<\alpha\}/\bar{F} = (d_{kn}^{({}^{I}\mathfrak{R})}\cdot\sum\{z_i : i<\alpha\})/\bar{F} = d_{kn}^{({}^{I}\mathfrak{R})}/\bar{F} = d_{kn}^{({}^{I}\mathfrak{R}/F)}$.

3. Case of Boolean operations: Recall that $E = \bar{F}^{\star}\circ h\circ d$. Clearly, $d\in\mathrm{Hom}({}^{\alpha}\mathfrak{R}, {}^{\alpha}({}^{I}\mathfrak{R}))$ and $\bar{F}^{\star}\in\mathrm{Hom}({}^{I}\mathfrak{R}, {}^{I}\mathfrak{R}/F)$. We show that $h\in\mathrm{Hom}(\mathfrak{Bl}\,{}^{\alpha}({}^{I}\mathfrak{R}), \mathfrak{Bl}\,{}^{I}\mathfrak{R})$. Let $\mathfrak{B} \overset{d}{=} \mathfrak{Bl}\,{}^{I}\mathfrak{R}$. By Claim 1, $z\in{}^{\alpha}B$ is an antichain in $\mathfrak{B}$ such that $\sum\{z_i : i<\alpha\} = 1^{\mathfrak{B}}$ and $\sum\{z_i\cdot y_i : i<\alpha\}$ exists in $\mathfrak{B}$ for every $y\in{}^{\alpha}B$. Then it is immediate by basic Boolean algebra theory that $h \overset{d}{=} \langle\sum\{z_i\cdot y_i : i<\alpha\} : y\in{}^{\alpha}B\rangle\in\mathrm{Hom}({}^{\alpha}\mathfrak{B}, \mathfrak{B})$ (e.g. it follows from [HMT]2.4.7 and 0.3.6(i)). Now ${}^{\alpha}\mathfrak{B} = {}^{\alpha}\mathfrak{Bl}\,{}^{I}\mathfrak{R} = \mathfrak{Bl}\,{}^{\alpha}({}^{I}\mathfrak{R})$ completes the proof.

We have seen that $E\in\mathrm{Hom}({}^{\alpha}\mathfrak{R}, {}^{I}\mathfrak{R}/F)$. Let $y\in{}^{\alpha}R$, $y \neq \langle 0 : i<\alpha\rangle$.

Then $(\exists i<\alpha) y_i \neq 0$. Let $\Delta \stackrel{d}{=} \Delta^{(\mathfrak{R})} y_i$. Then $|\Delta|<\omega$ and $c_{(\Delta)} y_i = 1$ by $\mathfrak{R} \in Cs_\alpha^{reg} \cap Lf_\alpha$. By Claim 1(ii) we have $z_i \cdot e(y) = z_i \cdot \sum\{z_j \cdot \overline{y_j} : j<\alpha\} = z_i \cdot \overline{y_i}$. Then by $\Delta(z_i/\bar{F}) = 0$ we have $c_{(\Delta)}(z_i/\bar{F} \cdot E(y)) = c_{(\Delta)}(z_i \cdot e(y)/\bar{F}) = c_{(\Delta)}(z_i/\bar{F} \cdot \overline{y_i}/\bar{F}) = z_i/\bar{F} \cdot c_{(\Delta)} \overline{y_i}/\bar{F} = z_i/\bar{F} \neq 0$. Thus $E(y) \neq 0$.

QED(Claim 2)

Claim 2 implies ${}^\alpha\mathfrak{R} \in \mathbf{SUp}\{\mathfrak{R}\}$ by ${}^I\mathfrak{R}/F \in \mathbf{Up}\{\mathfrak{R}\}$.

QED(Lemma 7.1.1.)

Lemma 7.1.2. ${}_\varkappa Gs_\alpha \subseteq \mathbf{SUp}\{\mathfrak{Rf}(\varkappa)\}$ for every $\varkappa \geq 2$.

Proof. First we prove ${}_\varkappa Gs_\alpha \subseteq \mathbf{SUp}({}_\varkappa Gs_\alpha \cap Lf_\alpha)$. Let $\varphi$ be any quantifier free formula in the discourse language of $CA_\alpha$-s. Assume ${}_\varkappa Gs_\alpha \nVdash \varphi$. Then there is $\mathfrak{A} \in {}_\varkappa Gs_\alpha$ such that $\mathfrak{A} \nVdash \varphi$. Let $\beta = \alpha+\alpha$. By [HMTI]8.5-8.6, there is $\mathfrak{B} \in \mathbf{I}\,{}_\varkappa Gs_\beta$ such that $\mathfrak{A} \subseteq \mathfrak{Nr}_\alpha \mathfrak{B}$ and $B = Sg^{(\mathfrak{B})} A$. Clearly, $\mathfrak{B} \nVdash \varphi$ since $\varphi$ contains no quantifier. Let $H \stackrel{d}{=} \{i \in \alpha : (\exists j)[c_i \text{ or } d_{ij} \text{ occurs in } \varphi]\}$. Then $|H|<\omega$. Let $L \stackrel{d}{=} (\beta \sim \alpha) \cup H$. Let $\rho : \alpha \rightarrowtail L$ be one-one and onto such that $H 1 \rho \subseteq Id$. Such a $\rho$ exists by $H \subseteq_\omega \alpha$ and $|L| = |\alpha| \geq \omega$. Then $\mathfrak{Rd}^{(\rho)} \mathfrak{B} \nVdash \varphi$ and $\mathfrak{Rd}^{(\rho)} \mathfrak{B} \in Lf_\alpha$ by $B = SgA$. By $\mathfrak{B} \in \mathbf{I}\,{}_\varkappa Gs_\beta$ and [HMTI]8.1 we have $\mathfrak{Rd}^{(\rho)} \mathfrak{B} \in \mathbf{I}\,{}_\varkappa Gs_\alpha$. We have seen that ${}_\varkappa Gs_\alpha \cap Lf_\alpha \nVdash \varphi$. Therefore ${}_\varkappa Gs_\alpha \subseteq \mathbf{SUp}({}_\varkappa Gs_\alpha \cap Lf_\alpha)$ by [HMT]0.3.83 and 0.3.70(i). Now, ${}_\varkappa Gs_\alpha \cap Lf_\alpha \subseteq \mathbf{SP}({}_\varkappa Ws_\alpha \cap Lf_\alpha) \subseteq \mathbf{SP}({}_\varkappa Cs_\alpha^{reg} \cap Lf_\alpha) \subseteq \mathbf{SP}\{\mathfrak{Rf}(\varkappa)\} \subseteq \mathbf{SUp}\{\mathfrak{Rf}(\varkappa)\}$, by [HMTI] 6.2, and by 3.15, 7.1.1.

QED(Lemma 7.1.2.)

Now we return to the proof of Theorem 7.1.

Let $\varkappa \geq \lambda \geq \omega$. We show that $\mathfrak{R} \stackrel{d}{=} \mathfrak{Rf}(\varkappa) \in \mathbf{SUp}\,{}_\lambda Gs_\alpha$. Let $I \stackrel{d}{=} \{\langle \Gamma, \mathfrak{B} \rangle : 2 \subseteq \Gamma \subseteq_\omega \alpha,\ \mathfrak{B} \in \mathbf{S}_\omega\, \mathfrak{Rd}_\Gamma \mathfrak{R}\}$. Let $F$ be an ultrafilter on $I$ such that $\{\langle \Delta, \mathfrak{D} \rangle \in I : \Gamma \subseteq \Delta,\ \mathfrak{B} \subseteq \mathfrak{Rd}_\Gamma \mathfrak{D}\} \in F$ for every $\langle \Gamma, \mathfrak{B} \rangle \in I$. Let $i \stackrel{d}{=} \langle \Gamma, \mathfrak{B} \rangle \in I$. Then $|B| \leq \omega$ and $\mathfrak{B} \in \mathbf{I}\,{}_\infty Gs_\Gamma$ by [HMTI]8.2 and by $\varkappa \geq \omega$. Therefore $\mathfrak{B} \in \mathbf{I}\,{}_\lambda Gs_\Gamma$ by [HMTI]3.18(iv), $\varkappa \geq \lambda \geq \omega$ and by $|\Gamma|<\omega$. Then $\mathfrak{B} \subseteq \mathfrak{Rd}_\Gamma \mathfrak{L}_i$ for some $\mathfrak{L}_i \in {}_\lambda Gs_\alpha$ since ${}_\lambda Gs_\Gamma \subseteq \mathbf{SRd}_\Gamma\, {}_\lambda Gs_\alpha$ by [HMTI]8.5, 8.7. Now

$\mathfrak{R} \cong \mathbf{I} \subseteq P_{i\in I}\,\mathfrak{L}_i/F$ can be seen similarly to the proofs of [HMT]0.3.71, 0.5.15.

Then $\mathfrak{Rf}(\kappa)\in\mathbf{SUp}\{\mathfrak{Rf}(\lambda)\}$ by 7.1.2. By [HMTI]7.25(ii) then $\mathbf{SUp}\{\mathfrak{Rf}(\kappa)\} = \mathbf{SUp}\{\mathfrak{Rf}(\lambda)\}$ for every $\kappa,\lambda\geq\omega$. Therefore $\mathbf{I}\,{}_{\infty}Cs_\alpha \subseteq \subseteq \mathbf{I}\,(\cup\{{}_\kappa Gs_\alpha : \kappa\geq\omega\}) \subseteq \mathbf{SUp}\{\mathfrak{Rf}(\kappa) : \kappa\geq\omega\} \subseteq \mathbf{SUp}\{\mathfrak{Rf}(\lambda)\} \subseteq \mathbf{SUp}({}_\lambda Cs_\alpha^{reg}\cap Lf_\alpha) \subseteq \subseteq \mathbf{SUp}\,{}_{\infty}Cs_\alpha = \mathbf{I}\,{}_{\infty}Cs_\alpha$ imply $\mathbf{I}\,{}_{\infty}Cs_\alpha = \mathbf{SUp}\{\mathfrak{Rf}(\omega)\} = \mathbf{SUp}({}_\lambda Cs_\alpha^{reg}\cap Lf_\alpha)$ for every $\lambda\geq\omega$. If $\kappa\in\omega\sim 2$ then $\mathbf{I}\,{}_\kappa Gs_\alpha$ is a variety by [HMTI]7.16 and [HMT]2.4.64, hence $\mathbf{I}\,{}_\kappa Gs_\alpha = \mathbf{SUp}\{\mathfrak{Rf}(\kappa)\}$ by Lemma 7.1.2. (i) of Theorem 7.1 is proved. (iv) follows from (i).

Let $K \stackrel{d}{=} \mathbf{I}\{\mathfrak{A}\in Gs_\alpha : \mathfrak{A}$ has characteristic $\kappa\neq 1$ or $|A|\leq 2\}$. Clearly, $Cs_\alpha \subseteq K$. For every $\kappa\in\omega$ let $a_\kappa \stackrel{d}{=} c_{(\kappa)}\bar{d}(\kappa\times\kappa)$. Let $Ax \stackrel{d}{=} \stackrel{d}{=} \{(a_\kappa=0 \lor a_\kappa=1) : 0<\kappa<\omega\} \cup \{\forall x(x\leq a_2 \lor x=1)\}$. Then $K = Md(Ax)\cap \mathbf{I}\,Gs_\alpha$ is easy to see by [HMT]2.4.63. Since $Ax$ consists of universal disjunctions of equations only, $Ax$ is preserved under $\mathbf{HSUp}$. By [HMTI]7.16 then $\mathbf{HSUp}\,K=K$. Now $K \subseteq \mathbf{I}\,{}_{\infty}Cs_\alpha \cup \mathbf{I}\,(\cup\{{}_\kappa Gs_\alpha : \kappa\in\omega\sim 2\})\cup \mathbf{I}\,{}_1Cs_\alpha \subseteq \mathbf{SUp}'\{\mathfrak{Rf}(\kappa) : : \kappa\leq\omega\} \subseteq \mathbf{SUp}'(Cs_\alpha^{reg}\cap Lf_\alpha) \subseteq \mathbf{SUp}\,Cs_\alpha \subseteq \mathbf{HSUp}\,Cs_\alpha \subseteq \mathbf{HSUp}\,K=K$ by [HMTI]7.21, Lemma 7.1.2, Theorem 7.1(i) and $\mathbf{I}\,{}_1Cs_\alpha = \mathbf{I}\{\mathfrak{Rf}(1)\}$. This, together with [HMTI]7.13 and $\mathbf{I}\,Cs_\alpha^{reg}\cap Lf_\alpha = (\mathbf{I}\,Ws_\alpha\cap Lf_\alpha)\cup \mathbf{I}\,{}_0Cs_\alpha$, see 3.15, completes the proof of Theorem 7.1.

QED(Theorem 7.1.)

Corollary 7.2. Let $\alpha\geq\omega$.

(i) Let $\kappa>1$ be a cardinal. Then $\mathbf{P}K \subseteq \mathbf{SUp}\,K$ for every $K \subseteq \subseteq {}_\kappa Gws_\alpha^{comp}$ such that $\mathfrak{Rf}(\kappa)\in\mathbf{SUp}\,K$. In particular $\mathbf{P}K \subseteq \mathbf{SUp}\,K$ for $K\in\{{}_\kappa Ws_\alpha, {}_\kappa Cs_\alpha^{reg}, {}_\kappa Cs_\alpha, {}_\kappa Ws_\alpha\cap Lf_\alpha, {}_\kappa Cs_\alpha^{reg}\cap Lf_\alpha, {}_\kappa Cs_\alpha\cap Lf_\alpha\}$.

(ii) $Lf_\alpha\cap\mathbf{SUp}\,Cs_\alpha = \mathbf{Ud}\,(\mathbf{I}\,Cs_\alpha\cap Lf_\alpha)$ but $\mathbf{SUp}\,Cs_\alpha \supsetneq \mathbf{UdI}\,Cs_\alpha$.

Proof: (i) follows from (1) in the proof of [HMTI]7.17 and from 7.1. (ii) follows from 4.15.

QED(Corollary 7.2)

Theorem 7.3. Let $1<\varkappa<\omega\leq\alpha$.

(i) $\mathbf{S}\,\mathbf{Up}\,Cs_\alpha^{reg} \neq \mathbf{Uf}\,\mathbf{Up}\,Cs_\alpha^{reg} \neq \mathbf{Uf}\,\mathbf{Up}\,Ws_\alpha \neq \mathbf{S}\,\mathbf{Up}\,Ws_\alpha$.

(ii) For any nondiscrete $CA_\alpha$ $\mathfrak{A}$ which is not of characteristic 0 statements a.-c. below hold.

a. $\mathbf{S}\,\mathbf{Up}\,\mathfrak{A} \not\subseteq \mathbf{Uf}\,\mathbf{Up}\,Gws_\alpha^{comp\ reg}$.

b. $\mathbf{S}\,\mathbf{Up}\,\mathfrak{A} \not\subseteq \mathbf{Uf}\,\mathbf{Up}\,Dind_\alpha$.

c. $\mathbf{S}\,\mathbf{Up}\,\mathfrak{A} \not\subseteq \mathbf{Uf}\,\mathbf{Up}\,(Ws_\alpha \cup Cs_\alpha^{reg})$.

(iii) ${}_\varkappa Cs_\alpha^{reg} \not\subseteq \mathbf{Uf}\,\mathbf{Up}\,(Ws_\alpha \cup {}_0Cs_\alpha)$ and ${}_\varkappa Cs_\alpha \not\subseteq \mathbf{Uf}\,\mathbf{Up}\,Dind_\alpha$.

(iv) $\mathbf{H}\ Ws_\alpha \not\subseteq \mathbf{Uf}\,\mathbf{Up}\,Ws_\alpha$, $\mathbf{H}Cs_\alpha^{reg} \not\subseteq \mathbf{Uf}\,\mathbf{Up}\,Cs_\alpha^{reg}$,

$\mathbf{H}_\varkappa Ws_\alpha \not\subseteq \mathbf{Uf}\,\mathbf{Up}\,Dind_\alpha$.

To prove Theorem 7.3 we shall use the following definitions and lemmas.

Definition 7.3.1. Let $\varkappa<\omega\leq\alpha$.

$a_\varkappa$ denotes the term $c_{(\varkappa)}\bar{d}(\varkappa\times\varkappa)$.

$at(x)$ denotes the formula $(\forall y[y\leq x \rightarrow (y=0 \vee y=x)] \wedge x\neq 0)$.

$supat(y)$ is the formula $\forall z[\forall x(at(x) \rightarrow x\leq z) \leftrightarrow y\leq z]$.

$\zeta_\varkappa$ is the formula $(a_\varkappa\neq 1 \rightarrow \forall y[supat(y) \rightarrow (y=0 \vee y=1)])$.

Note that for any $y\in A$, $\mathfrak{A}\in CA_\alpha$ we have $\mathfrak{A} \models supat(y)$ iff $y = \sum At\mathfrak{A}$.

Lemma 7.3.2. $Dind_\alpha \models \{\zeta_\varkappa : \varkappa<\omega\}$.

To prove Lemma 7.3.2 we shall need the following lemma.

Lemma 7.3.2.1. Let $\alpha\geq\omega$ and $\mathfrak{A}\in CA_\alpha$.

(i) Let $x\in At\mathfrak{A}$ be such that $x-a_\varkappa\neq 0$ for some $\varkappa$. Then $|\{z\in A : z\leq c_{(\Gamma)}x\}|<\omega$ for every $\Gamma\subseteq_\omega\alpha$. Hence $c_{(\Gamma)}x = \sum\{z\in At\mathfrak{A} : z\leq c_{(\Gamma)}x\}$.

(ii) Suppose $1^{\mathfrak{A}} = \sum\{-a_\varkappa : \varkappa<\omega\}$ and $\sum At\mathfrak{A}$ exists in $\mathfrak{A}$. Then $\Delta(\sum At\mathfrak{A})=0$.

Proof. Let $\alpha\geq\omega$ and $\mathfrak{A}\in CA_\alpha$. Proof of (i): It suffices to take

$2<\kappa<\omega$ and prove (i) for $\mathfrak{R} \stackrel{d}{=} \mathfrak{Rl}(-a_\kappa)\mathfrak{A}$ in place of $\mathfrak{A}$. Let $y \in \in At\,\mathfrak{R}$. Then, we claim, for each $i<\alpha$ the Boolean algebra $\mathfrak{Rl}(c_i y)\mathfrak{Bl}\,\mathfrak{R}$ is finite and has $<\kappa$ atoms (so (i) follows easily). Suppose not: then there exist non-zero pairwise disjoint $z_0,\ldots,z_{\kappa-1} \le c_i y$. Choose distinct $j_0,\ldots,j_\kappa<\alpha$ different from $i$ such that $y \le d_{js,jt}$ for all $s,t\le\kappa$. Let $\Gamma = \{j_0,\ldots,j_{\kappa-1}\}$ and set

$$w = \Pi\{c_i(c_{(\Gamma)}z_s \cdot d_{i,js}) : s<\kappa\}.$$

Note that $c_i y$ is an $\{i\}$-atom, and hence $c_i z_s = c_i y$ for all $s<\kappa$. Now

$$c_{(\Gamma)}w = \Pi\{c_{js}c_i(c_{(\Gamma)}z_s \cdot d_{i,js}) : s<\kappa\} =$$

$$= \Pi\{c_i c_{(\Gamma)}z_s : s<\kappa\} \ge y,$$

so $w\neq 0$. Now let $s,t<\kappa$, $s\neq t$. Let $\Delta=\{j_0,\ldots,j_\kappa\}$. Now for $\Theta\subseteq\Gamma$ and $j_\mu\in\Theta$ we have

$$c_{(\Theta)}z_s \cdot d_\Delta = c_{j\kappa}(c_{(\Theta\sim\{j\mu\})}z_s \cdot d_{j\mu,j\kappa}) \cdot d_{j\mu,j\kappa} \cdot d_\Delta = c_{(\Theta\sim\{j\mu\})}z_s \cdot d_\Delta .$$

It follows that $c_{(\Gamma)}z_s \cdot d_\Delta = z_s \cdot d_\Delta = z_s$. Hence $z_t \cdot c_{(\Gamma)}z_s = = z_t \cdot c_{(\Gamma)}z_s \cdot d_\Delta = 0$, so we infer in succession

$$c_{(\Gamma\sim\{jt\})}z_t \cdot c_{(\Gamma)}z_s \cdot d_{i,js} = 0,$$

$$c_{(\Gamma\sim\{jt\})}z_t \cdot d_{i,js} \cdot c_i(c_{(\Gamma)}z_s \cdot d_{i,js}) = 0,$$

$$c_{(\Gamma)}z_t \cdot d_{i,jt} \cdot c_i(c_{(\Gamma)}z_s \cdot d_{i,js}) \cdot d_{js,jt} = 0,$$

$$c_i(c_{(\Gamma)}z_t \cdot d_{i,jt}) \cdot c_i(c_{(\Gamma)}z_s \cdot d_{i,js}) \cdot d_{js,jt} = 0,$$

$$w \le -d_{js,jt}.$$

Thus $w \le \bar{d}(\Gamma\times\Gamma)=0$, a contradiction.

<u>Proof of (ii):</u> Let $1^{\mathfrak{A}} = \Sigma\{-a_\kappa : \kappa<\omega\}$ and suppose that $y \stackrel{d}{=} = \Sigma\, At\,\mathfrak{A}$ exists in $\mathfrak{A}$. Let $i\in\alpha$ and $x\in At\,\mathfrak{A}$. Then $(\exists\kappa<\omega)$ $x-a_\kappa\neq 0$. Thus $c_i x = \Sigma\{z\in At\,\mathfrak{A} : z\le c_i x\}$ by (i). Thus $c_i x\le y$. Now

$c_i y = \sum\{c_i x : x \in At\,\mathfrak{A}\} \leq y$ proves $c_i y = y$.

QED(Lemma 7.3.2.1.)

We return to the proof of Lemma 7.3.2. Let $\mathfrak{A} \in Dind_\alpha$ and $\varkappa < \omega$. If $a_\varkappa^{\mathfrak{A}} = 1$ then $\mathfrak{A} \models \zeta_\varkappa$. Assume $a_\varkappa^{\mathfrak{A}} \neq 1$. Since $a_\varkappa \in Zd\,\mathfrak{A}$, by $\mathfrak{A} \in Dind_\alpha$ then $a_\varkappa = 0$, i.e. $-a_\varkappa^{\mathfrak{A}} = 1$. Let $y \in A$ be such that $\mathfrak{A} \models supat(y)$. Then $y = \sum At\,\mathfrak{A}$, hence $y \in Zd\,\mathfrak{A} = \{0,1\}$ by 7.3.2.1(ii). Hence $\mathfrak{A} \models \zeta_\varkappa$.

QED(Lemma 7.3.2.)

Lemma 7.3.3. Let $\mathfrak{B} \in \mathbf{I}\,Gws_\alpha$ have characteristic $\varkappa > 1$. Let $B = Sg\{x,y\}$, $x \leq y \cdot d_{ij}$ for every $i,j \in \alpha$, $\Delta y = 0$, $y \neq 1$, $x \neq 0$. Then $\mathfrak{B} \not\models \zeta_n$ for every $\varkappa < n < \omega$.

Proof. Let $\mathfrak{B}$ satisfy the hypotheses of 7.3.3. Since $\mathfrak{B}$ is of characteristic $\varkappa > 0$, we have $a_n^{\mathfrak{B}} \neq 1$ for every $\varkappa < n$. We show that $\mathfrak{B} \models supat(y)$. First we show $x \in At\,\mathfrak{B}$. By $\mathbf{I}\,Gws_\alpha = \mathbf{I}\,Gws_\alpha^{wd}$ we may assume $\mathfrak{B} \in {}_\varkappa Gws_\alpha^{wd}$. Let $\bar{H} \stackrel{d}{=} \langle H : i < \alpha \rangle$ for every set $H$. Let $U \stackrel{d}{=} base(\mathfrak{B})$. By $\Delta y = 0$ there is $H \subseteq Subu(\mathfrak{B})$ such that $y = \cup \bar{H}$. By $x \leq y \cdot d_{ij}$ for $i,j \in \alpha$ there is $L \subseteq H$ and $b \in P_{v \in L}\, base(v)$ such that $x = \{\bar{b_v} : v \in L\}$. Assume $0 < z < x$ and $z \in B$. Then there are $v,w \in Subu(\mathfrak{B})$ such that $\bar{b_v} \in x - z$ and $\bar{b_w} \in z$. Let $f : U \rightarrowtail\!\!\!\rightarrow U$ be a permutation of $U$ such that $f^* base(v) = base(w)$, $f \circ f \subseteq Id$, $f(b_v) = b_w$ and $[U \sim base(v \cup w)]1f \subseteq Id$. By [HMTI]3.1 $f$ induces a base automorphism $\tilde{f} \in Is(\mathfrak{Sb}(1^{\mathfrak{B}}), \mathfrak{Sb}(1^{\mathfrak{B}}))$ of $\mathfrak{Sb}(1^{\mathfrak{B}})$. By $f(b_v) = b_w$ and $f(b_w) = b_v$ and $\bar{b_v}, \bar{b_w} \in x \subseteq y$ we have $\tilde{f}(x) = x$ and $\tilde{f}(y) = y$. Since $z \in Sg\{x,y\}$ then $\tilde{f}(z) = z$ contradicting $\bar{b_v} = f \bullet \bar{b_w} \notin z$ and $\bar{b_w} \in z$. This proves that $x \in At\,\mathfrak{B}$.

By $\Delta y = 0$ and $x \leq y$, $B = Sg\{x,y\}$ we have that $\mathfrak{Rl}(-y)\mathfrak{B} \in Mn_\alpha$, hence it is atomless by [HMT]1.10.5(ii). Hence $At\,\mathfrak{B} \subseteq \{z : z \leq y\}$. Let $\mathfrak{R} \stackrel{d}{=} \mathfrak{Rl}_y \mathfrak{B}$. Then $At\,\mathfrak{B} = At\,\mathfrak{R}$. We prove that $\sum At\,\mathfrak{R} = 1$ in $\mathfrak{R}$. Let $Q \stackrel{d}{=} \cup L$. Then $x \subseteq Q \subseteq y$ and $\Delta^{(\mathfrak{R})} Q = 0$. $R = Sg\{x\}$, $x$ is $Q$-wsmall and $x,Q$ satisfy the conditions of 4.7. Thus $rl_Q \in Is\,\mathfrak{R}$ since $\mathfrak{R}$ is of characteristic $\varkappa$ and $\Delta x = \alpha$. Let $\mathfrak{L} \stackrel{d}{=} \mathfrak{Rl}_Q \mathfrak{R}$. Then $1^{\mathfrak{L}} = Q$. Let $q \in Q$. Then $q \in V$ for some $V \in L$. Let $\Gamma \stackrel{d}{=} \{i \in \alpha : q_i \neq b_V\}$. Then $|\Gamma| < \omega$ since

$V$ is a $Ws_\alpha$ -unit and $\overline{b_V}\in V$, thus $q\in c_{(\Gamma)}x$. This proves $1^{\mathfrak{L}} = \cup\{c_{(\Gamma)}x : \Gamma \subseteq_\omega \alpha\}$. Thus $y = \sum\{c_{(\Gamma)}x : \Gamma\subseteq_\omega\alpha\}$, by $rl_Q\in Is\mathfrak{R}$ and $x\subseteq Q$. By 7.3.2.1(i) then $y=1^{(\mathfrak{R})} = \sum At\mathfrak{R}$, thus $y = \sum At\mathfrak{B}$ by $At\mathfrak{R} = At\mathfrak{B}$. I.e. $\mathfrak{B} \models \text{supat}(y)$. By $0\neq x\leq y\neq 1$ then $\mathfrak{B} \not\models \zeta_n$ for $n>\varkappa$.

QED(Lemma 7.3.3.)

Lemma 7.3.4. Let $1<\varkappa<n<\omega\leq\alpha$. Then ${}_\varkappa Cs_\alpha \not\models \zeta_n$ and $H_\varkappa Ws_\alpha \not\models \zeta_n$.

Proof. Let $\bar{0} \overset{d}{=} \langle 0 : i<\alpha\rangle$, $x \overset{d}{=} \{\bar{0}\}$, $y \overset{d}{=} {}^\alpha\varkappa^{(\bar{0})}$ and $\mathfrak{A} \overset{d}{=} \mathfrak{S}\mathfrak{g}^{(\mathfrak{Sb}\,{}^\alpha\varkappa)}\{x,y\}$. Then $\mathfrak{A} \not\models \zeta_n$ since $\mathfrak{A},x,y$ satisfy the conditions of 7.3.3. Thus ${}_\varkappa Cs_\alpha \not\models \zeta_n$.

Let $V \overset{d}{=} {}^\alpha\varkappa^{(\bar{0})}$, $y \overset{d}{=} \{q\in V : \cup\{n\in\omega : q_n\neq 0\}=2m$ for some $m\in\omega\}$, $x \overset{d}{=} \{\bar{0}_1^{2n} : n\in\omega\}$. Then $x\subseteq y$. $\mathfrak{L} \overset{d}{=} \mathfrak{Sb}V$ and $\mathfrak{A} \overset{d}{=} \mathfrak{S}\mathfrak{g}^{(\mathfrak{L})}\{x,y\}$. $I \overset{d}{=} \{z\in A : |\{\omega 1 q : q\in z\}|<\omega\}$. Then $I\in Il\mathfrak{A}$ and $x\notin I$, $-y\notin I$. Hence $0<x/I\leq y/I<1^{(\mathfrak{A})}/I$ in $\mathfrak{A}/I$. Let $i\in\omega$. Then $c_i y\cdot -y \subseteq \{q\in V : \omega\sim(i+1)1q\subseteq \bar{0}\}\in I$. Thus $\Delta(y/I)=0$ by $\varkappa<\omega$. Let $i,j\in\alpha$. Then $|x-d_{ij}|\leq 2$ and hence $x-d_{ij}\in I$ proving $x/I\leq d_{ij}$ for all $i,j\in\alpha$ (in $\mathfrak{A}/I$). Let $\mathfrak{B} \overset{d}{=} \mathfrak{A}/I$. Then $\mathfrak{B} \in I\,{}_\varkappa Gws_\alpha$ by [HMTI]7.15 and $\mathfrak{B}$, $x/I$, $y/I$ satisfy the hypotheses of 7.3.3. Thus $\mathfrak{B} \not\models \zeta_n$ for every $n>\varkappa$. Clearly $\mathfrak{B} \in H_\varkappa Ws_\alpha$.

QED(Lemma 7.3.4.)

Lemma 7.3.5. Let $\alpha\geq\omega$. Let $\mathfrak{A}$ be any nondiscrete $CA_\alpha$.

(i) ${}^I\mathfrak{A}/F \notin Dind_\alpha$ for any ultrafilter $F$ which is not $|\alpha|^+$ -complete.

(ii) Assume $(\exists\varkappa<\omega)\ \mathfrak{A} \models a_\varkappa\neq 1$. Then $\mathbf{S}\,\mathbf{Up}\,\mathfrak{A} \not\subseteq \mathbf{Uf}\,\mathbf{Up}\,Dind_\alpha$.

Proof. Let $\mathfrak{A}$ be any nondiscrete $CA_\alpha$. Let $F$ be any non-$|\alpha|^+$-complete ultrafilter on $I$. Let $\alpha=\lambda+m$ where $\lambda$ is a limit ordinal and $m\in\omega$. Since $F$ is not $|\alpha|^+$-complete, there is a function $h : I \to \lambda$ such that $(I / h|h^{-1})\cap F = 0$. Let $y \overset{d}{=} \langle d_{h(i),h(i)+1} : i\in I\rangle$. Then $y\in {}^I A$. Since $\mathfrak{A}$ is nondiscrete, we have that $0<y/F<1$. Let $\varkappa<\alpha$ be arbitrary. Then $\{i\in I : c_\varkappa y_i\neq y_i\} = \{i\in I : h(i)=\varkappa$ or $h(i)+1=\varkappa\}\notin F$

by the properties of $h$ and since $\Delta(d_{h(i),h(i)+1})=\{h(i),h(i)+1\}$. Therefore $\Delta(y/F)=0$ in ${}^I\mathfrak{A}/F$. Thus $|Zd\ {}^I\mathfrak{A}/F|>2$, i.e. ${}^I\mathfrak{A}/F\notin Dind_\alpha$.

Proof of (ii): Assume $\mathfrak{A}\models a_\varkappa\neq 1$. If $\mathfrak{A}\notin \mathbf{Uf\ Up}\,Dind_\alpha$ then we are done. Suppose $\mathfrak{A}\in\mathbf{Uf\ Up}\,Dind_\alpha$. Let $I\overset{d}{=}Sb_\omega\alpha$ and let $F$ be an ultrafilter on $I$ such that $(\forall\Gamma\in I)\{\Delta\in I:\Gamma\subseteq\Delta\}\in F$. Let $\mathfrak{L}\overset{d}{=}{}^I\mathfrak{A}/F$. Then $\mathfrak{L}\in\mathbf{Uf\ Up}\,Dind_\alpha$ is nondiscrete and $\mathfrak{L}\models a_\varkappa\neq 1$. Then $\mathfrak{L}$ is of characteristic $1<n<\varkappa$ since $Dind_\alpha\models\{(a_\lambda\neq 1\rightarrow a_\lambda=0):\lambda\in\omega\}$. Let $v,r\in P_{\Gamma\in I}(\alpha\sim\Gamma)$ be such that $(\forall\Gamma\in I)v_\Gamma\neq r_\Gamma$. Let $y\overset{d}{=}\langle d_{r\Gamma,v\Gamma}:\Gamma\in I\rangle/F$, $x\overset{d}{=}\langle d_{(\Gamma\cup\{r\Gamma,v\Gamma\})}:\Gamma\in I\rangle/F$, and $\mathfrak{B}\overset{d}{=}\mathfrak{Sg}^{(\mathfrak{L})}\{x,y\}$. Then $\Delta y=0$ since $(\forall i\in\alpha)(\forall\Gamma\in I)[i\in\Gamma\Rightarrow i\notin\{r\Gamma,v\Gamma\}]$. $y\neq 1$ since $\mathfrak{A}$ is nondiscrete. Similarly, $0\neq x\leq y\cdot d_{ij}$ for $i,j\in\alpha$. Thus $\mathfrak{B}\not\models\zeta_m$ for every $m>n$ by 7.3.3. Then $\mathfrak{B}\notin\mathbf{Uf\ Up}\,Dind_\alpha$ by 7.3.2. Now $\mathfrak{B}\in\mathbf{S\ Up}\,\mathfrak{A}$ completes the proof.

QED(Lemma 7.3.5.)

Definition 7.3.6. Let $\varkappa<\omega\leq\alpha$. Recall the formula $at(x)$ from 7.3.1. We define $\psi$ to be the formula

$$\forall x(at(x)\ \rightarrow\ \forall y[y\neq 0\rightarrow\exists z(at(z)\wedge z\leq y)]).$$

$\varphi_\varkappa$ is defined to be the formula $(a_\varkappa\neq 1\rightarrow\psi)$.
Note that $\mathfrak{A}\models\psi$ iff $\mathfrak{A}$ is either atomless or atomic.

Lemma 7.3.7. Let $1<\varkappa<\omega\leq\alpha$.

(i) ${}_\varkappa Cs_\alpha^{reg}\not\subseteq\mathbf{Uf\ Up}(Ws_\alpha\cup_0 Cs_\alpha)$, moreover
$\{\mathfrak{A}\in{}_\varkappa Cs_\alpha^{reg}:\mathbf{H}\mathfrak{A}\subseteq\mathbf{I}\,Cs_\alpha^{reg}\}\not\subseteq\mathbf{Uf\ Up}(Ws_\alpha\cup_0 Cs_\alpha)$.

(ii) $Ws_\alpha\models\{\varphi_n:n\in\omega\}$ and ${}_\varkappa Ws_\alpha\models\psi$.

(iii) $Cs_\alpha^{reg}\not\models\varphi_{\varkappa+1}$ and ${}_\varkappa Cs_\alpha^{reg}\not\models\psi$.

Proof. Let $1<\varkappa<\omega\leq\alpha$. Proof of (ii): It is enough to prove ${}_\varkappa Ws_\alpha\models\psi$ since if $\mathfrak{A}\models a_n\neq 1$ and $\mathfrak{A}\in Ws_\alpha$ then $\mathfrak{A}\in{}_\varkappa Ws_\alpha$ for some $\varkappa<n$. The case $\varkappa<2$ is obvious. Assume $\varkappa\geq 2$. Let $\mathfrak{A}\in{}_\varkappa Ws_\alpha$. Let $x\in At\,\mathfrak{A}$ and $y\in A\sim\{0\}$. By $\varkappa<\omega$ we have $|x|<\omega$ because $(\forall f,q\in z)ker(f)=ker(q)$ for any atom $z$ of any $Gws_\alpha$. Since $0\notin\{x,y\}$ there are $f\in x$ and $q\in y$.

Since $\mathfrak{A}\in Ws_\alpha$ there is a finite $\Gamma\subseteq\alpha$ such that $(\alpha\sim\Gamma)1f\subseteq q$. Then $w\overset{d}{=}c_{(\Gamma)}x\cap y\neq 0$. By $\varkappa+|x|<\omega$ we have $|c_{(\Gamma)}x|<\omega$ and hence $|w|<\omega$. Thus there is $z\in At\mathfrak{A}$ such that $z\leq w\leq y$, proving $\mathfrak{A}\models\psi$.

<u>Proof of (iii):</u> Let $H\subseteq\alpha$ be such that $|H|\cap|\alpha\sim H|\geq\omega$. For any set $s$ let $\bar{s}\overset{d}{=}\langle s : i<\alpha\rangle$. Let $x=\{\bar{0}\}$, $y\overset{d}{=}\{q\in{}^\alpha\varkappa : H1q\subseteq\bar{1}\}$ and $\mathfrak{A}\overset{d}{=}\mathfrak{Sg}^{(\mathfrak{G}\mathfrak{b}^\alpha\varkappa)}\{x,y\}$. Then $x\in At\mathfrak{A}$. We show that there is no atom below $y$. Assume $y\supseteq z\in At\mathfrak{A}$ for some $z$. Let $p\in z$ and $V\overset{d}{=}{}^\alpha\varkappa^{(p)}$. Then $rl_V\in Ho(\mathfrak{A},\mathfrak{B})$ for some $\mathfrak{B}\in Ws_\alpha$. Then $B=Sg\{V\cap y\}$ hence $\mathfrak{B}\in Dc_\alpha$ since $V\cap x=0$ by $|H|\geq\omega$. By $z\in At\mathfrak{A}$ we have $V\cap z\in\{0\}\cup At\mathfrak{B}$. By $\varkappa>1$ and $\mathfrak{B}\in{}_\varkappa Ws_\alpha$ we have $(\forall w\in At\mathfrak{B})\,\Delta w=\alpha$ (since either $w\leq d_{ij}$ or $w\leq -d_{ij}$ for all $i,j\in\alpha$). By $\mathfrak{B}\in Dc_\alpha$ we should have $V\cap z=0$. A contradiction, proving $\mathfrak{A}\models\neg\psi$. By Thm 1.3 $\mathfrak{A}\in{}_\varkappa Cs_\alpha^{reg}$ since $x,y\in Sm^{\mathfrak{A}}$.

<u>Proof of (i):</u> Let $\mathfrak{A},H,x,y$ be as in the proof of (iii). We have seen above that $\mathfrak{A}\in{}_\varkappa Cs_\alpha^{reg}$ is such that $\mathfrak{A}\not\models\psi$. We show that $\mathbf{H}\mathfrak{A}\subseteq$ $\subseteq\mathbf{I}\,Cs_\alpha^{reg}$. Then (i) will follow from (ii) and (iii).

Let $\mathfrak{B}\in\mathbf{H}\mathfrak{A}$. Then $\mathfrak{B}\cong\mathfrak{L}\in{}_\varkappa Gws_\alpha^{wd}$ by [HMTI]7.15. Let $h\in Ho(\mathfrak{A},\mathfrak{L})$, $a\overset{d}{=}h(x)$ and $b\overset{d}{=}h(y)$. Then

$$(*)\qquad (\forall i,j\in\alpha)a\leq d_{ij} \text{ and } (\forall i,j\in H)b\leq d_{ij} \text{ and } (\forall\theta\subseteq_\omega\alpha)a\cdot c_{(\theta)}b=0.$$

Let $V\in Subu(\mathfrak{L})$, $U=base(V)$. Then one of cases (i)-(iii) below holds.

(i) $(\exists u\in U)V\cap a=\{\bar{u}\}$ and $V\cap b=0$.

(ii) $(\exists w\in U)V\cap b=\{q\in V : H1q\subseteq\bar{w}\}$ and $V\cap a=0$.

(iii) $V\cap a=V\cap b=0$.

The fact that exactly one of (i)-(iii) holds for each $V\in Subu(\mathfrak{L})$ follows from $(*)$.

Let $W\overset{d}{=}\cup\{V\in Subu(\mathfrak{L}) : V\cap a=V\cap b=0\}$. Let $I\overset{d}{=}\{z\in C : z\subseteq W\}$. Clearly $I\in Il\mathfrak{L}$ and $\mathfrak{L}/I\cong\mathfrak{Rl}(-W)\mathfrak{L}$. Let $\mathfrak{N}\overset{d}{=}\mathfrak{Rl}_W\mathfrak{L}$. Let $z\in I$. Then $z\in N$. Clearly $N=Sg\{W\cap a, W\cap b\}=Sg\{0\}$. Hence $|\Delta^{(\mathfrak{N})}z|<\omega$. By $z\subseteq W$ and since $\Delta(W)=0$ in $1^{\mathfrak{L}}$ we have $\Delta^{(\mathfrak{L})}z=\Delta^{(\mathfrak{N})}z$. Thus $|\Delta^{(\mathfrak{L})}z|<\omega$. By 1.3.3 then $z\in Sg^{(\mathfrak{L})}\{0\}=Mn(\mathfrak{L})$ since $a,b\in Sm^{\mathfrak{L}}$ (and since $Dm_0^{\mathfrak{L}}$ is the greatest Lf-subuniverse of $\mathfrak{L}$). If $W=1^{\mathfrak{L}}$ then $\mathfrak{L}\in Mn_\alpha\cap{}_\varkappa Gws_\alpha$

and hence $\mathfrak{L} \in \mathbf{I}\, Cs_{\alpha}^{reg}$ and we are done. Assume therefore $W \neq 1^{\mathfrak{L}}$. Then $I \neq C$. By $\mathfrak{L} \in {}_{\varkappa}Gws_{\alpha}$, $\mathfrak{Mn}(\mathfrak{L})$ is simple and hence $I \cap Mn(\mathfrak{L}) = 0$. Thus $z=0$ by $z \in Mn(\mathfrak{L})$. Since $z \in I$ was chosen arbitrarily we proved $I=\{0\}$. Thus $rl_W \in Is\,\mathfrak{L}$. Hence we may assume $W=0$.

Let $V \in Subu(\mathfrak{L})$. Assume $V \cap b \neq 0$. Let $U = base(V)$. By (ii), $V \cap b = \{q \in V : H1q \subseteq \bar{w}\}$ for some $w \in U$ and $V \cap a = 0$. Let this $w \in U$ be fixed. Let $\mathfrak{R} \overset{d}{=} rl(V)^{*}\mathfrak{L}$. Let $\mathfrak{Y} \overset{d}{=} \mathfrak{Sg}^{(\mathfrak{U})}\{y\}$. By 3.14 there is $g \in Is(\mathfrak{R}, \mathfrak{Y})$ with $g(V \cap b) = y$ since $R = Sg\{V \cap b\}$. This proves (**) below.

(**) $(\forall V, Y \in Subu(\mathfrak{L}))[V \cap b \neq 0$ and $Y \cap b \neq 0 \Rightarrow (\exists h_{VY} \in Is(\mathfrak{Rl}_V \mathfrak{L}, \mathfrak{Rl}_Y \mathfrak{L}))$
$h_{VY}(V \cap b) = Y \cap b]$.

Let $T \overset{d}{=} \cup\{Y \in Subu(\mathfrak{L}) : Y \cap b \neq 0$ and $Y \neq V\}$. Let $z \in C$ with $0 < z \subseteq T$. Then $(\exists Y \in Subu(\mathfrak{L}))[Y \cap b \neq 0$ and $z \cap Y \neq 0]$. There is a term $\tau(x)$ such that $z = \tau^{(\mathfrak{L})}(b,a)$. Let $f \overset{d}{=} (h_{VY} \circ rl_V)$. Then $f \in Hom(\mathfrak{L}, \mathfrak{Rl}_Y \mathfrak{L})$ and $f(b) = rl(Y)b = Y \cap b$ and $f(a) = 0 = rl(Y)a$. Let $\mathfrak{R} \overset{d}{=} \mathfrak{Rl}_Y \mathfrak{L}$. Then $f(\tau^{\mathfrak{L}}(a,b)) = \tau^{\mathfrak{R}}(fa, fb) = \tau^{\mathfrak{R}}(rl(Y)a, rl(Y)b) = rl(Y)\tau^{\mathfrak{L}}(a,b) = Y \cap z \neq 0$. Hence $h_{VY}(V \cap z) \neq 0$, thus $V \cap z \neq 0$. We proved $rl(-T) \in Is\,\mathfrak{L}$. Hence we may assume $T=0$, i.e. $|\{V \in Subu(\mathfrak{L}) : V \cap b \neq 0\}| \leq 1$.

By a completely analogous argument we can prove $(\forall V \in Subu(\mathfrak{L}))$ $[V \cap a \neq 0 \Rightarrow rl(-\cup\{Y \in Subu(\mathfrak{L}) : Y \neq V$ and $Y \cap a \neq 0\}) \in Is\,\mathfrak{L}]$.

We have proved the existence of $f \in Is(\mathfrak{L}, \mathfrak{N})$ with $\mathfrak{N} \in {}_{\varkappa}Gws_{\alpha}^{wd}$ such that $|\{V \in Subu(\mathfrak{N}) : f(a) \cap V \neq 0\}| \leq 1$ and
$|\{V \in Subu(\mathfrak{N}) : f(b) \cap V \neq 0\}| \leq 1$ and
$(\forall V \in Subu(\mathfrak{N}))(f(a) \cap V \neq 0$ or $f(b) \cap V \neq 0)$.

<u>Case 1</u> Assume $b \neq 0$ and $a \neq 0$. Then $Subu(\mathfrak{N}) = \{V, Y\}$ with $f(a) \subseteq V$ and $f(b) \subseteq Y$. Above we proved the existence of $g \in Ism(\mathfrak{Rl}_Y \mathfrak{N}, \mathfrak{U})$ with $g(fb) = y$. Similarly there exists $k \in Ism(\mathfrak{Rl}_V \mathfrak{N}, \mathfrak{U})$ with $h(fa) = x$. Then $\mathfrak{N} \cong\, \mid\, \subseteq \mathfrak{Rl}_V \mathfrak{N} \times \mathfrak{Rl}_Y \mathfrak{N} \cong\, \mid\, \subseteq \mathfrak{U} \times \mathfrak{U}$ and there is $t \in Ism(\mathfrak{N}, {}^{2}\mathfrak{U})$ with $t(fa) = \langle x, 0 \rangle$ and $t(fb) = \langle 0, y \rangle$. Let $\mathfrak{Y} \overset{d}{=} \mathfrak{Sg}^{(\mathfrak{U} \times \mathfrak{U})}\{\langle x,0 \rangle, \langle 0,y \rangle\}$. Then $\mathfrak{N} \cong \mathfrak{Y}$. Moreover we have proved (***) below.

(***) $(\forall f \in Ho(\mathfrak{U}))[(fa \neq 0$ and $fb \neq 0) \Rightarrow f^{*}\mathfrak{U} \cong \mathfrak{Y}]$.

Since $Id\in Ho\mathfrak{A}$, by (***) we have $(\forall f\in Ho\mathfrak{A})[(fa\neq 0\neq fb) \Rightarrow f^*\mathfrak{A} \cong \mathfrak{A}]$.
Thus in the present case $\mathfrak{B}\in \mathbf{I}\mathfrak{A} \subseteq \mathbf{I}\ {}_{\kappa}Cs_{\alpha}^{reg}$ as it was desired.

Case 2 Assume $a=0$ and $b\neq 0$. Then by the proof preceding Case 1 above we have $\mathfrak{B} \cong \mathfrak{C} \subseteq \mathfrak{A} \in {}_{\kappa}Cs_{\alpha}^{reg}$, hence $\mathfrak{B} \in \mathbf{I}\ {}_{\kappa}Cs_{\alpha}^{reg}$.

Case 3 Assume $a\neq 0$ and $b=0$. We could treat this case analogously to Case 2 but we can also do more. By using parts of the above proof one easily proves $\mathfrak{B} \cong \mathfrak{N} \in {}_{\kappa}Cs_{\alpha}$ for some $\mathfrak{N}$. Then by 1.3 $\mathfrak{N}$ is regular.

We have proved $\mathbf{H}\mathfrak{A} \subseteq \mathbf{I}\ {}_{\kappa}Cs_{\alpha}^{reg}$. Actually we proved more than this, we have also proved (****) below

(****) $\mathbf{H}\mathfrak{A} = \mathbf{I}\{\mathfrak{Mn}(\mathfrak{A}), \mathfrak{Sg}^{(\mathfrak{A})}\{x\}, \mathfrak{Sg}^{(\mathfrak{A})}\{y\}, \mathfrak{A}\}$.

That is there are exactly 4 isomorphism types in $\mathbf{H}\mathfrak{A}$.

QED(Lemma 7.3.7.)

Now we turn to the proof of Theorem 7.3. $\mathbf{Uf\ Up}\, Cs_{\alpha}^{reg} \neq \mathbf{Uf\ Up}\, Ws_{\alpha}$ and the first part of (iii) follow from 7.3.7. The rest of (i), and (ii) follow from 7.3.5(ii), upon observing $Ws_{\alpha}\cup Cs_{\alpha}^{reg}\cup Gws_{\alpha}^{comp\ reg}\subseteq Dind_{\alpha}$. (iv) and the rest of (iii) follow from 7.3.4, using $Ws_{\alpha}\subseteq \mathbf{I}\, Cs_{\alpha}^{reg}$.

QED(Theorem 7.3.)

As a contrast to Proposition 7.4 below we note that $\mathbf{P}\ {}_{\kappa}Cs_{\alpha}^{reg}\subseteq \mathbf{S}\,\mathbf{Up}\, Cs_{\alpha}^{reg}$ and $\mathbf{P}\ {}_{\kappa}Ws_{\alpha}\subseteq \mathbf{S}\,\mathbf{Up}\, Ws_{\alpha}$ if $\kappa>1$, by Corollary 7.2.

Proposition 7.4. Let $0<\kappa<\omega\leq\alpha$. Then

(i) There are $\mathfrak{A}, \mathfrak{B} \in {}_{\kappa}Cs_{\alpha}^{reg}$ such that
$\mathfrak{A}\times\mathfrak{B}\notin \mathbf{Uf\ Up}\, Cs_{\alpha}^{reg}$ and $\mathbf{s}\ {}^{2}\mathfrak{A} \not\subseteq \mathbf{Uf\ Up}\, Cs_{\alpha}^{reg}$.

(ii) There are $\mathfrak{A}, \mathfrak{B} \in {}_{\kappa}Ws_{\alpha}$ such that
$\mathfrak{A}\times\mathfrak{B}\notin \mathbf{Uf\ Up}\, Gws_{\alpha}^{comp\ reg}$ and $\mathbf{s}\ {}^{2}\mathfrak{A} \not\subseteq \mathbf{Uf\ Up}\, Gws_{\alpha}^{comp\ reg}$.

(iii) $\mathbf{P}\ {}_{\kappa}Cs_{\alpha}^{reg} \not\subseteq \mathbf{Uf\ Up}\, Cs_{\alpha}^{reg}$ and $\mathbf{P}\ {}_{\kappa}Ws_{\alpha} \not\subseteq \mathbf{Uf\ Up}\, Gws_{\alpha}^{comp\ reg}$.

Proof. Let $0<\kappa<\omega\leq\alpha$. Let $U \stackrel{d}{=} (\kappa+\kappa)\sim\kappa$, and $\mathfrak{L} \stackrel{d}{=} \mathfrak{Sb}\,{}^{\alpha}\kappa$. $G \stackrel{d}{=} At\,\mathfrak{L}$, $\mathfrak{A} \stackrel{d}{=} \mathfrak{Sg}^{(\mathfrak{L})}G$ and $\mathfrak{B} \stackrel{d}{=} \mathfrak{Mn}(\mathfrak{Sb}\,{}^{\alpha}U)$. By Thm 1.3 then $\mathfrak{A},\mathfrak{B} \in {}_{\kappa}Cs_{\alpha}^{reg}$. If

$\varkappa=1$ then $\mathfrak{A}\times\mathfrak{B} \models \exists x(0<x<1 = d_{01})$ while $Gws_\alpha^{comp\ reg} \subseteq Dind_\alpha \models$ $\models \forall x \neg (0<x<1 = d_{01})$. Thus $\mathfrak{A}\times\mathfrak{B} \notin \mathbf{SUp}\,Dind_\alpha \supseteq \mathbf{Uf\,Up}\,Dind_\alpha$. Assume therefore $\varkappa>1$. Let $\mathfrak{N} = \mathfrak{A}\times\mathfrak{B}$. Then $At\,\mathfrak{N} = \{\langle x,0\rangle : x\in G\}$, since $At\,\mathfrak{A} = G$, $At\,\mathfrak{B} = 0$. Thus $\langle 1^{\mathfrak{A}},0\rangle = \sum At\,\mathfrak{N}$ proving $\mathfrak{N} \not\models \zeta_{n+1}$, see Def.7.3.1. This proves $\mathfrak{A}\times\mathfrak{B} \notin \mathbf{Uf\,Up}\,Dind_\alpha \supseteq \mathbf{Uf\,Up}\,Gws_\alpha^{comp\ reg}$, by Lemma 7.3.2. Let $\mathfrak{R} \overset{d}{=} \mathfrak{Sg}^{(\mathfrak{A}\times\mathfrak{A})}(\{\langle x,0\rangle : x\in G\}\cup\{\langle 1^{\mathfrak{A}},0\rangle\})$. Then one easily proves either $\mathfrak{R} \cong \mathfrak{A}\times\mathfrak{B}$ or directly $\mathfrak{R} \not\models \zeta_{n+1}$ as above. Both proofs show $\mathfrak{R} \notin \mathbf{Uf\,Up}\,Dind_\alpha$, proving the rest of (i). Let $\bar{s} \overset{d}{=}$ $\overset{d}{=} \langle s : i<\alpha\rangle$ for all sets s. Let $V \overset{d}{=} {}^\alpha{}_\varkappa(\bar{0})$ and $W \overset{d}{=} {}^\alpha{}_U(\bar{\varkappa})$. Let $G^+ \overset{d}{=} At\,\mathfrak{Sb}\,V$, $\mathfrak{A}^+ \overset{d}{=} \mathfrak{Sg}^{(\mathfrak{Sb}\,V)}G^+$, $\mathfrak{B}^+ \overset{d}{=} \mathfrak{Mn}(W)$. Replacing every occurrence of $G$, $\mathfrak{A}$ and $\mathfrak{B}$ in the above proof of (i) by $G^+$, $\mathfrak{A}^+$ and $\mathfrak{B}^+$ respectively, we obtain a proof for (ii). (iii) is an immediate corollary of the above.

QED(Proposition 7.4.)

Problem 7.5. Let $1<\varkappa<\omega\le\alpha$. Let $\mathfrak{A} \in {}_\varkappa Ws_\alpha$. Is ${}^2\mathfrak{A} \in \mathbf{Uf\,Up}\,Ws_\alpha$ true? What is the answer if $Ws_\alpha$ is replaced by $Cs_\alpha^{reg}$ in both places?

Problems 7.7. Let $1<\varkappa<\omega\le\alpha$.

(i) Is $\mathbf{Uf\,Up}\,Cs_\alpha = \mathbf{S\,Up}\,Cs_\alpha$?

(ii) Is $\mathbf{Uf\,Up}\,{}_\infty Ws_\alpha = \mathbf{S\,Up}\,{}_\infty Ws_\alpha$?

(iii) Is $\mathbf{Uf\,Up}\,Cs_\alpha^{reg} = \mathbf{Uf\,Up}\,Gws_\alpha^{comp\ reg}$?

(iv) Is $\mathbf{Uf\,Up}\,{}_\infty Cs_\alpha^{reg} = \mathbf{Uf\,Up}\,({}_\infty Ws_\alpha \cup {}_0 Cs_\alpha)$?

(v) Is $\mathbf{Uf\,Up}\,{}_\infty Cs_\alpha^{reg} = \mathbf{I}\,{}_\infty Cs_\alpha$?

(vi) Is $\mathbf{Uf\,Up}\,({}_\varkappa Gs_\alpha^{reg} \cap Lf_\alpha) = \mathbf{Uf\,Up}\,({}_\varkappa Cs_\alpha^{reg} \cap Lf_\alpha)$?

(vii) Is $\mathbf{Uf\,Up}\,(Dind_\alpha \cap Gws_\alpha) = \mathbf{Uf\,Up}\,Gws_\alpha^{comp\ reg}$?

Proposition 7.8. Let $\mathfrak{A}$ be any nondiscrete $CA_\alpha$, $\alpha\ge\omega$. Then $\mathbf{Up}\,\mathfrak{A} \not\subseteq \mathbf{SP}\,Dc_\alpha$.

Proof. Let $I \overset{d}{=} Sb_\omega\alpha$. $x \overset{d}{=} \langle \sum\{d_{0i} : i\in\Gamma\sim 1\} : \Gamma\in I\rangle$. Let F be an

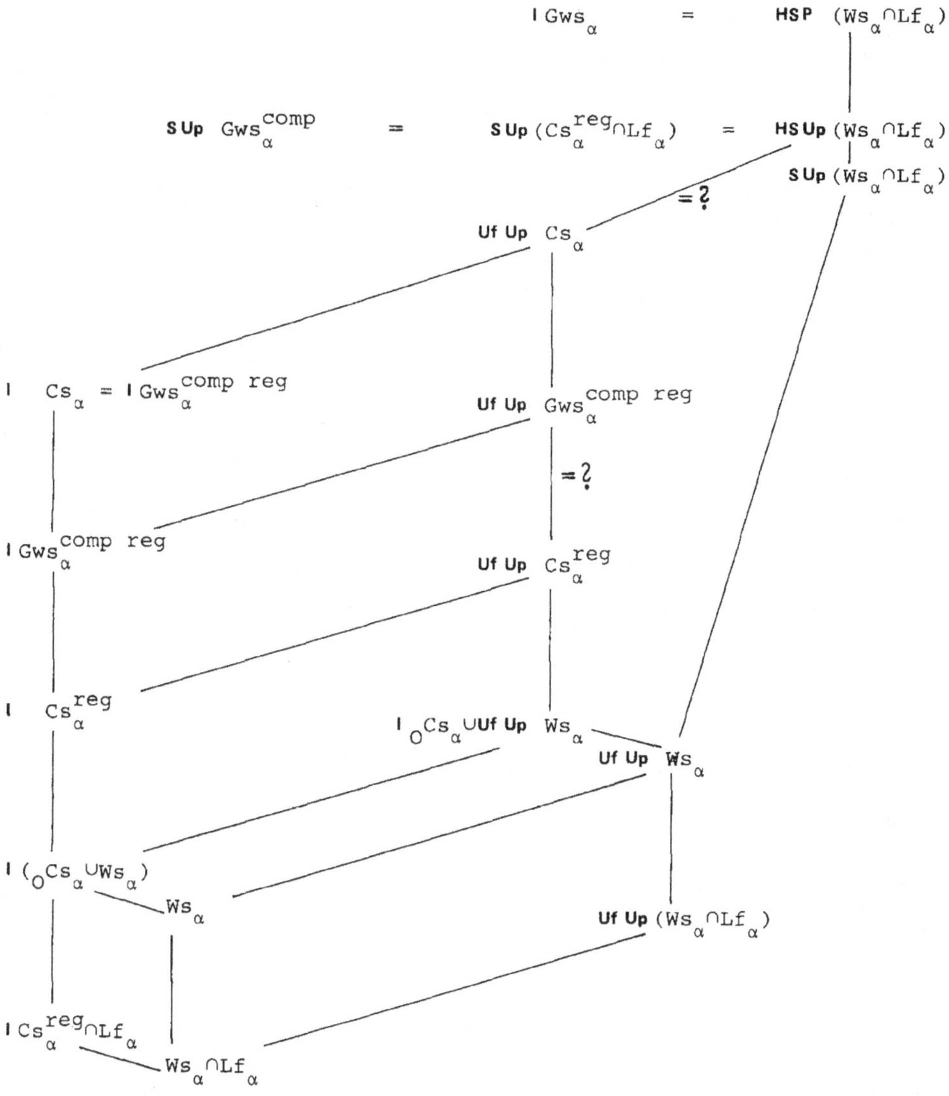


Figure 7.6. $(\alpha \geq \omega)$

The figure 7.6 gives all valid inclusions and equalities, except where =? appears we do not know about the equality, e.g. we do not know whether $\mathbf{Uf\,Up}\,Cs_\alpha = \mathbf{SUp}\,Cs_\alpha$ or not.

ultrafilter such that $(\forall\Gamma\in I)\{H\in I : \Gamma\subseteq H\}\in F$. Let $i\in\alpha$. Then $c_i(x/\bar{F}) = (c_i x)/\bar{F} = \langle c_i x_\Gamma : \Gamma\in I\rangle/\bar{F}$ and since $\{\Gamma\in I : i\in\Gamma\}\in F$ we have $c_i x/\bar{F} = 1$. Clearly $x/\bar{F} \neq 1$, by [HMT]. Then by $\mathbf{SP}Dc_\alpha \models [(\wedge\{c_i x=1 : i\in\alpha\}) \to x=1]$ we have ${}^I\mathfrak{A}/F \notin \mathbf{SP}Dc_\alpha$.

QED(Proposition 7.8)

Corollary 7.9. Let $\alpha\geq\omega$, $\varkappa\geq 2$. Then $\mathbf{HSP}Lf_\alpha \neq \mathbf{SP}Lf_\alpha$, $\mathbf{Up}({}_\varkappa Cs_\alpha^{reg}\cap Mn_\alpha) \not\subseteq \mathbf{SP}Dc_\alpha$.

Let $F$ be an ultrafilter on some set $I$. The notion of an $(F,\langle U_i : i\in I\rangle,\alpha)$-choice function $c$ and the function $Rep_c : P_{i\in I}(Sb^\alpha U_i)/F \to Sb^\alpha(P_{i\in I}U_i/F)$ associated with $c$ were defined in [HMTI] 7.1. By [HMTI]7.2, if $\mathfrak{A}\in{}^I Crs_\alpha$ and $(\forall i\in I)base(\mathfrak{A}_i)=U_i$ then $Rep_c\in Ho(P_{i\in I}\mathfrak{A}_i/F, \mathfrak{B})$ for some $\mathfrak{B}\in Crs_\alpha$. We shall use these notions without recalling their definitions.

Proposition 7.10 below is a generalization of [HMTI]7.3-6. We omit the immediate corollaries of Prop.7.10 analogous to [HMTI]7.8-10 concerning various closure properties of the classes $Gws_\alpha^{wd}$, $Gws_\alpha^{norm}$, ${}_\infty Gws_\alpha^{comp}$, ${}_\varkappa Gws_\alpha$ etc.

By (i) and (iii) of Prop.7.10 below regularity is the only property (among the ones investigated here) of $Gws_\alpha$-s which is destroyed under every nontrivial $Rep_c$.

Proposition 7.10. Let $\mathfrak{A}\in{}^I Gws_\alpha$ and let $F$ be any ultrafilter on $I$. Let $U \overset{d}{=} \langle base(\mathfrak{A}_i) : i\in I\rangle$. Assume $PU \neq 0$. Let $c$ be any $(F, U,\alpha)$-choice function. Then (i) - (viii) below hold.

(i) Let $\mathfrak{A}\in{}^I Cs_\alpha^{reg}$ be a system of nondiscrete algebras, $\alpha\geq\omega$. Then $Rep_c{}^*P\mathfrak{A}/F$ is regular iff $F$ is $|\alpha|^+$-complete.

(ii) Let $K\in\{Gws^{norm}, Gws^{comp}\}$. If $\mathfrak{A}\in{}^I K_\alpha$ then $Rep_c{}^*P\mathfrak{A}/F \in K_\alpha$.

(iii) Let $K\in\{Gws^{wd}, Gws^{norm}, Gws^{comp}, Gs, Cs, Ws\}$ and $\mathfrak{A}\in{}^I K_\alpha$. Then for every nonzero $x\in PA/F$ there exists an $(F,U,\alpha)$-choice function $e$ such that $\mathrm{Rep}_e{}^*P\mathfrak{A}/F\in K_\alpha$, $\mathrm{Rep}_e(x)\neq 0$ and $\mathrm{base}(\mathrm{Rep}_e{}^*P\mathfrak{A}/F) = PU/F$.

(iv) Assume that $F$ is $|\alpha|^+$-complete and let $K$ be as in (iii) or let $K = \mathrm{Gws}^{\mathrm{reg}}$. Then $\mathrm{base}(\mathrm{Rep}_c{}^*P\mathfrak{A}/F) = PU/F$ and if $\mathfrak{A}\in{}^IK_\alpha$ then $\mathrm{Rep}_c{}^*P\mathfrak{A}/F\in K_\alpha$.

(v) If $\mathfrak{A}\in{}^I\mathrm{Gws}_\alpha^{\mathrm{comp}}$ then $\mathrm{base}(\mathrm{Rep}_c{}^*P\mathfrak{A}/F)\in\{PU/F,0\}$.

(vi) Assume that $F$ is $|\alpha|$-regular. Then for every nonzero $x\in PA/F$ there is an $(F,U,\alpha)$-choice function $e$ such that $\mathrm{Rep}_e(x)\neq 0$ and $\mathrm{base}(\mathrm{Rep}_e{}^*P\mathfrak{A}/F) = PU/F$.

(vii) $\mathrm{Subb}(\mathrm{Rep}_c{}^*P\mathfrak{A}/F)\subseteq\{PY/\bar{F}^{(U)} : Y\in P_{i\in I}\,\mathrm{Subb}(\mathfrak{A}_i)\}$.

(viii) Let $\beta\overset{d}{=}|{}^I\varkappa/F|$. If $\mathfrak{A}\in{}^I{}_\varkappa\mathrm{Gws}_\alpha$ then $\mathrm{Rep}_c{}^*P\mathfrak{A}/F\in{}_\beta\mathrm{Gws}_\alpha$.

Proof. First we prove (i). Let $\alpha\geq\omega$ and $F$ be any ultrafilter on some set $I$. Let $\mathfrak{A}\in{}^I\mathrm{Cs}_\alpha$ be a system of nondiscrete $\mathrm{Cs}_\alpha$-s. Let $U\overset{d}{=}\langle\mathrm{base}(\mathfrak{A}_i) : i\in I\rangle$. Let $c$ be any $(F,U,\alpha)$-choice function. First we show that $\mathrm{Rep}_c{}^*P\mathfrak{A}/F$ is not regular if $F$ is not $|\alpha|^+$-complete. Suppose that $F$ is not $|\alpha|^+$-complete. We show that $\mathrm{Rep}_c{}^*P\mathfrak{A}/F\notin$ $\notin\mathrm{Dind}_\alpha$. Since $F$ is not $|\alpha|^+$-complete there are $\lambda=|\lambda|\leq\alpha$ and $Z\in$ $\in{}^\lambda F$ such that $(\forall i<j<\lambda)Z_i\supseteq Z_j$ and $\cap_{i<\lambda}Z_i\notin F$. This easily follows from Lemma 4.2.3 on p.180 in [CK]. Let $f:\lambda\rightarrowtail\lambda$ and $t:\lambda\rightarrowtail\lambda$ be two one-one functions such that $Rgf\cap Rgt = 0$ and $t(0)=0$. We may assume that $U_i$ is an ordinal for every $i\in I$. Let $\bar{0}\overset{d}{=}\langle 0 : i\in I\rangle$ and $\bar{1}\overset{d}{=}\langle 1 : i\in I\rangle$. Then $\bar{0},\bar{1}\in PU$ since $(\forall i\in I)2\subseteq U_i$ by the hypotheses. We define the function $Y:\lambda\to\mathrm{Sb}\,I$ such that $(\forall j\in\lambda)Y_j\overset{d}{=}\{i\in I : (\forall n<2)$ $c(f_j,\bar{n}/F)_i=c(t_j,\bar{n}/F)_i=n\}$. Let $Y^+\overset{d}{=}\langle\cap\{Y_j : j\leq\mu\}\cap Z_\mu : \mu<\lambda\rangle$. Let $\rho\overset{d}{=}\cap\{\beta\in(\lambda+1) : \cap_{i<\beta}Y^+_i\notin F\}$. Then $\rho$ is a limit ordinal, $\omega\leq\rho\leq\lambda$ by $\cap_{i<\lambda}Y^+_i\notin F$, and $(\forall i<\rho)Y^+_i=\cap\{Y_j : j\leq i\}\cap Z_i\in F$ because $\rho$ is a limit ordinal. Let $R\overset{d}{=}Y^+_0\sim\cap_{i<\rho}Y^+_i$ and $H\overset{d}{=}\langle Y^+_i\sim Y^+_{i+1} : i\in\rho\rangle$. Then $R\in F$ and $H:\rho\to\mathrm{Sb}R$ is a partition of $R$ such that $RgH\cap F = 0$ and $(\forall m<\rho)H_m\subseteq Y_m$. Let $z\in PA$ be such that $(\forall m<\rho)(\forall i\in H_m)z_i = d_{tm,fm}^{(\mathfrak{A}_i)}$.

Claim 1. $\Delta(z/F) = 0$ in $P\mathfrak{A}/F$.

Proof. Clearly, $\Delta(z/F) \subseteq t^*\rho \cup f^*\rho$. Let $j \in t^*\rho$. Let $j=tm$ for $m<\rho$. Let $T \overset{d}{=} \{i \in R : j \in \Delta(z_i)\}$. Then $T=H_m$ by $Rgt \cap Rgf = 0$ and since $t$ is one-one. Thus $j \notin \Delta(z/F)$ by $H_m \notin F$. The case $j \in f^*\rho$ is entirely analogous. QED.

Let $p,q \in {}^{\alpha}(PU/F)$ be such that $p \overset{d}{=} \langle \bar{0}/F : j<\alpha \rangle$ and $(\forall m<\rho)[q(tm)= =\bar{0}/F$ and $q(fm)=\bar{1}/F]$. Then in particular $p0=q0$ by $t0=0$.

Claim 2. $p \in Rep_c(z/F)$ and $q \notin Rep_c(z/F)$.

Proof. Let $m<\rho$ and $i \in H_m$. Then by $H_m \subseteq Y_m$ we have $c(tm,\bar{0}/F)_i = = c(fm,\bar{0}/F)_i = 0$ and $c(fm,\bar{1}/F)_i = 1$. Thus $(c^+p)_i = \langle c(j,p_j)_i : j<\alpha \rangle \in \in d_{tm,fm}^{(\mathfrak{A}_i)} = z_i$ and $(c^+q)_i \notin d_{tm,fm}^{(\mathfrak{A}_i)} = z_i$. By $\cup\{H_m : m<\rho\} \in F$ then $p \in Rep_c(z/F)$ and $q \notin Rep_c(z/F)$. QED.

Let $\mathfrak{P} \overset{d}{=} P\mathfrak{A}/F$. Let $\mathfrak{R} \overset{d}{=} Rep_c{}^*\mathfrak{P}$. Then $Rep_c \in Ho(\mathfrak{P},\mathfrak{R})$ and $\mathfrak{R} \in \in Cs_\alpha$ and $base(\mathfrak{R}) = PU/F$ by [HMTI]7.1-6. Therefore $p,q \in 1^{\mathfrak{R}}$ and hence $0<Rep_c(z/F)<1$ in $\mathfrak{R}$. By Claim 1 we have $\Delta^{(\mathfrak{R})}Rep_c(z/F) = 0$ proving $|Zd\,\mathfrak{R}|>2$ and hence $\mathfrak{R} \notin Dind_\alpha \supseteq Gws_\alpha^{comp\ reg}$. Thus $\mathfrak{R}$ is not regular by $\mathfrak{R} \in Cs_\alpha$. We have seen that $Rep_c{}^*P\mathfrak{A}/F$ is not regular if $F$ is not $|\alpha|^+$-complete. To prove the other direction, we shall use Lemma 7.10.1 below.

Lemma 7.10.1. Let $\mathfrak{A} \in {}^ICrs_\alpha$, $F$ any ultrafilter on $I$ and $c$ be any $(F,\langle base(\mathfrak{A}_i) : i \in I\rangle,\alpha)$-choice function. Let $x \in PA$, $Z \in F$ and $\Gamma \subseteq \alpha$ be such that $(\forall i \in Z)[x_i$ is $\Gamma$-regular in $\mathfrak{A}_i$ and $\Delta(x_i) \subseteq \Delta(x/\bar{F}) \cup \Gamma]$. Then $Rep_c(x/F)$ is $\Gamma$-regular in $Rep_c{}^*P\mathfrak{A}/F$.

Proof. The proof of [HMTI]7.6 proves the present lemma if we replace the statement "Let $\Gamma = 1 \cup \Delta f(a/\bar{F})$" with "assume all the hypotheses" and replace "$i \in I$" with "$i \in Z$" throughout the proof. QED(Lemma 7.10.1)

Suppose now $\mathfrak{A} \in {}^ICrs_\alpha^{reg}$ and $F$ is $|\alpha|^+$-complete. Let $\mathfrak{R} \overset{d}{=} \overset{d}{=} Rep_c{}^*P\mathfrak{A}/F$. We have to show that $\mathfrak{R}$ is regular. Let $y \in R$. Then $y = Rep_c(x/\bar{F})$ for some $x \in PA$. Let $\Delta \overset{d}{=} \Delta^{(\mathfrak{R})}y$. Then $\Delta = \Delta^{(P\mathfrak{A}/F)}(x/\bar{F})$

since $Rep_c$ is an isomorphism by [HMTI]7.3(ii). Let $Y_j \overset{d}{=}$ $\overset{d}{=} \{i\in I : j \notin \Delta^{(\mathfrak{A}_i)}x_i\}$ for every $j\in\alpha$. Then $Z \overset{d}{=} \cap\{Y_j : j\in\alpha\sim\Delta\}\in F$ by $|\alpha|^+$-completeness of F and by $\Delta = \Delta(x/\bar{F})$. We have $(\forall i\in Z)\Delta^{(\mathfrak{A}_i)}x_i\subseteq$ $\subseteq\Delta$. Thus $(\forall i\in Z)(x_i$ is $1\cup\Delta$-regular in $\mathfrak{A}_i)$, by $\mathfrak{A}_i\in Crs_\alpha^{reg}$. Then y is $1\cup\Delta$-regular in $\mathfrak{R}$ by Lemma 7.10.1. Then y is regular in $\mathfrak{R}$ by $\Delta^{(\mathfrak{R})}y=\Delta$. We have seen that $\mathfrak{R}$ is regular.

For the proofs of (ii) - (viii) assume the notation in the proof of [HMTI]7.4.

<u>To prove (ii)</u>, suppose that $K = Gws^{norm}$. By (2) and (3) it is enough to show that if $j,k\in PJ$ and $j/\bar{F} \neq k/\bar{F}$ then $Q_j\cap Q_k = 0$ or $Q_j=Q_k$. Indeed, we then have $\{i\in I : ji\neq ki\}\in F$. Clearly $\{i\in I : ji\neq ki\} \subseteq$ $\subseteq \{i\in I : Y_{i,ji}=Y_{i,ki}\} \cup \{i\in I : Y_{i,ji}\cap Y_{i,ki}=0\}$, so our desired conclusion is clear. The case $K = Gws^{comp}$ is similar, so (ii) holds.

The proof of [HMTI]7.4 gives <u>(vii) and (viii)</u> directly.

<u>For (v)</u>, suppose $base(Rep_c{}^*P\mathfrak{A}/F) \neq 0$, and let therefore $j\in PJ$ and $q\in W_j$. Take any $x\in X$. Then it is easily checked that $q_x^0\in W_j$ also, so that x is in the base. So $X = base(Rep_c{}^*P\mathfrak{A}/F)$, as desired.

<u>Proof of (iii):</u> The case $K = Gws^{comp}$ follows from 7.10(ii),(v) and [HMTI]7.2, the case $K = Ws$ follows from $Ws_\alpha\subseteq Gws_\alpha^{comp}$, 7.10(v) and [HMTI]7.5, and the case $K = Cs$ follows from [HMTI]7.2, 7.4(i).

Let $\mathfrak{A}\in {}^IGws_\alpha^{norm}$ and let $a\in PA$ be such that $a/\bar{F} \neq 0$. We may assume that $(\forall i\in I)p_{ir(i)}\in a_i$ for some $r\in PJ$. Let the equivalence $\equiv$ be defined on PJ by the following:

$(\forall z,w\in PJ)[z \equiv w$ iff $\{i\in I : Y_{iz(i)}=Y_{iw(i)}\}\in F]$.

Let $\rho : PJ/\equiv \to PJ$ be a choice function such that $\rho(r/\equiv) = r$. Let $\overset{d}{=} \langle\langle p_{i\rho(z)i} : i\in I\rangle : z\in PJ/\equiv\rangle$. Let $c : \alpha\times X \to PU$ be any choice function such that

(1) $c(\varkappa,pj_\varkappa\circ\pi(z)/\bar{F}) = pj_\varkappa\circ\pi(z)$ for all $\varkappa<\alpha$, $z\in PJ/\equiv$.

Such a c exists by the following. Let $z,w\in PJ/\equiv$ be such that $z\neq w$.

Then there is $Z \in F$ such that $(\forall i \in Z) Y_{i\rho(z)i} \neq Y_{i\rho(w)i}$ and hence $(\forall i \in Z)$ $Y_{i\rho(z)i} \cap Y_{i\rho(w)i} = 0$ by normality. Then $(\forall i \in Z)(\forall \varkappa < \alpha) \pi(z)_i \varkappa \neq \pi(w)_i \varkappa$. Thus $(\forall \varkappa < \alpha)(pj_\varkappa \circ \pi(z))/\bar{F} \neq (pj_\varkappa \circ \pi(w))/\bar{F}$.

We show that $c$ is a choice function with the desired properties. Let $f \overset{d}{=} Rep_c$. Then $f(V/\bar{F})$ is a $Gws_\alpha^{norm}$-unit, by 7.10(ii). $f(a/\bar{F}) \neq 0$ by [HMTI]7.2 since $c$ and $a$ satisfy the hypotheses of the last part of [HMTI]7.2 by $\rho(r/\equiv) = r$. It remains to show that $base(f(V/\bar{F})) = PU/\bar{F} = X$. Let $y = t/\bar{F} \in X$. Let $v \in PJ$ be such that $(\forall i \in I) t_i \in Y_{iv(i)}$. Let $z \overset{d}{=} v/\equiv$ and $q \overset{d}{=} \langle pj_\varkappa \circ \pi(z)/\bar{F} : \varkappa < \alpha \rangle (0/y)$. Then clearly $q \in f(V/\bar{F})$, and so $y \in$ $\in base(fV/\bar{F})$.

Suppose now that $\mathfrak{A} \in {}^I Gws_\alpha^{wd}$. We may assume that $(\forall i \in I)(\forall j \in J_i)$ $|Y_{ij}| > 1$. Then there is a choice function $c$ which, in addition to the above, satisfies

(2) $c(\varkappa, q)i \neq p_{i\rho(z)i}(\varkappa)$, for all $\varkappa < \alpha$, $i \in I$, $z \in PJ/\equiv$ and $q \in X$, $q \neq$ $\neq pj_\varkappa \circ \pi(z)/\bar{F}$.

Then it is easy to see that $f(V/\bar{F})$ is a $Gws^{wd}$-unit, see the proof of 7.14.2.

Suppose that $\mathfrak{A} \in {}^I Gs_\alpha$. Then there is a $c$ which, in addition to (1), satisfies (3) below.

(3) $c(\varkappa, q) \in P_{i \in I} Y_{i\rho(j/\equiv)i}$ for all $\varkappa < \alpha$, $j \in PJ$ and $q \in Q_j$.

Then it is not hard to see that condition (3) ensures that $f(V/\bar{F})$ is a $Gs_\alpha$-unit; see also (4) in the proof of [HMTI] 7.4.

Proof of (iv): Assume that $F$ is $|\alpha|^+$-complete. First we check that in any case, $base(f^* P\mathfrak{A}/F) = X$. Clearly $PU/F = \cup\{Q_j : j \in PJ\}$, so it suffices to show that for every $j \in PJ$ we have $W_j \neq 0$. Let $q\varkappa =$ $= \langle p_{i,ji} : i \in I \rangle / \bar{F}$ for all $\varkappa < \alpha$, and let $c'$ be an $(F, U, \alpha)$-choice function such that $c'(\varkappa, q\varkappa) = \langle p_{i,ji} : i \in I \rangle$ for all $\varkappa < \alpha$. Then $(c'^+ q)_i = p_{i,ji}$ for all $i \in I$, and so $q \in W_j$ using [HMTI]7.3.

Now for $K = Gws^{reg}$, (iv) follows from the above, from the proof of

(i) and from [HMTI]7.3(i). For the remaining choices of K, (iv) follows from (iii) and from [HMTI]7.3(i).

Proof of (vi): By (iv) we may assume that $\alpha\geq\omega$. Let F be an $|\alpha|$-regular ultrafilter on I. Then there is $h : I \to Sb_\omega\alpha$ such that $(\forall\varkappa<\alpha)\{i\in I : \varkappa\in h(i)\}\in F$. Let $\mathfrak{A}\in {}^I Gws_\alpha$ and $a\in PA$ be such that $Pa\neq 0$. Let $s\in Pa$ and let $q \stackrel{d}{=} \langle (pj_\varkappa\circ s)/\bar{F} : \varkappa<\alpha\rangle$. Then $q\in{}^\alpha X$. Let $w : X \to PU$ and $j : X \to PJ$ be such that $w(y)\in y$ and $w(y)\in P\langle Y_{i,j(y)i} : i\in I\rangle$ for all $y\in X$. Let the choice function $c : \alpha\times X \to PU$ be such that $c(\varkappa,q_\varkappa)=pj_\varkappa\circ s$ for all $\varkappa<\alpha$ and

$$c(\varkappa,y)_i = \begin{cases} p_{i,j(y)i} & \text{if } \varkappa\notin h(i) \\ w(y)_i & \text{if } \varkappa\in h(i) \end{cases} \quad \text{for all } \varkappa<\alpha,\ y\in X\sim Rgq, \text{ and } i\in I.$$

Such a c exists. Let $f = Rep(F,c)$. Now $f(a/\bar{F})\neq 0$ since $q\in f(a/\bar{F})$. Let $y\in X$ be arbitrary. If $y\in Rgq$ then $y\in base(f^*P\mathfrak{A}/F)$ since $q\in f(a/\bar{F})$. Let $y\in X\sim Rgq$. Let $p \stackrel{d}{=} \langle y : \varkappa<\alpha\rangle$. Let $i\in I$. Let $\gamma \stackrel{d}{=} j(y)i$. Then $(c^+p)_i = [(\alpha\sim h(i))1p_{i\gamma}\cup h(i)1w(y)_i]\in{}^\alpha Y_{i\gamma}{}^{(p_{i\gamma})} \subseteq V_i$, since $|h(i)|<\omega$. Therefore $p\in f(V/\bar{F})$, thus $y\in base(f^*P\mathfrak{A}/\bar{F})$.

QED(Proposition 7.10)

All conditions of Prop.7.10 above are needed. Part of this is formulated in Remark 7.11 below which is quoted in the second part of [HMTI]7.7.

Remark 7.11. (Discussion of Proposition 7.10.)

1. Statement (ii) of 7.10 does not generalize to any of the classes $Gws_\alpha^{wd}$, $Ws_\alpha$, $Gs_\alpha$, $Gws_\alpha^{reg}$.
2. Statement (iii) of 7.10 cannot be generalized to $Gws_\alpha$. Namely, let I and $\alpha\geq\omega$ be arbitrary. Assume $|I| < |\alpha|$. Then there exists $\mathfrak{A}\in Gws_\alpha$ such that for every nonprincipal ultrafilter F on I and every $(F,\langle base(\mathfrak{A}) : i\in I\rangle,\alpha)$ - choice function c we have $base(Rep_c{}^{*I}\mathfrak{A}/F) \neq {}^I(base(\mathfrak{A}))/\bar{F}$.
3. The condition $\mathfrak{A}\in{}^I Gws_\alpha^{comp}$ is needed in (v).: For any $\mathfrak{A}\in$

$I(Gws_\alpha \sim Gws_\alpha^{comp})$ and any not $|\alpha|^+$-complete ultrafilter $F$ on $I$ there is an $(F, \langle base(\mathfrak{A}_i) : i \in I\rangle, \alpha)$-choice function $e$ such that $base(Rep_e{}^* P\mathfrak{A}/F) \notin \{P_{i\in I} base(\mathfrak{A}_i)/\bar{F}, 0\}$.

4. For any nondiscrete $\mathfrak{A} \in Gws_\alpha$ with finite base $U$, for any $|\alpha|$--regular ultrafilter $F$ on $I$, for any $\mathfrak{B} \in Gws_\alpha \cap I\{{}^I\mathfrak{A}/F\}$ we have $base(\mathfrak{B}) \not\subseteq {}^IU/\bar{F}$ if $\alpha \geq \omega$.

5. Under the assumptions of (iii) there is a choice function $e$ for which the inclusion in (vii) can be replaced by equality and the conclusions of (iii) hold.

To save space, we omit the proofs of the above statements 1-5.

Theorem 7.12(ii) below implies that the condition $\varkappa = \varkappa^{|\alpha|}$ can be replaced with the weaker condition $\varkappa > \alpha$ in [HMTI]7.25(ii). $\varkappa = \varkappa^{|\alpha|}$ can be weakened to $\varkappa > \alpha$ in [HMTI]3.18(iii), too, if we require only isomorphism instead of ext-isomorphism.

<u>Theorem 7.12</u> Let $\varkappa$ and $\beta$ be two cardinals. Let $\alpha \geq \omega$, $\beta \geq \omega$ and $\mathfrak{A} \in {}_\beta Cs_\alpha$.

(i) If $\varkappa \geq |A| + |\alpha|^+$ then $\mathfrak{A} \in I\, {}_\varkappa Cs_\alpha$.

(ii) If $\varkappa \geq |A| + |\alpha|^+ + \beta$, or if $\varkappa \geq 2^{|\alpha\cup\beta|}$, then $\mathfrak{A}$ is strongly sub-isomorphic to a ${}_\varkappa Cs_\alpha$.

(iii) Let $\varkappa \geq 2^{|\alpha\cup\beta|}$. Then ${}_\beta Gws_\alpha \subseteq I\, {}_\varkappa Cs_\alpha$.

(iv) Assume the GCH. Let $\varkappa \geq \beta + |\alpha|^+$. Then ${}_\beta Cs_\alpha \subseteq I\, {}_\varkappa Cs_\alpha$, moreover every ${}_\beta Cs_\alpha$ is strongly sub-isomorphic to some ${}_\varkappa Cs_\alpha$.

Theorem 7.12(i)-(iii) follow from Lemma 7.12.1 below. In particular, 7.12(ii) follows from applying (ii) and (iii) of 7.12.1 together.

<u>Lemma 7.12.1</u> Let $\alpha \geq \omega$, $\mathfrak{A} \in {}_\infty Gws_\alpha$. Let $\varkappa \geq |A| \cap (2^{|\alpha|} + \Sigma\{2^{|Y|} : Y \in Subb(\mathfrak{A})\})$. Then (i)-(iii) hold for some $\mathfrak{B}, \mathfrak{L} \in I\mathfrak{A}$.

(i) $\mathfrak{B} \in {}_\varkappa Gws_\alpha^{comp}$ and $(\forall V \in Subu(\mathfrak{B}))V \cap 1^{\mathfrak{A}} \neq 0$.

(ii) If $\varkappa > \alpha$ then $\mathfrak{L} \in {}_\varkappa Cs_\alpha$.

(iii) Let $\varkappa \geq |base(\mathfrak{A})|$. Then $\mathfrak{A}$ is sub-isomorphic to both $\mathfrak{B}$ and $\mathfrak{L}$ and $rl(1^{\mathfrak{A}}) : Subu(\mathfrak{B}) \rightarrowtail\!\!\!\twoheadrightarrow Subu(\mathfrak{A})$. If $\mathfrak{A} \in Gws_\alpha^{comp}$ then $\mathfrak{A}$ is strongly sub-isomorphic to $\mathfrak{B}$.

Proof. Let $\alpha$, $\mathfrak{A}$ and $\varkappa$ be as in the hypotheses. We may assume that $\mathfrak{A}$ is nondiscrete, i.e. $|A| \neq 1$. Let $Y = \langle base(V) : V \in Subu(\mathfrak{A})\rangle$ and $p \in P\langle V : V \in Subu(\mathfrak{A})\rangle$. Let $I \overset{d}{=} Subu(\mathfrak{A})$. Then $(\forall i \in I) i = {}^{\alpha}Y_i^{(pi)}$. If $\varkappa \geq |\cup\{Rgp_i : i \in I\}|$ then let $J \overset{d}{=} I$. Otherwise let $J \subseteq I$ be such that $|J| \leq |A|$ and $(\forall a \in A \sim \{0\})(\exists j \in J) a \cap {}^{\alpha}Y_j^{(pj)} \neq 0$. Now $J$ is such that $\varkappa \geq$ $\geq |\cup\{Rgp_i : i \in J\}|$ and $J = I$ if $\varkappa \geq |base(\mathfrak{A})|$, since: Suppose $\varkappa < |\cup\{Rgp_i :$ $: i \in I\}|$. Then $\varkappa < |base(\mathfrak{A})|$ and $\varkappa < |I| \cdot |\alpha| \leq 2^{|\alpha|} + \sum\{2^{|Yi|} : i \in I\}$. Thus $\varkappa \geq |A| \geq |J|$ by the hypotheses on $\varkappa$. Since $\mathfrak{A}$ is nondiscrete we have $|A| \geq |\alpha|$ hence $\varkappa \geq |J| \cdot |\alpha|$.

Let $Z \overset{d}{=} \cup\{{}^{\alpha}Y_i^{(pi)} : i \in J\}$. Then $rl_Z^{\mathfrak{A}} \in Is(\mathfrak{A}, \mathfrak{B})$ for some $\mathfrak{B}$ with unit $Z$. Then by [HMTI]6.2, $\mathfrak{A} \cong | \subseteq P_{i \in J} \mathfrak{B}_i$ where each $\mathfrak{B}_i$ is a $Ws_\alpha$ with unit ${}^{\alpha}Y_i^{(pi)}$. We may assume $|B_i| \leq |A|$ for all $i \in J$. Then $|B_i| \leq \varkappa$ by $|B_i| \leq 2^{|\alpha \cup Yi|}$. By [HMTI]3.18(ii), and [HMTI]7.25(i) each $\mathfrak{B}_i$ is sub- or ext-isomorphic to a $\mathfrak{L}_i' \in Ws_\alpha$ with unit ${}^{\alpha}T_i^{(pi)}$ such that $\varkappa = |T_i| = |T_i \sim Rgpi|$. Let $r_i \in Is(\mathfrak{L}_i', \mathfrak{B}_i)$ be this sub- or ext-isomorphism. Let $U$ be a set such that $U \supseteq \cup\{Rgp_i : i \in J\}$ and $|U| \leq \varkappa$. Let $S$ be a set such that $U \subseteq S$ and $|S \sim U| = \varkappa$. Let $i \in J$. By $|T_i \sim$ $\sim Rgp_i| = |S \sim Rgp_i|$, there is a bijection $k_i : T_i \rightarrowtail\!\!\!\twoheadrightarrow S$ such that $(Rgp_i) 1 k_i \subseteq Id$. This $k_i$ induces a base-isomorphism $\tilde{k}_i \in Is(\mathfrak{L}_i', \mathfrak{L}_i)$ where $\mathfrak{L}_i$ is of unit ${}^{\alpha}S^{(pi)}$. Let $W \overset{d}{=} \cup\{{}^{\alpha}S^{(pi)} : i \in J\}$. By [HMTI] 6.2, $P_{i \in J} \mathfrak{L}_i \cong \mathfrak{P}$ where $\mathfrak{P}$ is of unit $W$. Hence $\mathfrak{P} \in {}_{\varkappa}Gws_\alpha^{comp}$ and $(\forall V \in Subu(\mathfrak{P})) V \cap 1^{\mathfrak{A}} \neq 0$. Now $\mathfrak{A} \cong | \subseteq P_{i \in J} \mathfrak{B}_i \cong P_{i \in J} \mathfrak{L}_i \cong \mathfrak{P}$ completes the proof of (i).

Proof of (ii): Let $\varkappa > \alpha$. By (i) we have that $\mathfrak{A} \cong \mathfrak{P}$ where $\mathfrak{P}$ has unit $W = \cup\{{}^{\alpha}S^{(pi)} : i \in J\}$. Let $Q \overset{d}{=} {}^{\alpha}S \sim W$ and let $Q = \cup\{{}^{\alpha}S^{(qi)} :$ $: i < \gamma\}$ be such that $(\forall i < j < \gamma) q_i \notin {}^{\alpha}S^{(qj)}$. $J \neq 0$ by $|A| \neq 1$. Let $e \in J$ be fixed. Then $\mathfrak{L}_e \in H\mathfrak{A}$ is of unit ${}^{\alpha}S^{(pe)}$ and $\varkappa = |S| \geq |C_e|$. Let $i < \gamma$. By $\varkappa > \alpha$ we have $|S \sim Rgq_i| = \varkappa$ and therefore by [HMTI]7.27,

$\mathfrak{L}_e \succcurlyeq \mathfrak{N}_i$ for some $\mathfrak{N}_i$ with unit ${}^{\alpha}S^{(qi)}$. By [HMTI]6.2, $\mathfrak{P}\times \times P_{i<\gamma}\mathfrak{N}_i \cong \mathfrak{R}$ where $\mathfrak{R}$ is of unit $W \cup \cup\{{}^{\alpha}S^{(qi)} : i<\gamma\} = {}^{\alpha}S$. Therefore $\mathfrak{R} \in {}_{\varkappa}Cs_{\alpha}$. By $\mathfrak{A} \cong \mathfrak{P}$ and $\{\mathfrak{N}_i : i<\gamma\} \subseteq \mathbf{H}\mathfrak{A}$ we have $\mathfrak{A} \cong \mathfrak{R}$. We have proved $\mathfrak{A} \in \mathbf{I}\,{}_{\varkappa}Cs_{\alpha}$.

Proof of (iii): Suppose $\varkappa \geq |base(\mathfrak{A})|$. We shall use the notations of the proof of (i). Then $J=I$. Let $i\in I$. We can choose $T_i$ such that $Y_i \subseteq T_i$ and $|T_i \sim Y_i| = \varkappa$. Choose $U$ to be $base(\mathfrak{A})$ and let $k_i : T_i \rightarrowtail\!\!\!\twoheadrightarrow S$ be such that $Y_i 1 k_i \subseteq Id$. $r_i \overset{d}{=} rl({}^{\alpha}Y_i) \in Is(\mathfrak{L}_i', \mathfrak{B}_i)$. Define $z_i \overset{d}{=} \tilde{k}_i \circ r_i^{-1}$. Then $z_i \in Is(\mathfrak{B}_i, \mathfrak{L}_i)$. Let $g \overset{d}{=} \langle \cup\{z_i(x \cap {}^{\alpha}Y_i^{(pi)}) : i\in I\} : x\in A\rangle$. Then $g \in Is(\mathfrak{A}, \mathfrak{P})$. Let $V = 1^{\mathfrak{A}}$. To show that $g$ is a sub-isomorphism it is enough to show that $(\forall x\in A) V\cap g(x) = x$. Let $x\in A$. Let $i\in I$ and let $x_i \overset{d}{=} x \cap {}^{\alpha}Y_i^{(pi)}$. Then $z_i x_i \in C_i \subseteq Sb\,{}^{\alpha}S^{(pi)}$, and therefore $V\cap z_i x_i = {}^{\alpha}Y_i \cap z_i x_i$. Then by $Y_i 1 k_i \subseteq Id$ and by the definition of $r_i$ we have ${}^{\alpha}Y_i \cap z_i x_i = {}^{\alpha}Y_i \cap \tilde{k}_i r_i^{-1} x_i = (r_i^{-1} x_i) \cap {}^{\alpha}Y_i = x_i$. We have seen that $(\forall i\in I) V\cap z_i x_i = x_i$. Now $V\cap g(x) = V \cap \cup\{z_i x_i : i\in I\} = \cup\{V\cap z_i x_i : i\in I\} = \cup\{x_i : i\in I\} = x$. We have seen that $g$ is a sub-isomorphism. By $1^{\mathfrak{P}} = W = \cup\{{}^{\alpha}S^{(pi)} : i\in I\}$ and $1^{\mathfrak{A}} = \cup\{{}^{\alpha}Y_i^{(pi)} : i\in I\}$ and $(\forall i\in I) Y_i \subseteq S$ we conclude $rl(1^{\mathfrak{A}}) : Subu(\mathfrak{B}) \rightarrowtail\!\!\!\twoheadrightarrow Subu(\mathfrak{A})$. Suppose $\mathfrak{A}$ is compressed. Then $(\forall i\in I) Y_i = U$. Then by the above, $V\cap z_i x_i = {}^{\alpha}Y_i \cap z_i x_i = {}^{\alpha}U \cap z_i x_i$. Therefore ${}^{\alpha}U \cap g(x) = x$ for every $x\in A$, showing that $g$ is a strong sub-isomorphism. Consider the proof of (ii): Let $h_i : \mathfrak{A} \to \mathfrak{N}_i$ be a homomorphism for every $i<\gamma$. Let $h \overset{d}{=} \langle g(x) \cup \cup\{h_i x : i<\gamma\} : x\in A\rangle$. Then $h \in Ism(\mathfrak{A}, \mathfrak{R})$. We show that $h$ is a sub-isomorphism. Since $h_i x_i \subseteq {}^{\alpha}S^{(qi)}$ and $q_i \notin W$, we have that $V\cap h_i x = 0$ for $i<\gamma$. Then $V\cap h(x) = V\cap g(x) = x$ shows that $h$ is a sub-isomorphism. This argument proves that $\mathfrak{A}$ is sub-isomorphic to a ${}_{\varkappa}Cs_{\alpha}$ if $\varkappa<\alpha$. If $\mathfrak{A}$ is a $Cs_{\alpha}$ then as before, ${}^{\alpha}U\cap h(x) = {}^{\alpha}U \cap g(x) = x$. Therefore $h$ is a strong sub-isomorphism.

QED(Lemma 7.12.1.)

Proof of Theorem 7.12(iv): Assume the hypotheses. By Theorem 7.12 (ii) we know that the conclusion holds if $\varkappa \geq 2^{|\alpha\cup\beta|}$ which is equal

to $|\alpha\cup\beta|^+$ by the GCH. Assume therefore $\varkappa \le |\alpha\cup\beta|$. Then $\alpha\cup\beta=\beta$ by $\varkappa>\alpha$. Thus $\varkappa=\beta$ by $\varkappa\ge\beta$. Then ${}_\varkappa Cs_\alpha = {}_\beta Cs_\alpha$ and we are done.
QED(Theorem 7.12.)

Remark 7.13. (Discussion of Theorem 7.12)

1. The condition "ext- or sub-isomorphic" does affect the cardinality conditions in Theorem 7.12, as the following Prop.7.13.1 shows.

Proposition 7.13.1. To every $\alpha\ge\omega$ there is a cardinal $\varkappa$ with the following property. Let $K = \{\mathfrak{A}\in{}_\infty Cs_\alpha : |A|\le\varkappa\}$. Then $K \subseteq I\,{}_\varkappa Cs_\alpha$ but some member of $K$ is not sub- or ext-isomorphic to any ${}_\varkappa Cs_\alpha$.
Proof. Let $\alpha\ge\omega$. Let $\varkappa$ be any cardinal such that $\alpha<\varkappa<\varkappa^{|\alpha|}$. Such a $\varkappa$ exists, e.g. $\varkappa \overset{d}{=} \aleph_{\alpha+\omega}$ is such since $cf\varkappa=\omega$. Then $(\forall\mathfrak{A}\in{}_\infty Cs_\alpha)$ $[|A|\le\varkappa \Rightarrow \mathfrak{A}\in I\,{}_\varkappa Cs_\alpha]$ by 7.12(i). In [HMTI]3.19, an $\mathfrak{A}\in{}_\infty Cs_\alpha\cap Lf_\alpha$ is constructed such that $|A|\le|\alpha|$ and $\mathfrak{A}$ is not sub- or ext-isomorphic to any ${}_\varkappa Cs_\alpha$. QED.

2. If $\varkappa<\beta$ then the condition $\varkappa\ge|A|+|\alpha|^+$ is needed in 7.12(i). For the necessity of $\varkappa\ge|A|$ see [HMTI]3.13, for the necessity of $\varkappa>\alpha$ see 7.13.2 below.

Proposition 7.13.2. Let $\alpha\ge\omega$. For every cardinal $\beta>\alpha$ there is $\mathfrak{A}\in{}_\beta Cs_\alpha$ such that $1<|A|\le\alpha$ and $(\forall\mathfrak{B}\in Cs_\alpha\cap H\mathfrak{A})$ $[|base(\mathfrak{B})|>\alpha$ or $|B|=1]$.
Proof. Let $\alpha\ge\omega$. Let $U = \beta\ge|\alpha|^+$. Let $X \overset{d}{=} \{q\in{}^\alpha U : q_0 \notin q^*(\alpha\sim1)\}$. Let $\mathfrak{A}$ be the $Cs_\alpha$ with base $U$ such that $A = Sg^{(\mathfrak{A})}\{X\}$. Then $|A|\le\alpha$, $(\forall i\in\alpha\sim1)X \subseteq -d^{\mathfrak{A}}_{0i}$ are obvious by definition, and $c_0X = 1^{\mathfrak{A}}$ is true because $|U|>\alpha$. Let $\mathfrak{B}$ be a $Cs_\alpha$ with base $W\neq0$ and assume $Hom(\mathfrak{A},\mathfrak{B})\neq0$. Then there is $h : \mathfrak{A} \to \mathfrak{B}$. Then $c^{\mathfrak{B}}_0 h(X)=1^{\mathfrak{B}}$ and $(\forall i\in\alpha\sim1)h(X) \subseteq -d^{\mathfrak{B}}_{0i}$ since these are preserved under homomorphisms. Let $\mathfrak{N}$ be a $Cs_\alpha$ of base $Z$ and let $y\in N$ be such that $(\forall i\in\alpha\sim1)y \subseteq -d_{0i}$. Then $(\forall q\in y)q_0\notin q^*(\alpha\sim1)$. Assume $0<|Z|\le\alpha$. Then there is $q\in{}^\alpha Z$ such that $q^*(\alpha\sim1)=Z$. Then $(\forall a\in Z)q^0_a\notin y$ and hence $q\notin c^{\mathfrak{N}}_0 y$. Thus $c^{\mathfrak{N}}_0 y \neq 1^{\mathfrak{N}}$ and therefore $\mathfrak{N}\neq\mathfrak{B}$. This proves that

$|W|>\alpha$.

QED(Proposition 7.13.2.)

3. If $\varkappa\geq\beta$ then we do not know how much the cardinality condition $\varkappa\geq|A|+|\alpha|^+$ is needed. See Remark 7.15 and Problems 7.16.

Theorem 7.14 below is a generalization of [HMTI]7.25 and of part of [HMTI]7.26. For example the cardinality conditions are improved and properties preserved under increasing the bases (or subbases) are investigated.

Theorem 7.14 Let $\varkappa, \beta$ be two cardinals, $\beta\geq\omega$. Let $\mathfrak{A}\in Gws_\alpha$ and $U \overset{d}{=} base(\mathfrak{A})$. Then (1)-(4) below hold.

(1) Assume $0<|U|\leq\varkappa$. Then $\mathfrak{A}$ is strongly sub-isomorphic to some $Crs_\alpha$ $\mathfrak{B}$ such that $\varkappa=|base(\mathfrak{B})|$ and (i)-(iii) below hold.

(i) If $\mathfrak{A}$ is regular then so is $\mathfrak{B}$.

(ii) If $\varkappa\geq\omega$ then for every $K\in\{Gws^{wd}, Gws^{norm}, Gws^{reg}, Gws, Gs\}$, if $\mathfrak{A}\in K_\alpha$ then $\mathfrak{B}\in K_\alpha$.

(iii) If $\alpha\geq\omega$ then $\mathfrak{B}\in Gws_\alpha$.

(2) Let $\varkappa\geq 2^{|\alpha\cup\beta|}$ or $\varkappa\geq|A|+\beta$. Let $K\in\{Gws^{wd}, Gws^{norm}, Gws^{comp}, Gws^{reg}$ $Gws, Ws\}$ and assume $\mathfrak{A}\in {}_\beta K_\alpha$. Then $\mathfrak{A}$ is strongly sub-isomorphic to some $\mathfrak{B}\in {}_\varkappa K_\alpha$ such that $rl(^\alpha U) : Subu(\mathfrak{B}) \rightarrowtail\!\!\!\twoheadrightarrow Subu(\mathfrak{A})$

(3) Assume $\varkappa\geq 2^{|\alpha\cup\beta|}$. Let $K\in\{Gws^{wd}, Gws^{norm}, Gws^{comp}, Gws^{reg}, Gws, Ws, Gs, Cs\}$. Then ${}_\beta K_\alpha \subseteq \mathbf{I}\, {}_\varkappa K_\alpha$.

(4) Assume $\varkappa=2^\delta$ for some cardinal $\delta$, and $\varkappa\geq\beta$. Let $K$ be as in (3) above. Then ${}_\beta K^{reg}_\alpha \subseteq \mathbf{I}\, {}_\varkappa K^{reg}_\alpha$ and ${}_\beta K_\alpha \subseteq \mathbf{I}\, {}_\varkappa K_\alpha$, moreover every ${}_\beta K^{reg}_\alpha$ or ${}_\beta K_\alpha$ is strongly sub-isomorphic to some ${}_\varkappa K^{reg}_\alpha$ or ${}_\varkappa K_\alpha$ respectively.

Proof. Proof of (1): Suppose $\varkappa\geq\omega$. Then the construction in the proof of [HMTI]7.25 proves the present statement (1), too. It is easy to check that the condition $|A|\leq\varkappa$ is not used in the quoted construction. Let $\varkappa<\omega$. Then, by induction, we may assume $\varkappa=|U|+1$. Let $r\in U$, $z\notin U$, $W \stackrel{d}{=} \{z\}\cup(U\sim\{r\})$, $f \stackrel{d}{=} (U1Id)^r_z$. Let $h \stackrel{d}{=} \langle x\cup\tilde{f}x : x\in A\rangle$. Then it can be proved that $h$ is a strong sub-isomorphism and $h$ preserves regularity. (1) is proved for finite $\alpha$. To see (iii), let $\alpha\geq\omega$. Let $\{{}^{\alpha}Y_i^{(pi)} : i<\rho\}$ be a partition of $1^{\mathfrak{A}}$. By $|U|+1=\varkappa<\omega$ we have $|Y_0|<\omega$, hence $|\{i\in\alpha : p_0(i)=r\}|\geq\omega$ for some $r\in Y_0$. Let $z\notin U$ and $f \stackrel{d}{=} (Y_0 1 Id)^r_z$. Let $h \stackrel{d}{=} \langle x\cup\tilde{f}(x\cap{}^{\alpha}Y_0^{(p0)}) : x\in A\rangle$. Then $h$ is a strong sub-isomorphism, $f^*\mathfrak{A}\in Gws_\alpha$ and $f^*\mathfrak{A}$ is regular if $\mathfrak{A}$ is so.

<u>Proof of (2):</u> First we prove a somewhat stronger version of (2) for $K\in\{Gws^{comp}, Ws\}$.

<u>Lemma 7.14.1.</u> Let $K\in\{Gws^{comp}, Ws\}$. Let $\beta\geq\omega$, $\mathfrak{L}\in{}_\beta K_\alpha$ and either $\varkappa\geq 2^{|\alpha\cup\beta|}$ or $\varkappa\geq|C|+\beta$. Let $|L|=\varkappa$, $U \stackrel{d}{=} base(\mathfrak{L})$. Then $\mathfrak{L}$ is strongly sub-isomorphic to some $\mathfrak{N}\in{}_\varkappa K_\alpha$ such that $base(\mathfrak{N}) = U\cup L$ and $rl({}^{\alpha}U) : Subu(\mathfrak{N}) \rightarrowtail\!\!\!\!\rightarrow Subu(\mathfrak{L})$.

<u>Proof</u>. Assume the hypotheses. Let $K=Gws^{comp}$. Then the cardinality conditions of 7.12.1(iii) are satisfied, since $|U|=\beta$. This proves all the present conclusions except $base(\mathfrak{N}) = U\cup L$. This follows by an easy base-isomorphism construction since $|L| = |base(\mathfrak{N})| = \varkappa\geq\beta = |U|$. Let $K=Ws$. By [HMTI]7.25(i) and [HMTI]3.18(ii) there is $\mathfrak{N}\in{}_\varkappa Ws_\alpha$ with $rl({}^{\alpha}U)\in Is(\mathfrak{N},\mathfrak{L})$ and $|base(\mathfrak{N})\sim U|=\varkappa$. The rest follows by an easy base-isomorphism construction.

<u>QED(Lemma 7.14.1)</u>

Assume the hypotheses of 7.14(2). Let $\mathfrak{A}\in{}_\beta Gws_\alpha^{norm}$. Let $I = Subb(\mathfrak{A})$. Then $(\forall Y\in I)\,\mathfrak{Rl}({}^{\alpha}Y)\mathfrak{A}\in{}_\beta Gws_\alpha^{comp}$. By 7.14.1 there are $\mathfrak{B}\in{}^{I}{}_\varkappa Gws_\alpha^{comp}$ and $\langle f_Y : Y\in I\rangle$ such that

(I) $(\forall Y,Z\in I)[f_Y = rl^{\mathfrak{B}_Y}({}^{\alpha}Y)\in Is(\mathfrak{B}_Y, \mathfrak{Rl}({}^{\alpha}Y)\mathfrak{A})$ and
$(Y\neq Z \Rightarrow base(\mathfrak{B}_Y)\cap base(\mathfrak{B}_Z)=0)$ and $base(\mathfrak{A})\cap base(\mathfrak{B}_Y)=Y]$.

Let $g \stackrel{d}{=} \langle \cup\{f_Y^{-1}(x\cap{}^{\alpha}Y) : Y\in I\} : x\in A\rangle$. Then $g\in Is(\mathfrak{A},\mathfrak{N})$ for some $\mathfrak{N}\in{}_\varkappa Gws_\alpha^{norm}$. Clearly $g$ is a strong sub-isomorphism. By 7.14.1 we have for all $Y \in I$ that

(II) $rl(^{\alpha}Y) : Subu(\mathfrak{B}_Y) \rightarrowtail Subu(\mathfrak{Rl}(^{\alpha}Y)\mathfrak{A})$, and

$Sb(1^{\mathfrak{B}Y}) 1 rl(^{\alpha}Y) \subseteq rl(^{\alpha}base(\mathfrak{A}))$.

This implies $rl(^{\alpha}base(\mathfrak{A})) : Subu(\mathfrak{N}) \rightarrowtail Subu(\mathfrak{A})$.

If $\mathfrak{A} \in Gws_{\alpha}^{wd}$ then by (II) above also $\mathfrak{N} \in Gws_{\alpha}^{wd}$. So far, 7.14(2) has been proved for $K \in \{Gws^{comp}, Ws, Gws^{norm}, Gws^{wd}\}$. It remains to consider the cases of Gws and $Gws^{reg}$.

To this end, let $\mathfrak{A} \in {}_{\beta}Gws_{\alpha}$ and assume the hypotheses of 7.14(2). Since $\mathfrak{A}$ may be not normal, we define our index set $I$ differently from the above $Gws^{norm}$ case. Let $I \overset{d}{=} Subu(\mathfrak{A})$. Then $\langle \mathfrak{Rl}_V \mathfrak{A} : V \in I \rangle \in$ $\in {}^{I}{}_{\beta}Ws_{\alpha}$. By 7.14.1 there are $\mathfrak{B} \in {}^{I}{}_{\varkappa}Ws_{\alpha}$ and $f \in P\langle Is(\mathfrak{B}_V, \mathfrak{Rl}_V\mathfrak{A}) : V \in I \rangle$ such that

(III) $(\forall V,W \in I)[f_V = rl^{\mathfrak{B}V}(V)$ and $(1^{\mathfrak{B}V}) \cap {}^{\alpha}base(\mathfrak{A}) = V$ and $(V \neq W \Rightarrow$

$\Rightarrow [base(\mathfrak{B}_V) \cap base(\mathfrak{B}_W) = base(V) \cap base(W)$, hence $1^{\mathfrak{B}V} \cap 1^{\mathfrak{B}W} = 0])]$.

Let $g \overset{d}{=} \langle \cup\{f_V^{-1}(x \cap V) : V \in I\} : x \in A \rangle$. Then $g \in Is(\mathfrak{A}, \mathfrak{N})$ for some $\mathfrak{N} \in$ $\in {}_{\varkappa}Gws_{\alpha}$ and $g^{-1} = rl^{\mathfrak{N}}(^{\alpha}base(\mathfrak{A}))$. $rl(^{\alpha}base(\mathfrak{A})) : Subu(\mathfrak{N}) \rightarrowtail$ $\rightarrowtail Subu(\mathfrak{A})$ follows from (III) and from $Subu(\mathfrak{N}) = \{1^{\mathfrak{B}V} : V \in I\}$. For the case $K = Gws$, 7.14(2) is proved.

Assume $\mathfrak{A} \in Gws_{\alpha}^{reg}$. We shall prove that $\mathfrak{N}$ is regular. Let $f \in 1^{\mathfrak{N}}$. Then there is $V \in Subu(\mathfrak{A})$ with $f \in 1^{\mathfrak{B}V}$. Let $T \overset{d}{=} base(\mathfrak{B}V)$ and $p \in V$. Then $1^{\mathfrak{B}V} = {}^{\alpha}T^{(p)}$ and $f \in {}^{\alpha}T^{(p)}$. Now we observe that

(**) $\{i \in \alpha : f_i \notin base(V)\}$ is finite.

Let $x \in A$, $k \in g(x)$, $f \in 1^{\mathfrak{N}}$ and assume $(1 \cup \Delta x) 1 k \subseteq f$. If $k^*(1 \cup \Delta x) \subseteq base(\mathfrak{A})$ then by (**) there are $f^+, k^+ \in 1^{\mathfrak{A}}$ such that $\{i \in \alpha : f_i \neq f_i^+$ or $k_i \neq$ $\neq k_i^+\} \subseteq [\alpha \sim (1 \cup \Delta x)]$ is finite hence $k^+ \in g(x) \cap 1^{\mathfrak{A}} = x$ and by regularity of $\mathfrak{A}$ $f^+ \in x \subseteq g(x)$ thus $f \in g(x)$. Assume therefore $k^*(1 \cup \Delta x) \nsubseteq base(\mathfrak{A})$. Then there is $i \in \alpha$ such that $k_i = f_i \notin base(\mathfrak{A})$. By definition of the unit of $\mathfrak{N}$ there is $V \in Subu(\mathfrak{A})$ such that $k_i = f_i \in T = base(\mathfrak{B}_V)$. By

(III) above then $k\in 1^{\mathfrak{B}V}$ and $f\in 1^{\mathfrak{B}V}$ (since clearly $k\in 1^{\mathfrak{B}V}$ and $f\in$ $\in 1^{\mathfrak{B}W}$ for some $V,W\in base(\mathfrak{U})$ hence $k_i=f_i\in([base(\mathfrak{B}V)\cap base(\mathfrak{B}W)]\sim$ $\sim base(\mathfrak{U}))$ and by (III), $[V\neq W \Rightarrow base(\mathfrak{B}V)\cap base(\mathfrak{B}W) \subseteq base(\mathfrak{U})]$ thus $V=W$). Then $k,f\in {}^{\alpha}T^{(p)}$ for some $p\in V$. Then $\Gamma \overset{d}{=} \{i\in\alpha : f_i\neq k_i\}$ is finite and $\Gamma\cap(1\cup\Delta g(x))=0$ since $\Delta gx=\Delta x$. Then $f\in c_{(\Gamma)}gx=gx$ proves $f\in gx$. We have proved that $\mathfrak{N}$ is regular. 7.14(2) is proved.

(3) is a corollary of (2) except for the cases of Gs and Cs. These follow from 7.12(iii).

Proof of (4): We shall need the following version of [HMTI]7.23.

**Lemma 7.14.2.** Let $\mathfrak{U}\in Gws_{\alpha}$, $U=base(\mathfrak{U})$, $V=1^{\mathfrak{U}}$ and let F be any ultrafilter on some set I. Let $\delta : A \rightarrowtail {}^{I}A/\bar{F}$ and $\varepsilon : U \rightarrowtail {}^{I}U/\bar{F}$ be as in [HMTI]7.12. Then there is an $(F,\langle U : i\in I\rangle,\alpha)$-choice function c such that $Rep_c\circ\delta : \mathfrak{U} \rightarrowtail \mathfrak{B}\in Gws_{\alpha}$ is a strong sub-base-isomorphism for some $\mathfrak{B}$ such that letting $f \overset{d}{=} Rep_c$ conclusions (i)-(iv) and (vi) of [HMTI]7.23 together with (vii) below hold.

(vii) Let K be as in 7.14(3). Let $L\in\{K,K^{reg}\}$. Then if $\mathfrak{U}\in L_{\alpha}$ then $(Rep_c\circ\delta)^{*}\mathfrak{U} \in L_{\alpha}$.

Proof. Assume the hypotheses. We use the notations of the proof of [HMTI]7.23. We may assume that $(\forall j\in J)|Y_j|\geq 2$. Let $v : X \rightarrow {}^{I}J$ be as in the proof of [HMTI]7.23, i.e.

(0) $(\forall y\in X)y\cap P_{i\in I}Y_{v(y)i}\neq 0$.

(1) If $r,s\in Rgv$ and $\{i\in I : Y_{ri}=Y_{si}\}\in F$ then $r=s$.

(2) $|Rg(v\varepsilon u)|=1$ for all $u\in U$.

Let $\pi(\varkappa,y) \overset{d}{=} \langle p_{v(y)i}(\varkappa) : i\in I\rangle$, for all $\varkappa<\alpha$, $y\in X$. Let $c : \alpha\times X \rightarrow PU$ be an $(F,\langle U : i\in I\rangle,\alpha)$-choice function such that for all $\varkappa<\alpha$ and $y\in X$ conditions (3)-(5) below hold.

(3) $c(\varkappa,\varepsilon u) = \langle u : i\in I\rangle$ for all $u\in U$.

(4) $c(\varkappa,y)\in P_{i\in I}Y_{v(y)i}$.

(5) $c(\varkappa,y)_i \neq \pi(\varkappa,y)_i$ for all $i\in I$ if $\pi(\varkappa,y)\notin y$.

It is easy to see that such a $c$ exists.

We show that (i)-(iv) and (vi)-(vii) hold for $c$. The hypotheses of [HMTI]7.12 hold by (3), so (i)-(iii) follow. Also, (3) implies (iv) by the following argument. Let $g \stackrel{d}{=} Rep_c \circ \delta$. It is enough to prove $g(V) \cap {}^{\alpha}(\varepsilon^* U) \subseteq \tilde{\varepsilon} V$. Let $q \in {}^{\alpha}(\varepsilon^* U)$. Then $q = \langle \varepsilon(k_j) : j<\alpha \rangle$ for some $k \in {}^{\alpha}U$. $(c^+ q)_i = \langle k_j : j<\alpha \rangle$ for all $i \in I$. Assume $q \in gV$. Then $(\exists i \in I)$ $(c^+ q)_i \in V$ and hence $k \in V$. Since $q = \varepsilon \circ k$, this implies $q \in \tilde{\varepsilon} V$. We have seen that (iv) holds. (iv) implies that $g \stackrel{d}{=} Rep_c \circ \delta : \mathfrak{A} \rightarrowtail \mathfrak{B}$ is a strong sub-base-isomorphism.

<u>Proof of (vi)-(vii)</u>: We have to show that $\mathfrak{A} \in K_\alpha \Rightarrow \mathfrak{B} \in K_\alpha$ for various choices of K. The cases K=Gws and K=Cs are taken care of by [HMTI]7.4. The cases $K \in \{Gws^{norm}, Gws^{comp}\}$ follow from 7.10(ii). If $\mathfrak{A}$ is regular then so is $g^* \mathfrak{A}$ by [HMTI]7.6 since $g$ is an isomorphism.

Suppose $\mathfrak{A} \in Gs_\alpha$. Then $g^* \mathfrak{A} \in Gs_\alpha$ follows from (1) and (4) exactly as in the proof of [HMTI]7.23, see the proofs of (5)-(8) there. In particular, using the notation $Q$ from the proof of [HMTI]7.4, $\mathfrak{A} \in$ $\in Gws_\alpha^{norm}$ implies $gV \subseteq \cup\{{}^{\alpha}Q_r : r \in Rgv\}$ and $(\forall r,s \in Rgv)[r \neq s \Rightarrow Q_r \cap Q_s = 0]$ by that proof.

Suppose $\mathfrak{A} \in Gws_\alpha^{wd}$. Let $y \in X$. Define $q \stackrel{d}{=} \langle \pi(\varkappa,y)/\bar{F} : \varkappa<\alpha \rangle$, and $r \stackrel{d}{=} v(y)$. By the above, and by $g^* \mathfrak{A} \in Gws_\alpha$ it is enough to prove $g(V) \cap {}^{\alpha}Q_r \subseteq {}^{\alpha}Q_r^{(q)}$. Let $k \in g(V) \cap {}^{\alpha}Q_r$. By $k \in {}^{\alpha}Q_r$ and by (1) we have $(\forall \varkappa<\alpha) v(k_\varkappa) = r$. Let $L \stackrel{d}{=} \{i \in I : (\exists j \in J \sim \{ri\})(c^+ k)_i \in {}^{\alpha}Y_j\}$. Then $L \notin F$. Thus $k \in gV$ implies that $(c^+ k)_i \in V \cap {}^{\alpha}Y_{ri}$ for some $i \in I$. Then by $\mathfrak{A} \in$ $\in Gws_\alpha^{wd}$ we have that $H \stackrel{d}{=} \{\varkappa<\alpha : (c^+ k)_{i\varkappa} \neq p_{ri}\varkappa\}$ is finite. By (5) we have $H \supseteq \{\varkappa<\alpha : k_\varkappa \neq q_\varkappa\}$. Thus $k \in {}^{\alpha}Q_r^{(q)}$ as desired.

The case K=Ws follows from the cases $K=Gws^{wd}$ and $K=Gws^{comp}$ since $Ws_\alpha = Gws_\alpha^{wd} \cap Gws_\alpha^{comp}$.

<u>QED(Lemma 7.14.2.)</u>

We turn to the proof of 7.14(4). Let $\varkappa = 2^\mu$, $\mu = |\mu|$, $\beta = |\beta|$, $\varkappa \geq \beta \geq \omega$. Let $\mathfrak{A} \in {}_\beta Gws_\alpha$ and $U = base(\mathfrak{A})$. Let $F$ be a $\mu$-regular ultrafilter on

$I \stackrel{d}{=} \mu$. We shall apply 7.14.2 to this $\mathfrak{A}$ and $F$. Let $c$ and $\delta$ be as in 7.14.2. Let $\mathfrak{B} \stackrel{d}{=} (Rep_c \circ \delta)^* \mathfrak{A}$. Then $\mathfrak{A}$ is strongly sub-base--isomorphic to $\mathfrak{B}$. We show that $\mathfrak{B} \in {}_\kappa Gws_\alpha$. Notation: for any set $s$ we define $\bar{s} \stackrel{d}{=} \langle s : i \in I \rangle$. Let $T \in Subb(\mathfrak{B})$. By 7.10(vii), there is $Y \in {}^I Subb(\mathfrak{A})$ such that $T = PY/\bar{F}^{(\bar{U})}$. By $|Y_i| = \beta$ and regularity of $F$ we have $|T| = \beta^\mu = 2^\mu$, since $2^\mu \geq \beta \geq 2$ was assumed. We have proved $\mathfrak{B} \in {}_\kappa Gws_\alpha$. Let $K$ be as in 7.14(3) and $L \in \{K, K^{reg}\}$. Assume $\mathfrak{A} \in {}_\beta L_\alpha$. Then by 7.14.2(vii) we have $\mathfrak{B} \in {}_\kappa L_\alpha$.

QED(Theorem 7.14.)

Remark 7.15. (Discussion of Theorem 7.14 and [HMTI]7.2, [HMTI]7.26.)

(i) Let $\kappa > \beta$ and $\alpha > 0$. Then $\mathbf{I}\,{}_\beta Gws_\alpha \neq \mathbf{I}\,{}_\kappa Gws_\alpha$.

Proof. $(\exists \mathfrak{A} \in {}_\kappa Gws_\alpha)(\exists x \in At\mathfrak{A})|\{z \in At\mathfrak{A} : z \leq c_0 x\}| = \kappa$. But $(\forall \mathfrak{A} \in {}_\beta Gws_\alpha)(\forall x \in At\mathfrak{A})|\{z \in At\mathfrak{A} : z \leq c_0 x\}| \leq \beta$ follows from $(\forall x \in At\mathfrak{A})(\forall z \leq c_i x)[z \neq 0 \Rightarrow c_i z = c_i x]$ which is proved by the following argument. Let $x \in At\mathfrak{A}$ and $0 < z \leq c_i x$. Then $0 < c_i z \leq c_i x$. By [HMT] 1.10.3(i), $c_i x \in At\mathfrak{Cl}_{\{i\}}\mathfrak{A}$. Since $c_i z \in (Cl_{\{i\}}\mathfrak{A}) \sim \{0\}$ we conclude $c_i z = c_i x$.

(ii) We do not know whether the cardinality conditions in 7.14(2) and (3) are needed. (Of course $\kappa \geq \beta \geq \omega$ is needed.) They are not the best possible since if $|\alpha| > \kappa = 2^\omega$ then ${}_\omega Cs_\alpha \subseteq \mathbf{I}\,{}_\kappa Cs_\alpha$ and, more generally, if $K$ is as in (3) then ${}_\omega K_\alpha \subseteq \mathbf{I}\,{}_\kappa K_\alpha$, by (4), but $\alpha, \kappa$ and $\beta = \omega$ do not satisfy the conditions of (3) and for some $\mathfrak{A} \in {}_\beta Cs_\alpha$ we have $|A| > \kappa$ and hence the conditions of (2) are not satisfied either.

(iii) The conditions of (1) are needed. More precisely, the conditions of (1) (ii) can be replaced by $[\kappa \geq \omega$ or $|U|$ divides $\kappa]$ but if $|U| < \kappa < \omega$ and $|U|$ does not divide $\kappa$ then there is a ${}_\beta Cs_\alpha$ not isomorphic to any $Gws_\alpha^{norm}$ with base of cardinality $\kappa$, if $\alpha > 1$. This can be proved by [HMTI]7.22 using the fact that finite dimensional $Cs_\alpha$ -s are simple. This also proves that the modified condition [$\varkappa+\alpha\geq\omega$ or $|U|$ divides $\varkappa$] is the best possible for (1)(ii) if $\alpha>1$, because if $\alpha<\omega$ then $Gws_\alpha = Gws_\alpha^{norm}$.

(iv) Let $\varkappa\geq\beta\geq\omega$ and $\varkappa>\alpha$. Let K be as in (3). Assume the GCH. Then ${}_\beta K_\alpha \subseteq \mathbf{I}\,{}_\varkappa K_\alpha$. (See the proof of 7.12(iv).) We do not know whether the condition $\varkappa>\alpha$ (or GCH) is needed here.

Problems 7.16. Let $\varkappa\geq\beta$ be two infinite cardinals.

1. Let $\varkappa\geq 2^{|\alpha\cup\beta|}$. Is ${}_\beta Cs_\alpha^{reg} \subseteq \mathbf{I}\,{}_\varkappa Cs_\alpha^{reg}$?
2. Let $\mathfrak{A}\in{}_\beta Cs_\alpha^{reg}$, $\varkappa\geq|\alpha|^+\cup(|A|\cap 2^{|\alpha\cup\beta|})$. Is then $\mathfrak{A}$ sub-isomorphic to some ${}_\varkappa Cs_\alpha^{reg}$?
3. How much are the cardinality conditions $\varkappa>\alpha$ and $\varkappa\geq(|A|\cap 2^{|\alpha\cup\beta|})$ needed for $\mathfrak{A}\in{}_\beta Cs_\alpha$ to be sub-isomorphic to a ${}_\varkappa Cs_\alpha$? (It is clear that these conditions are <u>not</u> the best possible.)
4. Is ${}_\beta Cs_\alpha \subseteq \mathbf{I}\,{}_\varkappa Cs_\alpha$ true?
5. How much are the cardinality conditions in Thm 7.14 needed? In (1) of Thm 7.14 they are the best possible. But how much are the cardinality conditions in (2), (3) needed? E.g. is ${}_\beta Gws_\alpha \subseteq \mathbf{I}\,{}_\varkappa Gws_\alpha$ true?

Problem 7.17. Let $\alpha\geq\omega$. Is $Ws_\alpha \subseteq \mathbf{I}\,Cs_\alpha^{reg}$ true without the Axiom of the existence of ultrafilters?

About Proposition 7.18 below we note that if $1<\alpha<\omega$ then there is a $Gs_\alpha$ $\mathfrak{A}$ with characteristic 0 such that $\mathfrak{A}\notin\mathbf{I}\,{}_\infty Gs_\alpha$ moreover for any $\varkappa>\alpha+1$, $\mathfrak{A}\notin\mathbf{I}\,{}_\varkappa Gs_\alpha$. Indeed, this $\mathfrak{A}$ is the ${}_{\alpha+1}Cs_\alpha$ constructed in [HMTI]7.22. Note also the contrast between (i) and (ii).

Proposition 7.18. Let $\varkappa<\omega$ and $\alpha>0$.

(i) $\mathbf{HSP}\,{}_\varkappa Gs_\alpha = \mathbf{I}\,{}_\varkappa Gs_\alpha$ and $\mathbf{HSP}\,{}_\infty Gs_\alpha = \mathbf{I}\,{}_\infty Gs_\alpha$ if $\alpha>1$.

(ii) $\mathbf{Up}'\,{}_\beta Gs_\alpha \neq \mathbf{I}\,{}_\beta Gs_\alpha$ if $\beta=|\beta|\geq\omega$.

(iii) $\mathbf{HSUp}\,{}_\infty Cs_\alpha = \mathbf{I}\,{}_\infty Cs_\alpha$.

Proof. The case $\alpha\geq\omega$ is proved in [HMTI]7.21 and 7.15. Assume therefore $\alpha<\omega$. By the first part of the proof of [HMTI]7.15 then $\mathbf{H}_\beta Gs_\alpha \subseteq \mathbf{SUp}_\beta Gs_\alpha$ for any cardinal $\beta$. By 7.10, $\mathbf{SUp}_\varkappa Gs_\alpha = \mathbf{I}_\varkappa Gs_\alpha$ and $\mathbf{SUp}_\infty Gs_\alpha = \mathbf{I}_\infty Gs_\alpha$. (iii) follows from $\mathbf{H}_\infty Cs_\alpha = \mathbf{I}_\infty Cs_\alpha$ (by simplicity of $Cs_\alpha$ -s and $\alpha>0$) and from [HMTI]7.4 which implies $\mathbf{SUp}_\infty Cs_\alpha = \mathbf{I}_\infty Cs_\alpha$. To see (ii), let $\mathfrak{A}\in{}_\beta Gs_\alpha$, $\beta\geq\omega$. Consider property (*) of $\mathfrak{A}$ below.

(*) $(\forall x\in At\,\mathfrak{A})\,|\{z\in At\,\mathfrak{A} : z\leq c_0x\}|\leq\beta$.

Clearly, (*) holds for any ${}_\beta Gs_\alpha$ but it is false for some $\mathbf{Up}'\,{}_\beta Gs_\alpha$ if $\beta\geq\omega$.

QED(Proposition 7.18.)

Geometrical representability as opposed to relational representability was discussed in [HMT]. The broadest possible version of geometrical representability (known to us) when applied to $CA_\alpha$-s yields the class $\mathbf{I}Crs_\alpha\cap CA_\alpha$. By [HMTI]2.14 this class is strictly larger than $\mathbf{I}Gws_\alpha$. By the proposition below this class is a variety.

Proposition 7.19. Let $\alpha>0$. Then $\mathbf{HSP}(Crs_\alpha\cap CA_\alpha) = \mathbf{I}Crs_\alpha\cap CA_\alpha$.

Proof. The proof uses [HMTI]7.2 similarly to the proof of [HMTI]7.15 but the construction is completely different. An outline of the proof is in [N]. To save space we omit the proof.

QED(Proposition 7.19.)

## 8. Reducts

Throughout, $\alpha$ and $\beta$ denote ordinals.

We shall use the functions $rb^\rho$ and $rd^\rho$ introduced in 4.7.1.1. Let $\alpha\leq\beta$. Then $rd_\alpha \stackrel{d}{=} rd^{(\alpha 1 Id)}$. By 4.7.1.2 $rd^\rho\in Is(\mathfrak{Rd}^\rho\,\mathfrak{A})$ for any $\mathfrak{A}\in Crs_\alpha$.

<u>Lemma 8.1.</u> Let $1<\alpha$ and let $\rho : \alpha \rightarrowtail \beta$ be one-one. Let $K\in\{Gws^{wd}, Gws^{norm}, Gws^{reg}, {}_{\aleph}Gws, {}_{\omega}Gws, Gws, Gs, Crs\}$. Then (i)-(iv) below hold.

(i) $\{rd^{\rho}{}^*\mathfrak{Rd}^{\rho}\mathfrak{A} : \mathfrak{A}\in K_{\beta}\} \subseteq K_{\alpha}$.

(ii) $\{rd_{\alpha}{}^*\mathfrak{Rd}_{\alpha}\mathfrak{A} : \mathfrak{A}\in Crs_{\beta}^{reg}\} \subseteq Crs_{\alpha}^{reg}$ if $\alpha\le\beta$ .

(iii) $\mathbf{I}K_{\alpha} = \mathbf{SRd}^{\rho}\mathbf{I}K_{\beta}$ and if $\alpha\le\beta$ then $\mathbf{I}K_{\alpha} = \mathbf{SNr}_{\alpha}\mathbf{I}K_{\beta}$.

(iv) $\mathbf{I}K_{\alpha} = \mathbf{Rd}_{\alpha}\mathbf{I}K_{\beta}$ iff $K = Crs$ or $\alpha=\beta$.

<u>Proof.</u> The proof is an easy extension of the proof of 4.7.1.2. It was proved there that if $\mathfrak{A}\in Crs_{\beta}$ then $rd^{\rho}\in Ism(\mathfrak{Rd}^{\rho}\mathfrak{A}, \mathfrak{Sb}\, rd^{\rho}1^{\mathfrak{A}})$ and if in addition $\mathfrak{A}\in Gws_{\beta}$ then $Subu(rd^{\rho}1^{\mathfrak{A}}) = \{{}^{\alpha}(base(V)\times\{(\beta\sim\sim Rg\rho)1q\})^{(rb^{\rho}q)} : q\in V\in Subu(\mathfrak{A})\}$. Let $\alpha,\rho$ and $K$ be as in the hypotheses. If $K\neq Gws^{reg}$ then it is easy to check, by the above, that if $1^{\mathfrak{A}}$ is a $K_{\beta}$-unit then $rd^{\rho}1^{\mathfrak{A}}$ is a $K_{\alpha}$-unit. Let $\mathfrak{A}\in Gws_{\beta}^{reg}$, $x\in A$. Let $\Gamma=\Delta rd^{\rho}(x)$. Clearly, $\Gamma=\rho^{-1}{}^*\Delta x$. By Lemma 1.3.4, $x$ is $\{\rho 0\}\cup\Delta x$-regular in $\mathfrak{A}$ and hence $rd^{\rho}x$ is $1\cup\Gamma$-regular, by the above. Thus $rd^{\rho}x$ is regular. This argument proves (i) for the case $K\neq Crs$. For $K=Crs$, (i) follows from 4.7.1.2(i) since $rd^{\rho}V$ is a $Crs_{\alpha}$-unit for any $Crs_{\beta}$-unit $V$. To prove (ii), let $\mathfrak{A}\in Crs_{\beta}^{reg}$, $x\in A$. Then $1\cup\Delta rd_{\alpha}x = \alpha\cap(1\cup\Delta x)$. Hence $rd_{\alpha}x$ is regular. This proves (ii). $\mathbf{Rd}^{\rho}K_{\beta} \subseteq \mathbf{I}K_{\alpha}$ is immediate by (i). $K_{\alpha} \subseteq \mathbf{SRd}^{\rho}\mathbf{I}K_{\beta}$ and $K_{\alpha} \subseteq \mathbf{SNr}_{\alpha}\mathbf{I}K_{\beta}$ if $\alpha\le\beta$ follow from [HMTI]8.1-3, [HMTI]8.5-6, and [HMTI]7.14. (iii) is proved. $\mathbf{I}Crs_{\alpha} = \mathbf{Rd}_{\alpha}\mathbf{I}Crs_{\beta}$ is proved in [N] Prop.8(ii). $Mn_{\alpha}\cap\mathbf{Rd}_{\alpha}CA_{\beta} = \mathbf{I}\,{}_1Cs_{\alpha}$ by [HMT]2.1.22, since $|\Delta(rd^{\rho}(d_{0\alpha}))|\neq 1$ implies $d_{0\alpha}=1$. This proves the negative part of (iv).

<u>QED(Lemma 8.1.)</u>

Statement (i) of 8.1 above does not extend to $Crs_{\alpha}^{reg}$ if $\alpha>3$ as it was proved in [N]Prop.15.

To investigate reducts, the notions introduced in Def.8.2 below (originating with Monk) are especially helpful. Recall from chap.1 that we apply the notions $\mathbf{Rd}_{\alpha}^{\rho}$, $\Delta$, Zd etc introduced for $CA_{\alpha}$ in [HMT] to

$Crs_\alpha$-s as well. In particular, $Rd_\alpha^\rho$ can be applied to $Bo_\beta$-s and more generally to arbitrary algebras similar to $CA_\beta$-s without any difficulty, since it only uses the similarity type of the algebra. Ord denotes the class of all ordinals. The following definition was first introduced in Monk[M1]. Definitions and results more general than the ones below can be found in [AN9],[AN2] and in [N].

Definition 8.2. By a system K of classes (of algebras) we understand $K = \langle K_\alpha : \alpha \in Ord \rangle$ such that $K_\alpha$ is a class of algebras similar to $CA_\alpha$-s, for all $\alpha \in Ord$. Let K be a system of classes.

(i) K is said to be definable by a scheme of equations iff
$K_\alpha = \mathbf{HSP}\ \mathbf{Rd}^\rho K_\beta$ for any $\alpha \geq \omega$ and one-one $\rho : \alpha \rightarrowtail \beta$.

(ii) K is strongly definable by a scheme of equations iff
$K_\alpha = \mathbf{HSP}\ \mathbf{Rd}^\rho K_\beta$ for any $\alpha \geq 2$ and one-one $\rho : \alpha \rightarrowtail \beta$.

The definitions given in [AN9] and in [AN2] for systems definable by schemes of equations are equivalent to the above one. This is proved in [A] and in [AN3].

We use the notation $CA = \langle CA_\alpha : \alpha \in Ord \rangle$, $\mathbf{I}Cs = \langle \mathbf{I}Cs_\alpha : \alpha \in Ord \rangle$ and similarly for all classes $K_\alpha$ which were defined simultaneously for all $\alpha \in Ord$. We note that CA is definable by a scheme of equations but is not strongly such, by [HMT]2.6.14(i) and [HMT]2.6.15.

Proposition 8.3. Let $1 < \varkappa < \omega$.

(i) Let $K \in \{\mathbf{I}Gs, \mathbf{I}\,{}_\varkappa Gs, \mathbf{I}\,{}_\infty Gs, \mathbf{I}Crs\}$. Then K is a system of classes strongly definable by schemes of equations.

(ii) $\langle \mathbf{I}Crs_\alpha \cap CA_\alpha : \alpha \in Ord \rangle$ is a system of classes definable by a scheme of equations.

Proof. Let $1 < \varkappa < \omega$. Let $1 < \alpha$. $\mathbf{I}Gs_\alpha$, $\mathbf{I}\,{}_\varkappa Gs_\alpha$ and $\mathbf{I}\,{}_\infty Gs_\alpha$ are varieties by 7.18(ii), [HMTI]7.16. Then $\mathbf{I}Gs$, $\mathbf{I}\,{}_\varkappa Gs$, $\mathbf{I}\,{}_\infty Gs$ are systems of classes definable by schemes of equations by 8.1(iii). An outline of the proof for the Crs case is in [N]. (ii) follows from 8.4(ii) below since I Crs and CA are systems of classes definable by schemes of equations. QED(Proposition 8.3.)

Lemma 8.4.

(i) Let $K$ be a system of classes. Then $K$ is definable by a scheme of equations iff $K_\alpha = \mathbf{Uf}\,\mathbf{Rd}^\rho K_\beta = \mathbf{HSP}\,K_\alpha$ for every $\alpha \geq \omega$ and one-one $\rho : \alpha \rightarrowtail \beta$.

(ii) If $K, L$ are systems of classes definable by schemes of equations then so is $\langle K_\alpha \cap L_\alpha : \alpha \in Ord \rangle$.

Proof. Notation: Let $\xi \in {}^\gamma\beta$ be one-one. Let $\varphi$ be a formula in the language of $CA_\gamma$. Then $\xi(\varphi)$ denotes the formula obtained from $\varphi$ by replacing in $\varphi$ the indices from $\gamma$ according to $\xi$. E.g. $\xi(c_i d_{ij}=1)$ is the equation $c_{\xi i} d_{\xi i \xi j}=1$. If $E$ is a set of formulas then $\xi(E) \stackrel{d}{=} \{\xi(\varphi) : \varphi \in E\}$.

Fact(*): $K_\beta \models \xi(\varphi)$ iff $\mathbf{Rd}^\xi K_\beta \models \varphi$ .

Construction 1. Let $\alpha \geq \omega$ and let $\rho : \alpha \rightarrowtail \beta$ be one-one. Let $I \stackrel{d}{=} Sb_\omega \alpha \times Sb_\omega \beta$. Let $F$ be an ultrafilter on $I$ such that $(\forall \langle \Gamma, \Delta \rangle \in I)$ $\{\langle \Gamma', \Delta' \rangle \in I : \Gamma \subseteq \Gamma', \Delta \subseteq \Delta'\} \in F$. Let $i \stackrel{d}{=} \langle \Gamma, \Delta \rangle \in I$. Let $Hi \stackrel{d}{=} \Delta \cup \rho^*\Gamma$ and let $\xi i : Hi \rightarrowtail \alpha$ be one-one such that $\rho^*\Gamma 1 \xi \subseteq \rho^{-1}$.

Let $\mathfrak{A}$ be any algebra similar to $CA_\alpha$-s. For every $i \in I$ let $\mathfrak{B}_i$ be an algebra similar to $CA_\beta$-s such that $\mathfrak{Rd}_{Hi} \mathfrak{B}_i = \mathfrak{Rd}^{\xi i} \mathfrak{A}$. Let $\mathfrak{B} \stackrel{d}{=} P_{i \in I} \mathfrak{B}_i / F$. Then $\mathfrak{A} \cong | \subseteq \mathfrak{Rd}^\rho \mathfrak{B}$, by the proofs of [HMT]0.3.71, 0.5.15.

Claim 1

(i) Let $K$ be a system of classes definable by a scheme of equations If $\mathfrak{A} \in K_\alpha$ then $\mathfrak{B} \in K_\beta$.

(ii) $\mathfrak{A}$ is elementarily equivalent to $\mathfrak{Rd}^\rho \mathfrak{B}$.

Proof. (i) Let $K_\beta \models e$ be any equation. Let $H$ be the set of indices occurring in $e$. Let $i = \langle \Gamma, \Delta \rangle \in I$ be such that $H \subseteq \Delta$. Let $\eta : \beta \rightarrowtail\!\!\!\!\rightarrow \beta$ be a permutation of $\beta$ such that $\eta \supseteq \xi i$. Then $\mathbf{Rd}^{\eta} K_\beta \models e$ by $\mathbf{Rd}^{\eta} K_\beta \subseteq K_\beta$, hence $K_\beta \models \eta(e)$ by Fact(*). Then $\mathbf{Rd}_\alpha K_\beta \models \eta(e)$ by $\eta^* H \subseteq \alpha$, hence $K_\alpha \models \eta(e) = \xi i(e)$. Thus $\mathfrak{A} \models$ $\models \xi i(e)$, i.e. $\mathfrak{Rd}^{\xi i} \mathfrak{A} \models e$ by Fact(*). Then $\mathfrak{B}_i \models e$. Now $\{\langle \Gamma, \Delta \rangle \in I : H \subseteq \Delta\} \in F$ finishes the proof.

Proof of (ii): Let $\varphi$ be any first order formula in the language of $CA_\alpha$-s. Let $H$ be the set of indices occurring in $\varphi$. Let $i = \langle \Gamma, \Delta \rangle \in I$ be such that $H \subseteq \Gamma$. Then $\mathfrak{A} \models \varphi$ iff $\mathfrak{Rd}^{\xi i} \mathfrak{A} \models \rho(\varphi)$ iff $\mathfrak{B}_i \models \rho(\varphi)$. Thus $\mathfrak{A} \models \varphi$ iff $\mathfrak{B} \models \rho(\varphi)$ iff $\mathfrak{Rd}^{\rho} \mathfrak{B} \models \varphi$ by $\{\langle \Gamma, \Delta \rangle \in I : H \subseteq \Gamma\} \in F$.

QED(Claim 1)

Now we turn to the proof of 8.4(i). Let $K$ be a system of classes. If $K_\alpha = \mathbf{Uf}\ \mathbf{Rd}^{\rho} K_\beta = \mathbf{HSP} K_\alpha$ for all $\alpha \geq \omega$ and one-one $\rho : \alpha \rightarrowtail \beta$ then $K$ is definable by a scheme of equations. Suppose $K$ is definable by a scheme of equations. Let $\alpha \geq \omega$ and let $\rho : \alpha \rightarrowtail \beta$ be one-one. We have to show $K_\alpha = \mathbf{Uf}\ \mathbf{Rd}^{\rho} K_\beta$. Clearly, $\mathbf{Uf}\ \mathbf{Rd}^{\rho} K_\beta \subseteq \mathbf{HSP}\ \mathbf{Rd}^{\rho} K_\beta \doteq K_\alpha$. Let $\mathfrak{A} \in K_\alpha$. Consider the algebra $\mathfrak{B}$ constructed from $\mathfrak{A}$ in Construction 1. Then $\mathfrak{B} \in K_\beta$ and $\mathfrak{A} \in \mathbf{Uf}\ \mathbf{Up}\ \mathfrak{Rd}^{\rho} \mathfrak{B}$ by Claim 1 and by the Keisler-Shelah ultrapower theorem (see [HMT]0.3.79). Thus $\mathfrak{A} \in$ $\in \mathbf{Uf}\ \mathbf{Up}\ \mathbf{Rd}^{\rho} K_\beta = \mathbf{Uf}\ \mathbf{Rd}^{\rho} K_\beta$ by [HMT]0.5.13(viii). Lemma 8.4(i) is proved. Let $K, L$ be systems of classes definable by schemes of equations. Let $\alpha \geq \omega$ and let $\rho : \alpha \rightarrowtail \beta$ be one-one. We have to show $K_\alpha \cap L_\alpha = \mathbf{HSP}\ \mathbf{Rd}^{\rho} (K_\beta \cap L_\beta)$. Clearly, $\mathbf{HSP}\ \mathbf{Rd}^{\rho}(K_\beta \cap L_\beta) \subseteq K_\alpha \cap L_\alpha$. Let $\mathfrak{A} \in K_\alpha \cap L_\alpha$. Consider the algebra $\mathfrak{B}$ constructed from $\mathfrak{A}$ in Construction 1. Then $\mathfrak{B} \in$ $\in K_\beta \cap L_\beta$ by Claim 1 and $\mathfrak{A} \cong | \subseteq \mathfrak{Rd}^{\rho} \mathfrak{B}$. Thus $\mathfrak{A} \in \mathbf{I\,S\,Rd}^{\rho}(K_\beta \cap L_\beta) \subseteq$ $\subseteq \mathbf{HSP}\ \mathbf{Rd}^{\rho}(K_\beta \cap L_\beta)$.

QED(Lemma 8.4.)

Corollary 8.5. Let $1<\alpha<\beta$ and $1<\varkappa<\omega$. Let $K\in\{\mathbf{I}\,Gs, \mathbf{I}_{\varkappa}Gs, \mathbf{I}_{\infty}Cs, CA\}$. Then (i)-(iii) below are equivalent.

(i) $K_\alpha = \mathbf{Uf}\ \mathbf{Rd}_\alpha K_\beta = \mathbf{Uf\,Up}\ \mathbf{Rd}_\alpha K_\beta$.

(ii) $Mn_\alpha \cap K_\alpha \subseteq \mathbf{Uf}\ \mathbf{Rd}_\alpha K_\beta$.

(iii) $\alpha\geq\omega$.

Proof. Let $\alpha,\beta,\varkappa$ and $K$ be as in the hypotheses. Then $\mathbf{Rd}_\alpha K_\beta = \mathbf{Rd}_\alpha \mathbf{Up} K_\beta = \mathbf{Up}\ \mathbf{Rd}_\alpha K_\beta$ by 8.3(i), 7.18(iii) and [HMT]0.5.13(viii). Now (iii) implies (i) by 8.4(i), 8.3 and [HMTI]7.21. Clearly, (i) implies (ii). (ii) ⇒ (iii) will be proved in Prop.8.10(2).
QED(Corollary 8.5)

By 8.5 and 8.1(iv) we have that $\mathbf{Rd}_\alpha \mathbf{I}\,Gs_\beta \neq \mathbf{Uf}\ \mathbf{Rd}_\alpha \mathbf{I}\,Gs_\beta$ for all $\omega\leq\alpha<\beta$. If $0<\alpha<\omega$ and $\beta\geq\omega_1$ then $\mathbf{Rd}_\alpha \mathbf{I}\,Gs_\beta \neq \mathbf{Uf}\ \mathbf{Rd}_\alpha \mathbf{I}\,Gs_\beta$, by [HMT]1.3.15 as the following argument shows. Let $\mathfrak{A}\in \mathbf{Rd}_\alpha Gs_\beta$ be nondiscrete. Then $|A|\geq\omega_1$. By the Löwenheim-Skolem-Tarski theorem there is an elementary subalgebra $\mathfrak{B}\subseteq\mathfrak{A}$ of $\mathfrak{A}$ with $|B|=\omega$ since $\alpha<\omega$ and $\mathfrak{A}\in CA_\alpha$. By the Keisler-Shelah isomorphic ultrapowers theorem $\mathfrak{B}\in\mathbf{Uf\,Up}\{\mathfrak{A}\}\subseteq \mathbf{Uf\,Up}\ \mathbf{Rd}_\alpha Gs_\beta$. By [HMT]0.5.13(viii), $\mathbf{Rd}_\alpha \mathbf{Up} K = \mathbf{Up}\ \mathbf{Rd}_\alpha K$ for all $K$, hence $\mathbf{Up}\ \mathbf{Rd}_\alpha Gs_\beta = \mathbf{Rd}_\alpha \mathbf{Up}\,Gs_\beta = \mathbf{Rd}_\alpha \mathbf{I}\,Gs_\beta$ by [HMTI]7.8. Thus $\mathfrak{B}\in \mathbf{Uf\,Up}\ \mathbf{Rd}_\alpha Gs_\beta = \mathbf{Uf}\ \mathbf{Rd}_\alpha \mathbf{I}\,Gs_\beta = \mathbf{Uf}\ \mathbf{Rd}_\alpha Gs_\beta$. $\mathfrak{B}$ is nondiscrete by $\mathfrak{B}\subseteq\mathfrak{A}$. Hence $\mathbf{Uf}\ \mathbf{Rd}_\alpha Gs_\beta$ has countable nondiscrete members but $\mathbf{Rd}_\alpha \mathbf{I}\,Gs_\beta$ does not. We have proved that $\mathbf{Rd}_\alpha \mathbf{I}\,Gs_\beta \neq \mathbf{Uf}\ \mathbf{Rd}_\alpha Gs_\beta$ whenever $\beta>\alpha>0$ and $\beta\geq\omega_1$ (or $\beta>\alpha\geq\omega$).

The corresponding question for $\mathbf{Nr}_\alpha$ and $\mathbf{Uf\,Nr}_\alpha$ seems to be much harder. (Cf. [HMT]2.11.) Thm 8.6 concerns the difference between $\mathbf{Nr}_\alpha$ and $\mathbf{Uf\,Nr}_\alpha$.

Theorem 8.6. Let $\beta>2$.

(i) $\mathbf{Uf\,Nr}_2({}_\infty Cs_\beta^{reg}\cap Lf_\beta) \not\subseteq \mathbf{Nr}_2 CA_\beta$.

(ii) $\mathbf{Nr}_2 K_\beta \neq \mathbf{Uf\,Nr}_2 K_\beta$ for $K\in\{\mathbf{I}\,Ws, \mathbf{I}\,Cs^{reg}, \mathbf{I}\,Gws^{comp\ reg}, \mathbf{I}_\infty Cs, \mathbf{I}\,Cs, \mathbf{I}\,Gs, CA\}$.

Proof. $R$ denotes the set of real numbers and $Q$ denotes the set of rational numbers. $U \stackrel{d}{=} (R\times\{0\})\cup(R\times\{1\})$. Let $u\in{}^2 2$ and $r\in R$. Then

$p(u,r) \stackrel{d}{=} \{s\in{}^2U : \langle s_0(1),s_1(1)\rangle=u$ and $s_0(0)=s_1(0)+r\}$,
$P \stackrel{d}{=} \{p(u,r) : u\in{}^2 2$ and $r\in R\}$. $\mathfrak{L} \stackrel{d}{=} \mathfrak{Sb}\,{}^2U$,
$\mathfrak{A} \stackrel{d}{=} \mathfrak{Sg}^{(\mathfrak{L})}P$ and $\mathfrak{B} \stackrel{d}{=} \mathfrak{Sg}^{(\mathfrak{L})}(P\sim\{p(\langle 0,1\rangle,r) : r\in R\sim Q\})$.

We shall prove that $\mathfrak{A}\in \mathbf{INr}_2\,{}_\infty Cs_\beta$, $\mathfrak{B}\notin \mathbf{Nr}_2\, CA_\beta$ and $\mathfrak{B}$ is an elementary submodel of $\mathfrak{A}$. Let $\beta>2$.

Claim 8.6.1 $\mathfrak{A} \in \mathbf{INr}_2\,{}_\infty Cs_\beta$.

Proof. Let $\underset{\sim}{R} \stackrel{d}{=} \langle R,+,-,r\rangle_{r\in R}$ denote the group of reals with constants. Let $S\in Sb_\omega\beta$. Let $u\in{}^S2$ and let $\varphi$ be any formula in the language of $\underset{\sim}{R}$ such that the variables occurring in $\varphi$ are among $\{x_i : i\in S\}$. Then we define

$E(u,\varphi) \stackrel{d}{=} \{s\in{}^SU : u=pj_1\circ s,\ \underset{\sim}{R} \models \varphi[pj_0\circ s]\}$.

Clearly, $E(u,\varphi) \subseteq {}^SU$ and $E(u,\varphi) = E(u,\psi)$ if $\underset{\sim}{R} \models (\varphi \leftrightarrow \psi)$.

$P(S) \stackrel{d}{=} \{E(u,x_i=x_j+r) : u\in{}^S2,\ r\in R,\ i<j$ and $i,j\in S\}$.
$\mathfrak{L}(S) \stackrel{d}{=} \mathfrak{Sb}\,{}^SU$ and $\mathfrak{A}(S) \stackrel{d}{=} \mathfrak{Sg}^{(\mathfrak{L}(S))}P(S)$.

In the sense of Def.6.0 we have $\mathfrak{A}(S), \mathfrak{L}(S)\in Cs_S$. Note that, by our previous notations, $\mathfrak{L} = \mathfrak{L}(2)$, $P=P(2)$ and $\mathfrak{A} = \mathfrak{A}(2)$. Some more definitions:

$F(S) \stackrel{d}{=} \{x_i=x_j+r,\ x_i\neq x_j+r : i,j\in S,\ r\in R\}$,
$F(S)^* \stackrel{d}{=} \{\ \wedge H : H \subseteq_\omega F(S)\}$, and
$F(S)^{**} \stackrel{d}{=} \{\ \vee H : H \subseteq_\omega F(S)^*\}$.
$G(S) \stackrel{d}{=} \{E(u,\varphi) : u\in{}^S2,\ \varphi\in F(S)^{**}\}$,
$G(S)^* \stackrel{d}{=} \{\ \sum H : H \subseteq_\omega G(S)\}$.

Lemma 8.6.1.1. $A(S) = G(S)^*$ for every $S\in Sb_\omega\beta$ if $2\leq|S|$.

Proof. Let $S\subseteq\beta$, $2\leq|S|<\omega$. First we prove $A(S) \subseteq G(S)^*$. By $P(S) \subseteq$

$\subseteq G(S)^*$ and $A(S) = SgP(S)$, it is enough to show that $G(S)^* \in Su\,\mathcal{L}(S)$. We shall need the following lemma.

Lemma 8.6.1.1.1.

(i) $(\forall\varphi,\psi\in F(S)^{**})(\exists\mu,\nu\in F(S)^{**})[\underset{\sim}{R} \models ((\varphi\wedge\psi) \leftrightarrow \mu)$ and $\underset{\sim}{R} \models ((\neg\varphi) \leftrightarrow \nu)]$.

(ii) $(\forall\varphi\in F(S)^{**})(\forall\varkappa\in\beta)(\exists\psi\in F(S\sim\{\varkappa\})^{**})\ \underset{\sim}{R} \models (\exists x_\varkappa\varphi \leftrightarrow \psi)$.

The proof of (ii) is a simple elimination of quantifiers argument. Since the proof of 8.6.1.1.1 is a straightforward computation, we omit it.

Now we show that $G(S)^*$ is closed under the operations of $\mathcal{L}(S)$.

TRUE stands for an arbitrary universally valid formula.

1. $G(S)^*$ is closed under multiplication: It is enough to show $(\forall g_1,g_2\in G(S))g_1\cdot g_2\in G(S)^*$. Let $u_1,u_2\in{}^S2$ and $\varphi_1,\varphi_2\in F(S)^{**}$.

$$E(u_1,\varphi_1)\cdot E(u_2,\varphi_2) = \begin{cases} 0 & \text{if } u_1\neq u_2 \\ E(u_1,\varphi_1\wedge\varphi_2) & \text{otherwise} \end{cases}.$$

$E(u_1,\varphi_1\wedge\varphi_2)\in G(S)$ by Lemma 8.6.1.1.1(i).

2. $G(S)^*$ is closed under negation: It is enough to show $(\forall g\in G(S))\,{}^SU{\sim}g\in G(S)^{**}$. Let $u\in{}^S2$, $\varphi\in F(S)^{**}$. Then ${}^SU{\sim}E(u,\varphi) = \sum\{E(u',\text{TRUE}) : u'\in{}^S2{\sim}\{u\}\} + E(u,\neg\varphi) \in G(S)^*$ by $|S|<\omega$ and by 8.6.1.1.1 (i).

3. $D_{ij}^{[{}^SU]}\in G(S)^{**}$ for $i,j\in S$ : Let $i,j\in S$. $D_{ij}^{[{}^SU]} = \sum\{E(u,x_i{=}x_j{+}0) : u\in{}^S2,\ u_i=u_j\}\in G(S)^*$ by $|S|<\omega$.

4. $G(S)^{**}$ is closed under cylindrifications: Let $j\in S$. It is enough to show $(\forall g\in G(S))C_j^{[{}^SU]}g\in G(S)^{**}$. Let $u\in{}^S2$ and $\varphi\in F(S)^{**}$. $C_j^{[{}^SU]}E(u,\varphi) = \sum\{E(u',\exists x_j\varphi) : u'\in C_j^{[{}^S2]}\{u\}\}$. Since $|S|<\omega$ we have $|C_j^{[{}^S2]}\{u\}|<\omega$, and therefore by 8.6.1.1.1(ii) we have $C_j^{[{}^SU]}E(u,\varphi)\in G(S)^{**}$.

By these statements we have seen that $G(S)^*\in Su\,\mathcal{L}(S)$. Therefore $A(S) \subseteq G(S)^*$. Next we prove $G(S)^* \subseteq A(S)$. It is enough to show $G(S) \subseteq A(S)$ since $A(S)\in Su\,\mathcal{L}(S)$. Let $u\in{}^S2$. By $E(u,\bigvee\{\varphi_i : i<n\}) = \sum\{E(u,\varphi_i) : i<n\}$ and $E(u,\bigwedge\{\varphi_i : i<n\}) = \prod\{E(u,\varphi_i) : i<n\}$ it is enough to show $(\forall\varphi\in F(S))E(u,\varphi)\in A(S)$.

<u>Case 1</u> $\varphi$ is $x_i=x_j+r$ for some $i,j\in S$ and $r\in R$. If $i<j$ then $E(u,\varphi)\in P(S)$. If $i>j$ then $E(u,\varphi) = E(u,x_j=x_i+(-r))\in P(S)$. If $i=j$ and $r\neq 0$ then $E(u,\varphi) = 0\in A(S)$. If $i=j$ and $r=0$ then $E(u,\varphi) =$ $= E(u,\mathrm{TRUE})$. Let $k\in S$, $k\neq i$. Such a $k$ exists by $|S|\geq 2$. Using the equalities obtained before, we have $C_m E(u,x_i=x_k) = \sum\{E(u',\mathrm{TRUE}) : u'\in$ $\in C_m\{u\}\}$ for $m\in\{i,k\}$, therefore $C_i E(u,x_i=x_k)\cdot C_k E(u,x_i=x_k) = \sum\{E(u',$ $\mathrm{TRUE}) : u'\in C_i\{u\}\cdot C_k\{u\}\} = E(u,\mathrm{TRUE})$. This shows $E(u,\mathrm{TRUE})\in A(S)$, since $E(u,x_i=x_k)\in P(S)$.

<u>Case 2</u> $\varphi$ is $x_i\neq x_j+r$ for some $i,j\in S$ and $r\in R$. We may suppose $i<j$ by the arguments of Case 1. Using the equalities obtained before we have $E(u,x_i\neq x_j+r) = E(u,\mathrm{TRUE})\cdot(\sum\{E(u',\mathrm{TRUE}) : u'\in {}^S2\sim\{u\}\} + E(u,x_i\neq$ $\neq x_j+r)) = E(u,\mathrm{TRUE}) - E(u,x_i=x_j+r)$. By $E(u,\mathrm{TRUE})\in A(S)$ and $E(u,x_i=$ $=x_j+r)\in P(S)$ then $E(u,x_i\neq x_j+r)\in A(S)$.

We have seen $G(S)^* \subseteq A(S)$.

<u>QED(Lemma 8.6.1.1.)</u>

Let $S,H\in(Sb_\omega\beta)\cup\{\beta\}$, and let $S\subseteq H$. We define

$$i(S,H) \overset{d}{=} \langle\{s\in{}^HU : S1s\in a\} \;:\; a\subseteq {}^SU\rangle,$$

$$\mathfrak{L}(\beta) \overset{d}{=} \mathfrak{Gs}\,{}^\beta U,\quad \mathfrak{A}(\beta) \overset{d}{=} \mathfrak{Sg}^{(\mathfrak{L}(\beta))}\cup\{i(S,\beta)^*A(S) : S\in Sb_\omega\beta\}.$$

Let $\mathfrak{D}\in CA_H$. Then $\mathfrak{Nr}_S\,\mathfrak{D}$ denotes the "neat S-reduct" of $\mathfrak{D}$, i.e. $Nr_S\mathfrak{D} \overset{d}{=} \{a\in D : (\forall\varkappa\in H\sim S)\, c_\varkappa^{\mathfrak{D}} a=a\}$.

<u>Lemma 8.6.1.2.</u> Let $S,H,Z\in(Sb_\omega\beta)\cup\{\beta\}$. Suppose $|S|\geq 2$ and $S\subseteq H\subseteq Z$. Then (1)-(3) below hold.

(1) $i(S,Z) = i(H,Z)\circ i(S,H)$.

(2) $i(S,H) : \mathfrak{A}(S) \rightarrowtail\!\!\twoheadrightarrow \mathfrak{Nr}_S\,\mathfrak{A}(H)$.

(3) $\mathfrak{A}(\alpha)\in \mathbf{I}\,\mathfrak{Nr}_\alpha\,{}_\infty Cs_\beta$ for every $\alpha\in\beta\cap(\omega\sim 2)$.

<u>Proof.</u> $i(H,Z)i(S,H)a = \{s\in{}^ZU : H1s\in i(S,H)a\} = \{s\in{}^ZU : S1(H1s)\in a\} =$ $= \{s\in{}^ZU : S1s\in a\} = i(S,Z)a$. This proves (1).

It is shown in [HMTI]8.5 that $i(S,H) : \mathfrak{A}(S) \rightarrowtail \mathfrak{Nr}_S\,\mathfrak{L}(H)$ is a one-one homomorphism (since $V={}^{S}U$ and $W={}^{H}S$ satisfy the conditions of [HMTI] 8.5.). By $\mathfrak{A}(H) \subseteq \mathfrak{L}(H)$ we have $Nr_S\,\mathfrak{A}(H) = A(H)\cap Nr_S\,\mathfrak{L}(H)$. Therefore to prove $i(S,H) : \mathfrak{A}(S) \rightarrowtail \mathfrak{Nr}_S\,\mathfrak{A}(H)$ it is enough to show $i(S,H)^*A(S)\subseteq \subseteq A(H)$. If $H=\beta$ then $i(S,H)^*A(S) \subseteq A(H)$ by definition of $A(\beta)$. Suppose $H\in Sb_\omega\beta$. Then $S\in Sb_\omega\beta$ too, and therefore by $A(S) = Sg^{(\mathfrak{L}(S))} P(S)$, it is enough to prove $i(S,H)^*P(S) \subseteq A(H) = G(H)^*$. Let $u\in{}^{S}2$ and $\varphi\in F(S)$. $i(S,H)E(u,\varphi) = \{s\in{}^{H}U : S1s\in E(u,\varphi)\} = \sum\{E(u',\varphi) : u'\in{}^{H}2, S1u'=u\}\in G(H)^*$, since $|H|<\omega$ (and since $F(S) \subseteq F(H)$). By these statements we have seen $i(S,H) : \mathfrak{A}(S) \rightarrowtail \mathfrak{Nr}_S\,\mathfrak{A}(H)$ for every $S,H \in \in(Sb_\omega\beta)\cup\{\beta\}$, if $S\subseteq H$. It remains to show $i(S,H)^*A(S) \supseteq Nr_S\,\mathfrak{A}(H)$. Suppose first $H\in Sb_\omega\beta$. Let $\Gamma \overset{d}{=} H\sim S$. Then $|\Gamma|<\omega$, and $Nr_S\,\mathfrak{A}(H) = \{c_{(\Gamma)}a : : a\in G(H)^*\}$ by 8.6.1.1. By additivity of $c_{(\Gamma)}$ it is enough to show $(\forall a\in G(H))(\exists g\in A(S))i(S,H)g=c_{(\Gamma)}a$. Let $u\in{}^{H}2$ and $\varphi\in F(H)^{**}$. $\exists x_{(\Gamma)}\varphi$ denotes $\exists x_{i1}\ldots\exists x_{in}\varphi$ where $\Gamma = \{i_1,\ldots,i_n\}$. $c_{(\Gamma)}E(u,\varphi) = \{s\in{}^{H}U : : (H\sim\Gamma)1pj_1\circ s\subseteq u,\ \underset{\sim}{R} \models \exists x_{(\Gamma)}\varphi[pj_0\circ s]\} = i(S,H)E(S1u,\exists x_{(\Gamma)}\varphi)$. Since $(\exists\psi\in \in F(S)^{**})$ $\underset{\sim}{R} \models (\exists x_{(\Gamma)}\varphi \leftrightarrow \psi)$ by 8.6.1.1.1(ii), this proves $c_{(\Gamma)}E(u,\varphi)\in \in i(S,H)^*A(S)$. By this, we have seen $i(S,H) : \mathfrak{A}(S) \twoheadrightarrow \mathfrak{Nr}_S\,\mathfrak{A}(H)$ for $S\subseteq H\in Sb_\omega\beta$. Next, $K \overset{d}{=} \{i(S,\beta)^*\,\mathfrak{A}(S) : S\in Sb_\omega\beta\}$ is directed by $\subseteq^r$, by 8.6.1.2(1). Then $\cup^r K$ exists by [HMT]0.5.11 and $\mathfrak{A}(\beta) = \cup^r K$ by [HMT]0.5.10. In particular, $A(\beta) = \cup\{i(S,\beta)^*A(S) : S\in Sb_\omega\beta\}$. Let $S\in \in Sb_\omega\beta$. Now we show $i(S,\beta)^*A(S) \supseteq Nr_S\,\mathfrak{A}(\beta)$. Suppose $a\in A(\beta)\sim i(S,\beta)^* A(S)$. We show that then $a\notin Nr_S\,\mathfrak{A}(\beta)$. Since $a\in A(\beta)$, there is $H\in Sb_\omega\beta$ such that $a\in i(H,\beta)^*A(H)$. By (1) of 8.6.1.2 we may suppose $S\subseteq H$. By $a\notin i(S,\beta)^*A(S)$ and by (1) then we have $a=i(H,\beta)g$ for some $g\in A(H)\sim \sim i(S,H)^*A(S)$. We have already seen that $i(S,H)^*A(S) = Nr_S\,\mathfrak{A}(H)$ (since $H\in Sb_\omega\beta$), and therefore $(\exists\varkappa\in H\sim S)$ $c_\varkappa^{\mathfrak{A}(H)}g \neq g$. Then $c_\varkappa^{\mathfrak{A}(\beta)}i(H,\beta)g = = i(H,\beta)\,c_\varkappa^{\mathfrak{A}(H)}g \neq i(H,\beta)g$, since $i(H,\beta)$ is a one-one homomorphism. I.e. $c_\varkappa^{\mathfrak{A}(\beta)}a \neq a$ for some $\varkappa\in\beta\sim S$, which shows $a\notin Nr_S\,\mathfrak{A}(\beta)$. Thus (2) is proved.

(3) is an immediate consequence of (2), since $\mathfrak{A}(\beta)\in{}_\infty Cs_\beta$.

<u>QED(Lemma 8.6.1.2.)</u>

By 8.6.1.2(3) we have $\mathfrak{A}(2)\in \mathbf{I}\,\mathbf{Nr}_2\,{}_\infty Cs_\beta$. Then $\mathfrak{A}=\mathfrak{A}(2)$ completes the proof.

QED(Claim 8.6.1)

Claim 8.6.2. $\mathfrak{B}\notin \mathbf{Nr}_2\,CA_\beta$.

Proof. Let $u\in{}^2 2$. $X_u \overset{d}{=} E(u,TRUE)$. $\tau(x)$ denotes the term $s^0_1 c_1 x\cdot$ $\cdot s^1_0 c_0 x$.

Lemma 8.6.2.1.

(i) $|\{b\in B : b\le X_{01}\}|\le\omega$ and $|\{b\in B : b\le X_{10}\}|>\omega$.

(ii) $\tau^{\mathfrak{B}}(X_{10}) = X_{01}$.

(iii) $CA_\beta \models {}_2s(0,1)c_2x\le\tau(c_2x)$.

(iv) $(\forall\mathfrak{N}\in CA_\beta)\ {}_2s(0,1)\in Ism(\mathfrak{Bl}\,\mathfrak{Nr}_2\,\mathfrak{N},\ \mathfrak{Bl}\,\mathfrak{Nr}_2\,\mathfrak{N})$.

Proof. Proof of (i): Let $K \overset{d}{=} P\sim\{p(\langle 0,1\rangle,r) : r\in R\sim Q\}$. Recall that $\mathfrak{B} = \mathfrak{Sg}^{(\mathfrak{L})}K$. Then clearly $|\{b\in B : b\le X_{10}\}| \ge |\{b\in K : b\le X_{10}\}|>\omega$. Now we show $|\{b\in B : b\le X_{01}\}|\le\omega$. Let $Z \overset{d}{=} \{c^{\mathfrak{L}}_0 a,\ c^{\mathfrak{L}}_1 a : a\in A\}\cup\{D^{\mathfrak{L}}_{01}\}$ and let $\mathfrak{D}$ $\overset{d}{=} \mathfrak{Sg}^{(\mathfrak{Bl}\,\mathfrak{L})}(Z\cup K)$. Then $D\supseteq B$ since $D\supseteq K$ and $D\in Su\,\mathfrak{A}\subseteq Su\,\mathfrak{L}$ by $Z\subseteq D$. Let $V \overset{d}{=} X_{01}$. Then $\{b\in B : b\le X_{01}\} = Rl_V\mathfrak{B}\subseteq Rl_V\mathfrak{D}$ by $B\subseteq D$, and therefore it is enough to show $|Rl_V\mathfrak{D}|\le\omega$. $rl_V$ is an endomorphism of the Boolean algebra $\mathfrak{D}$, since $V\in Z\subseteq D$. Therefore $Rl_V\mathfrak{D} = Sg^{(\mathfrak{D})} rl_V{}^*(Z\cup K)$. Thus it is enough to show $|rl_V{}^*(Z\cup K)|\le\omega$. By $rl_V{}^*K =$ $= \{p(\langle 0,1\rangle,r) : r\in Q\}$ we have $|rl_V{}^*K|\le\omega$. By By 8.6.1.1 we have $Z = \{\sum\{X_u : u\in S\} : S\subseteq_\omega {}^2 2\}\cup\{D_{01}\}$ and therefore $|rl_V{}^*Z|\le\omega$ by $|Z|\le\omega$. We have seen $|Rl_V\mathfrak{B}|\le\omega$.

Proof of (ii): $\tau^{\mathfrak{B}}(X_{10}) = c_0(D_{01}\cdot c_1X_{10})\cdot c_1(D_{01}\cdot c_0X_{10})$. $c_0(D_{01}\cdot c_1X_{10}) = c_0(D_{01}\cdot(X_{10}+X_{11})) = c_0(D_{01}\cdot X_{11}) = X_{01}+X_{11}$. Similarly, $c_1(D_{01}\cdot c_0X_{10}) = c_1(D_{01}\cdot(X_{00}+X_{10})) = c_1(D_{01}\cdot X_{00}) = X_{00}+X_{01}$. $(X_{01}+X_{11})\cdot$ $\cdot(X_{00}+X_{01})=X_{01}$.

Proof of (iv): Let $\mathfrak{N}\in CA_\beta$. Lemma 0 of [AN4] says that ${}_2s(0,1)$ is a complete and one-one endomorphism of $\mathfrak{Rl}_{\beta\sim 2}\mathfrak{N}$. Now $\mathfrak{Bl}\,\mathfrak{Nr}_2\,\mathfrak{N} =$ $= \mathfrak{Rl}_{\beta\sim 2}\mathfrak{N}$ completes the proof.

Proof of (iii): ${}_2s(0,1)c_2x \le {}_2s(0,1)c_1c_2x = s_0^2s_1^0s_2^1c_1c_2x = s_0^2s_1^0c_1c_2x =$ $= s_0^2s_1^0c_2c_1x = s_0^2c_2s_1^0c_1x = c_2s_1^0c_1x = s_1^0c_1c_2x$, by (iv) and by 1.5.12, 1.5.8 of [HMT]. Similarly (but not completely similarly), ${}_2s(0,1)c_2x \le$ $\le s_0^2s_1^0s_2^1c_0c_2x = s_0^2s_1^0c_0s_2^1c_2x = s_0^2s_2^1c_0c_2x = s_0^2s_0^1c_0c_2x = s_0^2c_2s_0^1c_0x =$ $= c_2s_0^1c_0x = s_0^1c_0c_2x$, by (iv) and by 1.5.12, 1.5.8 and 1.5.10(ii) of [HMT]. Thus ${}_2s(0,1)c_2x \le s_1^0c_1c_2x \cdot s_0^1c_0c_2x = \tau(c_2x)$.

QED(Lemma 8.6.2.1)

Suppose $\mathfrak{B} \in \mathbf{Nr}_2\, CA_\beta$. Then there is a $\mathfrak{N} \in CA_\beta$ such that $\mathfrak{B} =$ $= \mathfrak{Nr}_2\,\mathfrak{N}$. Let $N_u \overset{d}{=} \{b\in B : b\le X_u\} = \{n\in Nr_2\mathfrak{N} : n\le X_u\}$, for $u\in {}^22$. Then $|N_{01}|\le\omega$ and $|N_{10}|>\omega$ by 8.6.2.1(i). $X_{10} = c_2^{\mathfrak{N}}X_{10}$ by $X_{10}\in Nr_2\mathfrak{N}$ and therefore ${}_2s(0,1)X_{10} \le \tau^{\mathfrak{N}}(X_{10}) = \tau^{\mathfrak{B}}(X_{10}) = X_{01}$ by 8.6.1.2(ii)-(iii). By 8.6.2.1(iv) then ${}_2s(0,1)^*N_{10} \subseteq N_{01}$ and $|{}_2s(0,1)^*N_{10}|>\omega$ contradicting $|N_{01}|\le\omega$. Therefore $\mathfrak{B} \notin \mathbf{Nr}_2\, CA_\beta$.

QED(Claim 8.6.2)

Claim 8.6.3 $\mathfrak{B}$ is an elementary submodel of $\mathfrak{A}$ .

Proof. Let $P(0,1) \overset{d}{=} \{p(\langle 0,1\rangle,r) : r\in R\}$.

Lemma 8.6.3.1. Let $m : P(0,1) \rightarrowtail\!\!\!\!\rightarrow P(0,1)$ be any permutation of $P(0,1)$. Then $(\exists f\in Is(\mathfrak{A},\mathfrak{A}))[m\subseteq f$ and $(P\sim P(0,1))1f\subseteq Id]$.

Proof. Let $V \overset{d}{=} E(\langle 0,1\rangle,TRUE)$ and $u \overset{d}{=} \langle 0,1\rangle$. First we define an automorphism $h : \mathfrak{Bl}\,\mathfrak{Rl}_V\mathfrak{A} \rightarrowtail\!\!\!\!\rightarrow \mathfrak{Bl}\,\mathfrak{Rl}_V\mathfrak{A}$ such that $m\subseteq h$. Let $\mathfrak{I} \overset{d}{=}$ $\overset{d}{=} \mathfrak{Bl}\,\mathfrak{Rl}_V\mathfrak{A}$. The universe $D$ of $\mathfrak{I}$ is $D = Rl_V\mathfrak{A} = \{x\in A : x\le V\}$. By 8.6.1.1 we have $A = G(2)^*$ and then by $\sum\{E(u,\varphi_i) : i<n\} = E(u,\bigvee\{\varphi_i :$ $: i<n\})$ we have $D = \{E(u,\varphi) : \varphi\in F(2)^{**}\}$. Clearly, $At\,\mathfrak{I} = P(0,1)$. Then $E(u,x_0=x_1+r)\in P(0,1)$ for every $r\in R$, $E(u,x_0\ne x_1+r) = V-E(u,x_0=x_1+r)$, $E(u,\bigwedge\{\varphi_i : i<n\}) = \prod\{E(u,\varphi_i) : i<n\}$, $E(u, \bigvee\{\varphi_i : i<n\}) = \sum\{E(u,\varphi_i) :$ $: i<n\}$ show that $\mathfrak{I}$ is generated by the set $P(0,1)$ of its atoms. Lemma 8.6.3.1.1 below is known from BA-theory. Therefore we omit its proof. E.g. it is immediate by 12.3 of [S]p.34.

Lemma 8.6.3.1.1. Let $\mathfrak{D}$ be any BA generated by its atoms. Then every permutation of $At\mathfrak{D}$ can be extended to an automorphism of $\mathfrak{D}$ .

$\mathfrak{D} = \mathfrak{Bl}\,\mathfrak{Rl}_V\mathfrak{A}$ is generated by the set $P(0,1)$ of its atoms and $m$ is a permutation of $P(0,1)$. Then by 8.6.3.1.1 there is an automorphism $h$ of $\mathfrak{D}$ such that $m\subseteq h$. So far we have constructed an automorphism $h$ of $\mathfrak{Bl}\,\mathfrak{Rl}_V\mathfrak{A}$ such that $m\subseteq h$. Now we define $f : A \to A$ as $f \stackrel{d}{=}$ $\stackrel{d}{=} \langle h(y\cdot V)+y\cdot -V : y\in A\rangle$. We shall prove that $f\in Is(\mathfrak{A},\mathfrak{A})$ and $m\subseteq f$, $(P\sim P(0,1))\upharpoonright f\subseteq Id$. Let $y\in P(0,1)$. Then $y\subseteq V$ and therefore $f(y) = h(y) =$ $= m(y)$. Let $y\in P\sim P(0,1)$. Then $y\cdot V = 0$ and therefore $f(y) = y$. Next we prove that $f\in Is(\mathfrak{A},\mathfrak{A})$. It is easy to see that $f$ is one-one because $h$ is one-one, $f$ is onto since $h$ is onto and $f$ is a Boolean homomorphism since $h$ is a Boolean homomorphism. $f(d_{01}^{\mathfrak{A}}) = d_{01}^{\mathfrak{A}}$ since $d_{01}^{\mathfrak{A}}\cdot V = 0$. Next we shall use the following properties of $V =$ $= E(u,TRUE)$.

Fact 1 $(\forall\varkappa<2)(\forall y\in A)[0<y\leq V \Rightarrow c_\varkappa y=c_\varkappa V]$.

Fact 2 $(\forall\varkappa<2)(\forall y\in A)\ (c_\varkappa y)\cdot V \in \{0,V\}$.

Fact 2 is easy to see. It suffices to prove Fact 1 for $y=E(u,\varphi)$ where $u\in {}^2 2$ and $\varphi\in F(2)^{**}$, and $\varkappa=0$, $y\neq 0$. Clearly $C_0E(u,\varphi) =$ $= E(\langle 0,u_1\rangle,\exists x_0\varphi)\cup E(\langle 1,u_1\rangle,\exists x_0\varphi)$. Now by 8.6.1.1.1(ii) there is a $\psi\in$ $\in F(\{1\})^{**}$ such that $\underset{\sim}{R} \models \exists x_0\varphi \leftrightarrow \psi$. But clearly for any $\chi\in F(\{1\})^{**}$ we have $\underset{\sim}{R} \models \chi$ or $\underset{\sim}{R} \models \neg\chi$. Since $y\neq 0$ we must have $\underset{\sim}{R} \models \psi$. Hence $C_0E(u,\varphi) = E(\langle 0,u_1\rangle,TRUE)\cup E(\langle 1,u_1\rangle,TRUE) = C_0V$.

Let $\varkappa<2$ and let $y\in A$. $c_\varkappa f(y) = c_\varkappa(h(y\cdot V)+y-V) = c_\varkappa h(y\cdot V)+c_\varkappa(y-V) =$ $c_\varkappa(y\cdot V)+c_\varkappa(y-V) = c_\varkappa y$ since $c_\varkappa h(y\cdot V) = c_\varkappa(y\cdot V)$ by Fact 1. $f(c_\varkappa y) =$ $= h(V\cdot c_\varkappa y)+-V\cdot c_\varkappa y = V\cdot c_\varkappa y + -V\cdot c_\varkappa y$ since $h(V\cdot c_\varkappa y) = V\cdot c_\varkappa y$ by Fact 2 and since $h$ is an automorphism of $\mathfrak{Bl}\,\mathfrak{Rl}_V\mathfrak{A}$.

QED(Lemma 8.6.3.1.)

Now we prove that $\mathfrak{B}$ is an elementary submodel of $\mathfrak{A}$ . We shall show it by the Tarski-Vaught criterion, see e.g. Prop.19.16 in [M].

We have to show that for every $n\in\omega$, for every first order formula $\varphi(x_0,\dots,x_n)$ and for every $b_0,\dots,b_{n-1}\in B$ $\mathfrak{A}\models\exists x_n\varphi(b_0,\dots,b_{n-1},x_n)$ implies $(\exists b_n\in B)\ \mathfrak{A}\models\varphi(b_0,\dots,b_n)$. Let $\varphi(x_0,\dots,x_n)$ be any first order formula. Let $b_0,\dots,b_{n-1}\in B$ and assume $\mathfrak{A}\models\varphi(b_0,\dots,b_{n-1},a)$ for some $a\in A$. Since $A=\mathrm{Sg}P$, there is $H\subseteq_\omega P$ such that $a\in\mathrm{Sg}H$. Let this $H$ be fixed. Since $\{b_0,\dots,b_{n-1}\}\subseteq B=\mathrm{Sg}^{(\mathfrak{A})}(P\cap B)$, there is $Z\subseteq_\omega P\cap B$ such that $(H\cap B)\cup\{b_0,\dots,b_{n-1}\}\subseteq\mathrm{Sg}Z$. Let $m: P(0,1)\rightarrowtail\twoheadrightarrow P(0,1)$ be a permutation of $P(0,1)$ such that $(P(0,1)\cap Z)1m\subseteq Id$ and $m^*(H\sim B)\subseteq P(0,1)\cap B$. Such a permutation exists, since $|Z|<\omega$, $|H|<\omega$, $|P(0,1)\cap B|\geq\omega$ and $P\sim B\subseteq P(0,1)$. Let $f$ be an automorphism of $\mathfrak{A}$ such that $m\subseteq f$ and $P\sim P(0,1)1f\subseteq Id$. Such an automorphism exists by 8.6.3.1. Since $f$ is an automorphism of $\mathfrak{A}$ and $\mathfrak{A}\models\varphi(b_0,\dots,b_{n-1},a)$, we have $\mathfrak{A}\models\varphi(fb_0,\dots,fb_{n-1},fa)$. Since $Z1f\subseteq Id$ and $\{b_0,\dots,b_{n-1}\}\subseteq\mathrm{Sg}Z$ we have $(\forall i<n)fb_i=b_i$. Since $f^*H\subseteq B$ and $a\in\mathrm{Sg}H$ we have $fa\in B$. Therefore $\mathfrak{A}\models\varphi(b_0,\dots,b_{n-1},fa)$ and $fa\in B$ complete the proof.

QED(Claim 8.6.3)

By Claim 8.6.3 and by the Keisler-Shelah ultrapower theorem we have $\mathfrak{B}\in\mathbf{Uf\,Up}\,\mathfrak{A}$, and then by Claims 8.6.1-2 we have $\mathfrak{B}\in\mathbf{Uf\,Up\,Nr}_2\,{}_\infty Cs_\beta\sim\mathbf{I\,Nr}_2\,CA_\beta$. By 7.18(iii), 8.20 and by [HMTI]7.22 we have $\mathbf{I\,Nr}_2\,{}_\infty Cs_\beta=\mathbf{Nr}_2\,\mathbf{I}\,{}_\infty Cs_\beta=\mathbf{Nr}_2\,\mathbf{Up}\,{}_\infty Cs_\beta=\mathbf{Up\,Nr}_2\,{}_\infty Cs_\beta$. Therefore $\mathfrak{B}\in\mathbf{Uf\,Nr}_2\,{}_\infty Cs_\beta$. Then $\mathfrak{B}\in\mathbf{Uf\,Nr}_2\,\mathfrak{C}$ for some $\mathfrak{C}\in{}_\infty Cs_\beta$. Let $\mathfrak{N}\overset{d}{=}\mathfrak{Sg}^{(\mathfrak{C})}Nr_2\,\mathfrak{C}$. Then $\mathfrak{B}\in\mathbf{Uf\,Nr}_2\,\mathfrak{N}$ and $\mathfrak{N}\in{}_\infty Cs_\beta\cap Lf_\beta$. $|Zd\,\mathfrak{N}|=2$ by $|Zd\,\mathfrak{B}|=2$. Then $\mathfrak{N}\in\mathbf{I}\,{}_\infty Ws_\beta\cap Lf_\beta\subseteq\mathbf{I}\,{}_\infty Cs_\beta^{reg}\cap Lf_\beta$ by [HMTI]6.14 and by 3.15(a). (i) is proved.

(ii) follows from (i) by ${}_\infty Ws_\beta\cap Lf_\beta\subseteq K_\beta\subseteq CA_\beta$.

QED(Theorem 8.6.)

Problem 8.7. For which $2<\alpha<\beta$ does $\mathbf{Nr}_\alpha\,\mathbf{I}\,Gs_\beta=\mathbf{Uf\,Nr}_\alpha\,Gs_\beta$ hold?

Let $\alpha\geq\omega$. By 8.5 we have that $\mathbf{I}\,Gs_\alpha$ is the first order axiomatizable hull of $\mathbf{Rd}_\alpha Gs_\beta$. Theorem 8.8 below implies that the first order axiomatizable hull of $\mathbf{Nr}_\alpha \mathrm{Cs}_\beta$ is smaller than $\mathbf{I}\,\mathrm{Gs}_\alpha$ (i.e. there is a first order formula which is valid in $\mathbf{Nr}_\alpha \mathrm{Gs}_\beta$ but is not valid in $\mathrm{Gs}_\alpha$). As a contrast to Thm 8.8(iii) see Theorem 8.13.

Theorem 8.8. Let $1<\alpha<\beta$ and $\alpha\le\gamma$ be arbitrary.

(i) $\mathbf{Rd}_\alpha({}_\omega\mathrm{Ws}_\gamma) \not\subseteq \mathbf{Uf\,Up\,Nr}_\alpha \mathrm{CA}_\beta$.

(ii) $K \not\subseteq \mathbf{Uf\,Up\,Nr}_\alpha \mathrm{CA}_\beta$ if ${}_\omega\mathrm{Ws}_\alpha \subseteq K \subseteq \mathrm{CA}_\alpha$.

(iii) $\mathbf{I}\,\mathrm{Gs}_\alpha \ne \mathbf{Uf\,Nr}_\alpha \mathrm{Gs}_\beta = \mathbf{Uf\,Up\,Nr}_\alpha \mathrm{Gs}_\beta$ and $\mathbf{Uf\,Up}\,\mathrm{Cs}_\alpha \ne \mathbf{Uf\,Up\,Nr}_\alpha \mathrm{Cs}_\beta$.

Proof. First we define some formulas of the language of $\mathrm{CA}_2$. Let $\alpha>1$. The term $\tau(x)$ is defined to be $[(s^0_1 c_1 x)\cdot(s^1_0 c_0 x)]$. If $\mathfrak{A}$ is a $\mathrm{CA}_\alpha$ then $\tilde{\tau}^{\mathfrak{A}} : A \to A$ is the term-function defined by the term $\tau$ in $\mathfrak{A}$ see [HMT]p.4 3. $at(x)$ is the formula $\forall y[y\le x \to (y=0 \vee y=x)] \wedge x\ne 0$. Let $elem(x,y)$ be the formula $(at(\tau x) \wedge x\le y)$. We define $\psi(y_0,y_1)$ to be the formula

$$(\forall z[\forall x(elem(x,y_0) \to x\le z) \to z\ge y_0] \to [\forall x(elem(x,y_0) \to \tau x\le y_1) \wedge$$
$$\wedge\ \forall z(\forall x(elem(x,y_0) \to \tau x\le z) \to y_1\le z)]).$$

Claim 8.8.1. Let $\alpha\ge 2$ and $\mathfrak{A}\in\mathrm{CA}_\alpha$. Then conditions (i) and (ii) below are equivalent.

(i) $\mathfrak{A} \models \forall y_0 \exists y_1 \psi(y_0,y_1)$.

(ii) For all $X\subseteq A$ with $(\tilde{\tau}^{\mathfrak{A}})^* X \subseteq At\,\mathfrak{A}$, the existence of $\sum^{(\mathfrak{A})} X$ implies the existence of $\sum^{(\mathfrak{A})} (\tilde{\tau}^{\mathfrak{A}})^* X$.

The proof of Claim 8.8.1 is immediate by observing that $\psi(y_0,y_1)$ expresses that if $y_0 = \sum\{x : x\le y_0$ and $\tau(x)$ is an atom$\}$ then $y_1 = \sum\{\tau(x) : x\le y_0$ and $\tau(x)$ is an atom$\}$.

Let $1<\alpha<\beta$. Statement 1 of [AN4] says that every $\mathbf{Nr}_\alpha \mathrm{CA}_\beta$ satisfies (ii) of Claim 8.8.1 above. Therefore $\mathbf{Nr}_\alpha \mathrm{CA}_\beta \models \forall y_0 \exists y_1 \psi(y_0,y_1)$. Let $\gamma\ge\alpha$ be arbitrary. Statement 2 of [AN4] states the existence of $\mathfrak{A} \in \mathbf{S\,Nr}_\gamma \mathrm{CA}_{\gamma+\beta}$ such that $\mathfrak{A}$ does not satisfy (ii) of Claim 8.8.1.

The first few lines of the proof of the quoted Statement 2 show that $\mathfrak{A} \in {}_\omega Ws_\gamma$. By 8.8.1 we conclude that ${}_\omega Ws_\gamma \not\models \forall y_0 \exists y_1 \psi(y_0, y_1)$. Since $\alpha \geq 2$ the formula $\psi$ is in the language of $CA_\alpha$-s and therefore $\mathbf{Rd}_\alpha({}_\omega Ws_\gamma) \not\models \forall y_0 \exists y_1 \psi(y_0, y_1)$. Thus $\mathbf{Rd}_\alpha({}_\omega Ws_\gamma) \not\subseteq \mathbf{Uf\,Up\,Nr}_\alpha CA_\beta$. This proves (i). (ii) follows from (i) by choosing $\gamma = \alpha$. $\mathbf{I\,Nr}_\alpha Gs_\beta = \mathbf{Up\,Nr}_\alpha Gs_\beta$ will be proved in Thm 8.20. Now (iii) follows from (ii) since ${}_\omega Ws_\alpha \subseteq \mathbf{I}\,Cs_\alpha \subseteq$ $\subseteq \mathbf{I}\,Gs_\alpha \subseteq CA_\alpha$.

QED(Theorem 8.8.)

About Corollary 8.9 below note that if $\alpha \geq \omega$ then $Ws_\alpha \not\subseteq \mathbf{SP}(Cs_\alpha^{reg} \cap$ $\cap Lf_\alpha)$, moreover $Ws_\alpha \not\subseteq \mathbf{SP}Lf_\alpha$ by [HMTI]6.8(7). By [HMT]2.6.74 and 2.6.32, $\mathbf{S}$ cannot be omitted from Corollary 8.9 below.

Corollary 8.9. Let $1 < \alpha < \beta$. Then $\mathbf{SP}({}_\omega Cs_\alpha^{reg} \cap Lf_\alpha) \not\subseteq \mathbf{Nr}_\alpha CA_\beta$.

Proof. If $\alpha < \omega$ then we are done by 8.8.(ii). Assume $\alpha \geq \omega$. Let $\mathfrak{R}$ be the greatest $Cs_\alpha^{reg} \cap Lf_\alpha$-subalgebra of $\mathfrak{Sb}\,{}^\alpha\omega$. By 7.1 and [HMTI] 7.13 we have ${}_\omega Ws_\alpha \subseteq \mathbf{I}\,{}_\infty Cs_\alpha \subseteq \mathbf{S\,Up}\,\mathfrak{R}$. Thus $\mathbf{HSP}\,\mathfrak{R} \not\subseteq \mathbf{Nr}_\alpha CA_\beta$ by Thm 8.8. By Cor.8.20, $\mathbf{Nr}_\alpha CA_\beta = \mathbf{HNr}_\alpha CA_\beta$. Therefore $\mathbf{SP}\,\mathfrak{R} \not\subseteq \mathbf{Nr}_\alpha CA_\beta$.

QED(Corollary 8.9.)

Proposition 8.10(5) below is quoted in [HMTI]8.3.

Proposition 8.10. Let $1 < \alpha < \beta$ and $1 < \varkappa < \omega$.

(1) ${}_\varkappa Cs_1 \subseteq \mathbf{Rd}_1 \mathbf{I}\,{}_\varkappa Cs_2$ and ${}_\infty Cs_1 \not\subseteq \mathbf{Uf\,Rd}_1 CA_2$.

(2) $Mn_\alpha \cap \mathbf{Uf\,Rd}_\alpha CA_\beta = \mathbf{I}\,{}_1Cs_\alpha$ iff $\alpha < \omega$.

$Cs_\alpha \cap \mathbf{Uf\,Rd}_\alpha CA_\beta = {}_1Cs_\alpha$ iff $\alpha < \omega$ and $\beta > \alpha+1$.

(3) $Mn_\alpha \cap \mathbf{Rd}_\alpha CA_\beta = \mathbf{I}\,{}_1Cs_\alpha$.

$Cs_\alpha^{reg} \cap \mathbf{Rd}_\alpha CA_\beta = {}_1Cs_\alpha$ iff $\beta > \alpha+1$.

$Ws_\alpha \cap \mathbf{Rd}_\alpha CA_\beta = {}_1Ws_\alpha$ iff $\beta > \alpha+1$.

(4) $\mathbf{Rd}_\alpha(Cs_{\alpha+1} \cap Mn_{\alpha+1}) \subseteq \mathbf{I}\,Cs_\alpha^{reg}$.

$\mathbf{Rd}_\alpha(Cs_\beta^{reg} \cap Lf_\beta) \subseteq \mathbf{I}\,Cs_\alpha$ if $\omega \leq \alpha$ and $|\beta| \leq 2^{|\alpha|}$.

(5) $\mathbf{Rd}_\alpha Cs_\beta^{reg} \not\subseteq \mathbf{I}\,Cs_\alpha$.

$\mathbf{Rd}_\alpha(Cs_\beta^{reg} \cap Lf_\beta) \not\subseteq \mathbf{I}\, Cs_\alpha^{reg}$.

$\mathbf{Rd}_\alpha(Ws_\beta \cap Lf_\beta) \not\subseteq \mathbf{I}\, Gws_\alpha^{comp\ reg}$.

<u>Proof.</u> Let $1<\varkappa<\omega$. <u>Proof of (1):</u> Let $\mathfrak{A} \in {}_\varkappa Cs_1$ and $\mathfrak{B} \overset{d}{=} \mathfrak{Bl}\,\mathfrak{A}$. Then $\mathfrak{A} \models \forall x(x\neq 0 \to c_0x=1)$ and $\mathfrak{B}$ is a BA generated by $\lambda\leq\varkappa$ atoms. For every $i<\varkappa$ let $a_i \overset{d}{=} \{\langle j, j+i(\mathrm{mod}\ \varkappa)\rangle : j<\varkappa\}$. Let $\mathfrak{C}$ be the ${}_\varkappa Cs_2$ generated by $\{a_i : i<\lambda\}$. Then $\{a_i : i<\lambda\} \subseteq At\mathfrak{C}$ and $(\forall i<j<\lambda)a_i\neq a_j$ proves $\mathfrak{B} \cong \mathfrak{Bl}\,\mathfrak{C}$. Also, $(\forall x\in C\sim\{0\})c_0x=1$ by $(\forall i<\lambda)c_0a_i=1$. Thus $\mathfrak{A} \cong \mathfrak{Rd}_1\mathfrak{C}$. We have seen ${}_\varkappa Cs_1 \subseteq \mathbf{Rd}_1\mathbf{I}\,{}_\varkappa Cs_2$. Let $\beta\geq 2$. We prove ${}_\infty Cs_1 \not\subseteq \mathbf{Uf}\,\mathbf{Rd}_1 CA_\beta$. First we show $\mathbf{Rd}_1 CA_\beta \models (\forall x(x\neq 0 \to c_0x=1) \to \exists x\, at(x))$ where $at(x)$ is the formula $\forall y(y\leq x \to (y=0 \vee y=x)) \wedge x\neq 0$, cf. Def.7.3.1. $CA_\beta \models \forall x(0<x<d_{01} \to c_0x\neq 1)$ by (C7). Hence $CA_\beta \models [\forall x(x\neq 0 \to c_0x=1) \wedge 0\neq 1] \to at(d_{01})$, thus $\mathbf{Rd}_1 CA_\beta \models [\forall x(x\neq 0 \to c_0x=1) \wedge 0\neq 1] \to \exists x\, at(x)$. Therefore $Cs_1 \cap \mathbf{Uf}\,\mathbf{Rd}_1 CA_\beta \models \exists x\, at(x)$ by $Cs_1 \models \forall x(x\neq 0 \to c_0x=1)$. Hence $\mathfrak{C} \notin \mathbf{Uf}\,\mathbf{Rd}_1 CA_\beta$ for any atomless $Cs_1$ $\mathfrak{C}$.

<u>Lemma 8.10.1.</u> Let $0<\alpha<\beta$. Then $\mathbf{HRd}_\alpha CA_\beta \models (\exists x(c_0x\neq x) \to \exists x\Delta x=1)$, where $\Delta x=1$ is the ($\alpha$-ary) formula $c_0x\neq x \wedge \wedge\{c_ix=x : i\in\alpha\sim 1\}$.

<u>Proof.</u> Let $0<\alpha<\beta$ and $\mathfrak{A} \in \mathbf{HRd}_\alpha CA_\beta$. Then there are $\mathfrak{B} \in CA_\beta$ and $h\in Ho(\mathfrak{R}, \mathfrak{A})$ where $\mathfrak{R} \overset{d}{=} \mathfrak{Rd}_\alpha \mathfrak{B}$. Let $z \overset{d}{=} d_{0\alpha}^{\mathfrak{B}}$. Then $(\forall b\in B)[b\leq c_0^\partial z \Rightarrow \Delta b=0]$ by [HMT]1.3.19. Suppose $\mathfrak{A}$ is nondiscrete. We have $hz\neq 1$, for otherwise also $h(c_0^\partial z)=1$, and for any $a\in A$ and $i<\alpha$, say with $a=hb$, we have $c_ia = c_ih(b\cdot c_0^\partial z) = h(c_i(b\cdot c_0^\partial z)) = h(b\cdot c_0^\partial z) = a$, contradicting $\mathfrak{A}$ non-discrete. But $c_0h(z) = h(c_0d_{0\alpha}) = 1$ shows $0\in \Delta^{(\mathfrak{A})}(h(z))$. By $\Delta^{(\mathfrak{A})}z = 1$ this proves $\Delta^{(\mathfrak{A})}(h(z)) = 1$.

<u>QED Lemma(8.10.1.)</u>

By [HMT]2.1.22 we have $Mn_\alpha \models \neg\exists x\Delta x=1$. Let $\mathfrak{A} \in Mn_\alpha \cap \mathbf{Rd}_\alpha CA_\beta$. Then $\mathfrak{A} \models \forall x(c_0x=x)$ by 8.10.1. Hence $|A|\leq 2$ by $\mathfrak{A} \in Mn_\alpha$, thus $\mathfrak{A} \in \mathbf{I}\,{}_1Cs_\alpha$. Clearly ${}_1Cs_\alpha \subseteq Mn_\alpha \cap \mathbf{Rd}_\alpha CA_\beta$. We have seen $Mn_\alpha \cap \mathbf{Rd}_\alpha CA_\beta = \mathbf{I}\,{}_1Cs_\alpha$. Similarly, $Mn_\alpha \cap \mathbf{Uf}\,\mathbf{Rd}_\alpha CA_\beta = \mathbf{I}\,{}_1Cs_\alpha$ if $\alpha<\omega$, since $\Delta x=1$ is a first order formula if $\alpha<\omega$.

Lemma 8.10.2. Let $1<\alpha+1<\beta$. Then

$\mathbf{Rd}_{\alpha}CA_{\beta} \models \exists x(c_0x\neq x) \rightarrow \exists x(x\neq 0 \wedge x\neq 1 \wedge \Delta x=0)$

where $\Delta x=0$ is the formula $\wedge\{c_i x=x : i\in\alpha\}$.

Proof. Let $1<\alpha+1<\beta$. Let $\mathfrak{A} = \mathfrak{Rd}_{\alpha}\mathfrak{B}$ for some $\mathfrak{B}\in CA_{\beta}$. If $\mathfrak{A} \models$ $\models \exists x(c_0x\neq x)$ then $z \stackrel{d}{=} d^{\mathfrak{B}}_{\alpha,\alpha+1} \neq 1$ by [HMT]1.3.12. Clearly, $z\neq 0$ and $\Delta^{(\mathfrak{A})}z=0$.

QED(Lemma 8.10.2.)

Let $1<\alpha+1<\beta$. Let $\mathfrak{A}\in Dind_{\alpha}\cap\mathbf{Rd}_{\alpha}CA_{\beta}$. Then $\mathfrak{A}\models \forall x(c_0x=x)$ by 8.10.2, hence $|A|\leq 2$ by [HMT]1.3.12 and $\mathfrak{A}\in Dind_{\alpha}$. Thus $\mathfrak{A}\in \mathbf{I}\,{}_1Cs_{\alpha}$. Clearly, ${}_1Cs_{\alpha} \subseteq Dind_{\alpha}\cap\mathbf{Rd}_{\alpha}CA_{\beta}$. We have seen $Dind_{\alpha}\cap\mathbf{Rd}_{\alpha}CA_{\beta} = \mathbf{I}\,{}_1Cs_{\alpha}$. Similarly, if $\alpha<\omega$ then $Dind_{\alpha}\cap\mathbf{Uf}\,\mathbf{Rd}_{\alpha}CA_{\beta} = \mathbf{I}\,{}_1Cs_{\alpha}$, since $\Delta x=0$ is a first order formula if $\alpha<\omega$. Now ${}_1Cs_{\alpha} \subseteq Cs^{reg}_{\alpha} \subseteq Dind_{\alpha}$, $Ws_{\alpha} \subseteq Dind_{\alpha}$, and $Cs_{\alpha} \subseteq Dind_{\alpha}$ if $\alpha<\omega$ complete the proof of the "if-parts" of (2) and (3). If $\alpha\geq\omega$ then $\mathbf{I}Cs_{\alpha}\cap Mn_{\alpha}\cap\mathbf{Uf}\,\mathbf{Rd}_{\alpha}CA_{\beta} \subseteq \mathbf{I}Cs_{\alpha}\cap Mn_{\alpha} \not\subseteq \mathbf{I}\,{}_1Cs_{\alpha}$ by 8.5. Let $\beta=\alpha+1$. Then the negative parts of (2)-(3) will follow from (4), using [HMTI]7.13.

Proof of (4): Let $0<\alpha<\beta$. Let $\mathfrak{A}\in Cs_{\alpha+1}\cap Mn_{\alpha+1}$. Let $\mathfrak{R} \stackrel{d}{=} \mathfrak{Rd}_{\alpha}\mathfrak{A}$. Let $x\in A$ and $\Gamma \stackrel{d}{=} \alpha\cap\Delta^{(\mathfrak{A})}(x)$, Then $c^{\mathfrak{R}}_{(\Gamma)}x = c^{\mathfrak{A}}_{(\Gamma)}x$ and $\Delta^{(\mathfrak{A})}c_{(\Gamma)}x \subseteq$ $\subseteq \Delta^{(\mathfrak{A})}x\sim\Gamma$. Hence $|\Delta^{(\mathfrak{A})}c_{(\Gamma)}x|\leq 1$. Then $\Delta^{(\mathfrak{A})}c_{(\Gamma)}x=0$ by $\mathfrak{A}\in Mn_{\alpha+1}$ and [HMT]2.1.22. Then $c_{(\Gamma)}x\in\{0,1\}$ by [HMTI]5.3. We have seen $(\forall x\in A)$ $(\exists\Gamma \subseteq_{\omega} \alpha)c^{\mathfrak{R}}_{(\Gamma)}x\in\{0,1\}$. This $\mathfrak{R}$ is simple. Let $\mathfrak{B} \stackrel{d}{=} rd_{\alpha}{}^{*}\mathfrak{R}$. Then $\mathfrak{B}\in Gs^{reg}_{\alpha}$ by 8.1(i), since $\mathfrak{A}\in Cs^{reg}_{\alpha+1}$ by [HMTI]4,1, and $\mathfrak{B} \cong \mathfrak{R}$ by 4.7.1.2. Let $U\in Subb(\mathfrak{B})$. Then $rl({}^{\alpha}U)\in Is\mathfrak{B}$ since $\mathfrak{R}$ is simple and $rl({}^{\alpha}U)^{*}\mathfrak{B}\in Cs^{reg}_{\alpha}$ by [HMTI]3.16. Thus $\mathfrak{Rd}_{\alpha}\mathfrak{A} \in \mathbf{I}Cs^{reg}_{\alpha}$. We have seen $\mathbf{Rd}_{\alpha}(Cs_{\alpha+1}\cap Mn_{\alpha+1}) \subseteq \mathbf{I}Cs^{reg}_{\alpha}$.

Assume $\omega\leq\alpha$ and $|\beta|\leq 2^{|\alpha|}$. Let $\mathfrak{A}\in{}_{\lambda}Cs^{reg}_{\beta}\cap Lf_{\beta}$. If $\lambda\geq\omega$ then $\mathfrak{Rd}_{\alpha}\mathfrak{A} \in \mathbf{I}Cs_{\alpha}$ by 8.1(i) and [HMTI]7.21. Assume $\lambda<\omega$. Then $|A|\leq|\beta|\leq 2^{|\alpha|}$ by $\mathfrak{A}\in{}_{\lambda}Cs^{reg}_{\beta}\cap Lf_{\beta}$. $\mathfrak{Rd}_{\alpha}\mathfrak{A}$ has a characteristic and $\mathfrak{Rd}_{\alpha}\mathfrak{A} \in Lf_{\alpha}$. Thus $\mathfrak{Rd}_{\alpha}\mathfrak{A} \in \mathbf{I}Cs_{\alpha}$ by 4.17(i). We have seen that $\mathbf{Rd}_{\alpha}(Cs^{reg}_{\beta}\cap Lf_{\beta}) \subseteq$ $\subseteq \mathbf{I}Cs_{\alpha}$ if $\omega\leq\alpha$ and $|\beta|\leq 2^{|\alpha|}$.

Proof of (5): Let $1<\varkappa<\omega$. Let $\mathfrak{L} \overset{d}{=} \mathfrak{Fr}_{\beta}\varkappa$, $X \overset{d}{=} At\mathfrak{L}$ and $\mathfrak{A} \overset{d}{=} \mathfrak{Sg}^{(\mathfrak{L})}X$. Then $\mathfrak{A}$ is regular by 1.4. Let $\mathfrak{B} \overset{d}{=} \mathfrak{Rd}_{\alpha}\mathfrak{A}$. If $\alpha<\omega$ then $\mathfrak{B} \notin \mathsf{I}\,Cs_{\alpha}$ by $\mathfrak{B} \nvDash (x=0 \vee c_{(\alpha)}x=1)$. Suppose $\alpha\geq\omega$. By $X\subseteq At\mathfrak{B}$ there are more than $\varkappa$ atoms $a$ in $\mathfrak{B}$ such that $(\forall i<\alpha)\, a\leq d_{0i}$. Thus $\mathfrak{B} \notin \mathsf{I}\,{}_{\varkappa}Cs_{\alpha}$. Since $\mathfrak{B}$ is of characteristic $\varkappa\neq 0$, this means $\mathfrak{B} \notin \mathsf{I}\,Cs_{\alpha}$. We have seen $\mathsf{Rd}_{\alpha}Cs_{\beta}^{reg} \nsubseteq \mathsf{I}\,Cs_{\alpha}$.

Let $\mathfrak{L} \overset{d}{=} \mathfrak{Fr}_{\beta}\varkappa$, $x \overset{d}{=} \{s\in{}^{\beta}\varkappa : s(\alpha)=0\}$ and $\mathfrak{A} \overset{d}{=} \mathfrak{Sg}^{(\mathfrak{L})}\{x\}$. Then $\mathfrak{A} \in Cs_{\alpha}^{reg}\cap Lf_{\beta}$ by [HMTI]4.1. Let $\mathfrak{R} \overset{d}{=} \mathfrak{Rd}_{\alpha}\mathfrak{A}$. Then $\Delta^{(\mathfrak{R})}x=0$ by $\Delta^{(\mathfrak{A})}x=\{\alpha\}$. Thus $\mathfrak{R} \notin Dind_{\alpha}$ by $x \notin \{0,1\}$. This proves $\mathsf{Rd}_{\alpha}(Cs_{\beta}^{reg}\cap Lf_{\beta}) \nsubseteq Dind_{\alpha} \supseteq \mathsf{I}\,Cs_{\alpha}^{reg}$. Let $p\in{}^{\beta}\varkappa$ and $\mathfrak{B} \overset{d}{=} \mathfrak{Rl}({}^{\beta}\varkappa^{(p)})\mathfrak{A}$. Then $\mathfrak{B} \in {}_{\varkappa}Ws_{\beta}\cap Lf_{\beta}$ and $y \overset{d}{=} x\cap{}^{\beta}\varkappa^{(p)} \notin \{0,1\}$. Then $\Delta^{(\mathfrak{Rd}_{\alpha}\mathfrak{B})}y=0$ shows $\mathfrak{Rd}_{\alpha}\mathfrak{B} \notin Dind_{\alpha} \supseteq \mathsf{I}\,Gws_{\alpha}^{comp\ reg}$.

QED(Proposition 8.10.)

Problem 8.11. To what extent is the cardinality condition $|\beta|\leq 2^{|\alpha|}$ needed in 8.10(4)? (Obviously, $|\beta| \leq 2^{2^{|\alpha|}}$ is needed.)

Proposition 8.12. Let $\alpha\leq\beta$. Let $K = \langle \mathsf{I}\,Cs_{\alpha}\cap Lf_{\alpha} : \alpha\in Ord\rangle$.

(i) $K_{\alpha} = \mathbf{S}\mathsf{Rd}_{\alpha}K_{\beta}$ iff $2^{|\alpha|} = 2^{|\beta|}$.

(ii) $K_{\alpha} = \mathbf{S}\mathsf{Nr}_{\alpha}K_{\beta}$ iff either $\alpha+\beta<\omega$ or $2^{|\alpha|}=2^{|\beta|}$.

Proof. Let $\alpha\leq\beta$. Let $K = \langle \mathsf{I}\,Cs_{\alpha}\cap Lf_{\alpha} : \alpha\in Ord\rangle$. Suppose $2^{|\alpha|}=2^{|\beta|}$. If $\alpha<\omega$ then this implies $\alpha=\beta$ and we are done. Let $\omega\leq\alpha$. Let $\mathfrak{A} \in {}_{\varkappa}Cs_{\beta}\cap Lf_{\beta}$, $\mathfrak{B} \overset{d}{=} rd_{\alpha}^{*}\mathfrak{Rd}_{\alpha}\mathfrak{A}$. Then $\mathfrak{Rd}_{\alpha}\mathfrak{A} \cong \mathfrak{B} \in {}_{\varkappa}Gs_{\alpha}\cap Lf_{\alpha}$ by 4.7.1.2 and 8.1(i). If $\varkappa\geq\omega$ then $\mathfrak{B} \in \mathsf{I}\,Cs_{\alpha}$ by [HMTI]7.21. Suppose $\varkappa<\omega$. Then $\mathfrak{B}$ has $\leq|{}^{(\beta\sim\alpha)}2|$ subbases by 4.7.1.2(ii). Therefore $\mathfrak{B}$ has $\leq 2^{|\alpha|}$ subunits by our assumption $2^{|\beta|}=2^{|\alpha|}$ and by $\varkappa<\omega$. $\mathfrak{B} \in Lf_{\alpha}$ has non-zero characteristic by $\mathfrak{B} \in {}_{\varkappa}Gs_{\alpha}$, $\varkappa<\omega$. Hence $\mathfrak{B} \in \mathsf{I}\,Cs_{\alpha}$ by 4.16(i). We have seen that $\mathsf{Rd}_{\alpha}K_{\beta} \subseteq K_{\alpha}$. Thus $K_{\alpha} = \mathbf{S}\mathsf{Nr}_{\alpha}K_{\beta} = \mathbf{S}\mathsf{Rd}_{\alpha}K_{\beta}$ since $K_{\alpha} \subseteq \mathbf{S}\mathsf{Nr}_{\alpha}K_{\beta}$ by [HMTI]8.6. Assume $2^{|\alpha|} \neq 2^{|\beta|}$. Then $2^{|\beta|} > 2^{|\alpha|}$ by $\beta\geq\alpha$. If $\beta<\omega$ then $\mathsf{Rd}_{\alpha}K_{\beta} \nsubseteq K_{\alpha}$ by 8.10(5). Suppose $\beta\geq\omega$. Let $\mathfrak{L} \overset{d}{=} \mathfrak{Fr}_{\beta}2$, $z \overset{d}{=} Subu(\mathfrak{L})$, $\mathfrak{A} \overset{d}{=} \mathfrak{Sg}^{(\mathfrak{L})}z$.

Then $\mathfrak{A}\in Cs_\beta\cap Lf_\beta$ and $|Z|=2^{|\beta|}$. Assume $\mathfrak{B}\in Gws_\alpha\cap \mathbf{I}\,\mathfrak{Nr}_\alpha\mathfrak{A}$. Then $\mathfrak{B}\in$ $\in_2 Gws_\alpha$ and $\mathfrak{B}$ has an antichain of cardinality $2^{|\beta|}>2^{|\alpha|}$, since $Z\subseteq$ $\subseteq Zd\mathfrak{A}\subseteq Nr_\alpha\mathfrak{A}$. Therefore $\mathfrak{B}$ is not compressed. We have seen that $\mathbf{Nr}_\alpha K_\beta \not\subseteq K_\alpha$ if $\omega\le\alpha$ and $2^{|\alpha|}\neq 2^{|\beta|}$. Let $\alpha<\beta<\omega$. Then $K_\alpha = \mathbf{S}\mathbf{Nr}_\alpha K_\beta$ will be proved in 8.18(ii).

QED(Proposition 8.12.)

About the cardinality condition of 8.12 note that it is consistent with ZFC that $|\alpha|\neq|\beta|$ but $2^{|\alpha|}=2^{|\beta|}$, for some $\alpha,\beta$.

Theorem 8.13. Let $\alpha<\beta$.

(i) $\mathbf{Uf}\,\mathbf{Up}\,Cs_\alpha = \mathbf{Uf}\,\mathbf{Up}\,\mathbf{Rd}_\alpha\,Cs_\beta$ iff $\alpha\ge\omega$.

(ii) $\mathbf{Uf}\,\mathbf{Up}\,Gws_\alpha^{comp\ reg} = \mathbf{Uf}\,\mathbf{Up}\,\mathbf{Rd}_\alpha\,Gws_\beta^{comp\ reg}$ iff $\alpha\ge\omega$.

(iii) $\mathbf{Uf}\,\mathbf{Up}\,Ws_\alpha \supseteq \mathbf{Rd}_\alpha\,Ws_\beta$ iff $\alpha\ge\omega$.

To prove Theorem 8.13, we shall need the following definitions and lemmas.

Definition 8.13.1. By a Crs-structure we understand a four-sorted relational structure $\mathfrak{M}=\langle A,U,V,I,ext,E,C,D,+,-,1\rangle$ such that $A,U,V,$ $I$ are the four universes, and $ext : V\times I\to U$, $E\subseteq V\times A$, $C : I\times A\to A$, $D : I\times I\to A$, $+ : A\times A\to A$, $- : A\to A$ and $1\in A$.

Convention: Let $\mathfrak{M}$ be the above Crs-structure. Then $A^{\mathfrak{M}} = A$, $U^{\mathfrak{M}} =$ $= U$, etc. We shall omit the superscript if there is no danger of confusion.

Next we define some axioms in the discourse language of Crs-structures.

Definition 8.13.2. About the discourse language for Crs-structures we use the convention that $s,z$ denote variables of sort $V$; $i,j$ are variables of sort $I$; $x,y$ are of sort $A$ and $b,u,v$ are of sort $U$.

Consider axioms (S1)-(S9) below.

(S1) $(\forall s)(\forall z)[(\forall i)\mathrm{ext}(s,i)=\mathrm{ext}(z,i) \rightarrow s=z]$.

(S2) $(\forall x)(\forall y)[(\forall s)(E(s,x) \leftrightarrow E(s,y)) \rightarrow x=y]$.

(S3) $(\forall x)(\forall s)(\forall i)[E(s,C(i,x)) \leftrightarrow (\exists z)[E(z,x) \wedge (\forall j)(j\neq i \rightarrow \mathrm{ext}(s,j)= =\mathrm{ext}(z,j))]]$.

(S4) $(\forall s)(\forall i)(\forall j)[E(s,D(i,j)) \leftrightarrow \mathrm{ext}(s,i)=\mathrm{ext}(s,j)]$.

(S5) $(\forall x)(\forall y)(\forall s)[E(s,x+y) \leftrightarrow (E(s,x) \vee E(s,y))]$.

(S6) $(\forall x)(\forall s)[E(s,-x) \leftrightarrow \neg E(s,x)]$.

(S7) $(\forall s)E(s,1)$.

(S8) $(\forall s)(\forall i)(\forall u)(\exists z)[\mathrm{ext}(z,i)=u \wedge (\forall j)(j\neq i \rightarrow \mathrm{ext}(z,j)=\mathrm{ext}(s,j))]$.

(S9) $(\forall x)(\forall s)(\forall z)([(\forall i)(C(i,x)\neq x \rightarrow \mathrm{ext}(s,i)=\mathrm{ext}(z,i)) \wedge (\exists i)\mathrm{ext}(s,i)= =\mathrm{ext}(z,i)] \rightarrow (E(s,x) \leftrightarrow E(z,x)))$.

Now Crax is defined to consist of axioms (S1)-(S7), Cpax consists of (S1)-(S8) and Rgax is Crax$\cup\{$(S9)$\}$.

Definition 8.13.3.

(1) Let $\mathfrak{A}\in Crs_\alpha$. Then $str(\mathfrak{A}) \stackrel{d}{=} \langle A, base(\mathfrak{A}), 1^{\mathfrak{A}}, \alpha, ext, E, C, D, +^{\mathfrak{A}}, -^{\mathfrak{A}}, 1^{\mathfrak{A}}\rangle$ where $ext(s,i) \stackrel{d}{=} s(i)$, $E \stackrel{d}{=} \{\langle s,x\rangle\in 1^{\mathfrak{A}}\times A : s\in x\}$, $C(i,x) \stackrel{d}{=} c_i^{\mathfrak{A}}x$ and $D(i,j) \stackrel{d}{=} d_{ij}^{\mathfrak{A}}$ for every $i,j\in\alpha$, $s\in 1^{\mathfrak{A}}$ and $x\in A$. Clearly, $str(\mathfrak{A})$ is a Crs-structure.

(2) Let $\mathfrak{M}$ be a Crs-structure such that $\mathfrak{M}\models\{(S1),(S2)\}$. Let $\xi : \alpha \rightarrowtail\!\!\!\rightarrow I^{\mathfrak{M}}$ be one-one and onto. We define $Cy_\xi(\mathfrak{M}) \stackrel{d}{=} \langle A^{\mathfrak{M}}, +^{\mathfrak{M}}, 1^{\mathfrak{M}}, c_i, d_{ij}\rangle_{i,j\in\alpha}$ where $c_i \stackrel{d}{=} \langle C(\xi i,x) : x\in A\rangle$ and $d_{ij} \stackrel{d}{=} D(\xi i,\xi j)$. $cy_\xi \stackrel{d}{=} \langle\langle ext(s,\xi i) : i<\alpha\rangle : s\in V^{\mathfrak{M}}\rangle$ and $cyl_\xi \stackrel{d}{=} \langle\{cy_\xi(s) : \langle s,x\rangle\in E^{\mathfrak{M}}\} : x\in A^{\mathfrak{M}}\rangle$. Then $cy_\xi : V^{\mathfrak{M}} \rightarrowtail {}^\alpha U^{\mathfrak{M}}$ and $cyl_\xi : A^{\mathfrak{M}} \rightarrowtail Sb\,{}^\alpha U^{\mathfrak{M}}$ are one-one by axioms (S1) and (S2). We define $Cyl_\xi(\mathfrak{M}) \stackrel{d}{=} cyl_\xi{}^*Cy_\xi(\mathfrak{M})$.

We shall consider the similarity types of $CA_\alpha$-s to be identical with that of $Cyl(\mathfrak{M})$ because in any $Bo_\alpha$ the operations $\cdot$ and $0$ are definable by terms from the others. It is immediate by these definitions that $Cyl_\eta(str(\mathfrak{A})) = \mathfrak{A}$ for $\mathfrak{A}\in Crs_\alpha$ and $\eta=\alpha 1 Id$.

Lemma 8.13.4. Let $\mathfrak{M}$ be a Crs-structure. Let $\xi : \alpha \rightarrowtail\!\!\!\rightarrow I^{\mathfrak{M}}$ be one-one and onto. Let $\mathfrak{A}\in Crs_\alpha$.

(i) $\mathfrak{M} \models$ Crax implies $Cyl_\xi(\mathfrak{M}) \in Crs_\alpha$ and
$str(\mathfrak{A}) \models$ Crax.

(ii) $\mathfrak{M} \models$ Cpax implies $Cyl_\xi(\mathfrak{M}) \in Gws_\alpha^{comp}$ and
$str(\mathfrak{A}) \models$ Cpax if $\mathfrak{A} \in Gws_\alpha^{comp}$.

(iii) $\mathfrak{M} \models$ Cpax$\cup$Rgax implies $Cyl_\xi(\mathfrak{M}) \in Gws_\alpha^{comp\ reg}$ and
$str(\mathfrak{A}) \models$ Rgax if $\mathfrak{A} \in Gws_\alpha^{comp\ reg}$.

(iv) $\mathfrak{M} \models$ Rgax implies $Cyl_\xi(\mathfrak{M}) \in Crs_\alpha^{creg}$ and
$str(\mathfrak{A}) \models$ Rgax if $\mathfrak{A} \in Crs_\alpha^{creg}$.

<u>Proof.</u> Let $\mathfrak{M}$ be a Crs-structure and let $\xi : \alpha \rightarrowtail\!\!\!\twoheadrightarrow I^{\mathfrak{M}}$ be one-one and onto. Assume $\mathfrak{M} \models$ Crax. Then $\mathfrak{M} \models \{(S1),(S2)\}$. Let $\mathfrak{N} \overset{d}{=} Cy_\xi(\mathfrak{M})$ and $\mathfrak{B} \overset{d}{=} Cyl_\xi(\mathfrak{M})$. By Def.8.13.3, $cyl_\xi \in Is(\mathfrak{N},\mathfrak{B})$. Let $W \overset{d}{=} 1^{\mathfrak{B}}$. Then $W = \{cy_\xi(s) : s\in V\}$ by (S7).

<u>Convention:</u> In accordance with [HMT]p.28, instead of $\langle s,x\rangle \in E^{\mathfrak{M}}$ we sometimes write sEx.

(1) First we prove $\mathfrak{B} \in Crs_\alpha$. Let $x,y\in A$ and $i,j\in\alpha$. By (S6) and (S1) we have $-^{(\mathfrak{B})}cyl_\xi(x) = cyl_\xi(-^{(\mathfrak{N})}x) = cyl_\xi(-^{(\mathfrak{M})}x) = \{cy_\xi(s) : sE{-x}\} = \{cy_\xi(s) : s\in V$ and $\langle s,x\rangle\notin E\} = \{cy_\xi(s) : s\in V\}\sim\{cy_\xi(s) : sEx\} = W\sim cyl_\xi(x)$. Hence $-^{(\mathfrak{B})} = B1_{W}\sim$ is proved. By (S5) we have $cyl_\xi(x+y) = cyl_\xi(x)\cup cyl_\xi(y)$. By (S4) we have $sED(\xi i,\xi j)$ iff $cy_\xi(s)\in D_{ij}^{[W]}$, hence $d_{ij}^{(\mathfrak{B})} = cyl_\xi(D(\xi i,\xi j)) = \{cy_\xi(s) : sED(\xi i,\xi j)\} = D_{ij}^{[W]}$. By (S3) we have $sEC(\xi i,x)$ iff $cy_\xi(s)\in C_i^{[W]}cyl_\xi(x)$. Hence $c_i^{(\mathfrak{B})}cyl_\xi(x) = cyl_\xi(C(\xi i,x)) = C_i^{[W]}cyl_\xi(x)$. These statements prove that $\mathfrak{B} \in Crs_\alpha$.

(2) Assume $\mathfrak{M} \models$ Cpax. We prove $\mathfrak{B} \in Gws_\alpha^{comp}$. To this end, we have to prove that $W=cyl_\xi(1)$ is a $Gws^{comp}$-unit. Let $p\in W$ and $u\in U$. Then $p = cy_\xi(s) = \langle ext(s,\xi i) : i<\alpha\rangle$ for some $s\in V$. By (S8) there is $z\in V$ such that $ext(z,\xi i)=u$ and $(\forall j\in I)(j\neq\xi i \rightarrow ext(z,j) = ext(s,j))$. Let $q \overset{d}{=} cy_\xi(z)$. Then $q\in W$ and $q = p_u^i$. We have proved the following statement (*).

(*) $(\forall p\in W)(\forall u\in U)(\forall i\in\alpha)p_u^i\in W$ and $W\subseteq {}^\alpha U$.

Let $i,j\in\alpha$. Then $W\subseteq C_i^{[W]}(D_{ij}^{[W]}\cap W)$ since $(\forall p\in W)p(i/pj)\in W$ by (*).

We show $s^i_j W \subseteq W$ (operations in $\mathfrak{Gb}\,{}^{\alpha}U$). Let $p \in s^i_j W$, $p \in {}^{\alpha}U$. Then $p^i_{pj} \in W$, say $p^i_{pj} = cy_{\xi}(s)$, $s \in V$. By (S8) again, choose $z$ so that $ext(z,i) = pi$ and $(\forall j \neq i) ext(z,j) = ext(s,j)$. Clearly then $p = cy_{\xi}(z) \in W$. Thus $(\forall i,j \in \alpha) s^i_j W = W$ and hence $W$ is a $Gws_{\alpha}$-unit by [HMTI]2.1(ii)-(iii). Let $Y \in Subu(W)$. By (*) we have that $base(Y) = U$. Hence $W$ is a compressed $Gws_{\alpha}$-unit.

(3) $Crs^{creg}_{\alpha}$ was defined in 1.6.1. Assume $\mathfrak{M} \models Rgax$. We shall prove $\mathfrak{B} \in Crs^{creg}_{\alpha}$. Let $x \in B$ and $n \in \alpha$. Then $x = cyl_{\xi}(y)$ for some $y \in A^{\mathfrak{M}}$. Let $p,q \in 1^{\mathfrak{B}}$ be such that $(\{n\} \cup \Delta^{(\mathfrak{B})} x) 1 p \subseteq q \in x$. Then $p = cy_{\xi}(s)$, $q = cy_{\xi}(z)$ for some $s,z \in V$ such that $(\forall i \in \{n\} \cup \Delta^{(\mathfrak{B})} x) ext(s,\xi i) = ext(z,\xi i)$. In order to apply (S9) to $y,s$ and $z$, suppose $j \in I$ and $C(j,y) \neq y$. Let $j = \xi i$. Then $c^{\mathfrak{N}}_i y \neq y$ by $\mathfrak{N} = Cy_{\xi}(\mathfrak{M})$, hence $c^{\mathfrak{B}}_i x \neq x$ since $cyl_{\xi} \in Is(\mathfrak{N}, \mathfrak{B})$. Thus $ext(s,j) = ext(z,j)$. Clearly, $ext(s,\xi n) = ext(z,\xi n)$. By (S9) then $(sEy \longleftrightarrow zEy)$. By $q \in x$ we have $zEy$. Therefore $sEy$, which implies $p \in x$. We have seen that $x$ is cregular.

(4) Assume $\mathfrak{M} \models Cpax \cup Rgax$. Then $\mathfrak{B} \in Gws^{comp\ reg}_{\alpha}$, by (2)-(3). By 1.6.2(i), $Gws^{creg}_{\alpha} = Gws^{reg}_{\alpha}$ and hence $\mathfrak{B} \in Gws^{comp\ reg}_{\alpha}$.

By (1)-(4) we proved those parts of 8.13.4 which refer to $Cyl_{\xi}(\mathfrak{M})$. To prove the parts referring to $str(\mathfrak{A})$, let $\mathfrak{A} \in Crs_{\alpha}$ and $U \overset{d}{=} base(\mathfrak{A})$, $V \overset{d}{=} 1^{\mathfrak{A}}$. It is immediate by Definitions 8.13.2,3(1) that $str(\mathfrak{A}) \models Crax$, $str(\mathfrak{A}) \models$ (S8) if $\mathfrak{A} \in Gws^{comp}_{\alpha}$, and $str(\mathfrak{A}) \models$ (S9) if $\mathfrak{A} \in Crs^{creg}_{\alpha}$.

<u>QED(Lemma 8.13.4.)</u>

<u>Definition 8.13.5.</u> Let $X$ be a set of variables. Then $Fm(X,CA_{\alpha})$ (resp. $Tm(X,CA_{\alpha})$) denotes the set of first order formulas (resp. terms) in the discourse language of $CA_{\alpha}$-s using variables from $X$. Let elements of $X$ denote variables of sort $A$ in the discourse language of Crs-structures. Let $\xi : \alpha \rightarrowtail I$ be one-one. We define the "translation functions" $t_{\xi}$ and $tr_{\xi}$ on $Tm(X,CA_{\alpha})$ and $Fm(X,CA_{\alpha})$ respectively as follows. Let $x \in X$, $\tau,\delta \in Tm(X,CA_{\alpha})$ and $\varphi,\psi \in Fm(X,CA_{\alpha})$. Then $t_{\xi}x$ is $x$, $t_{\xi}(c_i\tau)$ is $C(\xi i, t_{\xi}\tau)$, $t_{\xi}d_{ij}$ is $D(\xi i,\xi j)$, $t_{\xi}(\tau+\delta)$ is

is $t_\xi\tau+t_\xi\delta$, $t_\xi-\tau$ is $-t_\xi\tau$. $tr_\xi(\tau=\delta)$ is $t_\xi\tau=t_\xi\delta$, $tr_\xi(\varphi\wedge\psi)$ is $tr_\xi\varphi\wedge$ $\wedge tr_\xi\psi$, $tr_\xi(\neg\varphi)$ is $\neg tr_\xi\varphi$ and $tr_\xi(\exists x\varphi)$ is $\exists x tr_\xi\varphi$. Let $\mathfrak{M}$ be a Crs-structure such that $I\subseteq I^{\mathfrak{M}}$. Clearly $tr_\xi\varphi$ and $t_\xi\tau$ are in the language of the expansion $\langle\mathfrak{M},i\rangle_{i\in I}$ of $\mathfrak{M}$. Hence $\langle\mathfrak{M},i\rangle_{i\in I}\models$ $\models tr_\xi\varphi$ is meaningful. Instead of this we shall write $\mathfrak{M}\models tr_\xi\varphi$. Similarly $(\widetilde{t_\xi\tau})^{\mathfrak{M}}$ denotes the term function $(\widetilde{t_\xi\tau})^{(\langle\mathfrak{M},i\rangle_{i\in I})}$.

Lemma 8.13.6. Let $\mathfrak{M}$ be a Crs-structure. Let $\xi:\alpha\rightarrowtail\!\!\!\!\twoheadrightarrow I^{\mathfrak{M}}$ be one-one and onto. Let $\varphi\in Fm(X,CA_\alpha)$. Then

$\mathfrak{M}\models tr_\xi(\varphi)$ iff $Cyl_\xi(\mathfrak{M})\models\varphi$.

Proof. Let $k:X\to A^{\mathfrak{M}}$ and $\tau\in Tm(X,CA_\alpha)$. Let $\mathfrak{L}\overset{d}{=}Cyl_\xi(\mathfrak{M})$. Then it is easy to check, by the definitions of $cyl_\xi$ and $t_\xi$, that $cyl_\xi$ $(\widetilde{t_\xi\tau})^{(\mathfrak{M})}k=\tilde{\tau}^{(\mathfrak{L})}(cyl_\xi\circ k)$. Therefore $\mathfrak{M}\models tr_\xi\varphi[k]$ iff $\mathfrak{L}\models$ $\models\varphi[cyl_\xi\circ k]$, for every $\varphi\in Fm(X,CA_\alpha)$. This implies $\mathfrak{M}\models tr_\xi\varphi$ iff $Cyl_\xi(\mathfrak{M})\models\varphi$ since $cyl_\xi:A^{\mathfrak{M}}\rightarrowtail\!\!\!\!\twoheadrightarrow C$ is one-one and onto.

QED(Lemma 8.13.6.)

For every $\alpha$, let $S_\alpha$ be the class of all Crs-structures $\mathfrak{M}$ for which $|I^{\mathfrak{M}}|=|\alpha|$.

Now we turn to the proof of Theorem 8.13. Let $K\in\{Cs, Gws^{comp\ reg}, Ws\}$. Let $Ax(Cs)\overset{d}{=}Ax(Ws)\overset{d}{=}Cpax$ and let $Ax(Gws^{comp\ reg})\overset{d}{=}Cpax\cup Rgax$. Let $\alpha\geq\omega$ and $\varphi\in Fm(X,CA_\alpha)$. Let $\Gamma$ be the set of all indices occurring in $\varphi$. Then $\Gamma\subseteq_\omega\alpha$. Suppose $K_\alpha\models\varphi$. We have to show $K_\beta\models\varphi$. Let $\mathfrak{A}\in K_\beta$. Then $\mathfrak{R}\overset{d}{=}str(\mathfrak{A})\in S_\beta\cap MdAx(K)$ by 8.13.4. By $\alpha\geq\omega$ and by the Löwenheim-Skolem theorem, there is an elementary submodel $\mathfrak{M}$ of $str(\mathfrak{A})$ such that $\mathfrak{M}\in S_\alpha\cap MdAx(K)$ and $\Gamma\subseteq I^{\mathfrak{M}}$. Let $\xi:$ $:\alpha\rightarrowtail\!\!\!\!\twoheadrightarrow I^{\mathfrak{M}}$ be one-one and onto such that $\Gamma 1\xi\subseteq Id$. Let $\mathfrak{L}\overset{d}{=}$ $\overset{d}{=}Cyl_\xi(\mathfrak{M})$. Then $\mathfrak{L}\in I\,K_\alpha$ by Lemma 8.13.4 and by (1) in the proof of [HMTI]7.17, in case $K\neq Ws$. Let $K=Ws$. Then $\mathfrak{L}\in Gws_\alpha^{comp}$ by 8.13.4. By $\mathfrak{A}\in Ws_\beta$ we have $(\forall s,z\in V^{\mathfrak{R}})|\{i\in I^{\mathfrak{R}}: ext(s,i)\neq ext(z,i)\}|<\omega$. Then the same holds for $\mathfrak{M}$ since $\mathfrak{M}\subseteq\mathfrak{R}$. Thus $\mathfrak{L}=Cyl_\xi(\mathfrak{M})\in Ws_\alpha$ Then $\mathfrak{L}\models\varphi$ by our assumption $K_\alpha\models\varphi$, i.e. $\mathfrak{M}\models tr_\xi\varphi$ by 8.13.6.

Then $\mathfrak{R}\models tr_\xi\varphi$ since $\mathfrak{M}$ is an elementary submodel of $\mathfrak{R}$. Let $\eta \overset{d}{=} \beta 1 Id$. Then $\Gamma 1\eta \subseteq \xi$, hence $tr_\xi\varphi$ equals $tr_\eta\varphi$ and $\mathfrak{R}\models tr_\eta\varphi$. Then $Cyl_\eta(\mathfrak{R}) \models \varphi$ by 8.13.6, thus $\mathfrak{A} \models \varphi$ since $Cyl_\eta(\mathfrak{R}) = Cyl_\eta(str(\mathfrak{A})) = \mathfrak{A}$.

We have seen $K_\alpha \models \varphi$ implies $K_\beta \models \varphi$. Let $K \neq Ws$. Suppose $K_\beta \models \varphi$. We have to show $K_\alpha \models \varphi$. Let $\mathfrak{A} \in K_\alpha$. Then $\mathfrak{R} \overset{d}{=} str(\mathfrak{A}) \in S_\alpha \cap \cap MdAx(K)$. Let $\mathfrak{R}'$ be an elementary submodel of $\mathfrak{R}$ with universes of cardinalities $\leq|\beta|$ and $I^{\mathfrak{R}'}=\alpha$. Let $\mathfrak{M}$ be an elementary extension of $\mathfrak{R}'$ such that $\mathfrak{M}\in S_\beta$. Let $\xi : \beta \rightarrowtail I^{\mathfrak{M}}$ be such that $\Gamma 1\xi\subseteq Id$. Then $\mathfrak{L} \overset{d}{=} Cyl_\xi(\mathfrak{M}) \models \varphi$, since $\mathfrak{L} \in K_\beta$ if $K\neq Cs$ and $\mathfrak{L} \in Gws_\beta^{comp} \subseteq \mathbf{I}\, Cs_\beta$ if $K=Cs$. Thus $\mathfrak{M} \models tr_\xi\varphi$, therefore $\mathfrak{R} \models tr_\eta\varphi$ where $\eta=\alpha 1 Id$, by $\Gamma 1\xi\subseteq\eta$ and since $\mathfrak{R}'$ is an elementary submodel both of $\mathfrak{R}$ and $\mathfrak{M}$. Hence $\mathfrak{A} \models \varphi$ since $Cyl_\eta(\mathfrak{R})=\mathfrak{A}$. $K_\alpha \models \varphi$ if $\mathbf{Rd}_\alpha K_\beta \models\varphi$ is proved. By [HMT]0.3.82, the "if-parts" of 8.13 are proved.

Let $1<\alpha<\omega$ and $\alpha<\beta$. Since $1<\alpha<\omega$, we have $Gws_\alpha^{comp}=Cs_\alpha$. Thus it is enough to prove $\mathbf{Rd}_\alpha Ws_\beta \nsubseteq \mathbf{Uf\,Up}\, Gws_\alpha^{comp}$. Let $\varphi$ be the formula $\forall x(c_{(\alpha)}x=x \rightarrow (x=0 \vee x=1))$. Then $\mathbf{Uf\,Up}\, Gws_\alpha^{comp} \models \varphi$. It is proved in 8.10(5) that $\mathbf{Rd}_\alpha (Ws_\beta \cap Lf_\beta) \not\models \varphi$.

QED(Theorem 8.13.)

By the above proof method more results can be obtained.

Prop.8.14 below is in contrast with 8.5.

Proposition 8.14. Let $1<\alpha<\beta$ and $1<\varkappa<\omega$.

(i) ${}_\varkappa Ws_\alpha \nsubseteq \mathbf{Uf\,Rd}_\alpha Ws_\beta$ if $|\alpha|=\omega$ or $|\beta\sim\alpha|<|\alpha|$.

(ii) ${}_\varkappa Ws_\alpha \nsubseteq \mathbf{Uf\,Rd}_\alpha Cs_\beta$ if $2^{|\beta\sim\alpha|}<|\alpha|+\omega$.

(iii) $Mn_\alpha \cap \mathbf{Uf\,Rd}_\alpha Ws_\beta = \mathbf{I}\, {}_1 Ws_\alpha$ if $|\alpha|=\omega$.

$\mathbf{Uf\,Rd}_\alpha Ws_\beta \models (\exists x(c_0x\neq x) \rightarrow \exists x\Delta x=1)$, if $|\alpha|=\omega$.

Proof. Let $1<\alpha<\beta$. If $\alpha<\omega$ then (i)-(iii) hold by Prop.8.10(2). Assume $\alpha\geq\omega$. Let $\varphi$ be the infinitary formula $\exists x(c_0x\neq x) \rightarrow \exists x\Delta x=1$.

Proof of (iii): Let $|\alpha|=\omega$. First we show $\mathbf{Uf\,Rd}_\alpha \mathrm{Ws}_\beta \models \varphi$. Let $\mathfrak{A}\in \mathrm{CA}_\alpha$ be such that $\mathfrak{A} \not\models \varphi$. Suppose $\mathfrak{B} \stackrel{d}{=} {}^I\mathfrak{A}/F \in \mathbf{Rd}_\alpha \mathrm{CA}_\beta$ for some ultrafilter $F$ on $I$. Then $\mathfrak{B} \models \varphi$ by 8.10.1. Therefore $F$ is not $|\alpha|^+$-complete, since $\neg\varphi$ is an $\alpha$-ary formula. Thus by $|\alpha|=\omega$ there is $H \in {}^\omega\mathrm{Sb}I$ such that $F\cap RgH=0$, $\cup RgH=I$ and $(\forall i<j<\omega)H_i\cap H_j=0$. Let $n : I \to \omega$ be such that $(\forall i\in I)i\in H_{ni}$. Let $x \stackrel{d}{=} \langle d^{\mathfrak{A}}_{n(i)} : i\in I\rangle$ and $y \stackrel{d}{=} \langle \Pi\{-d^{\mathfrak{A}}_{j,j+1} : j<ni\} : i\in I\rangle$. Then $x/\bar{F}$, $y/\bar{F} \in B\sim\{0\}$ since $\mathfrak{A}$ is nondiscrete by $\mathfrak{A} \not\models \varphi$. Let $j\in\omega$. Then $\{i\in I : x_i \not\leq d_{0j}\} \subseteq \{i\in I : j\geq ni\} = \cup\{H_m : m\leq j\} \notin F$ by $F\cap RgH=0$. Similarly, $\{i\in I : y_i \not\leq -d_{j,j+1}\} \subseteq \{i\in I : j\geq ni\} \notin F$. Thus $(\forall j<\omega)[x/F \leq d^{\mathfrak{B}}_{0j}$ and $y/F \leq -d^{\mathfrak{B}}_{j,j+1}]$. Therefore $\mathfrak{B} \notin \mathbf{IRd}_\alpha \mathrm{Ws}_\beta$ since for every $\mathfrak{C} \in \mathrm{Ws}_\beta$ we have that either $\cap\{d^{\mathfrak{C}}_{0j} : j<\omega\}=0$ or $\cap\{-d^{\mathfrak{C}}_{j,j+1} : j<\omega\}=0$. Thus $\mathfrak{A} \notin \mathbf{Uf\,Rd}_\alpha \mathrm{Ws}_\beta$. $\mathbf{Uf\,Rd}_\alpha \mathrm{Ws}_\beta \models \varphi$ is proved. Let $\mathfrak{A}\in \mathrm{Mn}_\alpha \sim \mathbf{I}\,{}_1\mathrm{Gws}_\alpha$. Then $\mathfrak{A} \not\models \varphi$ by [HMT]2.1.22. Thus $\mathrm{Mn}_\alpha \cap \mathbf{Uf\,Rd}_\alpha \mathrm{Ws}_\beta \subseteq \mathbf{I}\,{}_1\mathrm{Gws}_\alpha$. By Lemma 8.14.1 below we have $\mathbf{I}\,{}_1\mathrm{Gws}_\alpha \cap \mathbf{Uf\,Rd}_\alpha \mathrm{Ws}_\beta = \mathbf{I}\,{}_1\mathrm{Ws}_\alpha$ which completes the proof of (iii).

Lemma 8.14.1. Let $\beta>\alpha$ be arbitrary.
Then $\mathbf{I}\,{}_1\mathrm{Gws}_\alpha \cap \mathbf{Uf\,Up\,Rd}_\alpha \mathrm{Ws}_\beta = \mathbf{I}\,{}_1\mathrm{Ws}_\alpha$.

Proof. Assume $\alpha>0$. $\mathrm{Ws}_\beta \models (\forall x(c_0x=x) \to \forall x(x=0 \vee x=1))$. Then $\mathbf{Uf\,Up\,Rd}_\alpha \mathrm{Ws}_\beta \models (\forall x(c_0x=x) \to \forall x(x=0 \vee x=1))$. Since ${}_1\mathrm{Gws}_\alpha \models \forall x(c_0x=x)$ we have $\mathbf{I}\,{}_1\mathrm{Gws}_\alpha \cap \mathbf{Uf\,Up\,Rd}_\alpha \mathrm{Ws}_\beta \models \forall x(x=0 \vee x=1)$. Since $\mathbf{Rd}_\alpha \mathrm{Ws}_\beta \models (0\neq 1)$ we have $(\mathbf{I}\,{}_1\mathrm{Gws}_\alpha \cap \mathbf{Uf\,Up\,Rd}_\alpha \mathrm{Ws}_\beta) = \{\mathfrak{A}\in \mathrm{CA}_\alpha : |A|=2\} = \mathbf{I}\,\mathfrak{Gb}^\alpha 1 = \mathbf{I}\,{}_1\mathrm{Ws}_\alpha$. The case of $\alpha=0$ follows from ${}_1\mathrm{Gws}_0 = \{\mathfrak{Gb}0, \mathfrak{Gb}1\}$, $\mathbf{Rd}_0\mathrm{Ws}_\beta \models (0\neq 1)$, and ${}_1\mathrm{Ws}_0 = \mathrm{Ws}_0 = \{\mathfrak{Gb}\,1\}$.
QED(Lemma 8.14.1.)

Proof of (i)-(ii): If $|\alpha|=\omega$ then (i) is immediate by (iii). Let $1<\kappa<\omega$. Let $\bar{0} \stackrel{d}{=} \langle 0 : i<\alpha\rangle$, $V = {}^\alpha\kappa^{(\bar{0})}$ and $\mathfrak{A} \stackrel{d}{=} \mathfrak{Sg}^{(\mathfrak{Gb}V)}\{\{p\} : p\in V\}$. Then $(\forall x\in A)\Delta x\neq 1$ by 1.3.3 and by [HMT]2.1.22. We shall show that $\mathfrak{A} \notin \mathbf{Uf\,Rd}_\alpha \mathrm{Cs}_\beta$ if $2^{|\beta\sim\alpha|} < |\alpha|+\omega$ and $\mathfrak{A} \notin \mathbf{Uf\,Rd}_\alpha \mathrm{Ws}_\beta$ if $|\beta\sim\alpha| < |\alpha|+\omega^+$. Suppose $\mathfrak{B} \stackrel{d}{=} {}^I\mathfrak{A}/F \in \mathbf{Rd}_\alpha \mathrm{CA}_\beta$ for some ultrafilter $F$ on some I. Then F is not $|\alpha|^+$-complete since $\mathfrak{A}\nvDash\varphi$ and $\mathfrak{B}\vDash\varphi$ by 8.10.1. Then there is $\gamma\leq\alpha$ and $H:\gamma\to SbI$ such that $F\cap RgH=0$, $\cup RgH=I$ and $(\forall i<j<\gamma)H_i\cap H_j=0$. Let $n:I\to\gamma$ be such that $(\forall i\in I)i\in H_{ni}$.

Let $\delta=\begin{cases}2^\omega & \text{if } |\alpha|=\omega\\ |\alpha| & \text{if } |\alpha|>\omega\end{cases}$.

It is known that there is $G\subseteq{}^\gamma\alpha$ such that $|G|=\delta$ and $(\forall g,h\in G)[g\neq h\Rightarrow$ $\Rightarrow|g\cap h|<\omega]$ and $(\forall i\in\alpha)|g^{-1*}i|<\omega$, but we include here a short proof. If $|\alpha|=\omega$, let $\langle\Gamma_i:i<2^\omega\rangle$ be pairwise almost disjoint infinite subsets of $\alpha$, and for each $i<2^\omega$ let $g_i:\gamma\rightarrowtail\Gamma_i$. For $|\alpha|>\omega$ let $\alpha=\cup\{\Gamma_i:i<|\alpha|\}$ be a disjoint union with $|\Gamma_i|=\alpha$ for all $i<|\alpha|$, and let $g_i:\gamma\rightarrowtail\Gamma_i$. It is easy to check that these constructions really proved the claimed statements.

For every $j\in\alpha$ let $a(j)\overset{d}{=}\{\bar{0}^j_1\}$. Then $a(j)\in At\mathfrak{A}$. Let $g\in G$. Define $x(g)\overset{d}{=}\langle a(g(ni)):i\in I\rangle$. Then $x(g)/\bar{F}\in At\mathfrak{B}$ and $x(g)/\bar{F}\leq d^{\mathfrak{B}}_{0k}$ for every $k<\alpha$ since $\{i\in I:x(g)i\nleq d^{\mathfrak{A}}_{0k}\}\subseteq\{i\in i:g(ni)\in\{0,k\}\}=\cup\{H_m:$ $:g(m)\in\{0,k\}\}\notin F$ by $|(g^{-1*}0)\cup(g^{-1*}k)|<\omega$. Let $g,h\in G$ be such that $g\neq h$. Then $x(g)/F\neq x(h)/F$ since $\{i\in I:x(g)i=x(h)i\}=\{i\in I:gni=$ $=hni\}=\cup\{H_m:gm=hm\}\notin F$ by $|g\cap h|<\omega$. We have seen that $|\{x\in At\mathfrak{B}:$ $:(\forall j<\alpha)x\leq d_{0j}\}|\geq\delta$. Suppose $\mathfrak{B}\in\mathbf{I\,Rd}_\alpha Cs_\beta$. Then $\mathfrak{B}\in\mathbf{I\,Rd}_\alpha\,{}_\varkappa Cs_\beta$ since $\mathfrak{B}$ is of characteristic $\varkappa\neq0$. For every $\mathfrak{C}\in\mathbf{Rd}_\alpha\,{}_\varkappa Cs_\beta$ we have $|\{x\in At\mathfrak{C}:(\forall j<\alpha)x\leq d^{\mathfrak{C}}_{0j}\}|\leq\varkappa\cdot\varkappa^{|\beta\sim\alpha|}$. Therefore $\mathfrak{B}\notin\mathbf{I\,Rd}_\alpha\,{}_\varkappa Cs_\beta$ if $\varkappa^{|\beta\sim\alpha|}<|\alpha|+\omega^+\leq\delta$. Similarly, $\mathfrak{B}\notin\mathbf{I\,Rd}_\alpha Ws_\beta$ if $|\beta\sim\alpha|<|\alpha|+\omega^+$, since $(\forall\mathfrak{C}\in\mathbf{Rd}_\alpha\,{}_\varkappa Ws_\beta)[|\{x\in At\mathfrak{C}:(\forall j<\alpha)x\leq d_{0j}\}|=\rho\Rightarrow(\rho<\omega\ \vee\ \rho\leq|\beta\sim\alpha|)]$.

QED(Proposition 8.14.)

Proposition 8.15 below shows a difference between the behaviours of $\mathbf{Rd}_\alpha$ and $\mathbf{Nr}_\alpha$ since $\mathbf{HNr}_\alpha K_\beta=\mathbf{I\,Nr}_\alpha K_\beta$ for various classes $K_\beta$ will be proved in Corollary 8.20.

Proposition 8.15. Let $\omega\leq\alpha<\beta$ and $1<\varkappa$.

(i) $\mathbf{HRd}_\alpha\,{}_\varkappa Ws_\beta\nsubseteq\mathbf{Rd}_\alpha CA_\beta$.

(ii) $\mathbf{HRd}_\alpha K_\beta \neq \mathbf{IRd}_\alpha K_\beta$ for $K \in \{Ws, Cs, Gs, CA\}$.

We shall need the following lemmas.

<u>Lemma 8.15.1.</u> Let $1<\alpha+1<\beta$, $1<\varkappa$ and $\mathfrak{L} \in {}_\varkappa Ws_\beta$.

Then $\mathbf{H}\,\mathfrak{Rd}_\alpha \mathfrak{L} \not\subseteq \mathbf{Rd}_\alpha CA_\beta$.

<u>Proof.</u> Let $\alpha,\beta,\varkappa$ and $\mathfrak{L}$ be as in the hypotheses. Let $\mathfrak{R} \stackrel{d}{=} \mathfrak{Rd}_\alpha \mathfrak{L}$. By [HMTI]8.1-2 there exists an $\mathfrak{A} \in {}_\varkappa Ws_\alpha \cap \mathbf{H}\mathfrak{R}$. Then $\mathfrak{A} \notin \mathbf{Rd}_\alpha CA_\beta$ by Prop.8.10(3), $\varkappa>1$ and $\beta>\alpha+1$.

<u>QED(Lemma 8.15.1.)</u>

<u>Lemma 8.15.2.</u> Let $\alpha<\beta$. In any $\mathbf{Rd}_\alpha CA_\beta$ if $\Pi\{d_{0i} : i<\alpha\}$ exists and it is an atom then it is zero-dimensional.

<u>Proof.</u> Let $\mathfrak{A} = \mathfrak{Rd}_\alpha \mathfrak{B}$, $\mathfrak{B} \in CA_\beta$, $x=\Pi\{d_{0i} : i<\alpha\}$, $\alpha<\beta$. By [HMT] 1.2.10, $\alpha \notin \Delta^{(\mathfrak{B})}x$, hence $\Delta^{(\mathfrak{B})}x \neq \beta$. By $At\mathfrak{A} = At\mathfrak{B}$ and by [HMT] 1.10.5(ii), we have $\Delta x=0$ if $x \in At\mathfrak{A}$.

<u>QED(Lemma 8.15.2.)</u>

Now we turn to the proof of 8.15. Let $\omega\leq\alpha<\beta$ and let $1<\varkappa$ be any cardinal. If $\beta>\alpha+1$ then we are done by 8.15.1. Assume $\beta=\alpha+1$. Let $\bar{0} \stackrel{d}{=} \langle 0 : i<\beta\rangle$, $\mathfrak{L} \stackrel{d}{=} \mathfrak{Sb}^\beta{}_\varkappa(\bar{0})$, $y \stackrel{d}{=} \{\bar{0}\}$, $\mathfrak{B} \stackrel{d}{=} \mathfrak{Sg}^{(\mathfrak{L})}\{y\}$ and $\mathfrak{A} \stackrel{d}{=} \mathfrak{Rd}_\alpha \mathfrak{B}$. Then $\mathfrak{A} \in \mathbf{Rd}_\alpha\,{}_\varkappa Ws_\beta$. Let $z \stackrel{d}{=} (c_\alpha y)-y$, $J \stackrel{d}{=} Ig^{(\mathfrak{B})}\{z\}$ and $I \stackrel{d}{=} Ig^{(\mathfrak{A})}\{z\}$ Then $I\subseteq J$, $y\in J$ and $y\notin I$ by [HMT]2.3.10(i) since $(\forall\Gamma \subseteq_\omega \alpha) y \not\leq c_{(\Gamma)}z$. Let $\mathfrak{N} \stackrel{d}{=} \mathfrak{A}/I$. For every $x\in A$ we let $x^+ \stackrel{d}{=} x/I$. We show that $y^+ = \Pi\{d_{0i} : i<\alpha\}$ in $\mathfrak{N}$. Clearly, $y^+ \leq d_{0i}$ for every $i<\alpha$. Suppose $(\forall i<\alpha)x^+ \leq d_{0i}^{\mathfrak{N}}$. Let $\mathfrak{M} \stackrel{d}{=} \mathfrak{B}/J$. Then $(\forall i<\alpha)x/J \leq d_{0i}^{\mathfrak{M}}$, by $I\subseteq J$. $\mathfrak{M} \cong \mathfrak{Mn}(\mathfrak{B})$ by [HMTI]5.3, since $B = Sg\{y\}$, $y\in J$ and $J\neq B$. Thus $x/J= 0$ since $\Pi\{d_{0i} : i<\alpha\} = 0$ in $\mathfrak{Mn}(\mathfrak{B})$ by $\alpha\geq\omega$. Thus $x\in J$. By $\beta= \alpha+1$ and $J = Ig^{(\mathfrak{B})}\{z\}$ there is $\Gamma \subseteq_\omega \alpha$ such that $x\leq c_{(\Gamma)}c_\alpha z$. Then $x\leq c_{(\Gamma)}y + c_{(\Gamma)}z$ by $c_\alpha z = y+z$. This implies $x^+\leq c_{(\Gamma)}y^+$ by $z\in I$ and $\Gamma\subseteq\alpha$. Let $i\in\alpha\sim\Gamma$ and $\Delta \stackrel{d}{=} \Gamma\cup\{i\}$. Then $y = d_\Delta \cdot c_{(\Gamma)}y$ in $\mathfrak{B}$. By our assumptions, $x^+\leq d_\Delta$, hence $x^+\leq d_\Delta\cdot c_{(\Gamma)}y^+$. By $\Gamma\subseteq\Delta\subseteq\alpha$ we have

$d_\Delta \cdot c_{(\Gamma)} y^+ = y^+$, i.e. $x^+ \leq y^+$. We have proved that $y^+ = \Pi\{d_{0i} : i<\alpha\}$ in $\mathfrak{N}$. By $y \in At\mathfrak{B}$, $y \notin I$ and $(c_0 y - y) \notin I$ we have $y^+ \in At\mathfrak{N} \sim Zd\mathfrak{N}$. Thus $\mathfrak{N} \notin \mathbf{Rd}_\alpha CA_\beta$ by 8.15.2. So far, (i) is proved. (ii) follows from (i), using [HMTI]7.13.

QED(Proposition 8.15.)

Now we turn to neat reducts. A general algebraic definition of neat reducts is in [AN5] which does not even use the notions of systems of varieties or definability by schemes. Some of the results we use or prove here about neat reducts of $Gws_\alpha$-s are proved there as general algebraic theorems about neat reducts of arbitrary varieties.

For neat reducts of regular algebras, a representing function $rs_\alpha$ works with applications not shared by $rd_\alpha$. $rs_\alpha$ is much simpler than $rd_\alpha$. We shall see that many properties are preserved by $rs_\alpha$ which are not preserved by $rd_\alpha$. We denote this function by $rs_\alpha$ because it works by simply restricting all $\beta$-sequences in the unit of a $Gws_\beta$ to $\alpha$.

Definition 8.16. $rs_\alpha \overset{d}{=} \langle \{\alpha 1 q : q \in x\} : x$ is a set of functions$\rangle$. In particular, $rs_\alpha : Sb^\beta U \to Sb^\alpha U$ for any $\alpha \leq \beta$ and any $U$.

Lemma 8.17. Let $1<\alpha\leq\beta$. Let $K \in \{Ws, Cs, Gs, Gws^{comp}, Gws^{wd}, Gws^{norm}\}$.

(i) Let $\mathfrak{A} \in K_\beta^{reg}$. Then $rs_\alpha \in Is(\mathfrak{Nr}_\alpha \mathfrak{A}, \mathfrak{B})$ for some $\mathfrak{B} \in K_\alpha^{reg}$.

(ii) Let $\mathfrak{A} \in K_\beta$ be $\alpha$-regular. Then $rs_\alpha \in Is(\mathfrak{Nr}_\alpha \mathfrak{A}, \mathfrak{B})$ for some $\mathfrak{B} \in K_\alpha$.

Proof. Let $1<\alpha\leq\beta$. Let $V$ be a $Gws_\alpha^{norm}$-unit. Then $rs_\alpha V$ is a $Gws_\beta^{norm}$-unit and $Subb(rs_\alpha V) = Subb(V)$ and $Subu(rs_\alpha V) = rs_\alpha {}^* Subu(V)$. Let $K \in \{Ws, Cs, Gs, Gws^{comp}, Gws^{wd}\}$. Then $rs_\alpha V$ is a $K_\alpha$-unit if $V$ is a $K_\beta$-unit. Let $x \subseteq V$ be such that $\Delta^{[V]} x \subseteq \alpha$. If $x$ is regular in $V$ then clearly $rs_\alpha x$ is regular in $rs_\alpha V$. Let $\mathfrak{A} \in Gws_\beta$ be $\alpha$-regular.

Let $V \stackrel{d}{=} 1^{\mathfrak{A}}$, $\mathfrak{L} \stackrel{d}{=} \mathfrak{Sb}\, rs_\alpha V$ and $\mathfrak{N} \stackrel{d}{=} \mathfrak{Nr}_\alpha \mathfrak{A}$. We show that $rs_\alpha \in Ism(\mathfrak{N}, \mathfrak{L})$. Clearly, $rs_\alpha \in Hom(\langle SbV, \cup\rangle, \langle Sb(rs_\alpha V), \cup\rangle)$. Let $x \in N$. Then (*) below holds by $\alpha$-regularity of $x$ and by $\Delta x \subseteq \alpha$.

(*) $(\forall q \in rs_\alpha V)[q \in rs_\alpha x$ iff $(\forall f \in V)(q \subseteq f \Rightarrow f \in x)]$.

Now $rs_\alpha x \cap rs_\alpha(-x) = 0$ by (*). Then $rs_\alpha(-x) = rs_\alpha V \sim rs_\alpha x$ by $rs_\alpha x \cup rs_\alpha(-x) = rs_\alpha V$. We have seen $rs_\alpha \in Hom(\mathfrak{Bl}\,\mathfrak{N}, \mathfrak{Bl}\,\mathfrak{L})$. Let $i,j<\alpha$. Clearly, $rs_\alpha d^{\mathfrak{N}}_{ij} = d^{\mathfrak{L}}_{ij}$. Let $q \in c^{\mathfrak{L}}_i\, rs_\alpha x$. Then $q \subseteq g$ for some $g \in V$ and $q^i_a \in rs_\alpha x$ for some $a$. Let $q^i_a \subseteq f \in x$. There is $b \in Rg\, g \cap Rg\, f$ by $\alpha \geq 2$. Then $f^i_b, g^i_b \in V$ since $V$ is a $Gws_\beta$-unit. $f^i_b \in c_i x$ by $f \in x$ and then $g^i_b \in c_i x$ by $\alpha$-regularity of $c_i x$ and $\alpha 1 f^i_b \subseteq g^i_b$. Thus $g \in c_i x$ and therefore $q \in rs_\alpha c^{\mathfrak{N}}_i x$. We have seen $c^{\mathfrak{L}}_i\, rs_\alpha x \subseteq rs_\alpha c^{\mathfrak{N}}_i x$. Let $q \in rs_\alpha c_i x$. Then $q \subseteq f \in c_i x$. Then $f^i_a \in x$ for some $a$. Then $q^i_a \in rs_\alpha x$ hence $q \in c^{\mathfrak{L}}_i\, rs_\alpha x$. We have seen $c^{\mathfrak{L}}_i\, rs_\alpha x = rs_\alpha c^{\mathfrak{N}}_i x$. Thus $rs_\alpha \in Hom(\mathfrak{N}, \mathfrak{L})$. Then $rs_\alpha \in Ism(\mathfrak{N}, \mathfrak{L})$ since $rs_\alpha x \neq 0$ for all $x \neq 0$. (ii) is proved. (i) follows from (ii) since regularity implies $\alpha$-regularity by $\alpha > 1$.

QED(Lemma 8.17.)

Note that there is a $Gws^{reg}_\beta$ with unit $V$ such that $rs_\alpha V$ is not a $Gws_\alpha$-unit and there is an $\alpha$-regular $\mathfrak{A} \in Crs_\beta$ such that $rs_\alpha \notin Hom(\mathfrak{Nr}_\alpha \mathfrak{A}, \mathfrak{Sb}\, rs_\alpha 1^{\mathfrak{A}})$. Indeed, let $\alpha \stackrel{d}{=} \omega$, $\beta \stackrel{d}{=} \omega+\omega$, $p \stackrel{d}{=} \langle 0 : i<\omega\rangle \cup \langle 1 : \omega \leq i < \omega+\omega\rangle$, $q \stackrel{d}{=} \langle 0 : i<\omega\rangle \cup \langle 2 : \omega \leq i < \omega+\omega\rangle$, $V \stackrel{d}{=} {}^{\beta}2^{(p)} \cup {}^{\beta}\{0,2\}^{(q)}$ and $\mathfrak{A} \stackrel{d}{=} \mathfrak{Sb}\{p^0_1, q^0_2\}$. $\mathfrak{Mn}(\mathfrak{Sb}V)$ is regular by 4.2.

Corollary 8.18. Let $\beta > \alpha > 1$. Then (i)-(iv) below hold.

(i) $\mathsf{I}\,Ws_\alpha = \mathbf{SNr}_\alpha\, \mathsf{I}\,Ws_\beta$.

(ii) $\mathsf{I}\,Cs_\alpha = \mathbf{SNr}_\alpha\, \mathsf{I}\,Cs_\beta$ iff $\beta < \alpha+\omega$.

(iii) Let $K \in \{\mathsf{I}\,Cs^{reg}, \mathsf{I}\,(Gws^{comp})^{reg}\}$. Then

$K_\alpha \supseteq \mathbf{SNr}_\alpha K_\beta$ and if in addition $\alpha < \omega$ then

$K_\alpha = \mathbf{SNr}_\alpha K_\beta$.

(iv) Let $K$ be as in (iii). Then $K_\alpha \cap Lf_\alpha = \mathbf{SNr}_\alpha (K_\beta \cap Lf_\beta)$.

Proof. Let $K\in\{\mathbf{I}\,Ws, \mathbf{I}\,Cs^{reg}, \mathbf{I}\,Gws^{comp\ reg}\}$. Then $K_\alpha \supseteq \mathbf{SNr}_\alpha K_\beta$ follows from 8.17(i). Hence $\mathbf{I}\,Ws_\alpha = \mathbf{SNr}_\alpha \mathbf{I}\,Ws_\beta$ follows from [HMTI]8.6. By the above and by 4.1 and [HMTI]8.5 we have that $K_\alpha \cap Lf_\alpha = \mathbf{SNr}_\alpha (K_\beta \cap Lf_\beta)$. Thus if $\alpha<\omega$ then $K_\alpha = \mathbf{SNr}_\alpha K_\beta$. We have proved (i),(iii) and (iv).

Proof of (ii): Let $\beta<\alpha+\omega$. Then $\beta=\alpha+n$ for some $n\in\omega$, hence every $\mathfrak{A}\in Gws_\beta$ is $\alpha$-regular by 1.3.4(ii). Thus $\mathbf{Nr}_\alpha Cs_\beta \subseteq \mathbf{I}\,Cs_\alpha$ by 8.17(ii). Assume $\beta\geq\alpha+\omega$. Let $\mathfrak{N} \overset{d}{=} \mathfrak{Nr}_\alpha \mathfrak{Sb}^\beta 2$. Let $X \overset{d}{=} \{y\in N : (\forall i\in\alpha) y\leq \leq d_{0i}\}$. Then $|X|\geq\omega$ proving $\mathfrak{N}\notin \mathbf{I}\,Cs_\alpha$.

QED(Corollary 8.18.)

Problem 8.18.1. Let $K\in\{\mathbf{I}\,Cs^{reg}, \mathbf{I}\,Gws^{comp\ reg}\}$. Let $\beta>\alpha\geq\omega$. Is $K_\alpha = \mathbf{SNr}_\alpha K_\beta$?

We know that the answer is yes if $\alpha<\omega$, and that for all $\beta>\alpha>1$, $K_\alpha\cap Dc_\alpha \subseteq \mathbf{SNr}_\alpha K_\beta$. This can be proved from [HMT]1.11.9-12. So the question is whether $(K_\alpha\sim Dc_\alpha) \subseteq \mathbf{SNr}_\alpha K_\beta$.

Theorem 8.19 below will be used later. Concerning this theorem see [HMT]2.6.33. Thm.8.19 concernes the conditions under which the operator $\mathbf{Nr}_\alpha$ commutes with the other operators. Only the operators $\mathbf{S}$ and $\mathbf{P}$ are omitted. The reasons to omit them are the following two observations. $\mathbf{Nr}_\alpha \mathbf{S} K \neq \mathbf{SNr}_\alpha K$ and $\mathbf{Nr}_\alpha \mathbf{S} K \subset \mathbf{SNr}_\alpha K$, whenever ${}_\omega Ws_\beta \subseteq K \subseteq CA_\beta$ and $\beta>\alpha>1$, hold by 8.8, [HMTI]8.6 and by [HMT]2.6.29. $\mathbf{PNr}_\alpha K = \mathbf{Nr}_\alpha \mathbf{P} K$ for all $K\subseteq CA_\beta$ was proved in the proof of [HMT]2.6.33.

Theorem 8.19.

i) Let $0<\alpha\leq\beta$. The following conditions a.-g. are equivalent.

a. $\beta < \alpha+\omega$.

b. $\mathbf{Nr}_\alpha \mathbf{Up}\mathfrak{A} \subseteq \mathbf{HSP}\ \mathbf{Nr}_\alpha \mathfrak{A}$ for all $\mathfrak{A} \in Ws_\beta \cap Dc_\beta$.

c. $\mathbf{Uf}\,\mathbf{Nr}_\alpha \mathbf{Up}\mathfrak{A} \supseteq \mathbf{Up}\ \mathbf{Nr}_\alpha \mathfrak{A}$ for all $\mathfrak{A} \in Ws_\beta$.

d. $\mathbf{Nr}_\alpha \mathbf{Up} K = \mathbf{Up}\ \mathbf{Nr}_\alpha K$ for all $K \subseteq CA_\beta$.

e. $\mathbf{Nr}_\alpha \mathbf{H}\,\mathfrak{A} \supseteq \mathbf{H}\ \mathbf{Nr}_\alpha \mathfrak{A}$ for all $\mathfrak{A}\in Cs_\beta$.

f. $\mathbf{Nr}_\alpha \mathbf{H}\mathfrak{A} \subseteq \mathbf{H}\ \mathbf{Nr}_\alpha \mathbf{S}\mathfrak{A}$ for all $\mathfrak{A} \in Cs_\beta^{reg} \cap Dc_\beta$.

g. $\mathbf{Nr}_\alpha \mathbf{H}K = \mathbf{H}\ \mathbf{Nr}_\alpha K$ for all $K \subseteq CA_\beta$.

(ii) Let $\alpha \le \beta$ and $K \subseteq CA_\beta$. Then

$\mathbf{Up}\ \mathbf{Nr}_\alpha \mathbf{S}\mathbf{Up} K = \mathbf{Nr}_\alpha \mathbf{S}\mathbf{Up} K$ and $\mathbf{H}\ \mathbf{Nr}_\alpha \mathbf{S} K \subseteq \mathbf{Nr}_\alpha \mathbf{HS} K$.

(iii) There are $\alpha \le \beta$ and $K \subseteq CA_\beta$ such that

$\mathbf{Nr}_\alpha \mathbf{Uf} K \not\subseteq \mathbf{Uf}\ \mathbf{Up}\ \mathbf{Nr}_\alpha K$ and $\mathbf{Uf}\ \mathbf{Nr}_\alpha K \not\subseteq \mathbf{Nr}_\alpha \mathbf{HSP} K$.

<u>Proof.</u> Suppose $0<\alpha$ and $\alpha+\omega \le \beta$.

(1) We shall construct an $\mathfrak{A} \in Ws_\beta \cap Dc_\beta$ such that

$\mathbf{Nr}_\alpha \mathbf{Up}\mathfrak{A} \not\subseteq \mathbf{HSP}\ \mathbf{Nr}_\alpha \mathfrak{A}$.

<u>Case 1</u> $\alpha \ge 3$.

Let $\bar{0} = \langle 0 : i<\beta \rangle$, and let $Z$ be the set of all integers (both positive and negative). Let $H \subseteq (\alpha+\omega)\sim\alpha$ with $|H| = \omega \le |\beta\sim H|$. Let $X \stackrel{d}{=} \{q \in {}^\beta Z^{(\bar{0})} : q_0 < q_1$ and $H 1 q \subseteq \bar{0}\}$. Let $\mathfrak{L} \stackrel{d}{=} \mathfrak{Sb}\,{}^\beta Z^{(\bar{0})}$ and $\mathfrak{A} \stackrel{d}{=} \mathfrak{Sg}^{(\mathfrak{L})}\{X\}$. Then $\Delta X = 2 \cup H$. Clearly, $X$ is small in $\mathfrak{A}$ and hence, by Lemma 1.3.3, applied to $\mathfrak{Dm}^{(\mathfrak{A})}_{\beta\sim H}$, $A \subseteq Mn(\mathfrak{A}) \cup \{x \in A : |\Delta x \cap H| \ge \omega\}$. By $H \cap \alpha = 0$, then $\mathfrak{Nr}_\alpha \mathfrak{A} = \mathfrak{Nr}_\alpha \mathfrak{Mn}(\mathfrak{A}) \in Mn_\alpha$ by Cor.8.21(iii) since $\mathfrak{Mn}(\mathfrak{A}) \in Ws_\beta \cap Mn_\beta$. Next we define three unary terms $\vartheta(x)$, $\rho(x)$ and $\tau(x)$ below in the discourse language of $CA_3$.

$$\vartheta(x) = c^\partial_{(3)}[-(c_2x \cdot s^0_1 s^1_2 c_2 x) + s^1_2 c_2 x].$$

$$\rho(x) = c^\partial_{(2)} - (d_{01} \cdot c_2 x).$$

$$\tau(x) = \vartheta(x) \cdot \rho(x) \cdot c^\partial_0 c_1 c_2 x.$$

<u>Claim 8.19.1.</u> Let $\mathfrak{A} \in Gws_\alpha$, $x \in A$, $q \in V \in Subu(\mathfrak{A})$ and $U = base(V)$. Let $R(q,x) \stackrel{d}{=} \{\langle a,b \rangle \in {}^2U : q^{01}_{ab} \in c_2 x\}$. Then $q \in \tau(x)$ iff $R(q,x)$ is transitive antireflexive and $\mathcal{D}o R(q,x) = U$.

<u>Proof.</u> Let $\mathfrak{A}, x, U$ and $q$ be as in the hypotheses. Let $R \stackrel{d}{=} R(q,x)$. Suppose $\{\langle b,d \rangle, \langle d,e \rangle\} \subseteq R$. Then $q' \stackrel{d}{=} q^{012}_{bde} \in c_2 x \cdot s^0_1 s^1_2 c_2 x$. If $q \in \vartheta(x)$, then $q' \in s^1_2 c_2 x$ and hence $\langle b,e \rangle \in R$. Thus $q \in \vartheta(x)$ implies that $R$ is

transitive. The converse is easy, so $q\in\vartheta(x)$ iff $R$ is transitive. Let $b\in U$. Then $\langle b,b\rangle\in R$ implies $q^{01}_{bb}\in d_{01}\cdot c_2x$ implies $q\notin\rho(x)$. Conversely, $q\notin\rho(x)$ implies $\langle u,u\rangle\in R$ for some $u$. Thus $q\in\rho(x)$ iff $R$ is antireflexive. $DoR = U$ iff $(\forall b\in U)(\exists d)q^{01}_{bd}\in c_2x$ iff $q\in c^{\partial}_0 c_1 c_2 x$.

QED(Claim 8.19.1.)

Lemma 8.19.2. $Mn_\alpha \models \tau(x)=0$.

Proof. Let $\mathfrak{A}\in Mn_\alpha$. We may assume $\mathfrak{A}\in Gs_\alpha$ by [HMT]2.5.25. Suppose $\mathfrak{A} \not\models \tau(x)=0$. Then $\tau(x)\neq 0$ for some $x\in A$. Let $q\in\tau(x)$ and $R \overset{d}{=} R(q,x)$. Let $U\in Subb(\mathfrak{A})$ be such that $q\in{}^\alpha U$. By 8.19.1 we have that $R$ is a transitive, antireflexive relation on $U$ such that $DoR = U$. Thus there is $b\in{}^\omega U$ such that $(\forall i<j<\omega)\langle b_i,b_j\rangle\in R$. Then $|Rgb|\geq\omega$ by transitivity and antireflexivity of $R$. By [HMT]2.1.17 there is $\Gamma\subseteq_\omega\alpha$ such that $c_2x\in Sg^{(\mathfrak{Bl}\,\mathfrak{A})}G$ where $G \overset{d}{=} \{d_{ij} : i,j\in\Gamma\}\cup\{a_\varkappa : \varkappa<\alpha\cap\omega\}$ and $a_\varkappa = c_{(\varkappa)}\bar{d}(\varkappa\times\varkappa)$ for every $\varkappa<\alpha\cap\omega$. Let $H \overset{d}{=} q^*(\Gamma\cup 3)$. By $|H|<\omega$ and $|Rgb|\geq\omega$ there are $d,e\in U\sim H$ such that $\langle d,e\rangle\in R$. Let $i,j\in\Gamma$. If $i,j\in 2$ or $i,j\in\Gamma\sim 2$ then $q^{01}_{de}\in d_{ij}$ iff $q^{01}_{ed}\in d_{ij}$ and if $i\in 2$, $j\in\Gamma\sim 2$ then $\{q^{01}_{de}, q^{01}_{ed}\}\cap d_{ij} = 0$. Thus $(\forall g\in G)[q^{01}_{de}\in g$ iff $q^{01}_{ed}\in g]$ since $(\forall\varkappa<\alpha\cap\omega)$ $\Delta a_\varkappa=0$. By $c_2x\in Sg^{(\mathfrak{Bl}\,\mathfrak{A})}G$ then $q^{01}_{de}\in c_2x$ iff $q^{01}_{ed}\in c_2x$. By $\langle d,e\rangle\in R$ then $\langle e,d\rangle\in R$. THis is a contradiction since $R$ is transitive and antireflexive. Hence $\mathfrak{A}\models\tau(x)=0$.

QED(Lemma 8.19.2.)

Now we turn to our $\mathfrak{A}\in Ws_\beta\cap Dc_\beta$ generated by $\{X\}$. Let $F$ be a nonprincipal ultrafilter on $\omega$. Let $\mathfrak{B} \overset{d}{=} {}^\omega\mathfrak{A}/F$. Let $L \overset{d}{=} \langle H\cap(\alpha+n) : n\in\omega\rangle$ and $y \overset{d}{=} \langle c_{(Li)}X : i\in\omega\rangle$. Then $\Delta^{(\mathfrak{B})}(y/F)=2$ because $\Delta^{(\mathfrak{A})}X = 2\cup H$ and $(\forall i\in H)|\{j\in\omega : i\notin Lj\}|<\omega$ by $H\subseteq(\alpha+\omega)\sim\alpha$. Thus $y/F\in Nr_\alpha\mathfrak{B}$. Let $i\in\omega$. Then $\tau(y_i)\neq 0$ by 8.19.1 since $R(\bar{0},y_i) = \{\langle i,j\rangle\in{}^2Z : i<j\}$. Thus $\tau^{(\mathfrak{B})}(y/F)\neq 0$. Hence $\mathfrak{Nr}_\alpha\mathfrak{B}\notin$ **HSP** $Mn_\alpha$ by 8.19.2. Therefore $\mathfrak{Nr}_\alpha{}^\omega\mathfrak{A}/F\notin$ **HSP Nr**$_\alpha\mathfrak{A}$ since we have seen that $\mathfrak{Nr}_\alpha\mathfrak{A}\in Mn_\alpha$.

Case 2 $0<\alpha<\omega\leq\beta$.

We shall show **Nr**$_\alpha$ **Up**$(Ws_\beta\cap Mn_\beta) \not\subseteq$ **HSP Nr**$_\alpha$ $Mn_\beta$. Let $\mathfrak{A}\in Mn_\beta$ be nondiscrete and $F$ be a nonprincipal ultrafilter on $\omega$. Let $\mathfrak{B} \stackrel{d}{=} {}^{\omega}\mathfrak{A}/F$ and $y \stackrel{d}{=}$ $\stackrel{d}{=} \langle d_{0i}^{\mathfrak{A}} : i\in\omega\rangle/F$. Then $y\in B$ and $\Delta^{(\mathfrak{B})}y=1$. Let $\mathfrak{N} \stackrel{d}{=} \mathfrak{Nr}_{\alpha}\mathfrak{B}$. Then $y\in N$ by $\alpha>0$; and $c_{(\alpha)}y\neq y = c_{(\alpha\sim 1)}y$. But $\mathbf{Nr}_{\alpha} Mn_{\beta} \models c_{(\alpha)}x = c_{(\alpha\sim 1)}x$ by $\alpha\in\omega$ and [HMT]2.1.22. Hence $\mathfrak{N} \notin \mathbf{HSP}\ \mathbf{Nr}_{\alpha} Mn_{\beta}$. We have seen that $\mathbf{Nr}_{\alpha}\ \mathbf{Up}\mathfrak{A} \not\subseteq \mathbf{HSP}\ \mathbf{Nr}_{\alpha} Mn_{\beta}$ for all nondiscrete $\mathfrak{A}\in Mn_{\beta}$.

(2) We prove $(\exists\mathfrak{A}\in Ws_{\beta})\mathbf{Up}\,\mathbf{Nr}_{\alpha}\mathfrak{A} \not\subseteq \mathbf{Uf}\ \mathbf{Nr}_{\alpha}\ \mathbf{Up}\mathfrak{A}$.

Case 1 $\alpha\geq\omega$.

Lemma 8.19.3. Let $\omega\leq\alpha$ and $\alpha+\omega\leq\beta$. Then $\mathbf{Up}'\mathbf{Nr}_{\alpha} Ws_{\beta} \not\subseteq \mathbf{Uf}\ \mathbf{Nr}_{\alpha}\ \mathbf{Up} Ws_{\beta}$.

Proof. Recall the formula $at(x)$ from Def.7.3.1. Let $\varphi$ be the formula

$$\forall x\forall y(\wedge\{0<x\leq d_{0i} : i\in\omega\} \rightarrow \neg\wedge\{0<y\leq -d_{i,i+1} : i\in\omega\}).$$

Claim 8.19.3.1. Let $\alpha\geq\omega$ and $\beta\geq\alpha+\omega$. Then $\mathbf{Uf}\ \mathbf{Nr}_{\alpha}\ \mathbf{Up} Ws_{\beta} \models (\exists x\, at(x) \rightarrow \varphi)$.

Proof. Let $\mathfrak{A}\in\mathbf{Uf}\ \mathbf{Nr}_{\alpha}\ \mathbf{Up} Ws_{\beta}$. Then $\mathfrak{Nr}_{\alpha}P\mathfrak{R}/F \in \mathbf{Up}\mathfrak{A}$ for some $\mathfrak{R}\in{}^{I}Ws_{\beta}$ and some ultrafilter $F$ on $I$. Suppose $\mathfrak{A} \not\models \varphi$. Then $P\mathfrak{R}/F \not\models \varphi$, since $\varphi$ is a universal formula in the discourse language of $CA_{\omega}$-s and $\alpha\geq\omega$. Then $F$ is not $\omega^{+}$-complete by [CK] Thm.4.2.11, since $Ws_{\beta}\models$ $\models\varphi$. By $P\mathfrak{R}/F \not\models \varphi$ we have $P\mathfrak{R}/F$ is nondiscrete. Hence we may assume that $\mathfrak{R}_i$ is nondiscrete for every $i\in I$. Let $x/F \in Nr_{\alpha}P\mathfrak{R}/F$ be arbitrary. Let $Y \stackrel{d}{=} \langle\{i\in I : j\notin\Delta x_i\} : j\in\beta\rangle$. Then $(\forall j\in\beta\sim\alpha)\ Y_j\in F$. Let $\gamma \stackrel{d}{=} \alpha+\omega$. Then there is $Z\in{}^{(\gamma\sim\alpha)}F$ such that $\cap Rg Z = 0$ and $(\forall i\leq j\in\gamma\sim\alpha)$ $[Z_i \supseteq Z_j \subseteq Y_j]$. Let $H \stackrel{d}{=} \langle Z_j\sim Z_{j+1} : j\in\gamma\sim\alpha\rangle$. Let $h : I \rightarrow \gamma\sim\alpha$ be such that $(\forall i\in Z_{\alpha})\, i\in H(hi)$. Let $b \stackrel{d}{=} \langle d_{hi,hi+1} : i\in I\rangle$. Then $b/F \in Nr_{\alpha}P\mathfrak{R}/F$ because $\Delta(b/F)=0$ by $\{i\in I : n\in\Delta(b_i)\} \subseteq H_n\cup H_{n-1} \notin F$ for every $n\in\gamma\sim\alpha$. Let $i\in I$ be such that $x_i\neq 0$. Then $0<b_i\cdot x_i<x_i$ since $hi\notin\Delta x_i$ by $H(hi) \subseteq Y(hi)$ and $\mathfrak{R}_i$ is a nondiscrete $Ws_{\alpha}$. Thus $x/F \notin At\,\mathfrak{Nr}_{\alpha}P\mathfrak{R}/F$. Thus $\mathfrak{Nr}_{\alpha}P\mathfrak{R}/F \models \neg\exists x\, at(x)$, hence $\mathfrak{A} \models \neg\exists x\, at(x)$ by $\mathfrak{Nr}_{\alpha}P\mathfrak{R}/F\in\mathbf{Up}\mathfrak{A}$ since $at(x)$ is a first order formula.

QED(Claim 8.19.3.1.)

Let $\mathfrak{B}$ be any nondiscrete $Ws_\beta$ with $AtCl_{\beta\sim\alpha}\mathfrak{B} \neq 0$. $\mathfrak{A} \stackrel{d}{=} \mathfrak{Nr}_\alpha \mathfrak{B}$. Let F be any nonprincipal ultrafilter on $\omega$ and let $\mathfrak{P} \stackrel{d}{=} {}^\omega\mathfrak{A}/F$. Then $\mathfrak{P} \models \exists x\, at(x)$ by $\mathfrak{Nr}_\alpha\mathfrak{B} \models \exists x\, at(x)$. $\mathfrak{P} \models \neg\varphi$ since $\mathfrak{A}$ is nondiscrete (let $x \stackrel{d}{=} \langle d_n^{\mathfrak{A}} : n\in\omega\rangle/F$ and $y \stackrel{d}{=} \langle \prod\{-d_{j,j+1} : j<n\} : n\in\omega\rangle/F$) Hence $\mathfrak{P} \notin \mathbf{Uf\,Nr}_\alpha\,\mathbf{Up}\,Ws_\beta$ by 8.19.3.1.

QED(Lemma 8.19.3.)

Case 2 $0<\alpha<\omega\leq\beta$.

Let $\bar{0} \stackrel{d}{=} \langle 0 : i<\beta\rangle$, $V \stackrel{d}{=} {}^\beta\omega^{(\bar{0})}$, $\mathfrak{L} \stackrel{d}{=} \mathfrak{Sb}V$, $x \stackrel{d}{=} \langle \{q\in V : q_0=n\} : n\in\omega\rangle$ and $\mathfrak{A} \stackrel{d}{=} \mathfrak{Sg}^{(\mathfrak{L})}\{x_n : n\in\omega\}$. Let $\mathfrak{P} \in \mathbf{Up}\,\mathfrak{Nr}_\alpha\mathfrak{A}$ be such that $|P|>|A|$. Such a $\mathfrak{P}$ exists since $|Nr_\alpha\mathfrak{A}|\geq\omega$ by $\alpha>0$. By $\alpha<\omega$ we have $|Zd\,\mathfrak{P}|=2$, since $Zd\,\mathfrak{Nr}_\alpha\mathfrak{A} = Zd\mathfrak{A} \subseteq Zd\mathfrak{L}$, and $|Zd\mathfrak{L}|=2$ by [HMTI]6.13. Let $\mathfrak{B} = {}^I\mathfrak{A}/F$ for some ultrafilter F on I such that $|B|>|A|=|\beta|$. Then F is not $|\beta|^+$-complete. Let $H : \beta \rightarrow SbI$ be such that $(RgH)\cap F=0$, $\cup RgH=I$ and $(\forall i<j<\beta)H_i\cap H_j=0$. Let $b\in{}^IA$ be such that $(\forall j<\beta)[j+1\in\beta \Rightarrow (\forall i\in H_j)b_i=d_{j,j+1}]$. Then $\Delta^{(\mathfrak{B})}(b/F)=0$ and $b/F\notin\{0^{\mathfrak{B}},1^{\mathfrak{B}}\}$ since $\mathfrak{A}$ is nondiscrete. We have seen that $(\forall\,\mathfrak{B}\in\mathbf{Up}\,\mathfrak{A})[|B|>|A| \Rightarrow |Zd\,\mathfrak{Nr}_\alpha\mathfrak{B}|>2]$. Thus $\mathfrak{P} \notin \mathbf{Nr}_\alpha\,\mathbf{Up}\mathfrak{A}$. Moreover, $\mathbf{Up}\,\mathfrak{P} \cap \mathbf{Nr}_\alpha\,\mathbf{Up}\mathfrak{A}=0$ since $(\forall\,\mathfrak{G}\in\mathbf{Up}\,\mathfrak{P})$ $(|G|\geq|P|$ and $|Zd\,\mathfrak{G}|=2)$, by $\alpha<\omega$.

(3) We show $(\exists\,\mathfrak{A}\in Cs_\beta)\mathbf{HNr}_\alpha\mathfrak{A} \not\subseteq \mathbf{Nr}_\alpha\,\mathbf{H}\mathfrak{A}$.

Let $\varkappa$ be a cardinal such that $cf(\varkappa)>\omega$. (For example we may choose $\varkappa=\omega^+$.) Let $\mathfrak{L} \stackrel{d}{=} \mathfrak{Sb}\,{}^\beta\varkappa$ and $b \stackrel{d}{=} \{q\in{}^\beta\varkappa : q_0=0\}$. Let $\mathfrak{P} = {}^\omega\mathfrak{L}$, $w \stackrel{d}{=} \langle \{q\in{}^\beta\varkappa : q_0=\cup q^*((\alpha+\omega)\sim(\alpha+n))\} : n\in\omega\rangle$ and $x \stackrel{d}{=} \langle\langle 0 : i\in\omega\rangle_b^n : n\in\omega\rangle$. Let $\mathfrak{A} \stackrel{d}{=} \mathfrak{Sg}^{(\mathfrak{P})}\{w,x_n : n\in\omega\}$. Then $\mathfrak{A} \in \mathbf{SP}\,{}_\infty Cs_\beta \subseteq \mathbf{I}\,{}_\infty Cs_\beta$, using [HMTI]7.21.

Claim 8.19.4. $\mathbf{Nr}_\alpha\,\mathbf{H}\mathfrak{A} \models [0\neq1 \rightarrow \exists y(c_0^\partial y=0 \wedge \Delta y=1)]$.

Proof. Let $J\in Il\mathfrak{A}$ and $\mathfrak{R} = \mathfrak{Nr}_\alpha(\mathfrak{A}/J)$, $|R|>1$. Let $n\in\omega$. In $\mathfrak{A}$ we have $\Delta^{(\mathfrak{A})}x_n=1$ and $c_0^\partial x_n=0$. Thus $(x_n/J)\in R$ by $\alpha>0$ and $\Delta(x_n/J)=1$ if $x_n/J\neq0$. Therefore we are done if $Rgx \not\subseteq J$. Suppose $Rgx \subseteq J$. Then

$\Delta(w/J)\subseteq 1$ since $\Delta^{(\mathfrak{A})}w = 1\cup\{\alpha+n : n\in\omega\}$ and $(c_{\alpha+n}w-w) \subseteq c_0(x_0+\ldots+x_n)\in$ $\in J$ for every $n\in\omega$. (It is easily checked that $c_{\alpha+n}w_i-w_i = 0$ if $n<i$, while for $i\le n$, $1=c_0b=c_0x_{ii} \le c_0(x_0+\ldots+x_n)_i$.) Thus $w/J\in R$. In $\mathfrak{A}$ we have $c_0w=1$ and $c_0^\partial w=0$. Thus by $|R|>1$ we have $\Delta^{(\mathfrak{R})}(w/J)=1$ and $c_0^\partial(w/J)=0$.

QED(Claim 8.19.4.)

Let $I \stackrel{d}{=} Ig^{\mathfrak{A}}\{x_n : n\in\omega\}$ and $\mathfrak{B} \stackrel{d}{=} \mathfrak{Nr}_\alpha\mathfrak{A}/(I\cap Nr_\alpha\mathfrak{A})$.

Claim 8.19.5. $(\forall y\in B)[\Delta^{(\mathfrak{B})}y\ne 1 \lor c_0y\ne 0]$.

Proof. Let $v \stackrel{d}{=} \langle c_0x_n : n\in\omega\rangle$. Then $(\forall n\in\omega)v_n\in Zd\mathfrak{A}$, thus $I = \{a\in A :$ $: (\exists n\in\omega)a\le v_0+\ldots+v_n\}$. Note also that, since $c_0b=1$, $v_0+\ldots+v_n$ is the member $f$ of $P$ such that $f_i=1$ for all $i\le n$ and $f_i=0$ for all $i>n$. Let $y\in Nr_\alpha\mathfrak{A}$ and assume $\Delta(y/I)=1$ and $c_0^\partial(y/I)=0$. Then $y/I\ne 0$. By $y\in A$ there are $k\in\omega$ and $\Gamma\subseteq_\omega\beta$ such that $0\in\Gamma$ and $y\in Sg^{(\mathfrak{Rd}_\Gamma\mathfrak{A})}\{w,x_i : i<k\}$. By $\Delta(y/I)=1$ and $c_0^\partial(y/I)=0$ we have that $(c_0^\partial y+$ $+\Sigma\{c_iy-y : i\in\Gamma\sim 1\}) \le v_0+\ldots+v_{m-1}$ for some $m\in\omega$. Let $L \stackrel{d}{=} \{i\in\omega : y_i\ne 0\}$. Since $y/I\ne 0$, the above note concerning what $v_0+\ldots+v_n$ is shows that $|L|\ge\omega$. Let $r\in L$ be such that $r>m+k$ and $\Gamma\cap(\alpha+\omega)\subseteq\alpha+r$. Then $pj_r\in$ $\in Hom(\mathfrak{A},\mathfrak{L})$ is such that $pj_r(w)=w_r$ and $pj_r(v_i)=pj_r(x_i)=0$ for every $i<r$. Thus $y_r\in Sg^{(\mathfrak{L})}\{w_r\}$ since $k<r$. By $r>m$ and $r\in L$ we have $c_0^\partial y_r=0\ne y_r$ and $\Gamma\cap\Delta^{(\mathfrak{L})}(y_r)=1$. Thus there are $q\in y_r$ and $a\in\varkappa$ such that $q_a^0\notin y_r$. Let $u \stackrel{d}{=} q0$ and $H \stackrel{d}{=} (\alpha+\omega)\sim(\alpha+r)$. Then $\Gamma\cap H=0$. Let $t\in\varkappa$ be such that $t > a+u+\cup q^*H$. Such a $t$ exists by $cf(\varkappa)>\omega$. Let $i\in H$ and $p \stackrel{d}{=} (\beta\sim\Gamma)1q_t^i$. Let $h \stackrel{d}{=} \langle\{f\in{}^\Gamma\varkappa : (f\cup p)\in b\} : b\in C\rangle$. By [HMTI]8.1, $h\in$ $\in Hom(\mathfrak{Rd}_\Gamma\mathfrak{L}, \mathfrak{Gb}^\Gamma\varkappa)$. Let $\mathfrak{G} \stackrel{d}{=} \mathfrak{Gb}^\Gamma\varkappa$. By $\cup p^*H = t$ we have $h(w_r) =$ $= \{f\in{}^\Gamma\varkappa : f0=t\}$. Let $g \stackrel{d}{=} \Gamma 1q$. By $y\in Nr_\alpha\mathfrak{A}$ and $i\notin\alpha$ we have that $i\notin\Delta(y_r)$. Thus $q_t^i\in y_r$ and $q_{ta}^{i0}\notin y_r$. Hence $g\in h(y_r)$ and $g_a^0\notin h(y_r)$. By $\Gamma\cap\Delta(y_r)=1$ then $\Delta^{(\mathfrak{G})}h(y_r)=1$. Therefore $(\forall f\in{}^\Gamma\varkappa)[f_u^0\in h(y_r)$ and $f_a^0\notin$ $\notin h(y_r)]$.

Let $\pi \stackrel{d}{=} (\varkappa 1 Id)^{ua}_{au}$. Then $\pi$ is a permutation of $\varkappa$ interchanging $u$ and $a$ such that $\pi(t)=t$. By [HMTI]3.1, $\pi$ induces a base-

-automorphism $\tilde{\pi}\in Is(\mathfrak{G},\mathfrak{G})$. By $\pi(t)=t$ we have $\tilde{\pi}h(w_r) = h(w_r)$. Then $\tilde{\pi}h(y_r) = h(y_r)$ by $h(y_r)\in Sg\{h(w_r)\}$. This is a contradiction by the above, since $\pi(u)=a$. We have proved that $(\forall y\in Nr_\alpha \mathfrak{A})[\Delta(y/I)\neq 1 \vee$ $\vee\ c_0^\partial(y/I)\neq 0]$.

QED(Claim 8.19.5.)

By Claims 8.19.4-5 we have that $\mathfrak{B}\notin \mathbf{Nr}_\alpha \mathbf{H}\mathfrak{A}$. Then $\mathbf{HNr}_\alpha \mathfrak{A} \nsubseteq \mathbf{Nr}_\alpha \mathbf{H}\mathfrak{A}$ by $\mathfrak{B}\in \mathbf{HNr}_\alpha \mathfrak{A}$. Recall that $\mathfrak{A}\in \mathbf{I}\,{}_\infty Cs_\beta$. Thus (3) is proved.

(4) We show that $\mathbf{Nr}_\alpha \mathbf{H}\mathfrak{A} \nsubseteq \mathbf{HNr}_\alpha \mathbf{S}\mathfrak{A}$ for some $\mathfrak{A}\in Cs_\beta^{reg}\cap Dc_\beta$ as well as for some $\mathfrak{A}\in Ws_\beta\cap Dc_\beta$.

Recall that $\beta\geq\alpha+\omega$ and $\alpha>0$. Hence there is $H\subseteq\alpha\sim\beta$ with $|H|=\omega$ and $|\beta\sim H|\geq\omega$. Let $\varkappa\geq 2$ be a cardinal.

(4.1) The $Cs^{reg}\cap Dc$ case: Prop.4.11 states that there exist $\mathfrak{A}\in$ $\in {}_\varkappa Cs_\beta^{reg}\cap Dc_\beta$ and $\mathfrak{B}\in \mathbf{H}\mathfrak{A}$ with $|Zd\,\mathfrak{B}|>2$. It turns out from the proof of 4.11 that $A=Sg\{x\}$ with $\Delta x=H$ for some $x\in A$. Then $A=Dm_H$ by [HMT]1.6.4-8. Since $H\cap\alpha=0$, then $(\forall y\in A)|\alpha\cap\Delta y|<\omega$ and hence $(\forall y\in$ $\in Nr_\alpha A)|\Delta y|<\omega$. Therefore $\mathfrak{Nr}_\alpha \mathfrak{A}\in Lf_\alpha$. By Cor.8.18(iii) $\mathbf{Nr}_\alpha Cs_\beta^{reg} \subseteq$ $\subseteq \mathbf{I}\,Cs_\alpha^{reg}$. Hence $\mathfrak{Nr}_\alpha \mathfrak{A}\in \mathbf{I}\,Cs_\alpha^{reg}\cap Lf_\alpha$. By [HMT]2.6.29, $\mathbf{Nr}_\alpha \mathbf{S}\mathfrak{A} \subseteq \mathbf{SNr}_\alpha \mathfrak{A} \subseteq$ $\subseteq \mathbf{I}\,Cs_\alpha^{reg}\cap Lf_\alpha$. By [HMTI]5.2, $\mathbf{H}(Cs_\alpha^{reg}\cap Lf_\alpha) = \mathbf{I}\,Cs_\alpha^{reg}\cap Lf_\alpha$ and hence $\mathbf{HNr}_\alpha \mathbf{S}\mathfrak{A} \subseteq \mathbf{I}\,Cs_\alpha^{reg}\cap Lf_\alpha$. Thus $(\forall \mathfrak{C}\in \mathbf{HNr}_\alpha \mathbf{S}\mathfrak{A})|Zd\,\mathfrak{C}|\leq 2$. Above we have observed that $\mathfrak{B}\in \mathbf{H}\mathfrak{A}$ and $|Zd\,\mathfrak{B}|>2$. Let $\mathfrak{N} \stackrel{d}{=} \mathfrak{Nr}_\alpha \mathfrak{B}$. Then $|Zd\,\mathfrak{N}|=|Zd\,\mathfrak{B}|>2$. Thus $\mathfrak{N}\in \mathbf{Nr}_\alpha \mathbf{H}\mathfrak{A}$ and $\mathfrak{N}\notin \mathbf{HNr}_\alpha \mathbf{S}\mathfrak{A}$. We have proved for all $\varkappa\geq 2$ that $\mathbf{Nr}_\alpha \mathbf{H}\mathfrak{A} \nsubseteq \mathbf{HNr}_\alpha \mathbf{S}\mathfrak{A}$ for some $\mathfrak{A}\in {}_\varkappa Cs_\beta^{reg}\cap Dc_\beta$.

(4.2) The $Ws\cap Dc$ case: In [HMTI]6.16(2) an $\mathfrak{A}\in Ws_\beta$ and a $\mathfrak{B}\in \mathbf{H}\mathfrak{A}$ are constructed such that $|Zd\,\mathfrak{B}|>2$. By using the ideas of 4.12 this $\mathfrak{A}$ can be modified to a $Ws_\beta\cap Dc_\beta$ with $A=Dm_H$ as follows. Recall from the first lines of the proof of (4) that $H\subseteq\beta$, $|H|=\omega$ and $H\cap\alpha=0$ etc. Let $p\in{}^\beta\varkappa$ and $x\subseteq{}^\beta\varkappa$ be as in 4.12. Then $\Delta^{(\varkappa)}x=H$ was observed there. Let $V \stackrel{d}{=} {}^\beta\varkappa^{(p)}$ and $y \stackrel{d}{=} x\cap V$. Then $\Delta^{[V]}y=H$. Let $\mathfrak{C} \stackrel{d}{=} \mathfrak{Sb}\,V$ and $\mathfrak{A} \stackrel{d}{=} \mathfrak{Sg}^{(\mathfrak{C})}\{y\}$. Then $\mathfrak{A}\in {}_\varkappa Ws_\beta\cap Dc_\beta$ and $\Delta^{(\mathfrak{A})}y=H$. By repeating the proof of [HMTI]6.16(2) for this $\mathfrak{A}$ and H there exists $\mathfrak{B}\in \mathbf{H}\mathfrak{A}$ with

$|Zd\mathfrak{B}|>2$. Now the argument in (4.1) above can be repeated to prove the $Ws\cap Dc$-case as follows. Again $A=Dm_H$ as it was in (4.1) proving $\mathfrak{Nr}_\alpha\mathfrak{A}\in$ $\in Lf_\alpha$. By 8.18, $\mathbf{Nr}_\alpha Ws_\beta \subseteq \mathbf{I}\,Ws_\alpha$ hence $\mathfrak{Nr}_\alpha\mathfrak{A}\in \mathbf{I}\,Ws_\alpha\cap Lf_\alpha$. By 3.15(a), $\mathbf{H}(Ws_\alpha\cap Lf_\alpha) = \mathbf{I}\,Cs_\alpha^{reg}\cap Lf_\alpha$. Then the rest of the proof of (4.1) above works without any change to prove for all $\varkappa\geq 2$ that $\mathbf{Nr}_\alpha\mathbf{H}\mathfrak{A} \not\subseteq \mathbf{HNr}_\alpha\mathbf{S}\mathfrak{A}$ for some $\mathfrak{A}\in {}_\varkappa Ws_\beta\cap Dc_\beta$.

So far, the "only if-parts" of (i) have been proved.

(5) Now we prove (ii) and the "if-parts" of (i).

Lemma 8.19.6. Let $\beta\geq\alpha$.

(i) Let $\mathfrak{A}\in CA_\beta$ be such that $A = Dm_\alpha^{\mathfrak{A}}$. Let $\mathfrak{N} \overset{d}{=} \mathfrak{Nr}_\alpha\mathfrak{A}$ and $R\in Co\,\mathfrak{N}$. Then there is a unique $S\in Co\,\mathfrak{A}$ such that $R = S\cap {}^2N$ and for this $S$ we have $S^*\in Is((\mathfrak{Nr}_\alpha\mathfrak{A})/R,\ \mathfrak{Nr}_\alpha(\mathfrak{A}/S))$.

(ii) Let $K\subseteq CA_\beta$. Then $\mathbf{HNr}_\alpha\mathbf{S}K \subseteq \mathbf{Nr}_\alpha\mathbf{HS}K$ and if $|\beta\sim\alpha|<\omega$ then $\mathbf{HNr}_\alpha K = \mathbf{Nr}_\alpha\mathbf{H}K$.

Proof. Proof of (i): Assume the hypotheses. Let $I \overset{d}{=} 0^{\mathfrak{N}}/R$ and $J \overset{d}{=} Ig^{\mathfrak{A}} I$. By $A = Dm_\alpha^{\mathfrak{A}}$ we have $(\forall y\in A)|\Delta y\sim\alpha|<\omega$. Assume $I = E\cap N$. If $y\in E$ then $c_{(\Delta y\sim\alpha)}y \in E\cap N=I$ thus $y\in Ig^{\mathfrak{A}} I = J$. Thus $E\subseteq J$. By $I\subseteq E$, obviously $J\subseteq E$. This proves $I=E\cap N \Rightarrow E=J$. $I=J\cap N$ follows from $Ig^{\mathfrak{A}} I =$ $= \{x\in A : x\leq \Sigma I\}$ which holds by [HMT]2.3.8. So far we have seen that there is a unique extension $S\in Co\,\mathfrak{A}$ of $R\in Co\,\mathfrak{N}$, since 0-ideals function properly in CA-s. Let $S\in Co\,\mathfrak{A}$ be such that $R = S\cap {}^2N$. Then $0/S = J$. Let $h \overset{d}{=} (N/R)1S^*$. Then $h = \{\langle b/I, b/J\rangle : b\in N\}$. By [HMT] 0.2.23-27 we have $h\in Hom(\mathfrak{N}/R,\ \mathfrak{Nr}_\alpha(\mathfrak{A}/S))$. $h$ is one-one by $I = N\cap J$. Obviously, $Rgh \subseteq Nr_\alpha(\mathfrak{A}/S)$. Let $x\in Nr_\alpha(\mathfrak{A}/S)$ be arbitrary. Let $x =$ $= z/J$ and $\Gamma \overset{d}{=} \Delta z\sim\alpha$. Then $c_{(\Gamma)}z\in N$ and $x = c_{(\Gamma)}x = (c_{(\Gamma)}z)/J$. Thus $Rgh = Nr_\alpha(\mathfrak{A}/S)$. Therefore $h\in Is(\mathfrak{N}/R,\ \mathfrak{Nr}_\alpha(\mathfrak{A}/S))$.

Proof of (ii): Let $\mathfrak{A}\in CA_\beta$ and let $\mathfrak{L} \overset{d}{=} \mathfrak{Sm}_\alpha(\mathfrak{A})$. By (i) we have $\mathbf{HNr}_\alpha\mathfrak{L} = \mathbf{Nr}_\alpha\mathbf{H}\mathfrak{L}$. Thus $\mathbf{HNr}_\alpha\mathfrak{A} \subseteq \mathbf{Nr}_\alpha\mathbf{HS}\mathfrak{A}$ since $\mathfrak{L} \subseteq \mathfrak{A}$ and $\mathfrak{Nr}_\alpha\mathfrak{L} = \mathfrak{Nr}_\alpha\mathfrak{A}$. If $|\beta\sim\alpha|<\omega$ Then $\mathbf{HNr}_\alpha\mathfrak{A} = \mathbf{Nr}_\alpha\mathbf{H}\mathfrak{A}$ by $\mathfrak{A} = \mathfrak{L}$.

QED(Lemma 8.19.6.)

Lemma 8.19.7. Let $\beta\geq\alpha$.

(i) Let $\mathfrak{A}\in{}^{I}CA_{\beta}$, $\mathfrak{N} \overset{d}{=} \langle \mathfrak{Nr}_{\alpha}\mathfrak{A}_{i} : i\in I\rangle$ and $F$ be a filter on $I$. Then $\bar{F}^{(A)*} \in Is(P\mathfrak{N}/F, \mathfrak{Nr}_{\alpha}\mathfrak{B})$ for some $\mathfrak{B} \subseteq P\mathfrak{A}/F$.

(ii) Let $K \subseteq CA_{\beta}$. Then $\mathbf{Up\,Nr}_{\alpha}\,\mathbf{S\,Up}\,K = \mathbf{Nr}_{\alpha}\,\mathbf{S\,Up}\,K$ and if $|\beta{\sim}\alpha|<\omega$ then $\mathbf{Up\,Nr}_{\alpha}\,K = \mathbf{Nr}_{\alpha}\,\mathbf{Up}\,K$.

Proof. Notation: For any pair $\mathfrak{B}\subseteq\mathfrak{A}$ of algebras and $R\in Co\,\mathfrak{A}$ we define $\mathfrak{B}/R \overset{d}{=} (R^{\star})^{*}\mathfrak{B}$. Then $\mathfrak{B}/R \subseteq \mathfrak{A}/R$. Proof of (i): Assume the hypotheses. It is proved in the proof of [HMT]2.6.32 that $P\mathfrak{N} = \mathfrak{Nr}_{\alpha}P\mathfrak{A}$. Let $\mathfrak{L} = \mathfrak{Dm}_{\alpha}^{P\mathfrak{A}}$. Then $\mathfrak{Nr}_{\alpha}\mathfrak{L} = \mathfrak{Nr}_{\alpha}P\mathfrak{A} = P\mathfrak{N}$ and $\mathfrak{L}/F \subseteq P\mathfrak{A}/F$. By the notational conventions of [HMT], $N = Uv\circ\mathfrak{N}$ and $\bar{F}^{(N)} = \bar{F}^{(\mathfrak{N})}$. By definition, ${}^{2}(PN)\cap\bar{F}^{(A)} = \bar{F}^{(N)}\in Co\,P\mathfrak{N}$. Then by writing $F^{N}$ for $\bar{F}^{(N)}$, $\bar{F}^{(A)*} \in Is((\mathfrak{Nr}_{\alpha}\mathfrak{L})/F^{N}, \mathfrak{Nr}_{\alpha}(\mathfrak{L}/F^{A}))$, by 8.19.6(i). We have seen that

(*) $\bar{F}^{(A)*} \in Is(P\,\mathfrak{Nr}_{\alpha}\circ\mathfrak{A}/F, \mathfrak{Nr}_{\alpha}(\mathfrak{Dm}_{\alpha}^{P\mathfrak{A}}/F^{A}))$.

(i) is proved by $\mathfrak{Dm}_{\alpha}^{P\mathfrak{A}}/F^{A} \subseteq P\mathfrak{A}/F$.

Proof of (ii): Let $K \subseteq CA_{\beta}$. By (i) we have $\mathbf{Up\,Nr}_{\alpha}\,K \subseteq \mathbf{Nr}_{\alpha}\,\mathbf{S\,Up}\,K$, hence $\mathbf{Up\,Nr}_{\alpha}\,\mathbf{S\,Up}\,K \subseteq \mathbf{Nr}_{\alpha}\,\mathbf{S\,Up\,S\,Up}\,K = \mathbf{Nr}_{\alpha}\,\mathbf{S\,Up}\,K$. Suppose $|\beta{\sim}\alpha|<\omega$. Let $F$ be an ultrafilter on some $I$. Then $\mathfrak{Dm}_{\alpha}(P\mathfrak{A}) = P\mathfrak{A}$ for every $\mathfrak{A}\in{}^{I}K$, hence by (*) we have $\bar{F}^{(A)} \in Is(P\,\mathfrak{Nr}_{\alpha}\circ\mathfrak{A}/F, \mathfrak{Nr}_{\alpha}P\mathfrak{A}/F)$. This implies $\mathbf{Up\,Nr}_{\alpha}\,K = \mathbf{Nr}_{\alpha}\,\mathbf{Up}\,K$ as follows. Any element of $\mathbf{Nr}_{\alpha}\,\mathbf{Up}\,K$ is of the form $\mathfrak{Nr}_{\alpha}(P\mathfrak{A}/F)$ with $\mathfrak{A}\in{}^{I}K$ and $F$ an ultrafilter on $I$. By the above $\mathfrak{Nr}_{\alpha}P\mathfrak{A}/F \cong P(\mathfrak{Nr}_{\alpha}\circ\mathfrak{A})/F \in \mathbf{Up\,Nr}_{\alpha}\,K$. Conversely, any element of $\mathbf{Up\,Nr}_{\alpha}\,K$ is of the form $P(\mathfrak{Nr}_{\alpha}\circ\mathfrak{A})/F$ with $\mathfrak{A}\in{}^{I}K$ and $F$ an ultrafilter on $I$. By the above then $P(\mathfrak{Nr}_{\alpha}\circ\mathfrak{A})/F \cong \mathfrak{Nr}_{\alpha}P\mathfrak{A}/F \in \mathbf{Nr}_{\alpha}\,\mathbf{Up}\,K$.

QED(Lemma 8.19.7.)

(6) Now we prove (iii). Let $\alpha<\omega\leq\beta$, and $K = \{\mathfrak{A}\in\mathbf{I}\,Gs_{\beta} : |Zd\,\mathfrak{A}|\geq\omega\}$. Then $K=\mathbf{Up}\,K$ and $(\forall\mathfrak{B}\in\mathbf{Nr}_{\alpha}\,K)\,|Zd\,\mathfrak{B}|\geq\omega$. Therefore $(\forall\mathfrak{B}\in\mathbf{Uf\,Up\,Nr}_{\alpha}\,K)\,|Zd\,\mathfrak{B}|\geq\omega$, since $\mathbf{Nr}_{\alpha}\,K \models \exists x_{0}\ldots\exists x_{n-1}(\wedge\{c_{(\alpha)}x_{i}=x_{i} \wedge \wedge\{x_{i}\neq x_{j} : i<j<n\} : i<n\})$. By

$\beta\geq\omega$, ${}_2Ws_\beta \subseteq \mathbf{Uf}\,K$ since every ${}_2Ws_\beta$ is nondiscrete and hence has ultrapowers with infinitely many zero-dimensional elements, by the following. Let $\mathfrak{A}\in{}_2Ws_\beta$ and let $F$ be any nonprincipal ultrafilter on $\omega$. For every $i\in\omega$ let $f^{(i)} \stackrel{d}{=} \langle d_{j,i+j} : j\in\omega\rangle \cup \langle 0 : j\in\beta\sim\omega\rangle$. Then $\langle f^{(i)}/F : : i\in\omega\rangle$ is a system of distinct zero-dimensional elements in ${}^\omega\mathfrak{A}/F$. Thus ${}_2Ws_\beta \subseteq \mathbf{Uf}\,K$. But by $Ws_\beta \subseteq Dind_\beta$ we have $\mathbf{Nr}_\alpha({}_2Ws_\beta) \not\subseteq \mathbf{Uf\,Up\,Nr}_\alpha K$ and hence $\mathbf{Nr}_\alpha \mathbf{Uf}\,K \supseteq \mathbf{Nr}_\alpha({}_2Ws_\beta) \not\subseteq \mathbf{Uf\,Up\,Nr}_\alpha \mathbf{Up\,P}\,K$. The other part of (iii), $\mathbf{Uf\,Nr}_2 K \not\subseteq \mathbf{Nr}_2 \mathbf{HSP}\,K$, follows from 8.6.

QED(Theorem 8.19.)

Remark 8.19.8. (Discussion of 8.19.)

(1) By 8.19(ii) for all $\alpha\leq\beta$ and all $K \subseteq CA_\beta$ we have

$$\mathbf{HNr}_\alpha \mathbf{S}K \subseteq \mathbf{Nr}_\alpha \mathbf{HS}K.$$

(1.1) Here $\subseteq$ cannot be replaced by $=$ because of the following. Let $\beta>\alpha>0$ be such that $\beta>\alpha+\omega$. Then by 8.19(i)f there is $\mathfrak{A}\in Cs_\beta^{reg}\cap Dc_\beta$ with $\mathbf{HNr}_\alpha \mathbf{S}\mathfrak{A} \not\supseteq \mathbf{Nr}_\alpha \mathbf{H}\mathfrak{A}$. Let $K=\{\mathfrak{A}\}$. Then $K \subseteq CA_\beta$ and $\mathbf{HNr}_\alpha \mathbf{S}K \not\supseteq \mathbf{Nr}_\alpha \mathbf{H}K \subseteq \mathbf{Nr}_\alpha \mathbf{HS}K$. Thus whenever $\beta>\alpha+\omega$ there is $K \subseteq CA_\beta$ such that $\mathbf{HNr}_\alpha \mathbf{S}K \neq \mathbf{Nr}_\alpha \mathbf{HS}K$. B ut by 8.19(i)g the condition $\beta>\alpha+\omega$ is needed here, namely $|\beta\sim\alpha|<\omega$ iff [in 8.19(ii) $\subseteq$ can be replaced with $=$].

(1.2) If we delete all occurrence of $\mathbf{S}$ from this statement then we obtain $\mathbf{HNr}_\alpha K \subseteq \mathbf{Nr}_\alpha \mathbf{H}K$. This inequality is not valid because of the following. Let $\beta>\alpha+\omega$, $\alpha>0$. By 8.19(i)e then there exists $\mathfrak{A}\in Cs_\beta$ such that $\mathbf{HNr}_\alpha\mathfrak{A} \not\subseteq \mathbf{Nr}_\alpha \mathbf{H}\mathfrak{A}$. Hence the statement obtained by deleting $\mathbf{S}$ is false for some $\beta\geq\alpha$ and some $K \subseteq CA_\beta$. Again if $\beta\geq\alpha+\omega$ does not hold then $\mathbf{S}$ can be deleted as 8.19(i)e proves.

(2) By 8.19(ii) for all $\alpha\leq\beta$ and $K \subseteq CA_\beta$, $\mathbf{Uf\,Nr}_\alpha \mathbf{SUp}\,K = \mathbf{Nr}_\alpha \mathbf{SUp}\,K$. The obvious candidates to improve this equality would be (I)-(VI) below.

(I) $\mathbf{Up\,Nr}_\alpha \mathbf{Up}\,K = \mathbf{Nr}_\alpha \mathbf{Up}\,K$ (to delete all occurrences of $\mathbf{S}$)

(II) $\mathbf{Up\,Nr}_\alpha K \supseteq \mathbf{Nr}_\alpha \mathbf{Up}\,K$ ($\mathbf{Up}$ commutes one way)

(III) $\mathbf{Up\,Nr}_\alpha K \subseteq \mathbf{Nr}_\alpha \mathbf{Up}\,K$ ($\mathbf{Up}$ commutes the other way)

(IV) $\mathbf{SUp\,Nr}_\alpha K \supseteq \mathbf{Nr}_\alpha \mathbf{SUp}\,K$ ($\mathbf{SUp}$ commutes one way)

(V) $\mathbf{SUpNr}_\alpha K \subseteq \mathbf{Nr}_\alpha \mathbf{SUp} K$ ($\mathbf{SUp}$ commutes the other way)

(VI) $\mathbf{UpNr}_\alpha \mathbf{S} K = \mathbf{Nr}_\alpha \mathbf{SUp} K$ ($\mathbf{Up}$ commutes with $\mathbf{Nr}_\alpha \mathbf{S}$).

We prove that none of (I)-(VI) is valid.

(2.1) To disprove (V), let $\beta>\alpha>1$ and ${}_\omega Ws_\beta \subseteq K \subseteq CA_\beta$. By [HMTI]8.5-8.6, ${}_\omega Ws_\alpha \subseteq \mathbf{ISNr}_\alpha K$. By 8.8(ii), ${}_\omega Ws_\alpha \not\subseteq \mathbf{UfUpNr}_\alpha CA_\beta \supseteq \mathbf{Nr}_\alpha \mathbf{SUp} K$. Hence $\mathbf{SNr}_\alpha K \not\subseteq \mathbf{Nr}_\alpha \mathbf{SUp} K$.

(2.2) To disprove (IV) and (II) it is enough to show that $\mathbf{SUpNr}_\alpha K \not\supseteq \mathbf{Nr}_\alpha \mathbf{Up} K$. Let $\beta\geq\alpha+\omega$, $\alpha>0$. By 8.19(i)b there is $\mathfrak{A} \in Ws_\beta \cap Dc_\beta$ such that $\mathbf{HSP}\ \mathbf{Nr}_\alpha \mathfrak{A} \not\supseteq \mathbf{Nr}_\alpha \mathbf{Up} \mathfrak{A}$. Choosing $K=\{\mathfrak{A}\}$ completes the proof.

(2.3) To disprove (I) and (III) it is enough to find $\alpha,\beta$ and $K$ such that $\mathbf{UpNr}_\alpha K \not\subseteq \mathbf{Nr}_\alpha \mathbf{Up} K$ (since $\mathbf{UpNr}_\alpha \mathbf{Up} K \supseteq \mathbf{UpNr}_\alpha K$). Let $\beta\geq\alpha+\omega$, $\alpha>0$. By 8.19(i)c there is $\mathfrak{A}\in Ws_\beta$ such that $\mathbf{UpNr}_\alpha \mathfrak{A} \not\subseteq \mathbf{UfNr}_\alpha \mathbf{Up}\mathfrak{A} \supseteq \mathbf{Nr}_\alpha \mathbf{Up} \mathfrak{A}$. Choosing $K=\{\mathfrak{A}\}$ we have $\mathbf{UpNr}_\alpha K \not\subseteq \mathbf{Nr}_\alpha \mathbf{Up} K$ as it was desired.

(2.4) To disprove (VI) it is enough to show $\mathbf{UpNr}_\alpha \mathbf{S} K \not\supseteq \mathbf{Nr}_\alpha \mathbf{SUp} K$. Let $K$ be the class of all finite discrete $CA_\beta$-s. Then $(\forall \mathfrak{A}\in\mathbf{UpNr}_\alpha \mathbf{S}K)|A|\neq\omega$. But obviously $(\exists \mathfrak{B}\in\mathbf{Nr}_\alpha \mathbf{SUp}K)|B|=\omega$. This completes the proof. We note that there are nondiscrete counterexamples, too, e.g. the following one. Let $K \overset{d}{=} \{\mathfrak{A}\in CA_\beta : |Zd\mathfrak{A}|<\omega$ and $\mathfrak{A}$ is nondiscrete$\}$. Then $(\forall \mathfrak{A}\in \mathbf{UpNr}_\alpha \mathbf{S}K)|Zd\mathfrak{A}|\neq\omega$. (Hint: See e.g. the proof of 8.14(i)-(ii).) But obviously $(\exists \mathfrak{B}\in\mathbf{Nr}_\alpha \mathbf{SUp}K)|B|=\omega$. This proves $\mathbf{UpNr}_\alpha \mathbf{S} K \not\supseteq \mathbf{Nr}_\alpha \mathbf{SUp} K$ for the nondiscrete case too.

(3) In 8.19(i)e the class $Cs_\beta$ cannot be replaced with $Cs_\beta^{reg}$, moreover it cannot be replaced with $Dind_\beta$, since if $\alpha<\omega$ and $\mathfrak{A}\in Dind_\beta$ then $\mathfrak{Nr}_\alpha \mathfrak{A}$ is simple hence $\mathbf{HNr}_\alpha \mathfrak{A} = \mathbf{INr}_\alpha \mathfrak{A} \cup \mathbf{I}\,{}_0Gs_\alpha$.

<u>Corollary 8.20.</u> Let $\alpha\leq\beta$.

(i) Let $1<\varkappa<\omega$ and $K\in\{\mathbf{I}\,{}_\varkappa Gs, \mathbf{I}\,{}_\infty Gs, CA, \langle \mathbf{I}\,Crs_\gamma\cap CA_\gamma : \gamma\in Ord\rangle, \mathbf{I}\,Gs\}$.

Then $\mathbf{Nr}_\alpha K_\beta = \mathbf{HUp}\ \mathbf{Nr}_\alpha K_\beta$.

(ii) Let $K\in\{\mathbf{I}\,Ws, \mathbf{I}\,Cs^{reg}, \mathbf{I}\,Gws^{comp\ reg}, \mathbf{I}\,Cs\}$.

Then $\mathbf{Nr}_\alpha \mathbf{H}K_\beta = \mathbf{HNr}_\alpha \mathbf{H}K_\beta$. If $\beta<\alpha+\omega$ then $\mathbf{Nr}_\alpha \mathbf{H}K_\beta = \mathbf{HNr}_\alpha K_\beta$ and $\mathbf{Nr}_\alpha \mathbf{Up} K_\beta = \mathbf{UpNr}_\alpha K_\beta$.

2.(i) $\mathbf{Up\,Nr}_\alpha \mathrm{Ws}_\beta \subseteq \mathbf{Uf\,Nr}_\alpha \mathbf{Up}\,\mathrm{Ws}_\beta$ iff $\alpha\cap|\beta\sim\alpha|<\omega$.

(ii) $\mathbf{Uf\,Up\,Nr}_\alpha \mathrm{Ws}_\beta = \mathbf{Uf\,Nr}_\alpha \mathbf{Up}\,\mathrm{Ws}_\beta$ iff $\beta<\alpha+\omega$.

(iii) Let $K\in\{\mathbf{I}\,\mathrm{Ws}, \mathbf{I}\,\mathrm{Cs}^{reg}, \mathbf{I}\,\mathrm{Gws}^{comp\ reg}\}$. Then

$\mathbf{Nr}_\alpha K_\beta = \mathbf{Up\,Nr}_\alpha K_\beta$ iff $\alpha<\omega$.

(iv) Let $\alpha<\omega\leq\beta$. Then $\mathbf{Nr}_\alpha \mathbf{Up}\,\mathrm{Ws}_\beta \not\subseteq \mathbf{HS\,Up}\ \mathbf{Nr}_\alpha \mathrm{Gws}_\beta^{comp\ reg}$.

Proof. For 1(i), use 8.19(ii) and the fact that each indicated class is a variety (see, e.g., 8.3). For 1(ii), first part, we use 8.19(ii) again: $\mathbf{HNr}_\alpha \mathbf{H}K_\beta \subseteq \mathbf{HNr}_\alpha \mathbf{SHS}K_\beta \subseteq \mathbf{Nr}_\alpha \mathbf{HSHS}K_\beta = \mathbf{Nr}_\alpha \mathbf{H}K_\beta$. The second part of 1(ii) is clear by 8.19(i).

Proof of 2(iii): Let $K$ be as in the hypotheses. Let $\alpha<\omega$. Let $\mathfrak{A}\in\mathbf{Up\,Nr}_\alpha K_\beta$. By 1(i) then $\mathfrak{A} \cong \mathfrak{Nr}_\alpha \mathfrak{B}$ for some $\mathfrak{B}\in\mathrm{Gs}_\beta$. Let $\mathfrak{C} \overset{d}{=} \mathfrak{Dm}_\alpha(\mathfrak{B})$. Then $\mathfrak{A} \cong \mathfrak{Nr}_\alpha \mathfrak{C}$ and $\mathfrak{C}\in\mathrm{Lf}_\beta$ by $\alpha<\omega$. By 8.18 we have $\mathbf{Nr}_\alpha K_\beta \subseteq K_\alpha \subseteq \mathrm{Dind}_\alpha$, therefore $\mathfrak{A}\in\mathbf{Up\,Nr}_\alpha K_\beta \subseteq \mathrm{Dind}_\alpha$ by $\alpha<\omega$. Then $|\mathrm{Zd}\,\mathfrak{C}|\leq 2$ by $\mathfrak{A} \cong \mathfrak{Nr}_\alpha \mathfrak{C}$. Then [HMT]2.3.14 implies that $\mathfrak{C}$ is simple or $|C|=1$. Then $\mathfrak{C} \in \mathbf{I}\,\mathrm{Ws}_\beta \cup \mathbf{I}\,{}_0\mathrm{Gs}_\beta$ by [HMTI]6.14. Thus using [HMTI] 7.13 we have seen $\mathbf{Up\,Nr}_\alpha K_\beta \subseteq \mathbf{Nr}_\alpha K_\beta$ if $\alpha<\omega$. Suppose $\alpha\geq\omega$. Then clearly, $\mathbf{Nr}_\alpha K_\beta \neq \mathbf{Up\,Nr}_\alpha K_\beta$, since $\mathbf{Nr}_\alpha K_\beta \subseteq \mathrm{Dind}_\alpha$ while $\mathbf{Up\,Nr}_\alpha K_\beta \not\subseteq \mathrm{Dind}_\alpha$. ($\mathbf{Up\,Nr}_\alpha K_\beta \not\subseteq \mathrm{Dind}_\alpha$ can be proved analogously to the proof of ${}_2\mathrm{Ws}_\beta \subseteq \mathbf{Uf}\,K$ in part (6) of the proof of 8.19.) By this, 2(iii) is proved.

2(iv) holds since $\mathbf{Nr}_\alpha \mathrm{Gws}_\beta^{comp\ reg} \models \forall x(x=0 \vee c_{(\alpha)}x=1)$ and by $\beta\geq\omega$ members of $\mathbf{Nr}_\alpha \mathbf{Up}\,\mathrm{Ws}_\beta$ may contain infinitely many zero-dimensional elements. Now 2(i) follows from 2(iii), 8.19(i) and from 8.19.3. 2(ii) follows from 2(i), 2(iv) and 8.19(i)d.

QED(Corollary 8.20.)

Corollary 8.21. Let $0<\alpha<\beta$ and $K\in\{\mathrm{Ws}, \mathrm{Cs}^{reg}, \mathrm{Gws}^{comp\ reg}, \mathrm{Cs}, \mathrm{Gs}, \mathrm{CA}\}$. Let $L \overset{d}{=} \langle \mathbf{I}\,K_\alpha\cap\mathrm{Mn}_\alpha : \alpha\in\mathrm{Ord}\rangle$.

(i) $\mathbf{HS\,Up}\,L_\alpha \subset \mathbf{HSP}\,L_\alpha \subset \mathbf{SP}\,\mathrm{Cs}_\alpha$, in particular

$\mathbf{HSP}\,\mathrm{Mn}_\alpha \not\supseteq \mathrm{Ws}_\alpha\cap\mathrm{Lf}_\alpha$ and $\mathbf{HS\,Up}\,\mathrm{Mn}_\alpha \not\supseteq \mathbf{SP}(\mathrm{Ws}_\alpha\cap\mathrm{Mn}_\alpha)$.

(ii) $\langle \mathbf{HSP}\,L_\alpha : \alpha\in\mathrm{Ord}\rangle$ is a system of varieties definable by a scheme of equations, but is not strongly such.

(iii) $L_\alpha = \mathbf{Nr}_\alpha L_\beta$ iff $\mathbf{HS\,Up}\,L_\alpha = \mathbf{HS\,Up\,Nr}_\alpha L_\beta$ iff

iff (either $\alpha \geq \omega$ or $K \notin \{Gs, CA\}$).

(iv) Conditions a.-c. below are equivalent.

a. $\alpha \geq \omega$ or $\beta < \alpha + \omega$.

b. $\mathbf{Uf\,Up\,Nr}_\alpha L_\beta = \mathbf{Uf\,Up\,Nr}_\alpha \mathbf{Up}\,L_\beta$.

c. $\mathbf{HSP\,Nr}_\alpha L_\beta = \mathbf{HSP\,Nr}_\alpha \mathbf{Up}\,L_\beta$.

(v) Conditions a.-e. below are equivalent.

a. $\alpha \geq \omega$.

b. $\mathbf{Uf\,Up}\,L_\beta = \mathbf{Uf\,Up\,Rd}_\alpha L_\beta$.

c. $\mathbf{HSP}\,L_\alpha = \mathbf{HSP\,Rd}_\alpha L_\beta$.

d. $\mathbf{Uf\,Up}\,L_\alpha \neq \mathbf{S\,Up}\,L_\alpha$.

e. $\mathbf{HSP}\,L_\alpha \neq \mathbf{SP}\,L_\alpha$.

(vi) Let $\omega \leq \alpha \leq \beta$ and let $T \subseteq \mathbf{Up}\,Lf_\beta$. Then

$\mathbf{Uf\,Up\,Rd}_\alpha T = \mathbf{Uf\,Up\,Nr}_\alpha T = \mathbf{Uf\,Up\,Nr}_\alpha \mathbf{Up}\,T$.

To prove 8.21, we shall need the following lemmas.

<u>Lemma 8.21.0.</u> Let $\alpha, \beta$ and $L$ be as in the formulation of 8.21, Then $L_\alpha \subseteq \mathbf{S\,Nr}_\alpha L_\beta$.

<u>Proof.</u> First we prove

(*) $\mathbf{I}\,K_\alpha \cap Mn_\alpha \subseteq \mathbf{S\,Nr}_\alpha K_\beta$.

If $K \in \{Cs^{reg}, Gws^{comp\ reg}\}$ then (*) holds by 8.18(iv), to be precise this is not stated for the case $\alpha=1$ there but by $Gws_1^{comp\ reg} = Cs_1$ it follows from [HMTI]8.5-6. If $K \in \{Ws, Cs, Gs\}$ then (*) holds by [HMTI]8.6, and if $K=CA$ then (*) holds by [HMT]2.6.57, 2.6.31. Hence if $\mathfrak{A} \in K_\alpha \cap Mn_\alpha$ then $\mathfrak{A} \subseteq \mathfrak{Nr}_\alpha \mathfrak{B}$ for some $\mathfrak{B} \in \mathbf{I}\,K_\beta$. Let $\mathfrak{C} \overset{d}{=} \mathfrak{Sg}^{(\mathfrak{B})}A$. Then $\mathfrak{A} \in \mathbf{S\,Nr}_\alpha \mathfrak{C}$ and $\mathfrak{C} \in \mathbf{I}\,K_\beta \cap Mn_\beta$. This proves $K_\alpha \cap Mn_\alpha \subseteq \mathbf{S\,Nr}_\alpha (\mathbf{I}\,K_\beta \cap Mn_\beta)$ as desired.

<u>QED(Lemma 8.21.0.)</u>

Lemma 8.21.1. Let $\omega\le\alpha\le\beta$ and let $\mathfrak{A}\in\mathbf{Up}(Gs^{reg}_\beta\cap Lf_\beta)$.
Then $\mathfrak{Nr}_\alpha\mathfrak{A}$ is an elementary subalgebra of $\mathfrak{Rd}_\alpha\mathfrak{A}$.

Proof. Let $\alpha\ge\omega$. Let $F$ be an ultrafilter on $I$ and let $\mathfrak{A}=P\mathfrak{B}/F$ for some $\mathfrak{B}\in {}^I(Gs^{reg}_\beta\cap Lf_\beta)$. Let $\varphi(x,y_0,\dots,y_n)$ be a first order formula in the discourse language of $CA_\alpha$-s. Assume $\mathfrak{A}\models\varphi[a/F, b_0/F,\dots,b_n/F]$ for some $a\in PB$ and $b\in {}^{n+1}PB$ such that $(\forall k\le n)b_k/F\in\in Nr_\alpha\mathfrak{A}$. Let $\Gamma$ be the set of indices occurring in $\varphi$. Then $\Gamma\subseteq_\omega\alpha$. Let $i\in I$. Let $\Omega\overset{d}{=}\cup\{\Delta^{(\mathfrak{B}_i)}b_{ki} : k\le n\}$. Let $\xi:\beta\rightarrowtail\beta$ be a permutation of $\beta$ such that $(\Gamma\cup\Omega)1\xi\subseteq Id$, $\xi^{-1*}(\Delta(a_i)\sim\Omega)\subseteq\alpha$ and $\xi$ is a finite transformation of $\beta$. Such a $\xi$ exists by $\alpha\ge\omega$ and $\mathfrak{B}_i\in Lf_\beta$. Let $V$ be the unit of $\mathfrak{B}_i$ and let $\mathfrak{L}\overset{d}{=}\mathfrak{Sb}\,V$. Let $rs^\xi\overset{d}{=}\langle\{q\circ\xi : q\in x\} : x\in C\rangle$. By the proof of [HMTI]8.4 then $rs^\xi V=V$ and $rs^\xi\in\in Is(\mathfrak{Rd}^\xi\mathfrak{L},\mathfrak{L})$ since $V$ is a $Gs_\beta$-unit and $\xi$ is a permutation of $\beta$. Let $\tau\overset{d}{=}\xi^{-1}$. Then $(\forall b\in B_i)rs^\xi b=s_\tau b\in B_i$ by [HMT]1.11.10 (which clearly holds for $Gs^{reg}_\beta\cap Lf_\beta$) and by $\mathfrak{B}_i\in Gs^{reg}_\beta\cap Lf_\beta$. Thus $rs^{\xi*}B_i\subseteq\subseteq B_i$. Similarly, $rs^{\tau*}B_i\subseteq B_i$, thus $rs^\xi\in Is(\mathfrak{Rd}^\xi\mathfrak{B}_i,\mathfrak{B}_i)$ by $rs^\xi\circ\circ rs^\tau\subseteq Id$. By $\Omega1\xi\subseteq Id$ and $\mathfrak{B}_i\in Gs^{reg}_\beta$ we have $rs^\xi(b_{ki})=b_{ki}$ for every $k\le n$. Let $e_i\overset{d}{=}rs^\xi a_i$. Then $\Delta^{(\mathfrak{B}_i)}e_i\subseteq\alpha\cap\Omega$ by $\xi^{-1*}(\Delta(a_i)\sim\Omega)\subseteq\subseteq\alpha$. By $\Gamma1\xi\subseteq Id$ and by the above we have that $\mathfrak{B}_i\models\varphi[a_i,pj_i\circ b]$ iff $\mathfrak{Rd}^\xi\mathfrak{B}_i\models\varphi[a_i,pj_i\circ b]$ iff $\mathfrak{B}_i\models\varphi[e_i,pj_i\circ b]$. Let $j\in\beta\sim\alpha$. Then $[\wedge\{c_j(b_k/F)=b_k/F : k\le n\}\to c_j(e/F)=e/F]$ by $(\forall i\in I)$ $\Delta e_i\subseteq\alpha\cup\cup\{\Delta(b_{ki}) : k\le n\}$ where $e=\langle e_i : i\in I\rangle$. Thus $c_j(e/F)=e/F$ by $(\forall k\le n)$ $b_k/F\in Nr_\alpha\mathfrak{A}$. Therefore $e/F\in Nr_\alpha\mathfrak{A}$ and $\mathfrak{A}\models\varphi[e/F,b_0/F,\dots,b_n/F]$.
QED(Lemma 8.21.1.)

Lemma 8.21.2. Let $\alpha\ge\omega$ and $\varkappa=|\varkappa|>1$.
Then $\mathbf{SUp}'({}_\varkappa Ws_\alpha\cap Mn_\alpha)\not\models\forall x(at(x)\to c^\partial_0 d_{01}=x)$.

Proof. Let $\alpha\ge\omega$ and $\varkappa=|\varkappa|>1$. For any set $s$ we let $\bar{s}=\langle s : i<\alpha\rangle$. Let $\mathfrak{A}\overset{d}{=}\mathfrak{Mn}(\mathfrak{Sb}\,{}^\alpha\varkappa^{(\bar{0})})$. Then $\mathfrak{A}\in Mn_\alpha\cap{}_\varkappa Ws_\alpha$ and $\mathfrak{A}$ is nondiscrete. Let $I\overset{d}{=}Sb_\omega(\alpha)$ and $F$ be an ultrafilter on $I$ such that $(\forall\Gamma\in I)$

$\{\Delta\in I : \Delta\supseteq\Gamma\}\in F$. Let $x \overset{d}{=} \langle d_\Gamma^{(\mathfrak{A})} : \Gamma\in I\rangle$. Then $x/F > 0$ and $(\forall i,j\in\alpha)$ $x/F \le d_{ij}$ by the definition of F. Let $\mathfrak{C} \overset{d}{=} {}^I\mathfrak{A}/F$. Let $\mathfrak{B} \overset{d}{=}$ $\overset{d}{=} \mathfrak{Sg}^{(\mathfrak{C})}\{x/F\}$. Since $\mathfrak{B} \in \mathbf{S\,Up}\,\mathrm{Ws}_\alpha$ by 5.6(iii) and by [HMTI]7.16, $\mathfrak{B} \in$ $\in \mathbf{I}\,\mathrm{Gws}_\alpha^{wd}$. Since $\mathfrak{A}$ has a characteristic also $\mathfrak{B}$ has a characteristic. Hence by [HMTI]7.26, there exists a cardinal $\mu$ such that $\mathfrak{B} \in \mathbf{I}\,{}_\mu\mathrm{Gws}_\alpha^{wd}$. Then there are $h : \mathfrak{B} \rightarrowtail \mathfrak{N} \in {}_\mu\mathrm{Gws}_\alpha^{wd}$. Let $y \overset{d}{=} h(x/F)$. Then $N =$ $= Sg^{(\mathfrak{N})}\{y\}$ and $(\forall i,j\in\alpha) y\le d_{ij}^{(\mathfrak{N})}$ and $y\ne 0$. Let $U=base(\mathfrak{N})$. Then $y \subseteq \{\bar{u} : u\in U\}$. Thus there are $H \subseteq Subb(\mathfrak{N})$ and a choice function $n \in P(H1Id)$ such that $y = \{\overline{n_Y} : Y\in H\}$. Since $\mathfrak{N}$ is widely distributed, $(\forall Y\in H)\, 1^{\mathfrak{N}} \cap {}^\alpha Y = {}^\alpha Y(\langle nY : i<\alpha\rangle)$.

Assume $y \supset z \ne 0$. Then there are $q\in z$ and $g\in y\sim z$. By the above, there are $Q\in H$ and $G\in H$ such that $n(Q)=q$ and $n(G)=g$. Let $f :$ $: U \rightarrowtail U$ be such that $[U\sim(Q\cup G)]1f \subseteq Id$ and $f^*Q=G$ and $f = f^{-1}$ and $f\circ q = g$ and $f\circ g = q$. Let $\mathfrak{R} \overset{d}{=} \mathfrak{Gb}\,1^{\mathfrak{N}}$. Then $\tilde{f} : \mathfrak{R} \rightarrowtail \mathfrak{R}$ is a base-automorphism of $\mathfrak{R}$ by [HMTI]3.1. By $y = \{\overline{n_Y} : Y\in H\}$ and by $\tilde{f}\{g\} = q$, $\tilde{f}\{q\} = g$ and by $(\forall Y\in H\sim\{Q,G\})\tilde{f}\{\overline{n_Y}\}=\{\overline{n_Y}\}$ we have $\tilde{f}(y)=y$. By $N=Sg\{y\}$ therefore $N1\tilde{f} \subseteq Id$. By $g \in \tilde{f}\{q\} \subseteq \tilde{f}(z)$ and by $g\notin z$ we have $\tilde{f}(z)\ne z$ proving that $z\notin N$. We have proved that $(\forall z)$ $[0\ne z\subset y \Rightarrow$ $\Rightarrow z\notin N]$. This and $y\ne 0$ prove that $y \in At\,\mathfrak{N}$.

By $\varkappa>1$, we have $\Delta y=\alpha$ and hence $\mathfrak{N} \not\models \forall x(at(x) \rightarrow c_0^\partial d_{01}=x)$. Since $\mathfrak{N} \in \mathbf{S\,Up}\{\mathfrak{A}\} \subseteq \mathbf{S\,Up}'(\mathrm{Mn}_\alpha \cap {}_\varkappa\mathrm{Ws}_\alpha)$, we are done.

QED(Lemma 8.21.2.)

Now we turn to the proof of 8.21. Let $\alpha,\beta,K$ and L be as in the hypotheses.

Proof of (vi): Let $\omega\le\alpha<\beta$ and $T \subseteq \mathbf{Up}\,\mathrm{Lf}_\beta$. By [AN1] we have $\mathrm{Lf}_\beta \subseteq \mathbf{I}\,\mathrm{Gs}_\beta^{reg}$. Hence by $\mathbf{Up}\,T \subseteq \mathbf{Up}\,\mathrm{Lf}_\beta$ and by 8.21.1 we have $\mathbf{Uf\,Up\,Nr}_\alpha\, T =$ $\mathbf{Uf\,Up\,Rd}_\alpha\, T = \mathbf{Uf\,Up\,Rd}_\alpha\, \mathbf{Up}\,T = \mathbf{Uf\,Up\,Nr}_\alpha\, \mathbf{Up}\,T$.

Proof of (iv): Suppose $0<\alpha<\omega\le\beta$. Then $\mathbf{Nr}_\alpha\, \mathbf{Up}\,(\mathrm{Ws}_\beta\cap\mathrm{Mn}_\beta) \not\subseteq \mathbf{HSP}\,\mathbf{Nr}_\alpha\,\mathrm{Mn}_\beta$ s proved in Case 2 of step (1) in the proof of 8.19. If $\beta<\alpha+\omega$ then . holds by 8.19(i). If $\alpha\ge\omega$ then b. holds by (vi).

Proof of (v): Let $\alpha<\omega$. Then $\mathbf{Rd}_\alpha\,(\mathrm{Ws}_\beta\cap\mathrm{Mn}_\beta) \not\subseteq \mathbf{HSP}\,\mathrm{Mn}_\alpha$ since

$\mathsf{Rd}_\alpha(Ws_\beta \cap Mn_\beta) \not\models c_{(\alpha)}x = c_{(\alpha\sim 1)}x$. This proves $\mathsf{HSP}\,L_\alpha \neq \mathsf{HSP}\ \mathsf{Rd}_\alpha L_\beta$. Let $\mathfrak{A} \subseteq P\mathfrak{B}$ with $\mathfrak{B} \in {}^I CA_\alpha$ and $J \in Il\mathfrak{A}$. Then $J = A \cap E$ for some $E \in Il(P\mathfrak{B})$. Let $F \stackrel{d}{=} \{\{i \in I : x_i = 0\} : x \in E\}$. Then $F$ is a filter. Assume $\mathfrak{B} \in {}^I Dind_\alpha$. Then $E = 0/F$ and hence $P\mathfrak{B}/F = P\mathfrak{B}/E$. It is well known from model theory that for any class $M$ of algebras $\mathsf{SPUp}\,M$ is closed under taking reduced products. Therefore $P\mathfrak{B}/F \in \mathsf{SPUp}(Rg\,\mathfrak{B})$. Thus $\mathfrak{A}/J \in \mathsf{SPUp}(Rg\,\mathfrak{B})$. We have proved

(*) $\mathsf{HSP}\ T = \mathsf{SPUp}\ T$ for every $T \subseteq Dind_\alpha$, if $\alpha < \omega$.

By [HMT]2.5.25,28 we have $L_\alpha \subseteq \mathsf{P}(L_\alpha \cap Dind_\alpha)$ and $\mathsf{Uf\,Up}\,L_\alpha = L_\alpha$. Hence by (*) we have $\mathsf{HSP}\,L_\alpha = \mathsf{HSP}(L_\alpha \cap Dind_\alpha) = \mathsf{SPUp}(L_\alpha \cap Dind_\alpha) = \mathsf{SP}(L_\alpha \cap Dind_\alpha) = \mathsf{SP}\,L_\alpha$. $L_\alpha = \mathsf{HS\,Up}\,L_\alpha$ follows from [HMT]2.5.28. Let $\alpha \geq \omega$. Then b. holds by (vi) and (iii). $\mathsf{SP}\,Mn_\alpha \models \forall x(\Delta x \neq 1)$ but $\mathsf{Up}\,L_\alpha \not\models \forall x(\Delta x \neq 1)$; e.g. $\langle d_{0i} : i < \alpha \rangle / F$ has dimension set 1 in ${}^\alpha\mathfrak{A}/F$, if $\mathfrak{A}$ is a nondiscrete $Mn_\alpha$ and $F$ is a nonprincipal ultrafilter. Hence $\mathsf{Up}\,L_\alpha \not\subseteq \mathsf{SP}\,Mn_\alpha$. $\mathsf{Uf\,Up}\,Mn_\alpha \models \forall x(at(x) \rightarrow c_0^\partial d_{01} = x)$ by [HMT]2.1.20(ii). By Lemma 8.21.2 this formula fails in $\mathsf{SUp}(Ws_\alpha \cap Mn_\alpha)$ proving $\mathsf{Uf\,Up}\,L_\alpha \neq \mathsf{SUp}\,L_\alpha$. Actually, we have proved $\mathsf{Uf\,Up}\,L_\alpha \not\supseteq \mathsf{SUp}'({}_\varkappa Ws_\alpha \cap L_\alpha)$ for all $\varkappa > 1$ which is stronger than the claimed statement. Moreover we proved for every nondiscrete $\mathfrak{A} \in Ws_\alpha$ that $\mathsf{SUp}\,\mathfrak{A} \not\subseteq \mathsf{Uf\,Up}\,L_\alpha$.

<u>Proof of (iii):</u> Let $\mathfrak{A} \in L_\alpha$. By [HMT]2.5.25, $\nabla\mathfrak{A} \subseteq \alpha \cap \omega$. By Lemma 8.21.0, $\mathfrak{A} \cong \mathfrak{B} \subseteq \mathfrak{Nr}_\alpha \mathfrak{C}$ for some $\mathfrak{C} \in L_\beta$ and $\mathfrak{B}$. By $\mathfrak{C} \in Mn_\beta$ and by [HMT]2.5.25 we may assume $\mathfrak{C} \in Gws_\beta$ and $\{|Y| : Y \in Subb(\mathfrak{C})\} \subseteq \alpha \cup \{\omega\}$. Then $\nabla\mathfrak{B} = \nabla\mathfrak{C}$ and if $\alpha < \omega$ then $a_\alpha^{\mathfrak{B}} = \cup\{V : |base(V)| = \omega$ and $V \in Subu(\mathfrak{C})\} = a_\varkappa^{\mathfrak{C}}$ for all $\varkappa \in \omega \cap (\beta \sim \alpha)$, where $a_\varkappa \stackrel{d}{=} c_{(\varkappa)}\bar{d}(\varkappa \times \varkappa)$. Thus $\{a_\varkappa^{\mathfrak{B}} : \varkappa \in \omega \cap (\alpha+1)\} = \{a_\varkappa^{\mathfrak{C}} : \varkappa \in \omega \cap (\beta+1)\}$. By [HMT]2.1.17(ii) $Nr_\alpha \mathfrak{C} = B$ and hence $\mathfrak{B} = \mathfrak{Nr}_\alpha \mathfrak{C}$. This proves $L_\alpha \subseteq \mathsf{Nr}_\alpha L_\beta$ for all $\beta \geq \alpha$.

Let $\alpha \geq \omega$. Then $Mn_\alpha = \mathsf{Nr}_\alpha Mn_\beta$ by [HMT]2.1.17. By 8.18 and 8.1 we have $\mathsf{Nr}_\alpha K_\beta \subseteq \mathsf{I}\,K_\alpha$ for $K \neq Cs$. Thus $L_\alpha = \mathsf{Nr}_\alpha L_\beta$, since $\mathsf{I}\,Cs_\alpha \cap Mn_\alpha = \mathsf{I}\,Cs_\alpha^{reg} \cap Mn_\alpha$.

Let  $\alpha<\omega$.  1.) Assume  $K \notin \{Gs, CA\}$.  Let  $\mathcal{L} \in L_\beta$.  Then  $Zd\mathcal{L} =$
$= \{0^{\mathcal{L}}, 1^{\mathcal{L}}\}$  therefore by [HMT]2.1.17(ii) we have  $\mathcal{Nr}_\alpha \mathcal{L} \in Mn_\alpha$.  By 8.18,
$\mathcal{Nr}_\alpha \mathcal{L} \in L_\alpha$.  Thus  $L_\alpha = \mathbf{Nr}_\alpha L_\beta$.  2.) Let  $\mathcal{U} \in Gs_\beta \cap Mn_\beta$  and  $a_n$, $n \in \omega$  be
as in [HMT]2.1.17.  Let  $Subb(\mathcal{U}) = \{\alpha, (\alpha+1) \times \{\alpha\}\}$.  Then  $V \stackrel{d}{=} \beta_\alpha =$
$= -a_{\alpha+1}^{\mathcal{U}} \in Nr_\alpha \mathcal{U}$.  By [HMT]2.1.23,  $Mn_\alpha \vDash \forall x(a_\alpha \cdot c_{(\alpha)} x \in \{a_\alpha, 0\})$,  but
$a_\alpha^{\mathcal{U}} = 1^{\mathcal{U}}$   hence  $V \cdot a_\alpha \notin \{0,1\}$  in  $\mathcal{Nr}_\alpha \mathcal{U}$   proving  $\mathbf{Nr}_\alpha (Gs_\beta \cap Mn_\beta) \nsubseteq$
$\nsubseteq \mathbf{HS \, Up} \, Mn_\alpha$.

Proof of (i):   $\mathbf{HS \, Up} \, Mn_\alpha \nsupseteq \mathbf{SP} \, (Ws_\alpha \cap Mn_\alpha)$  is obvious for  $\alpha \leq 1$,  and
for  $\alpha>1$  it follows from  $Mn_\alpha \vDash \forall x(x \cdot c_0^\partial d_{01} \in \{0, c_0^\partial d_{01}\})$  (see [HMT]
2.1.20) which formula is not valid in  $\mathbf{SP} \, (Ws_\alpha \cap Mn_\alpha)$.  $\mathbf{HSP} \, Mn_\alpha \nsupseteq Ws_\alpha \cap Lf_\alpha$
follows from 8.19.1-2 for  $\alpha \geq \omega$,  and for  $\alpha<\omega$  from the fact that  $Mn_\alpha \vDash$
$\vDash c_{(\alpha)} x = c_{(\alpha \sim 1)} x$  for all  $\alpha \in \omega$.  $Mn_\alpha \subseteq \mathbf{I} \, Gs_\alpha$  is immediate by [HMT]
2.5.25. Hence  $\mathbf{HSP} \, Mn_\alpha \subseteq \mathbf{SP} \, Cs_\alpha$  by [HMTI]7.15.

(ii) is an immediate corollary of (v) by observing that for any  $\alpha$
and permutation  $\xi$  of  $\alpha$,  $\mathbf{Rd}^\xi L_\alpha = L_\alpha$  and since  $\mathbf{HSP} \, \mathbf{Rd}_\alpha T =$
$= \mathbf{HSP} \, \mathbf{Rd}_\alpha \mathbf{HSP} \, T$  by [HMT]0.5.4-13.

<u>QED(Corollary 8.21.)</u>

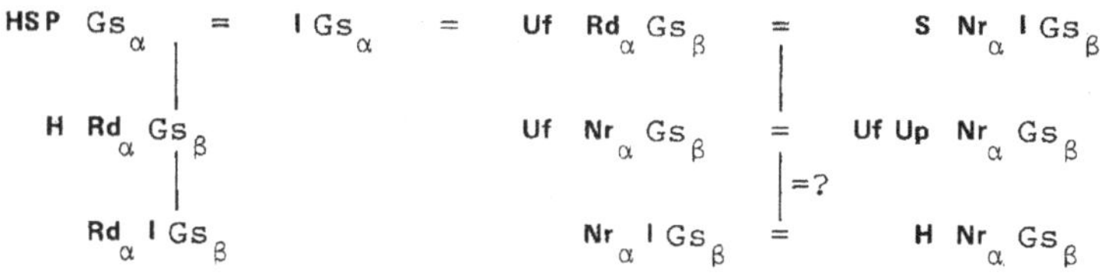

$$
\begin{array}{ccccccc}
\mathbf{HSP} \; Gs_\alpha & = & \mathbf{I} \, Gs_\alpha & = & \mathbf{Uf} \; \mathbf{Rd}_\alpha \, Gs_\beta & = & \mathbf{S} \; \mathbf{Nr}_\alpha \, \mathbf{I} \, Gs_\beta \\
| & & & & | & & | \\
\mathbf{H} \; \mathbf{Rd}_\alpha \, Gs_\beta & & & & \mathbf{Uf} \; \mathbf{Nr}_\alpha \, Gs_\beta & = & \mathbf{Uf} \; \mathbf{Up} \; \mathbf{Nr}_\alpha \, Gs_\beta \\
| & & & & & & |=? \\
\mathbf{Rd}_\alpha \, \mathbf{I} \, Gs_\beta & & & & \mathbf{Nr}_\alpha \, \mathbf{I} \, Gs_\beta & = & \mathbf{H} \; \mathbf{Nr}_\alpha \, Gs_\beta
\end{array}
$$

$(\omega \leq \alpha < \beta)$

<u>Figure 8.22.</u>

For any  $\omega \leq \alpha < \beta$,  the inclusions not indicated on the figure do not
hold with the only exception of the inclusion  $\mathbf{Uf} \; \mathbf{Nr}_\alpha \, Gs_\beta \subseteq \mathbf{Nr}_\alpha \, \mathbf{I} \, Gs_\beta$
which we do not know (cf. Problem 8.7).

Remark 8.23 (Discussion of Figure 8.22.)

(1) $\mathbf{Rd}_\alpha\, Gs_\beta \not\subseteq \mathbf{Uf}\ \mathbf{Nr}_\alpha\, Gs_\beta$ for all $1<\alpha<\beta$ follows from 8.8.

(2) Let $1<\alpha<\beta$. Then $\mathbf{Nr}_\alpha\, Ws_\beta \not\subseteq \mathbf{H}\ \mathbf{Rd}_\alpha\, CA_\beta$, moreover $\mathfrak{Nr}_\alpha \mathfrak{A} \notin \mathbf{H}\ \mathbf{Rd}_\alpha\, CA_\beta$ for any nondiscrete $\mathfrak{A}\in Mn_\gamma$, $\gamma\geq\alpha$.

Proof. Let $\alpha\leq\gamma$. By [HMT]2.1.22 we have $Mn_\gamma \models \forall x(\Delta x\neq 1)$. Thus $\mathbf{Nr}_\alpha\, Mn_\gamma \models \forall x(\Delta x\neq 1)$, by the definition of $\mathbf{Nr}_\alpha$. Let $1<\alpha<\beta$. Then every element of $\mathbf{Nr}_\alpha\, Mn_\gamma \cap \mathbf{HRd}_\alpha\, CA_\beta$ is discrete by 8.10.1. QED

(3) The rest of the inequalities in Figure 8.22 follow from 8.15 and (1)-(2) above. About $\mathbf{Nr}_\alpha\, \mathbf{I}\, Gs_\beta \subseteq \mathbf{Uf}\ \mathbf{Nr}_\alpha\, \mathbf{I}\, Gs_\beta$ see Theorem 8.6 and Problem 8.7.

(4) The positive statements of Figure 8.22 hold by 8.5, 8.1, 8.19 and by [HMTI]7.16.

We postponed one proof from section 3 to the present section because it uses tools developed here. Prop.8.24 below fills in this gap. It implies that the algebra $\mathfrak{A}$ constructed in the proof of Prop.3.7 is indeed regular as it was claimed (but not proved) there.

Proposition 8.24. Let $\alpha\geq\omega$, $\varkappa=|\varkappa|>\alpha$ and $H\subseteq\alpha\sim 1$ with $|H|\geq\omega$. Let $Z \stackrel{d}{=} \{q\in{}^\alpha\varkappa : q_0 \notin q^*H\}$. Let $\mathfrak{L} \stackrel{d}{=} \mathfrak{Sb}\,{}^\alpha\varkappa$. Then $\mathfrak{Sg}^{(\mathfrak{L})}\{Z\}$ is regular.

Proof. Assume the hypotheses. Let $\mathfrak{A} \stackrel{d}{=} \mathfrak{Sg}^{(\mathfrak{L})}\{Z\}$. Assume $\mathfrak{A}\notin Cs_\alpha^{reg}$. Then there is $y\in A$ such that $y$ is not regular. Since $y\in Sg\{Z\}$, there is $\Gamma\subseteq_\omega \alpha$ such that $0\in\Gamma$ and $y\in Sg^{(\mathfrak{Rd}_\Gamma\mathfrak{A})}\{Z\}$. Let $x \stackrel{d}{=} c_{(\Gamma\sim 1)}Z$. Clearly, $Z = \bar{d}(1\times(\Gamma\cap H))\cdot x$ and hence $y\in Sg^{(\mathfrak{Rd}_\Gamma\mathfrak{A})}\{x\}$. Let $\tau$ be a term in the discourse language of $\mathfrak{Rd}_\Gamma\mathfrak{A}$ such that $y = \tau^{\mathfrak{A}}(x)$. Since $Z$ is $H$-regular and $\Delta Z = 1\cup H$, by 1.3.5 we have that $y$ is $H$-regular. Hence $|H\sim\Delta y|\geq\omega$ by 1.3.4(ii), since $y$ is not regular. Let $f\in y$ and $k\in{}^\alpha\varkappa\sim y$ be such that $\Delta y 1 f\subseteq k$. Let $T \stackrel{d}{=} H\sim(\Gamma\cup\Delta y)$. Then $|T|\geq\omega$. Let $s \stackrel{d}{=} \Gamma 1 f$ and $Q \stackrel{d}{=} s^*\Gamma$. Let $p \in {}^\omega({}^\alpha\varkappa^{(f)})$ and $q \in {}^\omega({}^\alpha\varkappa^{(k)})$ be such that $s\subseteq p_n\cap q_n$, $\Delta y 1 p_n \subseteq f$, $\Delta y 1 q_n \subseteq k$, $Q\subseteq p_n{}^*T\cap q_n{}^*T$ and $|p_n{}^*T|\cap|q_n{}^*T|>n$, for every $n\in\omega$. Let

$n\in\omega$. Then $p_n\in y$ and $q_n\notin y$, since $p_n\in{}^{\alpha}\varkappa(f)$ and $q_n\in{}^{\alpha}\varkappa(k)$. Define $p'_n \stackrel{d}{=} (\alpha\sim\Gamma)1p_n$, $q'_n \stackrel{d}{=} (\alpha\sim\Gamma)1q_n$, $E_n \stackrel{d}{=} \varkappa\sim p_n{}^*(H\sim\Gamma)$ and $L_n \stackrel{d}{=} \varkappa\sim q_n{}^*(H\sim\Gamma)$. Then $|E_n|=|L_n|=\varkappa$ by $|\varkappa|>\alpha\geq\omega$. Let $F_n \stackrel{d}{=} \langle\{g\in{}^{\Gamma}\varkappa : (g\cup p'_n)\in a\} : a\in A\rangle$ and $G_n \stackrel{d}{=} \langle\{g\in{}^{\Gamma}\varkappa : (g\cup q'_n)\in a\} : a\in A\rangle$. Let $\mathfrak{G} \stackrel{d}{=} \mathfrak{Sb}\,{}^{\Gamma}\varkappa$. By [HMTI]8.1 we have $F_n, G_n \in \mathrm{Hom}(\mathfrak{Rd}_\Gamma\mathfrak{A}, \mathfrak{G})$, thus $F_n(y) = \tau^{(\mathfrak{G})}(F_n(x))$ and similarly for $G_n$. $s\in F_n(y)$ by $s\subseteq p_n\in y$ and $s\notin G_n(y)$ by $s\subseteq q_n\notin y$. By $s\subseteq p_n\cap q_n$, $\Omega\subseteq p_n{}^*(H\sim\Gamma)\cap q_n{}^*(H\sim\Gamma)$ and by the definitions of $x, Z$ we have $F_n(x) = \{g\in{}^{\Gamma}\varkappa : g_0\notin p_n{}^*(H\sim\Gamma)\} = \{g\in{}^{\Gamma}\varkappa : g_0\in E_n\}$ and $G_n(x) = \{g\in{}^{\Gamma}\varkappa : g_0\in L_n\}$. Let $D$ be a nonprincipal ultrafilter on $\omega$. Then by [HMTI]7.3-6 and by $|\Gamma|<\omega$ there are $\mathfrak{P}\in Cs_\Gamma$ and $h\in Is({}^{\omega}\mathfrak{G}/D, \mathfrak{P})$ such that $\varkappa\subseteq U \stackrel{d}{=} \mathrm{base}(\mathfrak{P})$ and by letting $\pi \stackrel{d}{=} \langle\langle F_n(a) : n\in\omega\rangle/D : a\in A\rangle$ we have $s\in\pi(y)$ and $\pi(x) = \{g\in{}^{\Gamma}U : g_0\in E'\}$ for some $E'\subseteq U$ with $|E'|\cap|U\sim E'|\geq\omega$ and $s\in{}^{\Gamma}(U\sim E')$. By the algebraic Downward Löwenheim-Skolem Theorem [HMTI]3.18, there is $Y\subseteq U$ such that $rl({}^{\Gamma}Y)\in Ho(\mathfrak{P},\mathfrak{B})$, $\mathfrak{B}\in Cs_\Gamma$, $s\in{}^{\Gamma}Y$ and $|Y\cap E'|=|Y\sim E'|=\omega$. Let $\rho \stackrel{d}{=} \langle{}^{\Gamma}Y\cap\pi(a) : a\in A\rangle$ and $M \stackrel{d}{=} Y\cap E'$. Then $s\in\rho(y)\cap{}^{\Gamma}(Y\sim M)$, $\rho(x) = \{g\in{}^{\Gamma}Y : g_0\in M\}$, $\rho(y) = \tau^{(\mathfrak{B})}(\rho(x))$ and $|M|=|Y\sim M|=\omega$. By applying the same argument to $\langle G_n : n\in\omega\rangle$ we obtain $\mathfrak{L}\in Cs_\Gamma$ with $W=\mathrm{base}(\mathfrak{L})$ and $N\subseteq W$, $\nu : A \to C$ such that $|N|=|W\sim N|=\omega$, $s\in{}^{\Gamma}(W\sim N)\sim\nu(y)$, $\nu(x) = \{g\in{}^{\Gamma}W : g_0\in N\}$ and $\nu(y) = \tau^{(\mathfrak{L})}(\nu(x))$. Let $b : W \rightarrowtail\!\!\!\twoheadrightarrow Y$ be one-one and onto such that $b\circ s = s$ and $b^*N = M$. By [HMTI]3.1, the base-isomorphism $\tilde{b}\in s(\mathfrak{Sb}\,W, \mathfrak{Sb}\,Y)$ is such that $\tilde{b}(\nu(x)) = \rho(x)$. Thus $\tilde{b}(\nu(y)) = \tilde{b}(\tau^{(\mathfrak{L})}(\nu(x))) = \tau^{(\mathfrak{B})}(\rho(x)) = \rho(y)$. This is a contradiction since $s\in\rho(y)$ while $b\circ s = s\notin\nu(y)$.

QED(Proposition 8.24.)

To save space, some of the open problems concerning reducts are stated in Section Problems only.

## 9. Problems

Problem 1 Let $V \subseteq {}^{\alpha}U$. What are the sufficient and necessary conditions on $V$ for $\mathfrak{Mn}(\mathfrak{Sb}\, V) \in CA_{\alpha}$? Cf. 0.3-0.4.

Problem 2 Let $K \in \{\mathbf{I}\,Crs_{\alpha}^{reg}, \mathbf{I}\,Crs_{\alpha}^{zdreg}, \mathbf{I}\,Crs_{\alpha}^{ireg}\}$. Is $\mathbf{I}\,Crs_{\alpha} \cap CA_{\alpha} = K \cap CA_{\alpha}$ or $\mathbf{I}\,Crs_{\alpha} = \mathbf{H}K$? Is $K \cap CA_{\alpha}$ a variety?

Note that $Crs_{\alpha} \not\subseteq \mathbf{H}Crs_{\alpha}^{oreg}$ for $\alpha \geq 2$ since if $\mathfrak{A} \in \mathbf{H}\mathfrak{B}$ and $\mathfrak{A} \models 0 < c_0 c_1 - d_{01} < 1$ then $\mathfrak{B} \models 0 < c_0 c_1 - d_{01} < 1$ and $(c_0 c_1 - d_{01}) \in Zd\mathfrak{B}$. Cf. 0.9 and 1.6.

Problem 3 Let $\alpha \geq \omega$. Are there two finitely generated base-minimal $Cs_{\alpha}^{reg} \cap Lf_{\alpha}$-s which are isomorphic but not base-isomorphic? Are there a finitely generated base-minimal $\mathfrak{A} \in Cs_{\alpha}^{reg} \cap Lf_{\alpha}$ and a $\mathfrak{B} \in Cs_{\alpha}^{reg} \cap \mathbf{I}\mathfrak{A}$ such that $\mathfrak{A}$ is not sub-base-isomorphic to $\mathfrak{B}$? Cf. 3.3-3.5.

Problem 4 Let $\alpha \geq \omega > \varkappa$. Are there two isomorphic countably generated ${}_{\varkappa}Gs_{\alpha}^{reg} \cap Lf_{\alpha}$-s which are not lower base-isomorphic? Cf. 3.4-3.6.

Problem 5 Does [HMTI]3.18 remain true if we replace the condition "$|A| \leq \varkappa$" with the weaker condition "$\mathfrak{A}$ can be generated by $\leq \varkappa$ elements and $(\forall x \in A)\varkappa \geq |\Delta x|$"? In particular, is the condition "$|\alpha| \leq \varkappa$" (which is implicit in the present wording of [HMTI]3.18) needed in [HMTI]3.18(i)c) and (iv) if we replace "$|A| \leq \varkappa$" by "$|G| \leq \varkappa$ and $A = SgG$"?

In this connection we note that 3.7 remains true if we delete the condition "$|\alpha| \leq \varkappa$" and replace (i) with (i') below.

(i') $A = SgG$ for some $|G| \leq \varkappa$ and $\lambda \geq \cup\{|\Delta x|^{+} : x \in A\}$.

This follows by replacing "$G \overset{d}{=} \{x\} \cup T \cup Y$" with "$G \overset{d}{=} \{x\} \cup T$" in the proof of 3.7. Cf. 3.7-3.9 and [HMTI]3.14.

Problem 6 Let $\alpha \geq \omega$. Let $\mathfrak{A}, \mathfrak{B} \in {}_{\infty}Cs_{\alpha}^{reg}$ and let $\mathfrak{A} \cong \mathfrak{B}$. Does there exist $\mathfrak{C} \in Cs_{\alpha}^{reg}$ ext-base-isomorphic to both $\mathfrak{A}$ and $\mathfrak{B}$? Cf. 3.11.

Problem 7 For any $\mathrm{Crs}_\alpha$ $\mathfrak{A}$, ultrafilter $F$ and $\langle F, \langle \mathrm{base}(\mathfrak{A}) : : i \in \cup F\rangle, \alpha\rangle$-choice function $c$ the homomorphism $ud^A_{cF} \in Ho\mathfrak{A}$ was introduced in Def. 3.12.

Let $\mathfrak{A}, \mathfrak{B} \in Cs_\alpha$ with $\mathfrak{A} \cong \mathfrak{B}$. Are there ultrafilters $F, D$ and an $\langle F, \langle \mathrm{base}(\mathfrak{A}) : i \in \cup F\rangle, \alpha\rangle$-choice function $c$ and a $\langle D, \langle \mathrm{base}(\mathfrak{B}) : : i \in \cup D\rangle, \alpha\rangle$-choice function $d$ such that $ud^A_{cF} {}^* \mathfrak{A}$ and $ud^B_{dD} {}^* \mathfrak{B}$ are base-isomorphic and $ud_{cF} \in Is\,\mathfrak{A}$, $ud_{dD} \in Is\,\mathfrak{B}$? Cf. 3.10-3.13.

Problem 8 Let $H \overset{d}{=} \omega \sim 2$. Let $q : H \times H \to \omega$ and $q^+ : H \to \omega$ be as in Problem 2 of [HMTI]. Let ${}_\beta K_\alpha \overset{d}{=} {}_\beta Cs_\alpha \cap Md\{\forall x(\bar{d}(\alpha\times\alpha) \geq x > 0 \to \bar{d}(\alpha\times\alpha) \leq \leq c_i x) : i \in \alpha\}$. Let $rq(\alpha,\beta) \overset{d}{=} \cap\{n \in \omega :$ every ${}_\beta K_\alpha$ can be generated by $n$ elements$\}$, and $rq^+(\alpha) \overset{d}{=} \cup\{\beta \in \omega : rq(\alpha,\beta) = 1\}$ for all $\alpha, \beta \in H$.

By [$P^3$], $rq^+(\alpha) = \alpha + 1 + |\alpha \cap \{4\}|$ for all $\alpha \in H$. The authors proved

$$(*) \quad rq(\alpha,\beta) \leq q(\alpha,\beta) \leq \lceil \log_2(|\beta \sim (\alpha+1)| + 2) \rceil \text{ for all } \alpha, \beta \in H.$$

Hence $q^+(\alpha) = \alpha + 1$ for $\alpha \in H \sim 5$, and $\alpha + 1 \leq q^+(\alpha) \leq \alpha + 2$ for all $\alpha \in H$. These motivate the questions:

Is $q^+(\alpha) = \alpha + 2$ for some $\alpha \in H$? Is $q^+(5) = \alpha + 2$? Is $q = rq$? Is there an approximation of $q$ better than (*)? Are $q$ and $q^+$ monotonic?

In this connection we note that [$P^3$] contains several results concerning the above problem as well as Problem 2 of [HMTI]. Cf.[HMTI] 4.5-4.8.

Problem 9 Let $\alpha \geq \omega$. Is $\{\mathfrak{A} \in {}_\kappa Gs_\alpha \cap Dc_\alpha : |Subb(\mathfrak{A})| < \omega$ and $1 < \kappa < \omega\} \subseteq \subseteq \mathbf{I}\,Cs_\alpha$? Cf. 4.14-4.18.

Problem 10 Is every epimorphism surjective in $CA_\alpha$ and in $Gs_\alpha$? Cf.5.11.

The $Gs_\alpha$-part of Problem 10 above is equivalent to the question whether or not the logic ${}_cL^t_F$ introduced in [AGN2]p.36 when restricted to $\alpha$ variables and with $t = \langle \alpha : i \in \omega \rangle$ has the Beth definability property. (By the notations of the quoted paper the set of all formulas of this restricted logic is $Fr_{\omega\ell_\alpha}$, and the class of its models is

$M_F^{\langle \alpha\ :\ i\in\omega\rangle}$, for any ordinal $\alpha$.)

Problem 11 Let $\alpha\geq\omega$. Does some of the equalities indicated by question marks on Fig.5.10 hold? How do Figures [HMTI] 5.7, 6.9, 6,10 look like if $\mathbf{I}\,Gws_\alpha^{comp\ reg}$ is included?

Problem 12 Let $2\leq\varkappa<\omega\leq\alpha$. Is $\mathbf{I}\,{}_\varkappa Gs_\alpha = \mathbf{HP}\,{}_\varkappa Cs_\alpha$? How will Fig.[HMTI]6.9 look like if for all $K\in\{Cs, Ws, Cs^{reg}, Gs, Gws\}$ we replace all occurrences of $K_\alpha$ with ${}_\varkappa K_\alpha$ in that figure? Is $\mathbf{HP}\,Cs_\alpha^{reg} =$ $= \mathbf{HP}\,Gws_\alpha^{comp\ reg}$? Cf. 5.6, 6.6 and [HMTI]6.8.

Problem 13 Let $wSdind_\alpha$ be the class of all weakly subdirectly indecomposable $CA_\alpha$-s. Let $\alpha\geq\omega$. Is $wSdind_\alpha\cap Gws_\alpha \subseteq \mathbf{I}\,Gws_\alpha^{comp\ reg}$, or $wSdind_\alpha\cap Cs_\alpha \subseteq \mathbf{I}\,Gws_\alpha^{comp\ reg}$, or $wSdind_\alpha\cap Gws_\alpha^{comp\ reg} \subseteq \mathbf{I}\,Cs_\alpha^{reg}$?

Note that the construction in the proof of 5.6(i) does not work to settle the last question since $wSdind_\alpha\cap Gws_\alpha^{comp\ reg} \models \forall x\forall y\ [(\Delta x)\cap\Delta y=0 \Rightarrow$ $\Rightarrow x\cap y\neq 0]$. Cf. 5.6, 6.3-6.6, [HMTI]6.13-16.

Problem 14 Let $1<\varkappa<\omega\leq\alpha$ and $K\in\{Ws, Cs^{reg}, Gws^{comp\ reg}\}$. Let $\mathfrak{A}\in{}_\varkappa K_\alpha$. Is then ${}^2\mathfrak{A}\in \mathbf{Uf\,Up}\,K_\alpha$? Is $\mathbf{Uf\,Up}\,({}_\varkappa Gs_\alpha\cap Lf_\alpha) = \mathbf{Uf\,Up}\,({}_\varkappa Cs_\alpha^{reg}\cap Lf_\alpha)$? Cf. 7.4-7.7.

Problem 15 Does some of the equalities (inclusions) indicated by questionmarks on Fig.7.6 hold?

Problem 16 Let $\alpha\geq\omega$. Which ones of the following equalities are true? $\mathbf{Uf\,Up}\,{}_\infty Ws_\alpha = \mathbf{S\,Up}\,{}_\infty Ws_\alpha$, $\mathbf{Uf\,Up}\,{}_\infty Cs_\alpha^{reg} = \mathbf{Uf\,Up}\,{}_\infty Ws_\alpha\cup\mathbf{I}\,{}_0Cs_\alpha$, $\mathbf{Uf\,Up}\,{}_\infty Cs_\alpha^{reg} = \mathbf{I}\,{}_\infty Cs_\alpha$, $\mathbf{Uf\,Up}\,(Dind_\alpha\cap Gws_\alpha) = \mathbf{Uf\,Up}\,Gws_\alpha^{comp\ reg}$, $(wSdind_\alpha\cap Gws_\alpha^{comp\ reg}) \subseteq \mathbf{Uf\,Up}\,Cs_\alpha^{reg}$ where $wSdind_\alpha$ was defined in Problem 13. Cf. 7.6-7.7.

Problem 17 Let $\varkappa\geq\beta$ be two infinite cardinals. Is ${}_\beta Cs_\alpha \subseteq \mathbf{I}\,{}_\varkappa Cs_\alpha$? Does $\varkappa\geq 2^{|\alpha\cup\beta|}$ imply ${}_\beta Cs_\alpha^{reg} \subseteq \mathbf{I}\,{}_\varkappa Cs_\alpha^{reg}$? How much are the cardinality conditions of 7.14(2),(3) needed? Cf. 7.14-17.

Problem 18 For which $\beta>\alpha>2$ is $\mathbf{Nr}_\alpha\,\mathbf{I}\,Gs_\beta = \mathbf{Uf}\ \mathbf{Nr}_\alpha\,Gs_\beta$? Cf. 8.6.

Problem 19 For which $|\beta|>2^{|\alpha|}$ is $\mathbf{Rd}_\alpha(Cs_\beta^{reg}\cap Lf_\beta) \subseteq \mathbf{I}\,Cs_\alpha$? Cf.8.10(4).

Problem 20 Let $\beta>\alpha\geq\omega$. If $\alpha\geq$ "the first uncountable measurable cardinal" then $\mathbf{Uf\,Up}\,Ws_\alpha = \mathbf{Uf\,Up\,Rd}_\alpha\,Ws_\beta$. Is this condition necessary? By the proof of 8.13 we have $|\beta|=|\alpha| \Rightarrow (\mathbf{Uf\,Up}\,K_\alpha = \mathbf{Uf\,Up\,Rd}_\alpha\,K_\beta)$ for $K\in$ $\in \{Ws, Cs^{reg}, Ws\cap Lf, Cs^{reg}\cap Dc\}$, so the question concerns the case $\beta>|\alpha|^+$.

Problem 21 By [AN8] and the proof of 8.13 and 8.4 we have $\mathbf{Uf\,Up}(Cs_\alpha\cap Lf_\alpha) = \mathbf{Uf\,Up}(Cs_\alpha\cap Dc_\alpha) = \mathbf{Uf\,Up\,Rd}_\alpha(Cs_\beta\cap Lf_\beta) =$ $= \mathbf{Uf\,Rd}_\alpha\,\mathbf{Uf\,Up}(Cs_\beta\cap Dc_\beta)$ for $\beta\geq\alpha\geq\omega$. Can $Cs$ be replaced with $Ws$ or $Cs^{reg}$ here?

Note that it can be replaced with CA, e.g. by using $Crax\cup tr_\xi\{C0-C7\}$ in the proof of 8.13. Cf. 8.13, 8.21.

Problem 22 Does 8.13.4 or the proof of 8.13 generalize to $\mathbf{I}\,Cs_\alpha^{reg}$ and $(S1)-(S9)\cup\Theta\rho\{str(\mathfrak{A}) : \mathfrak{A}\in Cs_\omega^{reg}\}$? Let

$Ex \overset{d}{=} \{[\forall i\exists! v\varphi(i,v) \rightarrow \exists z\forall i\varphi(i,ext(z,i))] : \varphi(i,v)$ is a formula $\varphi(i,v,x_m,u_m,s_m,j_m : m<n),\ n\in\omega$ in the discourse language of Crs-structures with free variables $i,v,x_m,u_m,s_m,j_m,m<n\}$.

What is the answer for $(S1)-(S9)\cup Ex$?

Problem 23 For which $\beta>\alpha\geq\omega$ is $\mathbf{Rd}_\alpha(Cs_\beta\cap Dc_\beta) \subseteq \mathbf{I}\,Cs_\alpha$ or $\mathbf{Rd}_\alpha(Cs_\beta^{reg}\cap$ $\cap Dc_\beta) \subseteq \mathbf{I}\,Cs_\alpha$?

Problem 24 To what extent are the cardinality conditions needed in 8.14? How much is the condition $|\alpha|=\omega$ needed? The existence of an uncountable measurable cardinal implies $Ws_\alpha \subseteq \mathbf{Uf\,Rd}_\alpha\,Ws_\beta$ for some $|\beta|>\alpha>\omega$, by the proof method of 8.13 but it is consistent with ZFC that the condition $|\alpha|=\omega$ can be omitted from 8.14.

Problem 25 In 8.21(vi) $\mathbf{Uf}\,Lf_\beta$ cannot be replaced neither with $\mathbf{Uf\,Up}\,Lf_\beta$ nor with $Dc_\beta$ if $\beta\geq\alpha+\omega$ (by [AN8]). Then what are the necessary conditions?

Problem 26 In 8.14(iii) the condition "$|\alpha|=\omega$" can be replaced with the new condition "there is no uncountable measurable cardinal $\leq \alpha$". Is this new condition necessary?

Problem 27 Let $K \in \{\mathbf{I}\,Cs^{reg}, \mathbf{I}\,Gws^{comp\ reg}\}$. Let $\beta>\alpha\geq\omega$. Is $K_\alpha = \mathbf{SNr}_\alpha K_\beta$? Cf. 8.18.1 and 8.18(iii).

Problem 28 Let $\beta>\alpha>0$. Let $\mathfrak{A}\in Gws_\alpha^{reg}$. Does there exist $\mathfrak{B} \in \mathbf{SNr}_\alpha Gws_\beta^{reg}$ such that $rs_\alpha \in Is(\mathfrak{B},\mathfrak{A})$?

Problem 29 Let $\beta>\alpha>0$. Let $\mathfrak{A}\in Gws_\beta$ and $\mathfrak{B} \subseteq \mathfrak{Nr}_\alpha \mathfrak{A}$. Assume $(\forall x\in B)[x$ is regular in $\mathfrak{A}]$. Is then $\mathfrak{Sg}^{(\mathfrak{A})}B$ regular?

## REFERENCES


[A] Andréka,H., Universal Algebraic Logic, Dissertation, Hungar. Acad. Sci.Budapest 1975. (In Hungarian)

[AGN1] Andréka,H. Gergely,T. and Németi,I., Purely algebraical construction of first order logics, Publications of Central Res.Inst. for Physics, Hungar. Acad. Sci., No KFKI-73-71, Budapest, 1973.

[AGN2] Andréka,H. Gergely,T. and Németi,I., On universal algebraic constructions of logics, Studia Logica XXXVI, 1-2(1977), pp.9-47.

[AN1] Andréka,H. and Németi,I., A simple, purely algebraic proof of the completeness of some first order logics, Algebra Universalis 5 (1975), pp.8-15.

[AN2] Andréka,H. and Németi,I., On universal algebraic logic and cylindric algebras, Bulletin of the Section of Logic, Vol.7, No.4 (Wroclaw, Dec. 1978), pp.152-158.

[AN3] Andréka,H. and Németi,I., On universal algebraic logic,Preprint, 1978.

[AN4] Andréka,H. and Németi,I., The class of neat-reducts of cylindric algebras is not a variety but is closed w.r.t. **HP**, Math.Inst.Hungar. Acad. Sci., Preprint No 14/1979, Budapest 1979. Submitted to The J. of Symb. Logic.

[AN5] Andréka,H. and Németi,I., Neat reducts of varieties, Studia Sci. Math. Hungar. 13(1978), pp.47-51.

[AN6] Andréka,H. and Németi,I., $\mathbf{I}Crs$ is a variety and $\mathbf{I}Crs^{reg}$ is a quasivariety but not a variety, Preprint Math. Inst. Hungar. Acad.Sci.,

Budapest, 1980.

[AN7] Andréka,H. and Németi,I., On the number of generators of cylindric algebras, Preprint, Math.Inst. Hungar. Acad.Sci., Budapest 1979.

[AN8] Andréka,H. and Németi,I., Dimension complemented and locally finite dimensional cylindric algebras are elementarily equivalent, Algebra Universalis, to appear.

[AN9] Andréka,H. and Németi,I., Varieties definable by schemes of equations, Algebra Universalis 11(1980), pp. 105-116.

[CK] Chang,C.C. and Keisler,H.J., Model Theory, North-Holland, 1973.

[G] Gergely,T., Algebraic representation of language hierarchies, Working Paper, Research Inst. for Applied Computer Sciences (Hungary), Budapest 1981. To appear in Acta Cybernetica.

[H] Hausdorff,F., Über Zwei Sätze von G. Fichtenholz und L. Kantorovitch, Studia Math. 6(1936), pp.18-19.

[HMT] Henkin,L. Monk,J.D. and Tarski,A., Cylindric Algebras Part I, North-Holland, 1971.

[HMTI] Henkin,L. Monk.J.D. and Tarski,A., Cylindric set algebras and related structures, this volume.

[M] Monk,J.D., Mathematical Logic, Springer Verlag, 1976.

[M1] Monk,J.D., Nonfinitizability of classes of representable cylindric algebras, The J. Symb. Logic 34(1969), pp. 331-343.

[N] Németi,I., Connections between cylindric algebras and initial algebra semantics of CF languages, In: Mathematical Logic in Computer Science (Dömölki,B. Gergely,T. eds.) Colloq.Math.Soc.J.Bolyai Vol.26, North-Holland, 1981, pp. 561-606.

[N1] Németi,I., Some constructions of cylindric algebra theory applied to dynamic algebras of programs, Comput. Linguist. Comput. Lang. (Budapest) Vol.XIV(1980), pp.43-65.

[$P^3$] Pálfy,P.P., On the chromatic number of certain graphs, Math.Inst. Hungar. Acad.Sci. Preprint No 17/1980. To appear in Discrete Mathematics.

[S] Sikorski,R., Boolean Algebras, Springer Verlag, 1960.

# INDEX OF SYMBOLS

This list should be used together with the index of symbols in the book [HMT]. For [HMT] see any one of the lists of references in this volume.

| | |
|---|---|
| $V^{\sim}$ | $\langle V\sim x$ : x is a set$\rangle$ ; [HMT] |
| $A\sim B$ | $\{a\in A : a\notin B\}$ ; [HMT] |
| $c^{+}$ | defined if c is an $\langle F,U,\alpha\rangle$-choice function; 86(I.7.1) |
| Rep(c), Rep(F,U,$\alpha$,A,c), Rep | representing function of ultraproducts of $Crs_{\alpha}$-s; 86(I.7.1) |
| $Rep_c$ | $Rep_c \overset{d}{=} Rep(c)$ ; 244, 86 |
| $ud_F^A$, $ud_F$ | diagonal ultrapower sub-base-isomorphism; 162(3.5.1) |
| $ud_{cF}^A$, $ud_c$ | diagonal ultrapower homomorphism if c is an $\langle F,U,\alpha\rangle$-choice function; 181(3.12) |
| $\twoheadrightarrow$ | onto function; 132 |
| $\rightarrowtail$ | one-one function; 132 |
| $\rightarrowtail\!\!\!\!\twoheadrightarrow$ | one-one and onto function; 132 |
| $\subseteq_{\omega}$ | "finite subset of" relation; 132 |
| $Sb_{\omega}$ V | $\{X : X \subseteq_{\omega} V\}$ ; 132 |
| Sb V | powerset of V; [HMT] |
| $\mathfrak{Cs}$ V | full $Crs_{\alpha}$ with unit V; 132 |
| $\mathfrak{Mn}(\mathfrak{A})$ | minimal subalgebra of $\mathfrak{A}$ ; 132 |
| $Mn(\mathfrak{A})$ | universe of $\mathfrak{Mn}(\mathfrak{A})$; 132 |
| Zd $\mathfrak{A}$ | $\{x\in A : \Delta^{(\mathfrak{A})}x=0\}$ ; 133, 262-3 |
| $\mathfrak{Zd}\ \mathfrak{A}$ | $\mathfrak{Sg}^{(\mathfrak{Rd}_0\mathfrak{A})}Zd\mathfrak{A}$ ; 133, 262-3 |
| ZdA, AtA, etc. | the corresponding notions for $\mathfrak{Sg}^{(\mathfrak{Cs}\cup A)}A$; 133 |
| base(V), base($\mathfrak{A}$) | base of; 133(0.1) |
| Subu(V), Subu($\mathfrak{A}$) | set of subunits of; 133(0.1) |
| Subb(V), Subb($\mathfrak{A}$) | set of subbases of; 133(0.1) |
| $\bar{s}$ | $\langle s : i\in\alpha\rangle$ ; 141 |
| $Sm^{\mathfrak{A}}$, Sm | set of small elements of $\mathfrak{A}$ ; 146(1.2) |
| $I_H^{\mathfrak{A}}$, $I_H$ | 146(1.3.1) |
| $Dm_H^{\mathfrak{A}}$, $Dm_H$, $Dm_H(\mathfrak{A})$ | $\{x\in A : \lvert\Delta x\sim H\rvert<\omega\}$ ; 146(1.3.1) |
| $\mathfrak{Dm}_H^{\mathfrak{A}}$, $\mathfrak{Dm}_H$, $\mathfrak{Dm}_H(\mathfrak{A})$ | subalgebra of $\mathfrak{A}$ with universe $Dm_H$; 146(1.3.1) |
| H-$dim^{\mathfrak{A}}$, H-dim | set of almost H-dimensional elements of $\mathfrak{A}$; 195(4.7.2.1) |

| | |
|---|---|
| $f^{AB}$ | base-relation induced by f on $A \times B$; 170 |
| **Ud** K | class of directed unions of members of K; 203 |
| **Uf** K | class of ultraroots of members of K; 229 (7.0), [HMT] |
| **Up**′K | class of ultrapowers of members of K; 229(7.0), [HMT] |
| $\mathrm{Dind}_{\alpha}$ | $\{\mathfrak{A} \in CA_{\alpha} : \lvert Zd\,\mathfrak{A}\rvert \leq 2\}$; 210 |
| $rb^{\rho}$, $rb^{(\rho)}$ | $rb^{(\rho)} \overset{d}{=} rb^{\rho}$ ; 191(4.7.1.1) |
| $rd^{\rho}$, $rd^{(\rho)}$ | $rd^{(\rho)} \overset{d}{=} rd^{\rho}$ ; 191(4.7.1.1) |
| $rd_{\alpha}$ | $rd^{(\alpha 1 Id)}$ ; 261 |
| $rs_{\alpha}$ | 289(8.16) |
| $\mathbf{Rd}_{\alpha}^{(\rho)} Bo_{\beta}$ | 263 |
| Ord | class of all ordinals; 263 |
| $CA_H$, $Cs_H$, etc. | H-dimensional CA-s; 222(6.0) |
| $Rd_H \mathfrak{A}$, $\mathfrak{Rd}_H \mathfrak{A}$ | defined for $\mathfrak{A} \in CA_S$, $S \supseteq H$; 222(6.0), [HMT] |
| $Nr_H \mathfrak{A}$, $\mathfrak{Nr}_H \mathfrak{A}$ | defined for $\mathfrak{A} \in CA_S$, $S \supseteq H$; 269, 222(6.0), [HMT] |
| $\mathfrak{Rf}(\varkappa)$ | greatest $Cs_{\alpha}^{reg} \cap Lf_{\alpha}$ with base $\varkappa$; 230(7.1) |
| CA, ICs, etc. | systems of classes of algebras; 263(8.2) |
| (S1) - (S9) | 281(8.13.1) |
| Crax, Cpax, Rgax | sets of axioms; 281(8.13.2) |
| $str(\mathfrak{A})$ | Crs-structure associated to $\mathfrak{A}$; 281(8.13.3) |
| $Cyl_{\xi}(\mathfrak{M})$ | $Crs_{\alpha}$ associated to $\mathfrak{M}$; 281(8.13.3) |
| Id | identity relation; [HMT] |
| Do R , *Do*R | domain of R; [HMT] |
| Rg R , *Rg*R | range of R; [HMT] |
| A1R | R domain-restricted to A; [HMT] |
| $R^{*}A$ | R-image of A; [HMT] |
| $R^{\star}x$ | $R^{*}\{x\}$; [HMT] |
| $pj_a$ | a-th projection; [HMT] |
| Θρ K | theory of K; [HMT] |
| $Md\varphi$ | class of models of $\varphi$; [HMT] |

| | |
|---|---|
| $Su\,\mathfrak{A}$ | set of subuniverses of $\mathfrak{A}$ ; [HMT] |
| $\mathbf{S}K$, $\mathbf{S}\mathfrak{A}$ | class of subalgebras; [HMT] |
| $Sg^{(\mathfrak{A})}X$, $SgX$ | subuniverse of $\mathfrak{A}$ generated by X; [HMT] |
| $\mathfrak{Sg}^{(\mathfrak{A})}X$, $\mathfrak{Sg}X$ | subalgebra of $\mathfrak{A}$ generated by X; [HMT] |
| $Ho\,\mathfrak{A}$ | class of homomorphisms on $\mathfrak{A}$; [HMT] |
| $Is\,\mathfrak{A}$ | class of isomorphisms on $\mathfrak{A}$ ; [HMT] |
| $h^{*}\mathfrak{A}$ | h-image of $\mathfrak{A}$ ; [HMT] |
| $Ho(\mathfrak{A},\mathfrak{B})$ | set of homomorphisms from $\mathfrak{A}$ onto $\mathfrak{B}$ ; [HMT] |
| $Is(\mathfrak{A},\mathfrak{B})$ | set of isomorphisms from $\mathfrak{A}$ onto $\mathfrak{B}$ ; [HMT] |
| $Hom(\mathfrak{A},\mathfrak{B})$ | set of homomorphisms from $\mathfrak{A}$ into $\mathfrak{B}$ ; [HMT] |
| $Ism(\mathfrak{A},\mathfrak{B})$ | set of isomorphisms from $\mathfrak{A}$ into $\mathfrak{B}$; [HMT] |
| $\mathfrak{A} \preccurlyeq \mathfrak{B}$ | $\mathfrak{A}$ is a homomorphic image of $\mathfrak{B}$ ; [HMT] |
| $\mathbf{H}K$, $\mathbf{H}\mathfrak{A}$ | class of homomorphic images; [HMT] |
| $\mathbf{I}K$, $\mathbf{I}\mathfrak{A}$ | class of isomorphic images; [HMT] |
| $Co\,\mathfrak{A}$ | set of congruence relations on $\mathfrak{A}$ ; [HMT] |
| $Il\,\mathfrak{A}$ | set of ideals of $\mathfrak{A}$ ; [HMT] |
| $Ig^{(\mathfrak{A})}X$, $Ig\ X$ | ideal generated by X; [HMT] |
| $P\mathfrak{B}$, $P_{i\in I}\mathfrak{B}_i$ | direct product of $\mathfrak{B}$ ; [HMT] |
| $\mathbf{P}K$ | class of isomorphic images of direct products; [HMT] |
| $\mathbf{Up}\ K$ | class of isomorphic images of ultra-products; [HMT] |
| $\mathfrak{B} \subseteq^{r} \mathfrak{A}$ | $\mathfrak{B}$ is a subreduct of $\mathfrak{A}$ ; [HMT] |
| $\cup^{r} K$ | reduct union of K; [HMT] |
| $\langle A,+,\cdot,-,0,1\rangle$ | Boolean algebra; [HMT] |
| BA | class of all Boolean algebras; [HMT] |
| $\Sigma^{(\mathfrak{A})}$, $\Sigma$ | sup; [HMT] |
| $\Pi^{(\mathfrak{A})}$, $\Pi$ | inf; [HMT] |
| $\langle A,+,\cdot,-,0,1,c_{\kappa},d_{\kappa\lambda}\rangle_{\kappa,\lambda<\alpha}$ | cylindric algebra; [HMT] |
| $\mathfrak{Bl}\,\mathfrak{A}$ | $\mathfrak{Rd}_{0}\mathfrak{A}$ ; 263, [HMT] |
| $c^{\partial}_{\kappa}$ | dual cylindrification; [HMT] |
| $s^{\kappa}_{\lambda}$ | substitution operation, $\lambda$ for $\kappa$; [HMT] |

| | |
|---|---|
| ${}_{\mu}s(\varkappa,\lambda)$ | substitution operation, interchanging $\varkappa$ and $\lambda$; [HMT] |
| $Cl_{\Gamma}\mathfrak{A}$, $\mathfrak{Cl}_{\Gamma}\mathfrak{A}$ | BA of $\Gamma$-closed elements of $\mathfrak{A}$ ; [HMT] |
| $c_{(\Gamma)}$ | generalized cylindrification; [HMT] |
| $d_{\Gamma}$ | generalized diagonal element; [HMT] |
| $\bar{d}\ R$ | generalized co-diagonal element; [HMT] |
| $Lf_{\alpha}$ | class of all locally finite $CA_{\alpha}$-s; [HMT] |
| $Dc_{\alpha}$ | class of all dimension-complemented $CA_{\alpha}$-s; [HMT] |
| $Mn_{\alpha}$ | class of all minimal $CA_{\alpha}$-s; [HMT] |
| $\mathfrak{Rd}^{(\rho)}\mathfrak{B}$, $\mathfrak{Rd}^{\rho}\mathfrak{B}$ | $\rho$-reduct of $\mathfrak{B}$ ; 263, [HMT] |
| $\mathbf{Rd}_{\alpha}^{(\rho)}K$, $\mathbf{Rd}_{\alpha}K$ | class of reducts; [HMT] |
| $\mathbf{Nr}_{\alpha}K$ | class of neat-reducts; [HMT] |
| $Bo_{\alpha}$ | class of all BA-s with operators; [HMT] |
| $At\,\mathfrak{A}$ | set of atoms of $\mathfrak{A}$ ; p.225 of [HMT] |
| $\oplus$ | symmetric difference; [HMT] |
| $R\|S$ | relative product of the relations R and S; [HMT] |
| $a_{\varkappa}^{\mathfrak{A}}$ , $a_{\varkappa}$ | $c_{(\varkappa)}\bar{d}(\varkappa\times\varkappa)$; 234(7.3.1), [HMT] |
| $at(x)$ | formula; 234(7.3.1) |
| $Cr_{\alpha}$ | $\mathbf{R}\ell\ CA_{\alpha}$ ; [HMT] |
| $\mathfrak{B}\subseteq_{d} P\mathfrak{A}$ | $\mathfrak{B}$ is a subdirect product of $\mathfrak{A}$; [HMT] |
| $ker(f)$ | $\{\langle x,y\rangle : \exists z(\langle x,z\rangle\in f \text{ and } \langle y,z\rangle\in f)\}$ ; 238 |

# INDEX OF DEFINED TERMS

This list should be used together with the "index of names and subjects" of the book [HMT].